Lecture Notes in Computer Science 2001

Edited by G. Goos, J. Hartmanis and J. van Leeuwen

Lecture Notes in Computer Science
Edited by G. Goos, J. Hartmanis and J. van Leeuwen

Springer
Berlin
Heidelberg
New York
Barcelona
Hong Kong
London
Milan
Paris
Singapore
Tokyo

Gul A. Agha   Fiorella De Cindio
Grzegorz Rozenberg (Eds.)

Concurrent
Object-Oriented Programming
and Petri Nets

Advances in Petri Nets

Springer

Series Editors

Gerhard Goos, Karlsruhe University, Germany
Juris Hartmanis, Cornell University, NY, USA
Jan van Leeuwen, Utrecht University, The Netherlands

Volume Editors

Gul A. Agha
University of Illinois at Urbana-Champaign
Department of Computer Science
1304 West Springfield Avenue, 61801-2987 Urbana, Illinois, USA
E-mail: agha@cs.uiuc.edu

Fiorella De Cindio
University of Milan, Computer Science Department
via Comelico 39/41, 20135 Milan, Italy
E-mail: fiorella.decindo@unimi.it

Grzegorz Rozenberg
Leiden University, Leiden Institute of Advanced Computer Science (LIACS)
Niels Bohrweg 1, 2333 CA, Leiden, The Netherlands
E-mail: rozenber@liacs.nl

Cataloging-in-Publication Data applied for

Die Deutsche Bibliothek - CIP-Einheitsaufnahme

Concurrent object oriented programming and Petri nets : advances in
Petri nets / Gul A. Agha ... (ed.). - Berlin ; Heidelberg ; New York ;
Barcelona ; Hong Kong ; London ; Milan ; Paris ; Singapore ; Tokyo :
Springer, 2001
  (Lecture notes in computer science ; 2001)
  ISBN 3-540-41942-X

CR Subject Classification (1998):F.1, D.1-3, F.3, C.2, J.1, G.2

ISSN 0302-9743
ISBN 3-540-41942-X Springer-Verlag Berlin Heidelberg New York

Springer-Verlag Berlin Heidelberg New York
a member of BertelsmannSpringer Science+Business Media GmbH

http://www.springer.de

' Springer-Verlag Berlin Heidelberg 2001
Printed in Germany

Typesetting: Camera-ready by author, data conversion by Steingräber Satztechnik GmbH, Heidelberg
Printed on acid-free paper   SPIN: 10782248   06/3142   5 4 3 2 1 0

Preface

There are two aspects to object-oriented programming: encapsulation of state and procedures to manipulate the state; and inheritance, polymorphism, and code reuse. The first aspect provides a correspondence with the natural separation of physical objects, and the second aspect is a way of organizing into categories and sharing by abstracting commonalities. Although much of the early research in object-based programming on sequential systems, in recent years concurrency and distribution have become the dominant concern of computer science. In fact, objects are a natural unit of distribution and concurrency – as elucidated quite early on by research on the Actor model.

It is therefore natural to look at models and theories of concurrency, the oldest of these being Petri nets, and their relation to objects. There are a number of directions in which models of concurrency need to be extended to address objects – these include the semantics of concurrent object-oriented programming languages, object-oriented specification languages, and modeling methods. Developments in these areas will lead to better system development and analysis tools. Among the issues that must be addressed by formal models of concurrency are inheritance, polymorphism, dynamically changing topologies, and mobility. These present deep challenges for the modeling capabilities of the existing formalisms and require investigation of ways to extend these formalisms.

For these reasons, we organized a series of two workshops on Models of Concurrency and Object-Oriented Programming which were held in conjunction with the Annual Petri Net Conference in 1995 (Turin, Italy) and in 1996 (Osaka, Japan). Subsequently, a subset of authors at these workshops, as well as a few others, were invited to submit papers for this volume. The papers then underwent a referee cycle with revisions.

The papers in this volume are organized into three sections. The first consists of long papers, each of which presents a relatively detailed approach to integrating Petri nets and object orientation – i.e. defining a syntax and its semantics, and illustrating it with examples. Section II includes shorter papers where the emphasis is on concrete examples to demonstrate an approach. Finally, Section III includes papers which significantly build on the Actor model of computation.

In order to provide an opportunity to more easily understand and compare the various approaches, we solicited some sample problems at the first workshop, which could serve as canonical case studies. In preparation of the second workshop, two case studies were then suggested to the authors – namely, the Hurried Philosophers problem, proposed by C. Sibertin-Blanc, and the specification of a Cooperative Petri Net Editor proposed by R. Bastide, C. Lakos, and P. Palanque. Many papers in this collection deal with these two problems and the original text describing the problems is included in the last section of this volume.

We are grateful to the referees and to the authors for their patience in the production of this volume. We would like to thank the organizers of the Petri Net conference for their help in organizing the workshops. Finally, we are also very grateful to Alfredo Chizzoni for his support in organizing the two workshops, and to Nadeem Jamali and Patricia Denmark for their assistance in preparing the manuscript.

January 2001   G. Agha
                                                                          F. De Cindio
                                                                        G. Rozenberg

Table of Contents

Section I

Section II

Section III

Section IV: Case Studies

Object Oriented Modelling with Object Petri Nets

Charles Lakos

Computer Science Department,
University of Adelaide,
Adelaide, SA, 5005,
Australia.
`Charles.Lakos@adelaide.edu.au`

Abstract. This paper informally introduces Object Petri Nets (OPNs) with a number of examples. OPNs support a thorough integration of object-oriented concepts into Petri Nets, including inheritance and the associated polymorphism and dynamic binding. They have a single class hierarchy which includes both token types and subnet types, thereby allowing multiple levels of activity in the net. The paper discusses some theoretical issues pertinent to the analysis of OPNs, and compares the provisions of OPNs with those of other Concurrent Object-Oriented Programming Languages.

The paper then considers a case study of using OPNs to model a cooperative editor for hierarchical diagrams. This extended example demonstrates the applicability of OPNs to the modelling of non-trivial concurrent systems. The methodology for deriving a Petri Net model is to adapt an object-oriented design methodology: the Object Model is prepared in Rumbaugh's OMT notation; the Dynamic Model is then prepared in the form of lifecycles, following the Shlaer-Mellor methodology; and finally these models are mapped into an OPN model. This approach has the advantage of guiding the development with well-accepted methodologies, and demonstrates the generality and flexibility of the OPN formalism.

1 Introduction

This paper informally introduces *Object Petri Nets* (OPNs) and considers their application to object-oriented modelling. The goal of OPNs is to achieve a complete integration of object-orientation into the Petri Net formalism, and thereby reap the complementary benefits of these two paradigms. Petri Nets are used for the formal specification of concurrent systems. They have a natural graphical representation, which aids in the understanding of such formal specifications, together with a range of automated and semi-automated analysis techniques. Object-oriented technology, on the other hand, has become extremely popular because of its provision of powerful structuring facilities, which stress encapsulation and promote software reuse. This addresses a traditional weakness of Petri Net formalisms, namely the inadequate support for compositionality.

The most distinctive feature of OPNs is the single unified class hierarchy. This has not been evident in previous object-oriented Petri Net approaches (see the review

G. Agha et al. (Eds.): Concurrent OOP and PN, LNCS 2001, pp. 1–37, 2001.

in [17]). Even our own *Language for Object-Oriented Petri Nets* (LOOPN) had separate class hierarchies for token types and subnet types. This meant that there was always a clean separation between tokens and subnets, resulting in the traditional style of Petri Nets with the control aspects of the system being coded as a global net structure.   This is not the flexibility one is accustomed to in object-oriented programming languages, where objects can contain attributes which are themselves objects, and so on.

The provision of a single unified class hierarchy in OPNs means that both token types and subnet types are classes, and these classes can be intermixed. A token can thus encapsulate a subnet. This implies that OPNs support the possibility of multiple levels of activity in the net and the modeller is free to choose how various activities are to be composed, i.e. whether a particular object should be active, passive, or both (depending on the focus of attention). This capability can address the increasing demand to model complex hierarchical systems (as in [9], [11]).

This paper reports on the progress towards achieving the goal of integrating the two technologies or paradigms. In doing so, the first part of the paper addresses the various aspects which have been identified above. Firstly, OPNs are informally introduced in §2, together with their graphical representation. This extends the usual Petri Net conventions. The formal definition of OPNs has been published elsewhere [17], [19], [22] and is not repeated here. Analysis capabilities for OPNs have not yet been implemented, but §3 introduces some theoretical issues which impact on the analysis of OPNs. In assessing the extent to which OPNs provide the flexible structuring possibilities promised by object-orientation, §4 compares the capabilities of OPNs with other concurrent object-oriented programming languages (COOPLs).

The suitability of a formalism is often determined by its applicability to non-trivial problems. To this end, the second part of this paper presents the application of OPNs to a non-trivial problem – the modelling of a cooperative editor for hierarchical diagrams. The basic issues of object-oriented modelling and the benefits offered by an integrated environment based on OPNs are introduced in §5. The (static) object model for the case study is presented and discussed in §6, while the dynamic model is considered in §7. The transformation of these models into an OPN model is considered in §8. Finally, the conclusions and the directions for further work are given in §9.

2 Object Petri Nets (OPNs)

Traditionally, Petri Nets are drawn with ovals representing places (which hold tokens), rectangles representing transitions (which change the distribution of the tokens and hence the state of the net), and directed arcs (which indicate how transitions affect neighbouring places). For Coloured Petri Nets (CPNs), tokens have an associated data value. Places are then annotated to indicate what type of value they hold, arcs indicate which values are consumed or generated (possibly with the use of variables), and transitions can have a guard to further constrain their enabling.

For example, the traditional dining philosophers problem can be modelled as the CPN of Fig. 1. All the places contain tokens with numeric values in the range *1..n*.

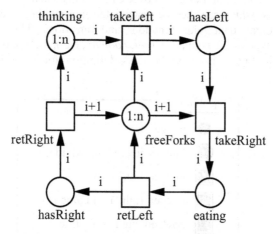

Fig. 1. Coloured petri net of the dining philosophers problem

The inscription of arcs with variables shows how the token values are matched from the various places. So, in order to fire the transition takeRight, for example, a token for philosopher i must be present in the place hasLeft, a fork numbered i+1 must be available in the place freeForks (with addition modulo n assumed). On firing this transition, a token for philosopher i is added to the place eating. In this example, all the transition guards can have the default value of true.

Note that some of the annotations have been omitted from the diagram, such as the type of each variable and the type of tokens for each place. This avoids cluttering the diagram and such selective display of annotations is normally supported by a graphical editor for Petri Nets.

OPNs extend CPNs in a number of significant ways, as considered in the following subsections. The extensions are demonstrated in the development of a solution to the *Russian philosophers* problem, an enhanced version of the dining philosophers problem. The name is meant to conjure up the image of the Russian matrioshka dolls which can be pulled apart to reveal other dolls inside. The Russian philosophers are like the dining philosophers, except that each Russian philosopher *is thinking about* the dining philosophers problem. Only when such an imagined dining philosophers problem deadlocks, will the corresponding Russian philosopher stop thinking and try to start eating. When such a Russian philosopher stops eating, he or she starts thinking about a fresh dining philosophers problem. Naturally, this system can be nested arbitrarily deeply. This problem is considered to be typical of a multi-level system where one part is performing some task which is being monitored by another.

2.1 Subnets as Classes

The first significant extension of OPNs over CPNs is that each Petri Net (or subnet) is defined by a *class*, which can then be instantiated in a number of different contexts. Each class can contain data fields, functions, and transitions. The type of a data field may be a simple type (such as *integer, real, boolean,* etc.), a class, or a multiset of the above.   Given in Fig. 2 is an OPN class definition for identifying each dining philosopher.

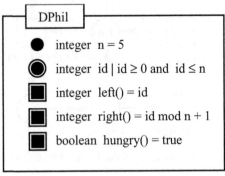

Fig. 2. Dining philosophers token type

Note that data fields generalise Petri Net places and are therefore drawn as ovals. Their annotation indicates the type, the identifier, an optional initial value (preceded by a character '=') and an optional guard (preceded by a character '|'). The guard must evaluate to true in a valid instance. It is therefore like a conjunct of a class invariant [26], a database integrity constraint, or a *fact* of earlier Petri Net formalisms [29]. Functions are drawn as rectangles to emphasise that they share transition semantics – they have bindings of variables and are evaluated at a particular point in time. The exporting of fields and functions is indicated by drawing them with a double outline. A data field of multiset type (written *type**) corresponds to a traditional Petri Net place, and is called a *simple place* as opposed to a *super place* (see §2.5).

2.2 Functions for Read-Only Access to the Subnet State

Using the normal conventions, an OPN solution to the dining philosophers problem would then appear as in Fig. 3. Note that places are annotated with their (multiset) type, and optionally specify an initial value or marking. Thus, the place *thinking* holds tokens of type *DPhil*, one of them being initialised with an *id* field of 1.

As with fields, the guard for a transition is indicated by a vertical bar followed by a boolean expression. For the transition to be enabled, the guard must evaluate to true. In Fig. 3, the transition *takeLeft* has a guard which requires that the function *hungry* for the chosen philosopher should evaluate to true.

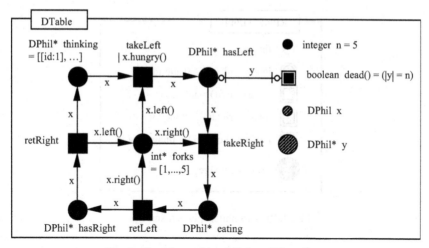

Fig. 3. Class for a table of dining philosophers

There is an attractive symmetry about the use of guards for class components and for transitions. In fact, the formal definition for the binding of a transition (such as *takeLeft*) is really an instance of a class, with fields for the variables (in this case *x*) and a constraint to restrict the valid bindings [19]. It would be possible to define each kind of binding with a separate class diagram, but this would be overly cumbersome. Therefore, we allow the shorthand of annotating a transition (externally) with a guard, and of graphically representing transition variables, such as *x* and *y*. A striped shading is used for variables to highlight that they are really local to a transition, and their value exists only for the duration of transition firing. (Note that the fact that transition bindings can be defined by classes is exploited for super transitions in §2.6.)

In this example, unlike that of Fig. 2, we define a function *dead()* which is state-dependent, i.e. it determines whether the table of dining philosophers is deadlocked. This dependence of the function on the place *hasLeft* is explicitly shown with an equal arc (which is a compound arc consisting of a test arc and an inhibitor arc, each having the same inscription) [15]. This arc has the effect of binding the variable *y* to the marking of the place *hasLeft*. The function then returns *true* if the place *hasLeft* holds *n* tokens.

The ability to define functions with read-only access to the current state of a subnet is another significant extension of OPNs over CPNs. While such access functions are commonly supported by COOPLs, this is not the case for Petri Net formalisms, even though such functions can significantly enhance the encapsulation of subnets [23].

2.3 Inheritance

In order to qualify for the title *object-oriented* [36], OPNs support the definition of classes with inheritance. In order to demonstrate this, we consider the OPN class definition used to instantiate each Russian philosopher (Fig. 4).

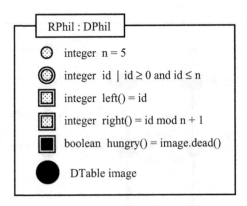

Fig. 4. Russian philosophers token type

The label of the class frame indicates that the class *RPhil* inherits the features or components of the class *DPhil*. The grey shading identifies those features or components which are inherited without change, while the black shading indicates components which have been introduced by this class or which override inherited components. Thus, class *RPhil* modifies class *DPhil* by augmenting it with *image* (the image of the table of dining philosophers), and by overriding the function *hungry*, so that Russian philosophers are hungry only when *image* deadlocks.

It is an open question as to the appropriate behavioural relationship between a parent and a child subclass [18]. We have adopted the approach of the Eiffel language [26], in requiring that the type of an overriding field must be a subclass of the type of the overridden field (in the parent class), and that the guards in a subclass should be stronger than those of the parent class. This ensures that an instance of the subclass also qualifies as an instance of the parent. It is essentially a pragmatic solution, but does have the benefit of being the foundation of a programming and design methodology, namely programming by contract [35].

It is worth noting that we have adopted the convention that all inherited components are displayed graphically (with the appropriate shading). Experience has shown that if the inherited components are not displayed at all (analogous to the conventions of object-oriented programming languages), then it is difficult to determine the role of the subclass since many of the visual cues are now missing. This is especially the case for Fig. 5. Nevertheless, we would expect a graphical editor for OPNs to allow the display of inherited components to be toggled on and off.

2.4 Arbitrary Instantiation of Subnets and Polymorphism

Having defined subnets as classes, it is natural to conclude that they can be instantiated. One of the most significant extension of OPNs over CPNs is that OPNs allow classes to be instantiated as tokens. In Fig. 3, this can be seen in the places holding tokens which are instances of class *DPhil*. This is not so startling since the class *DPhil* contains no actions, and therefore can be considered as a data structure with no encapsulated activity. This kind of instantiation is already possible in CPNs [12]. Similarly, the instantiation of the table of dining philosophers in the class *RPhil* of Fig. 4 is akin to the instantiation of pages by substitution transitions in HCPNs [12].

However, OPNs also allow tokens to be arbitrary subnets. This can be demonstrated with the class for a table of Russian philosophers, as shown in Fig. 5. Note that here the places holding *RPhil* tokens are holding tokens which have their own encapsulated activity. The current state of one of these tokens can be interrogated using the defined access function *hungry()*.

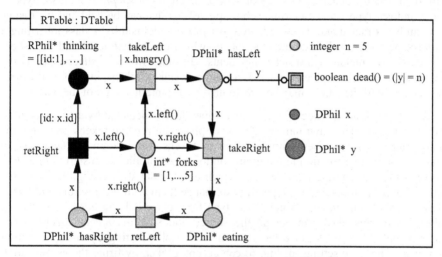

Fig. 5. Class for a table of Russian philosophers

This example also demonstrates the use of polymorphism in OPNs since *RPhil* (or subclass) instances may occur in *DPhil* (or superclass) contexts. Thus, the class *RTable* needs to override the place *thinking* so that it can initialise it to hold *RPhil* tokens. However, transitions which have been written to fetch and store *DPhil* tokens will also work with *RPhil* tokens and hence do not need to be overridden. Similarly, places defined to hold *DPhil* tokens can also store *RPhil* tokens. The only other change required to the *DTable* class is to override the transition *retRight*, so as to ensure that each Russian philosopher, on returning their right fork, starts thinking about a fresh dining philosophers problem rather than the same deadlocked one which led to eating.

2.5 Simple and Super Places

The previous subsection has shown that OPN classes can be arbitrarily instantiated. As well as implying that OPNs can directly model complex systems with multiple levels of activity (like the Russian philosophers problem), it also gives freedom to the modeller to choose the encapsulation which best suits the problem at hand. For example, the above solutions to the dining philosophers and Russian philosophers problems are derived from a standard Petri Net solution (Fig. 1). However, it is not a natural object-oriented solution because the philosophers are passive components while the table is active. A preferred solution would be to model each philosopher as an active component which interacts with the table setting (which contains the forks). The table setting could also store the number of philosophers once, instead of duplicating this for every philosopher and every table.

This leads naturally to the notion of a *super place*, which is another significant extension provided by OPNs. OPNs generalise the notion of a place to include any data field which can act as a source or sink of tokens. For *simple places*, the type is some multiset type which determines both the tokens that can be held and the tokens that can be extracted and deposited. A *super place* is defined by a class inheriting a multiset type, which then defines the tokens which can be extracted and deposited. The class may include extra net components. In other words, a super place is the instantiation of a class which can act as a source or sink of tokens, but where the offer or acceptance of the tokens may be constrained by the internal logic of the place.

The offer and acceptance of tokens by a superplace is depicted by arcs incident on the class frame. The environment of a super place can remove tokens provided this action can be synchronised with the offer of tokens from within the super place. Thus, an alternative solution to the dining philosophers problem can start with a class to act as the table setting (as in Fig. 6) which is instantiated as in Fig. 9. The table setting holds the forks and is prepared to exchange them with the environment. This class inherits from *integer** which is the type for a place holding *integer* tokens. In other words, this class can act as the source and sink of *integer* tokens. The environment (of Fig. 9) can offer a token to the table setting by firing transition *putdown*. The table setting (of Fig. 6) can accept a token by firing the *ret* transition (and similarly for the *pickup* and *get* transitions). The class also supplies the unique definition of *n* (the number of seats at the table) and the functions *left* and *right*.

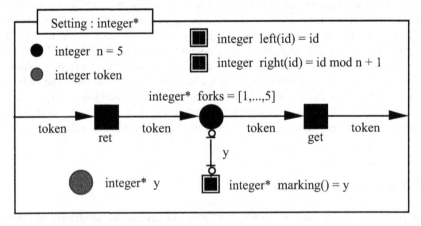

Fig. 6. Table setting class for active dining philosophers

The above approach embodies the formal definitions of [17], [19] by matching the external token interaction with the internal activity. These definitions do not constrain the annotations of the arcs incident on the class boundary. However, if these arcs were allowed to be inscribed by arbitrary multiset expressions, then there would be significant performance penalties in matching the internal and external interaction. In practice, it seems reasonable to constrain these arc inscriptions to specify a single token. Then each token specified by an external arc (which may have an arbitrary multiset inscription) simply needs to be matched by the enabling of one of the boundary transitions of the superplace. This clearly leads to an efficient implementation and conceptually captures the notion that the tokens in a place or superplace are essentially independent, i.e. they can be accessed individually.

The formal definitions also mean that the logic of the superplace can determine when tokens are accepted or offered. This makes it easy, for example, to design a superplace which acts as a capacity place, i.e. one that will not accept more than a predetermined multiset of tokens. An alternative approach to the definition of super places, based on the notion of net morphisms, is presented in [22]. This considers the notion of abstract places and transitions in the context of CPNs, but it is equally applicable to OPNs. In the simplest case, an abstract place is a place-bordered subnet, as illustrated in Fig. 7.

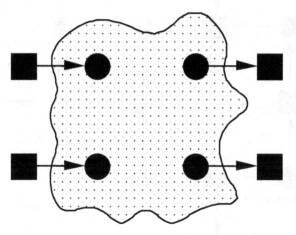

Fig. 7. An abstract place as a place-bordered subnet

In this case, the abstract place is largely passive, i.e. it cannot constrain when tokens are deposited or extracted (provided that they are available in border places). Thus, it is not possible to design an abstract place which acts as a capacity place. Accordingly, we prefer the former approach (as in Fig. 6) for providing interaction with the superplace.

The paper on abstraction [22] also argued that the notion of a place abstraction should not be merely conceptual or structural (i.e. a place-bordered subnet), but should include behavioural properties. Specifically, a place abstraction should include the notion of an abstract marking which is a projection of the state of the subnet, and which should be invariant over the firing of internal transitions but not over the firing of interface transitions. (Fig. 6 only has interface transitions.) This notion of a marking can be provided in various ways. One possibility is to require the modeller to supply a function called *marking* which returns the appropriate value. Such a function is shown in Fig. 6. Alternatively, the implementation could associate a redundant place with each super place, which could record the projected marking and which would only be affected by the transfer of tokens to and from the super place. Then the extraction of a token would require the token to be present in the redundant place and also for one of the interface transitions to be enabled for that token.

Using the table setting of Fig. 6, the active dining philosophers can be modelled by the class of Fig. 8, where the state of the philosopher is indicated by the position of the single *null* (or colourless) token. This class is also a superplace, which exchanges fork requests (−1 for a left fork, +1 for a right fork) with its environment.

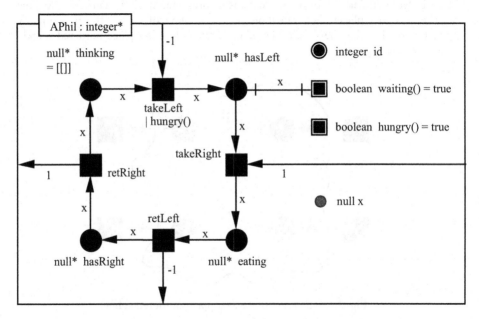

Fig. 8. Class definition for active dining philosophers

Finally, the above components can be combined into the class of Fig. 9 for a table of active dining philosophers. The class has a super place for the table setting and a place holding the active philosophers. It has two transitions which allow the philosophers to interact with the table setting. We have found that this pattern of transitions is common to interaction across the multiple levels of containment in OPNs (see §4.2). Here, the selection of token x (for an active philospher) from place *phils* is synchronised with interaction with that token (by exchanging fork request j) and with access to fork i.

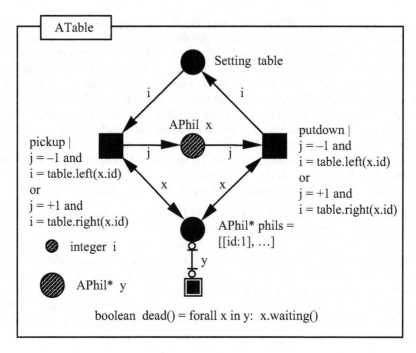

Fig. 9. Revised class definition for the table of dining philosophers

The thoughtful reader will note that the transitions in class *ATable* mean that the active philosophers do not encapsulate all the activity of the system. There are difficulties in having philosophers refer directly to the table setting via pointers, for example (see §3.1). Further, it is a convention of Petri Nets to constrain interaction to neighbouring components. We consider that it is reasonable to show explicitly by a net (as in Fig. 9) the environment in which active objects interact.

It is left as an exercise for the reader to derive a solution to the Russian philosophers problem using active philosophers.

2.6 Simple and Super Transitions

Just as it is valuable to build abstractions of places (called *super places*), it is also valuable to build abstractions of transitions (called *super transitions*). A number of Petri Net formalisms, including HCPNs [12], provide this capability in the form of *substitution transitions*. Here, an instance of a subnet or page is substituted for a transition in a net, and the component places of the subnet instance can be fused with the neighbouring places of the transition (as specified by a port assignment). As discussed in [22], this mechanism is quite general but it cannot be considered to be a transition abstraction since it provides no notion of firing for the substitution transition. The only semantics is in terms of expanding out the subnet in place of the substituted transition. If a subnet like that of Fig. 10 were used in this way (without the arcs incident on the class frame), then places *p1* and *p2* would be exported (or declared as port places). When the subnet substitutes a transition, the places *p1* and *p2* would be fused with the neighbouring places of the transition. Then the transfer of tokens into *p1* and out of *p2* would be determined solely by external transitions, and the transfer of tokens from *p1* and to *p2* would be solely determined by the internal transition *xfer*. There would be no guarantee (or requirement) that even if tokens were deposited into place *p1*, that there would ever be tokens extracted from place *p2*. In other words, there is no sense in which the actions associated with the arcs incident on the substituted transition should be synchronised in any way. Consequently, the semantics of such a substitution transition (in terms of its interaction with its environment) can only be understood from the documentation or by expanding out all the levels of the hierarchy. This is not helpful in practical applications where an HCPN model may have three or more hierarchy levels.

Arising from this critique, it is proposed [22] that super transitions should support the notion of firing, namely that all the effects of the incident arcs are synchronised in some sense. Specifically, the firing of the super transition will normally correspond to a multiset of internal actions, determined from the abstract firing mode. This set of internal actions will determine the interaction of the super transition with its environment, i.e. the effect of *firing* the super transition. However, The fact that it is a set of actions will not be visible to the environment (in the same way that the details of the state of a super place are not visible to its environment). In this sense, the arc effects are synchronised (as far as the environment is concerned).

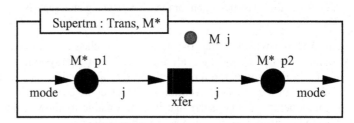

Fig. 10. A simple supertransition

The set of internal actions is governed by the transfer of tokens to and from the subnet, as indicated in Fig. 10 (the dual of that for super places in Fig. 6). The class inherits from the (pseudo) class *Trans*, thus indicating that it is a super transition. It also inherits from M^*, thus indicating that the firing modes of the transition are given by class M. (Recall from §2.2 that the bindings or firing modes of a transition can be defined by classes, just like tokens.) If the super transition fires with a particular mode (or binding), then that mode must be transferred on every arc incident on the class frame. In effect, this means that the *start* of firing of the super transition is achieved by activating the arcs *from* the class boundary, and the *end* of firing is achieved by activating the arcs *to* the class boundary. In our simple example, this means that the firing of the super transition can commence with the mode as a token being deposited in the place *p1*. This means that the transition *xfer* can then fire. Finally, the firing of the super transition can be completed by extracting the mode as a token from place *p2*.

The formal definition of OPNs [17], [19] does not support such a flexible form of super transition. In fact, they require that the internal and external effects of a super transition must match exactly and be exactly synchronised. This was recognised as being too restrictive in practice and was one of the motivations for considering a different form of abstraction [22]. Just as for place abstractions, it was argued that transition abstractions should not be merely conceptual or structural (i.e. a transition-bordered subnet), but should include behavioural properties. Specifically, a transition abstraction should indicate which internal actions correspond to an abstract firing mode. In our example, this is determined by the tokens deposited and extracted from the subnet.

3 Theoretical Aspects of OPNs

As already noted, the formal definition of OPNs has been published elsewhere [17], [19] and is not reproduced here. Instead, we highlight certain aspects of the formal definition which impact on the traditional Petri Net properties and analysis.

At the time of writing, the analysis capabilities discussed below have not been implemented. However, it should be noted that foundation for the analysis of OPNs is provided by the ability to map OPNs into behaviourally equivalent CPNs [16].

3.1 Reachability Analysis and the Use of Object Identifiers

The formal definition of OPNs consistently uses object identifiers to refer to class instances [19]. Thus, when a token is extracted from a place, it is the object identifier which is specified, and not the whole object (with potentially many levels of nested components). If such an object identifier is extracted from a place and not deposited into another place, then the whole object (including the nested components) is discarded.

This approach makes it possible for a transition to synchronise activity across different levels of containment in an object. For example, transition *pickup* in Fig. 9 removes token x for an active philosopher from place *phils* and then returns it. At the same time, a fork request j is added to the philosopher x. Thus, the token x returned to place *phils* has the same object identifier, but modified contents. This is a common requirement for such multi-level nets, and is clumsy to achieve without the use of object identifiers.

The possibility of reachability analysis for OPNs is affected by the use of object identifiers, since the number of these is potentially infinite. For this kind of analysis to be feasible, the number of object identifiers will need to be finite (as is assumed in [4], for example). Alternatively, some symmetry analysis will be required to determine if two objects are identical apart from relabelling of object identifiers. We anticipate that in practical cases, the number of objects will be finite.

Another aspect of the formal definition related to object identifiers is that tokens are required to be self-contained — it is not possible for some components of a token to be referenced by other net components *outside* the token. This is a constraint on the use of references and was imposed because of the perceived demands of invariant analysis. As a result, memory management for an OPN implementation is explicit, rather than being implicit using garbage collection. However, more recent work indicates that a less constrained solution (without requiring self-contained tokens) may also be possible.

3.2 Place Invariants for OPNs

It is possible to define place invariants directly for OPNs, in a similar manner to that of [10]. For a monolithic, non-hierarchical CPN, this means that a weight function is defined which, for each place, maps the (multiset) marking of that place to a common multiset type. For a place invariant, this projection of the marking of the net will be invariant for all possible firing sequences of the net. For a modular CPN, a family of weight functions is defined, one for each module or page instance. The local weight function for a module maps the marking of each module place to a multiset (or weighted set) over some common colour set. For such a family of weight functions to define a place flow (or a place invariant), each weight function has to be a local place flow, and the weight functions for different modules must coincide on fused places.

Given the dynamic allocation and deallocation of subnets in OPNs, it is not feasible to define a different weight function for each page instance (as for modular CPNs). Instead, we define one weight function per class. The contribution of a class instance to the weighted marking of the net is then given by the local weight function for the class combined with a weight determined by the context of the instance within the net.

Thus, if we wish to define a place invariant with range given by the colour set R (which may not be a set of object identifiers), then for each class s, we define a *local weight function* W_s, satisfying the following properties:

(a) W_s is a function mapping object identifiers of class s (and hence the instances of the class) into weighted sets over R.

(b) W_s is defined by mapping the markings of the data fields in the class instance into weighted sets over R.

Note that having defined the weight functions for individual object identifiers, they can be uniquely extended to linear functions over the domain of sets (or multisets) of object identifiers. The set of all local weight functions is called a *family of weight functions* $W = \{ W_s \}$.

Given a family of weight functions W and a marking M of the OPN, a *weighted marking* is given by $W(oid_0)$, where oid_0 is the object identifier of the instance of the root class (which indirectly instantiates all other net components). The family of weight functions applied to oid_0 applies weight functions to the data field values in the instance of the root class which, in turn, may apply other weight functions to the component instances. In other words, the contribution of a class instance to the weighted marking is given by the local weight function applied to the instance together with the composition of weight functions for the component instances. The composition of the the weight functions for the component instances is defined as a *context weight function W'*, and is defined by the derivation of each class instance from the root class instance.

This can be illustrated by the Russian philosophers example of Fig. 5. Suppose we wish to prove that for each Russian philosopher the forks in their dining philosophers image are conserved. Consider Russian philosopher with identity i, as follows:

(a)   define a local weight function for the root class *RTable* which determines the result from the root class instance in terms of the object identifiers for the appropriate Russian philosopher i:

$$W_{RTable}\,(x) = W_{RPhil}\,(W_1\,(M(thinking_x) + M(hasLeft_x) + M(eating_x) + M(hasRight_x)))$$

where p_x is the instance of place p in the class instance x and where

$$W_1\,(x) = if\,x.id = i\,then\,x\,else\,\emptyset$$

In other words, the contribution of instance x to the forks for Russian philosopher i is determined by the markings of places *thinking, hasLeft, eating* and *hasRight* for Russian philosopher i.

(b)   define a local weight function for the class *RPhil* which determines the contribution of its instances to the result, in this case the contribution from the image of the dining philosophers:

$$W_{RPhil}\,(x)\,=\,W_{DTable}\,(M(image_x))$$

In other words, the contribution of Russian philosopher x to the forks is given by the marking of *image*, the image of the table of dining philosophers.

(c)   define a local weight function for the class *DTable* which returns the weighted sum of forks according to the state of the various philosophers:

$$W_{DTable}\,(x) =\,W_{DPhil}\,(M(hasLeft_x) + M(eating_x)) + M(forks_x)\,+ W_2\,(W_{DPhil}\,(M(eating_x) + M(hasRight_x)))$$

where

$$W_2\,(x) = x+1$$

In other words, the contribution of a table of dining philosophers to the forks is determined from the philosopher tokens in places *hasLeft, hasRight* and *eating* (with philosopher tokens in place *eating* contributing both left and right forks), together with the forks still unused in place *forks*.

(d)   define a local weight function for the class *DPhil* which returns the identity of the particular dining philosopher:

$$W_{DPhil}\,(x) = x.id$$

In other words, the contribution of a dining philosopher instance to the weighted marking is given by the local weight function W_{DPhil} composed with the context weight function derived from the composition of functions W_{RTable}, W_{RPhil}, and W_{DTable}. The invariance of this family of weight functions would then be proved in the usual way, by showing that each transition conserves the weighted marking. Note that this assumes that when a token is consumed, all its components are consumed as well (see §3.1).

4 Discussion of Concurrent Object-Oriented Features of OPNs

Having introduced OPNs, we now compare the provisions of OPNs with key aspects of other concurrent object-oriented programming languages (COOPLs).

4.1 The Actor Model

The work on Actors has been a catalyst for many ideas related to COOPLs. In a recent paper [1], Agha et.al. have noted that the transition from sequential to parallel and distributed computing has been widely accepted as a major paradigm shift occurring in Computer Science. They argue that the Actor paradigm with objects as concurrent computational agents is much more natural than that found in sequential object-oriented languages.

Fundamental to the notion of Actor systems is the notion of a computational agent which is able to respond to the reception of a message by sending another message, by creating a new actor with some specified behaviour, or by replacing the current behaviour. Behaviour replacement *gives actors a history-sensitive behaviour necessary for shared mutable objects ... in contrast to a purely functional model* [1].

Under these criteria, OPNs qualify as a formal model for actor systems. OPNs naturally model concurrent, distributed systems. Each instance of an OPN class can be an independent computational agent, and message-passing between OPN objects is captured by token passing. New actors can be created simply by creating new tokens which are instances of the appropriate class. The ability of an actor to replace its behaviour with another, is not directly supported, but is available indirectly, since OPN class instances are mutable objects which can change state.

4.2 Communication Abstractions

Agha et.al. [1] note that while asynchronous message sending is the most efficient form of communication in a distributed system, and the form originally assumed by the Actor model [2], they argue that concurrent languages should provide a number of flexible communication abstractions to simplify the task of programming. They argue for supporting call/return communication, pattern-directed communication, synchronisation constraints, and synchronisers.

The standard form of message-passing in the Actor model is asynchronous. The actor which initiates the message continues with its own computation rather than waiting for a reply. *Call/return communication* has both synchronous and asynchronous dimensions – it is synchronous in that the message originator waits for a reply before continuing, but it is asynchronous in that the sending of the call is not blocked. Call/return communication can be implemented by *remote procedure call*. OPNs can model call/return communication directly by super transitions (as proposed in §2.6).

Pattern-directed communication removes the constraint for an actor to specify its communication partner explicitly. The usual point-to-point communication allows locality to be directly expressed and optimised. However, in some situations it is preferable for an actor to specify communication with any one of a set of actors, known as an *actorSpace*. An *actorSpace is a computationally passive container of actors which acts as a context for matching patterns* [1].

A number of features of OPNs support the possibility of pattern-directed communication. The unified class hierarchy allows places to be declared to contain a set of active objects, which can then qualify as an actorSpace. Agha et. al. allow for actorSpaces to overlap, which is not supported by OPNs since a token always resides in one place at a time. (Removing the restriction on self-contained tokens (see §3.1) would get over this limitation.) A transition can select a token from the place which is an actorSpace, and then interact with the token, as in Fig. 11. Note that the actor x is selected from actorSpace $s1$, is sent the message y, and then is deposited in the actor space $s2$ (which could optionally be the same as $s1$). This is simply the multi-level interaction of §2.5.

Note also that, just as with actorSpaces, the selection of x can be dependent on a transition guard which is external to the actor but which may examine some of the actor's attributes. Alternatively, since an actor (in this case) is a super place, the reception of the message y may be constrained by the internal logic of x, in which case the transition is only enabled if an x can be selected which is prepared to accept y, i.e. on a condition internal to the actor.

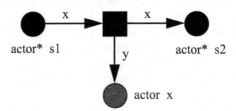

Fig. 11. Pattern-directed communication of message y to actor x from actorSpace $s1$

Agha et.al. [1] also consider the provision of *synchronisation constraints*, i.e. where a concurrent object can constrain the invocation of its methods. For example, a bounded buffer object should not accept a put action unless there is space in its buffer. It is argued that synchronisation constraints should be separated from the representation of methods.

The OPN notion of a super place supports synchronous interaction (see §2.5). The transition which deposits (or extracts) a token at a super place must synchronise with an internal transition of the superplace which accepts (or produces) a matching token. The internal transition (as well as the external transition) can constrain the interaction with a suitable guard. This has been demonstrated in the table setting object for the active dining philosophers (in Fig. 6) and the pattern-directed communication (of Fig. 11). The other activity of the superplace (corresponding to method bodies) can be represented by other internal transitions. In other words, OPNs directly support the notion of synchronisation constraints and the separation of synchronisation and method bodies.

Synchronisers relate to multiple actor coordination patterns while synchronisation constraints (considered above) relate to the acceptability of communication with a single actor. Petri Nets have always supported the synchronisation of multiple interactions with a single transition (such as a dining philosopher picking up two forks at once), and so the extension of synchronisation constraints to synchronisers follows directly for OPNs. Furthermore, the step semantics of Petri Nets directly supports truly concurrent interactions.

4.3 Compositionality

Nierstrasz [27] lists a number of basic requirements on the operational view of concurrent objects that affect compositionality. Firstly, in order to guarantee object autonomy, every object ought to be active and have control over its synchronisation of concurrent requests. Secondly, it should be possible to replace a sequential implementation of an active object with a concurrent one without affecting clients. Thirdly, scheduling should be transparent, both for requests and replies.

We have already noted that each OPN object is the instance of some class which may encapsulate a subnet, and thus be active. Furthermore, super places have synchronous interaction with their environment and can thus control the synchronisation of concurrent requests. It ought to be noted that the normal firing rules of Petri Nets allow for true concurrency, so that concurrent requests can be handled concurrently (assuming that the appropriate enabling conditions are satisfied), rather than forcing some sequential interleaving. Similarly the firing rules for Petri Nets depend on the availability of tokens rather than on any distinction between sequential or concurrent subsystems. This also means that transition enabling is not constrained by some predetermined delays, and thus scheduling is transparent.

Nierstrasz also notes the need for composable synchronisation policies. He proposes an Object Calculus for combining components. It is a matter for further study whether such composition policies can be emulated with OPNs, perhaps with super places and super transitions, as in [22].

4.4 Inheritance Issues

There has been much work on the so-called inheritance anomaly which is the interaction between inheritance and synchronisation policies in COOPLs [24]. Because of it, some COOPL designers have chosen *not* to support inheritance as a fundamental language feature. It is our view (supported by [6]) that much of the problem stems from the way that method definitions in traditional programming languages tend to combine the synchronisation code with the internal state transformation instead of separating these concerns, as recommended in the Actor model.

In the same way, the inheritance anomaly can be signficantly reduced for OPNs and for super places in particular, since it is easy to identify the internal transitions which control the acceptance and supply of tokens, and hence which embody the synchronisation constraints. Furthermore, OPNs naturally support synchronisation constraints by method guards, which is the technique with the least problems as considered by Matsuoka and Yonezawa [24]. Method guards are only deficient for history-sensitive synchronisation, but we argue that it is unrealistic to demand history-sensitive synchronisation without first storing the necessary history information in the object.

A more difficult issue concerning inheritance is the extent to which the behaviour of a parent class should be retained in a subclass. Some have advocated a bisimilarity or labelled transition sequences, while others have proposed the maintenance of labelled state sequences [4]. We have found that both of these appear to be too restrictive in practice [18].

5 Object-Oriented Modelling

Having considered the elements of OPNs and compared their facilities with those of other COOPLs, we now consider the application of OPNs to object-oriented modelling.

Mellor and Shlaer [25] assert: „The ability to execute the application analysis models is a *sine qua non* for any industrial-strength method, because analysts need to be able to verify the behaviour of the model with both clients and domain experts.“ On the other hand, a recent assessment of CASE tools [28] has noted: „The biggest mistake was to overestimate the ease with which the early CASE tools would evolve into the kinds of truly integrated CASE systems that users want. It is a long way from diagramming tools to generated code.“ Similarly, it has been observed [14]: „Tool integration therefore is an important issue. Lack of integration is a major criticism regarding CASE tools because it is detrimental to acceptance of CASE by the users.“

This paper now examines whether the above deficiencies can be addressed by employing OPNs as the formal foundation for an integrated CASE environment. OPNs are formally defined, allowing them to be used for software specifications; and they are executable, thus allowing OPN software designs to be used as prototypes.

In order for OPNs (or any other proposal to integrate object orientation and Petri Nets [3], [4], [5], [8], [32]) to qualify as a formalism suitable as the foundation for an integrated CASE environment, it must be able to model realistic case studies. We therefore consider such case study – an editor for hierarchical diagrams in some cooperative working environment (see the appendix.)

In line with the problem specification, our solution concentrates on the locking mechanisms required to allow cooperative editing. Hence the precise nature of the diagrams – whether Petri Net diagrams, Circuit Diagrams, Object-Oriented Designs, etc. – is irrelevant. It should be possible to refine the basic framework presented here for any of these specific diagrams. The essential aspect of the diagrams is that they are hierarchical, i.e. one graphic component may be an abstraction of a diagram which can then be opened and edited.

Again in line with the problem specification, our solution ignores display issues such as windows, icons, menus, etc. We are interested in the underlying objects, not their graphical representation, nor the way that the user communicates with the system. (This is akin to the distinction made in the context of graphical user interfaces (GUIs) between *subjects*, *views* and *editors* [34].) However, since the underlying objects of our case study are diagrams, there will be some graphics-related attributes which will need to be stored. For example, the position of a graphic in a diagram is relevant to the underlying object since it is an attribute that may be modified by the user via the GUI. This can be accommodated under the general notion of editing an object's attributes.

This paper derives a solution to the problem by preparing an object-oriented design and then mapping it into an OPN. This has the significant advantage of building on existing object-oriented methodologies which are widely used, admittedly with many variations. It will also have the effect of constraining the possible solutions. An advantage of this is that it will force us into a particular style of solution which will therefore test the flexibility of OPNs. A disadvantage is that it may exclude natural Petri Net solutions, especially those with enhanced concurrency. Another significant advantage is that this approach can provide a formal semantics for informal design methodologies.

The design methodology adopted in the subsequent section is a hybrid which uses the Rumbaugh notation [30] to capture the static object model, and the Shlaer-Mellor approach to capture the dynamic models [31]. This approach has been chosen because the Rumbaugh notation is well-accepted but is vague about the integration of static and dynamic models. On the other hand, the Shlaer-Mellor approach provides clear guidelines on the integration of static and dynamic models, but has a somewhat unconventional approach to inheritance. This hybrid approach is not completely unusual since it is allowed by the Unified Method [7].

Because the problem specification is incomplete, it will be necessary to make assumptions about the case study. We normally choose simplifying assumptions but note the situations where these have significant ramifications.

6 The Object Model

From the problem description, we derive the Object Model of Fig. 12, given in the OMT notation of Rumbaugh [30].

For those unfamiliar with the notation, the classes are denoted by labelled rectangles, the associations by lines. Lines with no terminator indicate that the class participates once in the association, a solid dot as terminator indicates that the class may participate zero or more times, while a hollow dot as terminator indicates that the class may participate zero or one times. The diamond terminator indicates an aggregation relationship, while a triangle indicates an inheritance relationship.

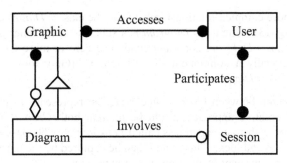

Fig. 12. Object Model for the Cooperative Editor

Thus, in Fig. 12, each *User* participates in a number of editing *Session*s, and each *Session* has a number of participating *User*s. Similarly, each *Graphic* is accessed by a number of *User*s, and each *User* accesses a number of *Graphic*s. Each *Session* involves one *Diagram*, and each *Diagram* may be the subject of an editing *Session*.

Note that while the identifier *Graphic* is used, it is *not* meant to imply that this class holds information relevant to the graphical representation or display of the component. It simply indicates the underlying object for a diagram component. (It is a *subject* in the terminology of [34].) As already noted in §5, a *Graphic* may need to store attributes which affect the graphical display, but these are treated uniformly as editable attributes.

The problem specification suggests the use of some locking mechanism so that certain kinds of accesses to a *Graphic* will preclude other accesses (by a different *User*) to the same *Graphic*. This could be made explicit by introducing a class *Lock* as in the alternative object model of Fig. 13.

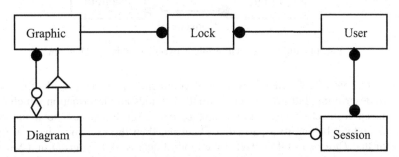

Fig. 13. Alternative Object Model for the Cooperative Editor, with explicit locks

Each instance of *Lock* would capture the interaction between a single *Graphic* and a single *User*, with the kind of locking being an attribute of *Lock*. This approach has not been followed since it would preclude other forms of interaction, such as version control. By simply having an association between *Graphic* and *User* (as in Fig. 12), it is possible to map the model into a solution with locking or version control, or some other form of interaction. Certainly, locking is possible, with the kind of lock now being an attribute of the association.

There is a more complex relationship between the classes *Diagram* and *Graphic* in Fig. 12 (and Fig. 13). The class *Diagram* is a subclass of (or inherits from) the class *Graphic*. Further, a *Diagram* is an aggregation of zero or more *Graphics*, and each *Graphic* is optionally a component of a *Diagram*. (Only root *Diagrams* are not components of other *Diagrams*.)

This relationship between *Graphic* and *Diagram* represents an interesting design choice. An alternative approach would be to treat both *Graphic* and *Diagram* as subclasses of some abstract superclass called *Lockable*, i.e. the class of objects which the user can access and lock. This would then be captured by the partial model of Fig. 14. Now, a *Diagram* may or may not be an expansion of a *Graphic*, depending on whether it is a root diagram. Similarly, a *Graphic* may or may not have an expansion, depending on whether it is a simple or a complex graphic. The original solution of Fig. 12 was considered to be simpler and therefore preferable.

Another implication of the approach in Fig. 14 is that every *Graphic* belongs to a *Diagram*, which is not required by the original solution. The original solution therefore leaves open the option of a *Graphic* being accessible in some other context, such as an icon on the workstation desktop.

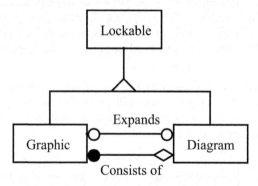

Fig. 14. Alternative relationship between Graphic and Diagram

Finally, the role of a *Session* is worthy of comment. This class is identified in the problem description, but it is not clear whether it adds any information which cannot be deduced from elsewhere. For example, if a *User* has access to a number of *Graphics* which belong to a particular *Diagram*, then the user must have a *Session* open for that *Diagram*. Of course, if the kind of access to a *Diagram* in a *Session* is different to the kind of access to a *Graphic*, then it will be helpful to include the notion of a *Session*. In order to cater for this possibility, we have chosen to include this class.

In developing the object model, the attributes and operations pertinent to each class have not been considered. The reasons for this are two-fold. Firstly, the choice of these components is usually a simpler process than the choice of classes and associations. (We introduce some attributes in specifying the OPN model in §8.) Secondly, the Shlaer-Mellor methodology emphasises the design of the dynamic model in terms of lifecycles, to the exclusion of the choice of operations.

7 The Dynamic Model

In order to capture the dynamics of the case study, we adopt the *Object Lifecycle* approach of Shlaer and Mellor [31]. This determines the lifecyles for each relevant class in the form of a finite state machine (FSM). The choice of such an FSM model has the advantage of building on a well-accepted methodology, but has the disadvantage that it may exclude a range of Petri Net solutions, particularly those with high levels of concurrency. Since each object lifecycle is a finite state machine, concurrency within an object is not considered.

In a lifecycle, the states are drawn as labelled rectangles, while the transitions between the states are drawn as directed arcs. Each arc is annotated by the event which can cause the transition between the states. (The events may originate either internally or externally to the object.) Each state can be annotated with an action which occurs on entry to the state. That action may involve the generation of new events and the modification of the local state. Where an action is non-trivial, it is described by an *Action Dataflow Diagram* (ADFD).

Parts of the lifecycles for *User* and *Graphic* are given in Figs. 15 and 16. To avoid intersecting lines, the state *Idle* has been duplicated in each case, a facility we would expect to be supported by a graphical editor.

Thus, for *User*, the initial (and final) state is *Dormant*. A *User*, through the GUI, may request the opening of a *Diagram* (event *U3*). The appropriate event (*D6*) is sent to the chosen *Diagram*, which will respond in due course with acceptance (event *U1*) or rejection (event *U0*). Once a diagram is open, the *User* will be in the *Idle* state. From there, it is possible to respond to display updates (event *U2*) and to perform various operations, as dictated by user actions in the GUI. One possibility is for the *User* to select a *Graphic* for inspection (event *U7*), causing a change of state from *Idle* to *RQ inspect*. On entry to the new state, an event (*G3*) is sent to the chosen *Graphic* in order to convey the request. The *Graphic* may respond with rejection (event *U0*) or acceptance (event *U1*). We have adopted the convention of using three states for each such operation – one to indicate that a request has been made, one to indicate that the operation is being performed, and one to indicate that it is completed.

Note that we have distinguished between an *open* and an *expand* operation. An *open* event (*D6*) can only be sent to a *Diagram* while an *expand* event (*G5*) can be sent to a *Graphic* or a *Diagram*. If an *expand* event is sent to a *Graphic* which is *not* a *Diagram*, it requests the conversion from a simple *Graphic* to a *Diagram*. In both cases an *open* event will eventually be processed.

In the lifecycle for a *Graphic*, the initial (and final) state is *Idle*. Again, we have used three states for each operation. At the request state (after receiving the relevant request event), a decision is made whether to accept or reject the request and the corresponding event (*U1* or *U0*) is sent to the *User*, and a similar event to the *Graphic*. (Note that the logic for making the choice will need to be given by an ADFD interacting with the current state of the object.) On returning to the *Idle* state, a *U2* event is sent to all relevant users to ensure that all views of the diagram are updated.

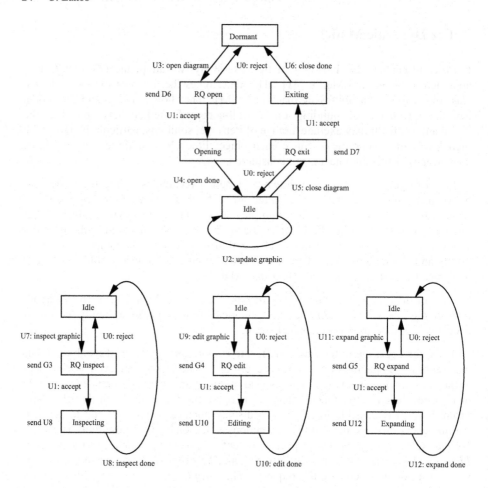

Fig. 15. Lifecycle for the class *User*

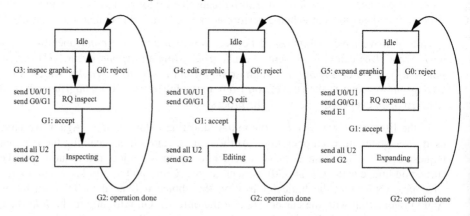

Fig. 16. Lifecycle for the class *Graphic*

It is worth noting that an expand request (event *G5*), if it is accepted, is handled by sending the appropriate event (*E1*) to the environment to convert the *Graphic* into a *Diagram*.

The above lifecycle diagrams reflect the simple locking scheme described in the problem definition. In a more realistic scheme, there would be a need for some kind of transactions, so that if the user had a complex sequence of related activities to perform, it would be possible to establish the locks on all the relevant graphics first, before any of the editing operations were performed. At the completion of editing, all the locks would be released. In this case, each cycle of three states for an operation would need to be replaced by three cycles – one to establish a lock, one to perform the operation (for which the appropriate lock has already been established), and one to release the lock. The locking scheme could be two- or three-phase. Alternatively, some form of version control could be used, which would allow the possibility of multiple copies of each diagram and graphic, and would specify the merging of different versions. In contrast, the above scheme assumes that the lock needs to be established in one place, wherever the relevant component is stored.

An interesting issue is the relationship between the lifecycle of a class and its subclass. In our Object Model, *Diagram* is a subclass of *Graphic*. As described above, only a *Diagram* responds to *open* and *close* events, and hence the lifecycle of *Diagram* will need to augment the lifecycle of *Graphic*. On the other hand, both *Diagram* and *Graphic* respond to *expand* events, but with different logic, and hence the lifecycle of *Diagram* will need to override part of the lifecycle of *Graphic*. These components of *Diagram* are shown in Fig. 17. The modified response to an *expand* request is relatively minor – an event *E2* is sent to the environment instead of an event *E1*. Given the close relationship between a state transition and the action on entry to the new state, both of these have been shown as being overridden in the subclass.

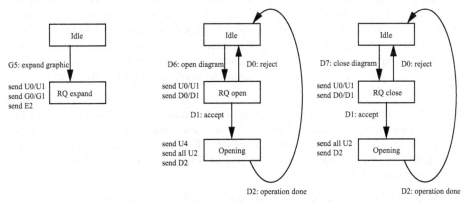

Fig. 17. Additional lifecycle components for the class *Diagram*

In addition to the above minor change, a *Diagram* exhibits considerable additional behaviour. An interesting question then arises as to the allowable ways in which a subclass can change the behaviour of its parent while still being considered to be a specialisation of the parent (see §4.4).

This question of the relationship between the lifecycle of a class and its subclass is ignored by most object-oriented modelling schemes. It is considered by Shlaer and Mellor [31], though in a somewhat confusing manner, possibly because the work products of the methodology can be mapped into a non-object-oriented language. It therefore does not speak of inheritance in the same way as traditional object-oriented programming languages. Instead, a subclass is split up into two parts – one containing the components of the superclass (called the *supertype*), and one containing the components unique to the subclass (called the *subtype*). An instance of the subclass thus involves an instance of both the supertype and the subtype. A lifecycle can optionally be associated with each of the two parts. If a lifecycle is only associated with the supertype, then all the subclass instances have the same behaviour. If lifecycles are only associated with the subtypes, then the behaviour is different for each subclass instance. Alternatively, both supertype and subtypes may have lifecyles, in which case the lifecycle of a subclass instance is formed by splicing together one of each to form a composite lifecycle. Shlaer and Mellor even allow for an instance to migrate between different subtypes over its lifetime. We consider that this approach is unfortunate since it muddies the water about the nature of inheritance. Instead, we affirm that each object is an instance of a single class. Further, a subclass will inherit both the data and the lifecycle(s) of its parent(s), and will have the option of augmenting or overriding both. In other words, any lifecycle splicing will be done explicitly by the modeller, as we have done in Fig. 17 – the states *Idle* and *RQ expand* would override the same states of Fig. 16, while state *RQ open* would become a new state, etc.

Similarly, Shlaer and Mellor speak of polymorphic events, but do not treat polymorphism in the usual object-oriented way. In the above example, the event *G5* is polymorphic since it can be received both by a *Graphic* and by a *Diagram*. The labelling of events is the one suggested by Shlaer and Mellor, namely a letter for the class together with a unique number. If one sticks rigidly to this convention, a table is required to indicate how such a polymorphic event can be translated into other events, e.g. event *G5* for graphics may be translated into an (equivalent) event *D5* for diagrams. Again, we feel that this requirement is unfortunate, and propose that a subclass (in this case *Diagram*) can respond to events defined for the superclass (in this case *Graphic*).

With its complete integration of object-oriented structuring, OPNs support both the augmentation and overriding of parts of a net. As a result, we can support a more natural form of inheritance and polymorphism. Consequently, the augmentation and overriding of lifecycle components (as in Fig. 17) can be mapped directly into the corresponding OPNs.

As already noted, the Shlaer-Mellor methodology specifies that complex actions (which occur on entry to states) are modelled by *Action Dataflow Diagrams* (ADFDs). In Fig. 16, for example, a *Graphic*, on receiving a request for an edit operation (event *G4*), needs to decide whether that operation will be allowed, and to respond with acceptance or rejection (events *U1* or *U0*). That choice will depend on the current locks being held on the *Graphic*. The details of this decision-making process would be represented as an ADFD.

Finally, the Shlaer-Mellor methodology captures the flow of events between objects and with external entities by the *Object Communication Model* (OCM). Part of the OCM for the cooperative editor is shown in Fig. 18.

The external components of the system are drawn as rectangles. For example, the external user generates events via the GUI. The objects of the system are shown as rounded rectangles. Furthermore, Shlaer and Mellor recommend that the entities which have more global knowledge should be drawn at the top of the diagram.

To a large extent, the OCM can be deduced from the lifecycle diagrams. However, the provision of both gives a pleasing separation of concerns – the lifecycle diagrams capture the response to events *within* an object, while the OCM captures the flow of events *between* objects. Note, however, that the OCM does *not* capture the handling of events such as creation events which require a response from the run-time environment.

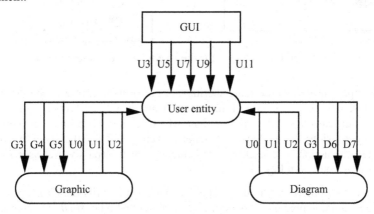

Fig. 18. Part of the Object Communication Model for the Editor

8 The Object Petri Net Model

The Object Model of §6 and the Dynamic Model of §7 are now integrated into an Object Petri Net model. The semantics of OPNs means that the resulting model can be simulated (as a prototype) or analysed (for erroneous situations). While the class attributes were not considered in §6, the Object Model would determine the data for each class as follows:

Attribute	=	Identifier × Value
Event	=	GraphicId × UserId × EventKind
EditEvent	=	Event × Attribute
Lock	=	GraphicId × UserId × Action
Graphic	=	GraphicId × GraphicId × Attribute* × Lock*
Diagram	=	Graphic × GraphicId*
User	=	UserId × Lock* × ...

Thus, a (graphic) *Attribute* consists of an *Identifier* and a *Value* (both of which are treated as primitive values). Examples of an *Attribute* could be a description suitable for a data dictionary, a name, the colour and location of the graphic, etc. An *Event* consists of a *GraphicId* and a *UserId* (to identify the relevant *Graphic* and *User*) and an *EventKind* (to identify the particular event). Further data can be attached to *Event*s. Thus, an *EditEvent* is a subclass of an *Event*, with the addition of a new *Attribute* value.

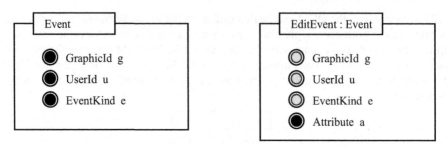

Fig. 19. OPN classes for the Event and Edit-Event classes

It is important to observe that inheritance plays a natural role in the definition of events, just as in the modelling of inheritance from the Object Model. Polymorphism is also significant since the environment which exchanges events between objects need only consider the exchange of generic events, and not the specific events for particular interactions. The OPN representations of the (passive) classes *Event* and *EditEvent* are shown in Fig. 19.

Thus, class *EditEvent* inherits from class *Event* – it inherits the *GraphicId*, *UserId*, and *EventKind* components, and augments the parent with an *Attribute* component.

A *Graphic* consists of its own *GraphicId*, the *GraphicId* of its enclosing diagram, the multiset of associated *Attribute*s and the multiset of current *Lock*s. As before, multisets are indicated by a suffix asterisk and identify traditional Petri Net places in the relevant class. Thus each *Graphic* will be represented by a class containing (at least) two constant fields holding *GraphicId*s and two places holding the current *Attribute*s and *Lock*s (see 20). The *GraphicId* of the enclosing diagram is used to implement one side of the association between *Graphic* and *Diagram* of Fig. 12. The multiset of *Lock*s serves to implement one side of the association between *Graphic* and *User*. Note that the attribute of the association, i.e. the kind of lock, is stored as part of the *Lock*.

It is worth noting that the OPN model stores some kind of identification for related objects, here *GraphicId*s and *UserId*s. It does *not* store the object identifiers or pointers to those objects, since tokens must be self-contained (see §3.1). It is always possible to refer to other objects indirectly via some value managed by the net (as in [33]). However, this means that OPNs directly support inheritance and aggregation, but only indirectly support other class relationships.

The lifecycles from the Dynamic Model of §7 are mapped in a fairly obvious way into the corresponding Petri Nets. Each possible state is mapped into a place which can hold a *null* token. (A *null* token has no attached attribute information.) The current state is indicated by the presence of a single *null* token in the appropriate place. The initial state (of 20) is given by a token in place *Idle*. Each state transition is mapped into a Petri Net transition with input and output arc to the preceding and following state, respectively.

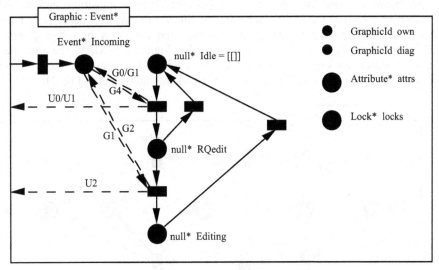

Fig. 20. Part of the OPN class for a graphic

The transition also has an input arc indicating the event which causes the transition, and output arcs indicating the resultant generated events. (An alternative approach would be to map each state transition from the lifecycle into two Petri Net transitions and an intermediate place, the first transition responding to the input event and the second generating the output events. As we shall see below, this is not necessary since the Petri Net transitions will commonly be refined into a subnet.)

Since each object interacts by exchanging events, the most natural representation of the object as a whole is as an OPN superplace which can accept and offer tokens which are event notices. Thus, 20 shows part of the net for a *Graphic*. It inherits from *Event** to indicate that it exchanges tokens of type *Event*. The state components for *Graphic* are shown on the right hand side and the event interaction on the left hand side. Incoming events are buffered in the place *Incoming*, in line with the Shlaer-Mellor approach of handling events asynchronously [31]. (The OPN formalism also supports synchronous interaction, which might be appropriate for the Shlaer-Mellor Object Access Model, which we do not consider. In fact, an OPN super place could be written to handle some events synchronously, and some asynchronously.)

In this case of *Graphic*, the transition from state *Idle* to state *RQedit* is triggered by the reception of event *G4*, and results in the generation of events *U0* or *U1* and *G0* or *G1*. Similarly, a transition from state *RQedit* to state *Editing* is enabled on reception of event *G1* and coincides with the generation of events *U2* and *G2*. The net of 20 indicates the possible transitions and should be a recognisable representation of part of the lifecyle of Fig. 16. Clearly, it does not show all the interactions in order not to clutter the diagram. For example, the transitions from state *RQedit* to *Idle*, and from *Editing* to *Idle*, do not show the consumption and generation of events.

Furthermore, it does not show the interaction between the transitions and the state components (such as the locks), nor does the lifecyle of Fig. 16. As already noted, this form of interaction is modelled by Shlaer and Mellor with *Action Dataflow Diagrams* (ADFDs). In order to represent this logic in an OPN, the transitions between states are expanded into super transitions. For example, the transition between the states *Idle* and *RQedit* could be expanded into the net of Fig. 21.

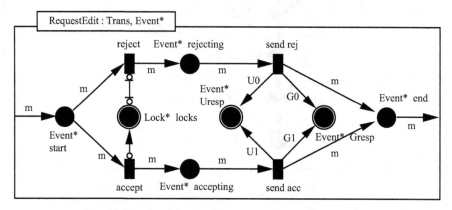

Fig. 21. Details of the Graphic response to an edit request

In this case, the response could be determined by a simple transition, without the need for a computation sequence, but we have chosen to expand it into a subnet so as to demonstrate the capabilities of supertransitions. While not necessary to specify this, we anticipate that this supertransition will be instantiated with arcs incident on places *idle*, *RQedit* and *Incoming*. The firing mode will be determined by the *G4* event (which includes a specification of the user and the graphic). The firing of the supertransition commences with the deposit of this mode in place *start*. Then transition *reject* or *eject* fires depending on whether there is already a conflicting lock or not. Then the relevant transition (*send rej* or *send acc*) fires to generate events *U0* and *G0*, or *U1* and *G1* respectively. Note that we do allow the subnet to include component places which are exported and thus can be fused to other places in the environment. However, the set of transitions to be fired is determined by the abstract firing mode (as already discussed in §2.6).

It is worth emphasising that the mapping of Shlaer-Mellor work products into OPNs naturally requires both superplaces and supertransitions. It is common in Petri Net formalisms to supply only one of these mechanisms, but it is obviously beneficial to have both. It is also worth noting that the more relaxed requirements for the synchronisation of super transition actions (as discussed in §2.6) is appropriate here.

Another work product of the Shlaer-Mellor methodology is the *Object Communication Model* (OCM). While the object lifecycles show the reception and generation of events internal to an object, the OCM shows the external flow of events between objects and external entities. It could therefore be considered as the environment for the objects. Unfortunately, it does not specify the complete environment. For example, events which result in the creation and destruction of objects are not shown. Similarly, when an object migrates from one subtype to another (as discussed in §7), the relevant events are not shown.

In OPNs, with their support for multiple levels of activity, it is a simple matter to capture the behaviour of the environment by a net, in which the objects previously considered become tokens. Given that the number of instances of *Graphic* and *User* varies dynamically during the lifetime of a session, it is appropriate that they be represented as tokens in places holding all the relevant instances. It is then necessary to allow these objects to exchange events.

This leads to the general pattern of interaction given in Fig. 22, which shows the exchange of events between existing *Graphic* and *User* instances. Thus, the transition *touser* has the effect of selecting (and replacing) a graphic *g* from place *graphics* and a user *u* from place *users*. An event *x* is then extracted from *g* and deposited into the user *u*. The transition *touser* thus synchronises activity across multiple levels, as discussed in §2.5.

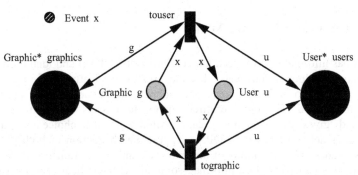

Fig. 22. Interaction between graphics and users

Other forms of interaction between graphics and users are shown in 23. These forms of interaction are not shown in the Shlaer-Mellor Object Communication Diagram since they do not reflect a simple exchange of events. It shows that a user can create a graphic at any time with transition *newgraphic*, and that a user may delete a graphic with transition *delgraphic*, provided that the graphic accepts the event (i.e. the appropriate locks are already in place). It shows that a user can be generated at any time by transition *newuser* or can be deleted by transition *deluser*, provided that the user accepts the event (i.e. it holds no outstanding locks). Note that the transitions *delgraphic* and *deluser* both require synchronous event interaction (as considered in §8) in order to constrain the acceptance of the event by a *Graphic* and *User*, respectively. Shlaer and Mellor assume that such creation and deletion events are handled by the environment, but how that is accomplished is not shown in their work products.

Similarly, Shlaer and Mellor do not explicitly consider how an instance can be converted from one subclass to another, even though they speak of subtype migration in relation to subtype lifecycles. In 23, we show how the expansion of a graphic is achieved with the *expandgraphic* transition. A graphic g is extracted from place *graphics* and an expand event x is received from it. A new diagram object d is created with state information copied from the graphic g (as indicated by the guard $d = g$).

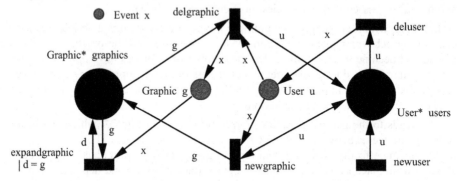

Fig. 23. Other forms of interaction with the environment

9 Conclusions and Further Work

This paper has informally introduced Object Petri Nets (OPNs) and identified the key extensions over Coloured Petri Nets. The most distinctive feature is the unified class hierarchy which allows both tokens and subnets to be defined by classes and arbitrarily intermixed. This provides direct support for models with multiple levels of activity, and gives the modeller great flexibility in the choice of modelling components as active or passive. OPNs support the abstraction of both places and transitions, leading to the notions of super places and super transitions. While the emphasis of this paper has been informal, some of the theoretical issues pertinent to the analysis of OPNs were discussed, particularly the implications of using object identifiers. These simplify access to class instances, but require special handling if reachability analysis is to be feasible. The features of OPNs were compared with those provided by other concurrent object-oriented programming languages, and the general conclusion was that OPNs provide a natural representation for many of the proposals introduced by the Actor model.

A more extensive examination of the modelling capabilities of OPNs was conducted by applying them to the case study of a cooperative editor for hierarchical diagrams. The OPN model was derived by adapting existing object-oriented modelling techniques. Firstly, a static object model was built using OMT notation [30], since this provides explicit support for inheritance (as traditionally understood in object-oriented languages). Secondly, a dynamic model was built using the Shlaer-Mellor approach [31], since this provides a clean integration of the static and dynamic models – there is a clean separation of the activity within an object (captured by lifecycles) and the communication between objects (captured in the Object Communication Model).

The paper demonstrated that OPNs provide a natural target formalism for such a methodology. Each class from the static object model can be mapped naturally into an OPN super place class which contains the appropriate attributes and object identifiers, and exchanges events with its environment. Each OPN class can be augmented with Petri Net transitions corresponding to the state transitions from the object lifecycles. The details of each action executed on entry to a state are readily represented by the refinement of the above Petri Net transitions into super transitions. In other words, the transitions show the event flow, while their refinement into subnets shows the detailed logic of the associated actions. Given that the unified class hierarchy of OPNs readily supports multiple levels of activity, the Object Communication Model can also be mapped into an OPN subnet. Further, this subnet can be extended to reflect issues which are not explicitly represented in the Shlaer-Mellor methodology, such as the response to creation and destruction events and the possibility of converting an object from one class to another.

We claim that the mappings from the work products of the Shlaer-Mellor methodology into OPNs are very natural, and possibly amenable to automation. The ability to refine both places and transitions (into super places and super transitions) is an important part of this process, as is the ability of OPNs to model systems with multiple levels of activity. Thus the OCM has components which are themselves active objects.

We consider that OPNs are therefore ideal as the underlying semantic model for an integrated CASE environment based on this methodology. The fact that a well-defined formalism is a target of this methodology means that some aspects can be more precisely defined, such as the relationship between the lifecycles of a class and its subclass and the response to polymorphic events. Finally, the fact that OPNs are executable means that the models produced with this methodology can serve as prototypes.

OPNs have been formally defined [17], [19], and have been proved to be behaviourally equivalent to Coloured Petri Nets [16]. This provides a basis for adapting analysis techniques for Coloured Petri Nets [12] for use with Object Petri Nets. Significant theoretical effort for OPNs should now be directed towards the development of analysis techniques, taking into account the considerations of §3. An important issue here will be the possibility of incremental analysis which can build on the incremental specifications possible in OPNs.

A textual form of OPNs in the language LOOPN++ has been implemented [20] and has been considered for the modelling of information systems [13]. Currently, the compiler for this language produces either C++ or Java code, the latter raising the possibility of using OPNs to control Java applets and distributed applications.

While there is great scope for further work in extending the formal foundations for OPNs, their analysis, and the associated tool implementations, this paper makes a significant contribution to justifying the claim that OPNs embody a thorough integration of object-orientation into Petri Nets.

Acknowledgements

The author is pleased to acknowledge the stimulating discussions and encouragement of the Networking Research Group of the University of Tasmania, Australia.

References

[1]   G. Agha, S. Frølund, W.Y. Kim, R. Panwar, A. Patterson, and D. Sturman *Abstraction and Modularity Mechanisms for Concurrent Computing* Research Directions in Concurrent Object-Oriented Programming, G. Agha, P. Wegner, and A. Yonezawa (eds.), pp 3-21, MIT Press (1993).

[2]   G.A. Agha *Actors: A Model of Concurrent Computation in Distributed Systems* The MIT Press series in artificial intelligence, MIT Press (1986).

[3]   M. Baldassari and G. Bruno *An Environment for Object-Oriented Conceptual Programming Based on PROT Nets* Advances in Petri Nets 1988, G. Rozenberg (ed.), Lecture Notes in Computer Science 340, pp 1–19, Springer Verlag (1988).

[4]   E. Battiston, A. Chizzoni, and F. de Cindio *Inheritance and Concurrency in CLOWN* Proceedings of Workshop on Object-Oriented Programming and Models of Concurrency, Torino, Italy (1995).

[5]   E. Battiston, F. de Cindio, and G. Mauri *OBJSA Nets: A Class of High-level Nets having Objects as Domains* Advances in Petri Nets 1988, G. Rozenberg (ed.), Lecture Notes in Computer Science 340, pp 20–43, Springer-Verlag (1988).

[6]   M.Y. Ben-Gershon and S.J. Goldsack *Using inheritance to build extendable synchronisation policies for concurrent and distributed systems* Proceedings of TOOLS Pacific 1995, pp 109-122, Melbourne, Australia, Prentice-Hall (1995).

[7]   G. Booch and J. Rumbaugh *Unified Method for Object-Oriented Development* Version 0.8, Rational Software Corporation (1995).

[8]   D. Buchs and N. Guelfi *CO-OPN: A Concurrent Object Oriented Petri Net Approach* Proceedings of 12th International Conference on the Application and Theory of Petri Nets, Gjern, Denmark (1991).

[9]   L. Cherkasova, V. Kotov, and T. Rokicki *On Net Modelling of Industrial Size Concurrent Systems* Proceedings of 15th International Conference on the Application and Theory of Petri Nets – Case Studies, Zaragoza (1994).

[10]   S. Christensen and L. Petrucci *Towards a Modular Analysis of Coloured Petri Nets* Application and Theory of Petri Nets, K. Jensen (ed.), Lecture Notes in Computer Science 616, pp 113-133, Springer-Verlag (1992).

[11]   P.A. Fishwick *Computer Simulation: Growth Through Extension* Proceedings of Modelling and Simulation (European Simulation Multiconference), pp 3-20, Barcelona, Society for Computer Simulation (1994).

[12]   K. Jensen *Coloured Petri Nets: Basic Concepts, Analysis Methods and Practical Use – Volume 1: Basic Concepts* EATCS Monographs in Computer Science, Vol. 26, Springer-Verlag (1992).

[13]   C.D. Keen and C.A. Lakos *Information Systems Modelling using LOOPN++, an Object Petri Net Scheme* Proceedings of 4th International Working

Conference on Dynamic Modelling and Information Systems, pp 31-52, Noordwijkerhout, the Netherlands, Delft University Press (1994).

[14]   K. Kurbel and T. Schnieder *Integration Issues of Information Engineering Based I-CASE Tools* Proceedings of 4th International Conference on Information Systems Development, pp 431-441, Bled, Slovenia, Moderna Organizacija, Kranj (1994).

[15]   C. Lakos and S. Christensen *A General Systematic Approach to Arc Extensions for Coloured Petri Nets* Proceedings of 15th International Conference on the Application and Theory of Petri Nets, Lecture Notes in Computer Science 815, pp 338-357, Zaragoza, Springer-Verlag (1994).

[16]   C.A. Lakos *Object Petri Nets – Definition and Relationship to Coloured Nets* Technical Report TR94-3, Computer Science Department, University of Tasmania (1994).

[17]   C.A. Lakos *From Coloured Petri Nets to Object Petri Nets* Proceedings of 16th International Conference on the Application and Theory of Petri Nets, Lecture Notes in Computer Science 935, pp 278-297, Torino, Italy, Springer-Verlag (1995).

[18]   C.A. Lakos *Pragmatic Inheritance Issues for Object Petri Nets* Proceedings of TOOLS Pacific 1995, pp 309-321, Melbourne, Australia, Prentice-Hall (1995).

[19]   C.A. Lakos *The Consistent Use of Names and Polymorphism in the Definition of Object Petri Nets* Proceedings of 17th International Conference on the Application and Theory of Petri Nets, Lecture Notes in Computer Science 1091, pp 380-399, Osaka, Japan, Springer-Verlag (1996).

[20]   C.A. Lakos *The LOOPN++ User Manual* Technical Report R96-1, Department of Computer Science, University of Tasmania (1996).

[21]   C.A. Lakos *Towards a Reflective Implementation of Object Petri Nets* Proceedings of TOOLS Pacific 1996, pp 129-140, Melbourne, Australia, Monash Printing Services (1996).

[22]   C.A. Lakos *On the Abstraction of Coloured Petri Nets* Proceedings of 18th International Conference on the Application and Theory of Petri Nets, Lecture Notes in Computer Science 1248, pp 42-61, Toulouse, France, Springer-Verlag (1997).

[23]   C.A. Lakos and C.D. Keen *Modelling a Door Controller Protocol in LOOPN* Proceedings of 10th European Conference on the Technology of Object-oriented Languages and Systems, Versailles, Prentice-Hall (1993).

[24]   S. Matsuoka and A. Yonezawa *Analysis of Inheritance Anomaly in Object-Oriented Concurrent Programming Languages* Research Directions in Concurrent Object-Oriented Programming, G. Agha, P. Wegner, and A. Yonezawa (eds.), pp 107-150, MIT Press (1993).

[25]   S.J. Mellor and S. Shlaer *A deeper look ... at execution and translation* Journal of Object-Oriented Programming, 7, 3, pp 24-26 (1994).

[26]   B. Meyer *Object-Oriented Software Construction* Prentice Hall (1988).

[27]   O. Nierstrasz *Composing Active Objects* Research Directions in Concurrent Object-Oriented Programming, G. Agha, P. Wegner, and A. Yonezawa (eds.), pp 151-171, MIT Press (1993).

[28]   G. Philipson *CASE technology's mid-life crisis* Informatics, pp 34-36 (1993).

[29]   W. Reisig *Petri nets : An Introduction* EATCS Monographs on Theoretical Computer Science, Vol. 4, Springer-Verlag (1985).

[30]   J. Rumbaugh and et al *Object-oriented modeling and design* Prentice-Hall (1991).

[31]   S. Shlaer and S.J. Mellor *Object Lifecycles – Modeling the World in States* Yourdon Press, Prentice Hall (1992).

[32]   C. Sibertin-Blanc *Cooperative Nets* Proceedings of 15th International Conference on the Application and Theory of Petri Nets, Lecture Notes in Computer Science 815, pp 471-490, Zaragoza, Spain, Springer-Verlag (1994).

[33]   P.A.C. Verkoulen *Integrated Information Systems Design: An Approach Based on Object-Oriented Concepts and Petri Nets* PhD Thesis, Technical University of Eindhoven, the Netherlands (1993).

[34]   J.M. Vlissides *Generalized Graphical Object Editing* Technical Report CSL-TR-90-427, Stanford University (1990).

[35]   K. Waldén and J. Nerson *Seamless Object-Oriented Software Architecture* Prentice-Hall (1995).

[36]   P. Wegner *Dimensions of Object-Based Language Design* Proceedings of OOPSLA 87, pp 168-182, Orlando, Florida, ACM (1987).

Appendix: Problem Specification
for a Cooperative Diagram Editor

Introduction

The aim of this case study is to serve as a common example for formal specification approaches that combine a formal model of concurrency (such as Petri nets) and the object-oriented approach. You are expected to exercise your formalism of choice on this problem, in order to demonstrate how it may deal with the specification of a software of reasonable size.

It is desirable to highlight or emphasize the way that the structure of the problem can be modelled in your formalism, and the way that reuse of components can be incorporated.

Statement of the Problem

The system to be studied is software allowing for cooperative editing of hierarchical diagrams. The diagrams may be the work products of some Object-Oriented Design methodology, Hardware Logic designs, Petri Net diagrams, etc. (Note that if the diagrams happen to coincide with the formalism you are proposing, be careful to distinguish clearly between the two.)

One key aspect of this problem is that the editor should cater for several users, working at different workstations, and cooperating in constructing the one diagram. In the Computer Supported Cooperative Work (CSCW) vocabulary, such a tool could be ranked amongst synchronous groupware (each user is informed in real time of the actions of the others) allowing for relaxed WYSIWIS (What You See Is What I See) :

each user may have his own customized view of the diagram under design, viewing different parts of the drawing or examining it at a different level of detail.

A second key aspect of this problem is that the editor should cater for hierarchical diagrams, i.e. components of the diagram can be exploded to reveal subcomponents.

A simple coordination protocol is proposed to control the interactions between the various users:

- Users may join or leave the editing session at will, and may join with different levels of editing priveleges. For example, a user may join the session merely to view the diagram, or perhaps to edit it as well. (See below.) The current members of the editing session ought to be visible to all, together with their editing privileges.
- Graphical elements may be free or owned by a user. Different levels of ownership should be supported, including ownership for deletion, ownership for encapsulation, ownership for modification, and ownership for inspection. (The ownership must be compatible with the user's editing privileges.)
- Ownership for deletion requires that no-one else has any ownership of the component - not even for inspection.
- Ownership for encapsulation requires that only the owner can view the internal details of the component - all other users can only view the top level or interface to the component.
- Ownership for modification allows the user to modify attributes, but not to delete the component.
- Ownership for inspection only allows the user to view the attributes.
- Only ownership for encapsulation can persist between editing sessions. (Note that this ownership is tied to a particular user, not a particular workstation.) All other ownership must be surrendered between sessions.
- Ownership for inspection is achieved simply by selecting the component.
- Other forms of ownership (and release) are achieved by an appropriate command, having first selected the component.
- The level of ownership of a component is visible to all other users, as is the identity of the owner.
- The creator of an element owns it for deletion until it is explicitly released.

Treatment of the Case Study

The above text states the basic requirements of the case study in an informal, incomplete (and maybe even inconsistent) way. One of the purposes of the case study is to explore the extent to which a formal specification may further the completeness of these requirements. You may deviate from those requirements if necessary, to highlight some feature of the approach you propose, but in this case you are expected to state precisely the reason for this deviation.

You are free to undertake this case study at whichever level is appropriate for the approach (e.g. to deal only with requirements engineering, or with the software architecture), or to focus only on one precise aspect of the problem that you feel most relevant (e.g. with the technical details of the necessary network communications, or on the high-level protocol between users). However, please try to make precise the scope of your treatment of the case study.

Using Petri Nets for Specifying Active Objects and Generative Communication

Tom Holvoet* and Pierre Verbaeten

Department of Computer Science, K.U.Leuven,
Celestijnenlaan 200A
B-3001 Leuven Belgium
Tom.Holvoet@cs.kuleuven.ac.be

Abstract. A majority of the research committee dealing with the development of advanced future software systems appears to agree upon the *requirements* that a software modelling approach should meet. In short, it should enable a "cost-effective software development for open parallel and distributed systems". One concept that is intrinsicly involved is *agent-based modelling techniques*, which combine object technology with concurrency.

Most current approaches to combining objects and concurrency fail to provide a well-founded integration of the conflicting dimensions in advanced object-oriented environments, such as classes, inheritance, concurrency and types.

This paper first elaborates on the set of requirements of an advanced software development approach. Thereafter, it discusses an approach for modelling, implementing and reasoning upon such systems. It is based on principles of concurrent object-orientation, generative communication – in particular the coordination language called Objective Linda – and Petri nets. The main contribution of this paper is to provide formal definitions of a Petri net formalism that is used to provide (1) semantics for the Objective Linda language, and (2) modelling the internal behaviour of concurrent objects.

The approach is illustrated through a case study, proposed by C. Sibertin-Blanc. The case is a variant of the well known dining philosophers problem, called "the Hurried Philosophers" problem.

1 Introduction / Motivation

In this introduction, we elaborate on the background from which the research presented in this paper evolved. We feel that this is necessary mainly because, unlike most other researchers interested in combining objects and Petri nets, we start from an innovative advanced software development technique – using a coordination model based on *generative communication* as the basic mechanism for agent interaction – into which we integrate Petri nets. We introduce the basics of this technique by presenting a short overview of related research domains.

---

* Research Assistant of the Belgian Fund for Scientific Research

G. Agha et al. (Eds.): Concurrent OOP and PN, LNCS 2001, pp. 38–72, 2001.

Since a few years, there has been a considerable shift on the concept of a software system. Formerly, a software system was "just" a program, possibly consisting of several modules, which runs on a computer, and sometimes on a parallel or distributed system. In the latter case, the programs use some kind of communication libraries in order to allow the distributed or parallel entities to cooperate, such as PVM, TCP, and so on. Some advanced object-oriented approaches provide "transparent remote object invocation", a means to hide distribution of objects for system designers.

However, a number of new factors make this static view of a program untenable. First, future software systems are bound to run on top of a network system (either local or wide-area networks). Besides practical issues, such as the reliability of the communication media and computer failure, a software development paradigm should allow systems to be *open*. The openness of systems is the most essential characteristic of future software systems. These are the main requirements for covering open systems:

- **distributed**: several entities, which may reside on any computer in a network system, constitute an application; one cannot rely on a notion of a global state;
- **concurrency**: the software entities perform concurrent activities; two kinds of concurrency may be involved: concurrency between entities, and multiple concurrent activities within one entity; in an object environment, one talks of inter-object and intra-object concurrency;
- **autonomy**: there is no a priori master/slave or client/server relation between the concurrent entities; entities possess the ability to proceed with their activities as it decides to;
- **evolution**: in such a concurrent and distributed environment, entities may *dynamically* join or leave the system, or may be replaced by other entities;
- **heterogeneity**: not only the computer architecture, involved networks and operating systems may be heterogeneous, the programming languages for realising software components may vary.

Second, the development of new applications is still too often started from (almost) scratch. Object-oriented techniques offer a number of concepts to enhance reusability (such as inheritance and polymorphism).

Petri Nets. Petri nets are undoubtedly the most attractive formalism for modelling concurrent systems. They have a well-founded, mathematical foundation, which has been studied for many years now. Petri nets are above all a visual formalism, which makes them accessible for constructing and understanding complex concurrent system models. Their main drawback is their lack of thorough modularization techniques. Several extensions to standard Petri nets have been proposed in order to deal with this deficiency. Still, these approaches basically offer what "step-wise refinement" offered to structured programming. Though these are of practical use for particular closed systems, they certainly cannot be applied to tackle the advanced software systems as mentioned above. This can easily be shown by pointing out that the result of any Petri net model is one overall net, modelling the concurrent behaviour of a concurrent system. During the execution of the modelled software system, the Petri net cannot cope with the addition of new, previously unknown entities.

Object-Orientation. The complexity of several applications and the quest for concepts to enable reusability naturally led to the ideas of object-orientation. This drift is called naturally, since it has been accepted that: the closer the software model is related to the real world, the easier it is to "control" the complexity of the system. Objects model real world entities.

Object-orientation offers a number of features that can be considered a first step towards an approach to tackle open systems:

- **encapsulation**: an object can be seen as a service provider to other objects; its internal state can only be inspected through its interface; hence, a client does not have to deal with the (complexity of the) the internals of other objects;
- **information hiding**: from the server object point of view, information hiding allows one to change the implementation of an object (data structures and implementation of interface operations) without having to change its clients;
- **classes and inheritance**: these are basic compositional features; classes represent sets of similar objects; inheritance is often called an "incremental modification"[11] technique, allowing the reuse of previously defined classes by extending them in a number of possible ways.

Object-Orientation and Concurrency: Agents. Yet one step closer to the requirements of future systems is the notion of active objects or agents[7]. In this paper, we will consider agents to be objects which have an autonomous activity. They are no longer purely reactive (only responding to external stimuli, i.e. invocations of other objects), they are pro-active (they can decide for themselves what actions they perform, e.g. based on their own internal state). The resulting design model is yet more natural, since real world entities are intrinsically concurrent objects. Agents meet our requirement of concurrency in future systems retaining the advantages of (sequential) object-orientation.

PINOLI: An Approach for Advanced Open System Development. Agent-orientation as shortly described above still does not meet the requirements of *open* systems. One of the main missing parts is a thorough *coordination model for open systems*. Moreover, for a number of reasons, it is important to have a formal instrument for reasoning upon the behaviour of an agent, and also upon the behaviour of a set of cooperating agents, called a configuration of agents. Based on such a notion of behaviour, *types of agents* and *types of configurations of agents* can then be defined.

In this paper, we overview the approach called PINOLI. It is first presented in [2]. We introduce the fundamentals of its coordination model, which is generative communication[1], show how this model is extended to meet the requirements of open systems, resulting in the coordination model Objective Linda[5]. We define a new high-level Petri net formalism that can be used to define semantics for the operations of this coordination language, and illustrate how this formalism can be used as an implementation language for active objects.

2   Generative Communication

Generative communication[1] aims to provide a dynamic communication forum between entities. It is based on a so-called *blackboard* that is used as a shared data space. Objects communicate by putting data items (or messages or ...) into the data space, which can be retrieved later on by other objects. An object can retrieve data items by asking the data space for an item *matching* a certain template item.

Generative communication has two fundamental characteristics. The first can be described as "communication orthogonality". The receiver of a message does not have any prior knowledge about the sender, and the sender has no prior knowledge about the receiver. Communication orthogonality has two important consequences:

- space uncoupling: there is no explicit need for a distributed naming scheme; the shared data space does not reveal the location of the sender and receiver of a message; this is sometimes called *anonymous communication*;
- time uncoupling: a data item which is put into the data space can be retrieved at any time by other entities, independently of whether the sender even still exists.

This uncoupling appears to be necessary for dealing with the openness of systems. It reduces the shared knowledge between different entities to a minimum, and entities can be replaced or added without adapting or even inform other entities.

Let us consider a simple example: a producer/consumer problem (see Figure 1). This can be modelled by two active objects, sharing a data space. Each produced item is put into the data space. The consumer continuously retrieves an item from the data space for consuming it.

Linda[1] is probably the best known example of a coordination model based on generative communication. It offers a set of primitive operations on a shared data space. The operations are blocking; they return only when their execution has terminated successfully.

- **in** (parameters): retrieves an item from the data space matching the actual and/or formal parameters of the operation, and removes it;
- **rd** (parameters): retrieves an item from the data space matching the actual and/or formal parameters of the operation, but does not remove it;

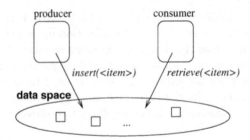

Fig. 1. The producer/consumer model using generative communication.

- **out** (parameters): puts the item, given through the parameters, into the data space.

Linda-like communication is a very powerful tool, allowing a large degree of flexibility by enforcing as few restrictions on communicators as possible. The weaknesses of Linda are that there is no support for *modularization* of the software system built around the shared data space, and that *dynamic process creation* is not considered in its first introduction[1]. While the latter has soon been resolved by an additional operation, **eval** (which creates a new active entity which blocks its originator until it delivers its results), the former has been a highly researched subject. Objective Linda, which is presented in the next section, provides by far the most elegant and natural solution to this problem.

3   Objective Linda

In [5], **Objective Linda** is proposed. Objective Linda is designed to meet the open system requirements of a coordination model. Its primary contribution is the support for multiple blackboard abstractions in a structured way. The data space is replaced by a hierarchy of object spaces.

Objective Linda aims at enhancing Linda mainly in three ways:

- *object matching*: contrary to Linda, where matching of objects is based on the contents of variables denoting the object's state, object matching in Objective Linda is based on object *types* and the *predicates* defined by the corresponding type interfaces; the items that entities (agents) can put and retrieve from object spaces are (passive) objects;
- *non-blocking multi-set operations on object spaces*: the operations defined on the object spaces (mainly insertion and retrieval of objects) are annotated with a time-out value; this is necessary to cope with system failures, which are not unlikely in open distributed systems; moreover, the operations are extended such that multi-sets of matching objects can be inserted in or retrieved from object spaces atomically.
- *multiple object spaces*: having only one object space is not feasible mainly for two reasons: scalability (for a huge set of agents, one global shared

blackboard is untenable from a practical point of view) and communication structure (using one blackboard as the communication medium for multiple groups of closely related agents imposes undesirable complexity); since object spaces are the forum for communication between objects, multiple object spaces may be used for separate goals, i.e. closely related objects should use the same object space (see Figure 2); in Objective Linda, each newly created active object knows of two object spaces: the object space in which it has been created (the *context* object space), and a newly created object space (the *self* object space); also, objects may create new object spaces, and object spaces may be communicated between active objects, using particular passive objects, called object space logicals (OS_Logicals); this allows agents to offer their object spaces to other agents.

At the abstract view, there is also a notion of *AgentSpace*, denoting the set of all active objects in the system.

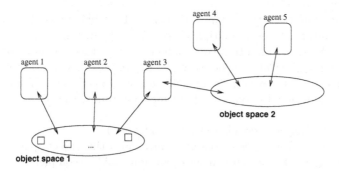

Fig. 2. Object spaces are the communication forum for closely related agents. Agents can be related to more than one object space.

Objective Linda allows to build open systems by modelling systems as dynamically changing sets of entities communicating in a decoupled manner, and by introducing multiple object spaces with enhanced operations (allowing multisets), reflecting the requirements of openness of the coordination model. The reader is referred to [5] for more details about Objective Linda.

We now present the Objective Linda operations on object spaces: out, in, rd, eval and attach. The operations in and rd specify multisets of matching objects to be retrieved by two parameters, namely min and max; min gives the minimal number of objects to be found in order to successfully complete the operation whereas max denotes an upper bound allowing to retrieve (small) portions of all objects of a kind. An infinite value for max allows to retrieve all currently available objects of a kind. While multisets of objects are necessary for in and rd, they have no substantial benefits for out or eval, but for consistency and simplicity reasons, Objective Linda also uses multisets of objects for out and eval. In the following, Objective Linda's operations on object spaces are summarized.

The notation is based on a binding to the C++ language, and thus the interface of the class Object_Space is shown. In order to simplify code, default values are assigned to the min, max, and timeout parameters causing Objective Linda's operations to behave in the default case analogous to the corresponding Linda operations.

bool out (Multiset *m , double timeout = inf_time)

> Tries to move the objects contained in m into the object space. Returns true if the operation completed successfully; returns false if the operation could not be completed within timeout seconds.

Multiset *in (OIL_Object *o , min = 1 , max = 1 , double timeout = inf_time)

> Tries to remove multiple objects $o_1 \ldots o_n$ matching the template object o from the object space and returns a multiset containing them if at least min matching objects could be found within timeout seconds. In this case, the multiset contains at most max objects, even if the object space contained more. If min matching objects could not be found within timeout seconds, the result has a NULL value.

Multiset *rd (OIL_Object *o , min = 1 , max = 1 , double timeout = inf_time)

> Tries to return clones of multiple objects $o_1 \ldots o_n$ matching the template object o and returns a multiset containing them if at least min matching objects could be found within timeout seconds. In this case, the multiset contains at most max objects, even if the object space contained more. If min matching objects could not be found within timeout seconds, the result has a NULL value.

bool eval (Multiset *m , double timeout = inf_time)

> Tries to move the objects contained in m into the object space and starts their activities. Returns true if the operation could be completed successfully; returns false if the operation could not be completed within timeout seconds.

Object_Space *attach (OS_Logical *o , double timeout = inf_time)

> Tries to get attached to an object space for which an OS_Logical matching o can be found in the current object space. Returns a valid reference to the newly attached object space if a matching object space logical could be found within timeout seconds; otherwise the result has a NULL value.

int inf_matches

> Returns a constant value which is interpreted as infinite number of matching objects when provided as min or max parameter to in and rd.

double inf_time

> Returns a constant value which is interpreted as infinite delay when provided as timeout parameter to out, in, rd, eval, and attach.

4   PINOLI: Petri Net Objects in Objective Linda

As most existing languages and systems, Objective Linda suffers from two aspects: programs are hardly readable (i.e. execution and communication scenarios are sometimes hard to track down), and it does not provide any formal framework for reasoning upon the behaviour, either of isolated entities, configurations of closely-related entities or entire systems. As such, the designer falls back on ad hoc methods for e.g. finding out whether the system may deadlock. This is often bound to result in erroneous conclusions. Moreover, the distributed and highly concurrent nature of the environment we consider makes this problem even more pronounced.

In this section, we first define a Petri net formalism. We show how this Petri net formalism can be considered as adding "syntactic sugar" to a well-known and widely used high-level Petri net formalism, Coloured Petri Nets. This allows us to use established formal analysis techniques defined for CPNs, as well as existing tools. The main contribution of this paper is to provide a formal definition of this high-level Petri net formalism, which has already been used on a more intuitive basics.

Then, we use the formalism to provide formal semantics for the Objective Linda operations.

We also recall an algorithm for translating an application consisting of a set of agent classes which use Objective Linda operations, passive object classes and a particular configuration, into one CPN. This CPN can then be used for formally analysing, simulating or implementing the application.

Figure 3 illustrates the basics of these ideas for the producer/consumer problem mentioned earlier. There are two (active) Petri net objects, a producer and a consumer, each specified by a simple C/E-net. These communicate through a common object space, which is modelled as a place, the arcs of which are annotated similar as in high-level nets.

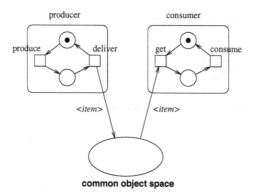

Fig. 3. The producer/consumer model using Petri nets for specifying individual object behaviour, and representing object spaces as places for relating nets of communicating objects.

4.1 Functional Passive Objects

Passive objects that are used in Objective Linda applications are considered record structures, including object data representation and operations. Object operations are defined as *functions* over a set of parameters, and yield an operation result. These passive objects will constitute the structured tokens for the high-level Petri net we propose in Section 4.2.

In this section, we propose a limited set of notations that are used for defining and using passive objects, rather than an entire programming language calculus. In particular, we define object types and subtypes, variable declarations and operation invocations.

Passive object types. We distinguish between primitive types and user-defined object types. Instances of primitive types as well as the most straightforward operations (summation for integers and reals, logical expressions for booleans, and so on) are understood without further explanation. The set of primitive types is denoted by $\mathcal{PT}$.

User-defined object types, or object types for short, include all types of passive objects which are described by their data representation and operations. $\mathcal{OT}$ denotes the set of object types.

- The set of all types (primitive and user-defined) is denoted by $\mathcal{T}$, $\mathcal{T} = \mathcal{PT} \cup \mathcal{OT}$
- **Primitive types** $\mathcal{PT}$ include   Boolean   .
                                                   Integer
                                                   Real
                                                   Char
- **User-defined object types** $\mathcal{OT}$:

Definition 1. *Object types A type $T \in \mathcal{OT}$ is a set of records, representing the object data representation as well as the object operations:*

$$
T = \left\{ \left(\begin{array}{|l|} \hline \textbf{Data} \\ \hline label_1 \ : \ T_1 \\ label_2 \ : \ T_2 \\ \ldots \\ label_n \ : \ T_n \\ \hline \end{array} ; \begin{array}{|l|} \hline \textbf{Operations} \\ \hline oper_1 \ : \ T \times T_{1,1} \times \ldots \times T_{1,n_1} \longrightarrow T_{1,res} \\ oper_2 \ : \ T \times T_{2,1} \times \ldots \times T_{2,n_2} \longrightarrow T_{2,res} \\ \ldots \\ oper_m \ : \ T \times T_{m,1} \times \ldots \times T_{m,n_m} \longrightarrow T_{m,res} \\ \hline \end{array} \right) \right\}
$$

where
- *$label_i$ and $oper_j$ are strings over some alphabet Σ ;*
- *$T_k \in \mathcal{T}$;*
- *$n, m, n_l \in I\!\!N$*

$\forall i, j, k, l \in I\!\!N$.

The data and operations part of object type T are referred to as $T.Data$ and $T.Operations$ respectively; the data items and operations are denoted by $T.Data.label_i$ and $T.Operations.oper_j$ respectively.

Contravariant subtyping. Types can be related through a subtype relationship. A subtype S of type T is denoted by $S \preceq T$; symmetrically, T is called a supertype of S, $T \succeq S$. A subtype S of type T extends this type T through

- additional data items,
- additional operations,
- (possibly) redefined operations from *T.Operations*.

The redefinition of the operations from *T.Operations* is restricted by the mechanism of contravariant specialization, in order to ensure the "principle of substitutability" [11].

Definition 2. *Contravariant subtyping Given two functional object types, S and T, $S \preceq T$. The subtype relation is contravariant if*

$$\forall\ oper \in S.Operations,\ oper\ :\ S \times S_1 \times \ldots \times S_n \longrightarrow S_{res}\ :$$
$$oper\ \in\ T.Operations$$
$$oper\ :\ T \times T_1 \times \ldots \times T_m \longrightarrow T_{res}$$

$$\Downarrow$$

$$(n = m) \wedge (S_1 \preceq T_1) \wedge \ldots \wedge (S_n \preceq T_n) \wedge (T_{res} \preceq S_{res})$$

Definition 3. *Type compatibility An object type S is called* type compatible *with type a T if $S \preceq T$.*

Definition 4. *Equivalent types If $S \preceq T$ and $T \preceq S$, S and T are called equivalent types $(S \equiv T)$.*

Declaration. A declaration $\boxed{var\ o : T}$ $(T \in \mathcal{T})$ defines a name of a variable, o, creates an object of type T by recursively declaring all data items of the object and yielding a record consisting of these data items and the operation functions defined for this type, and assigns this object to the variable. The data items are initialized as follows:

- $T \in \mathcal{PT} \Rightarrow o$ contains a random value $\in T$;
- $T \in \mathcal{OT} \Rightarrow o$ is a record as described above, where each label is declared recursively.

Reversely, when the scope in which the variable is declared reaches its point of termination, objects referred to by object variables become unreachable. A garbage collector can be responsible for regularly deleting unreachable objects.

Operation invocation. An operation invocation $\boxed{o.oper(par_1, \ldots, par_n)}$ invokes the function *oper* as defined in *o.Operations* with parameters $o, par_1, \ldots, par_n$. If *o.Operations.oper* does not exist, or if the number of invocation parameters does not match the number of parameters from the *oper* function signature, or if the type of one of the invocation parameters is not type compatible with the type of the corresponding formal parameter from the function signature, a terminal error occurs.

The result of the invocation is the result of the function call.

Polymorphism and dynamic binding. A similar type compatibility restriction as explained for operation invocations is imposed on variable bindings to objects, which we denote as an assignment at this time: $\boxed{o_1 = o_2}$ is legal only if the type of o_2 is compatible with the type of o_1. The result of this binding is that o_1 refers to an object of type $T(o_2)$, including its data and operations.

Dynamic binding bases the selection of the operation in an invocation on the actual type of the object rather than the declaration type, i.e. the type in the declaration of the object variable. Since in our type model, objects carry their entire set of interface operations in them, this selection is achieved straightforward by searching in the *Operations* part of the object directly.

4.2   OPN: A High-Level Petri Net Formalism

The formalism we propose, which is called OPN (Object Petri Net), is based on Coloured Petri Nets (CPNs) as defined in [4], and we adopt time annotation syntax from Time Petri nets (TPNs) [6] for introducing timeout exception handling.

In Sections 4.2 through 4.2, we informally overview all features of OPN. We show how these features are merely "syntactic sugar" for CPNs, and how they translate to CPN features. The choice of CPNs as the basic Petri net formalism is motivated by the fact that (1) CPNs are a well-known class of Petri nets which also serve as a reference for ongoing efforts for defining a standard high-level Petri net formalism [9], and (2) we intend to exploit existing CPN formal analysis techniques as well as tools for analyzing OPNs.

We provide a formal definition for OPN in Section 4.2.

Timeout Transitions.   The first "extension" of OPN compared to CPNs is a timeout mechanism. We illustrate this using anonymous tokens, without loss of generality. We introduce this timeout mechanism for transitions in order to model timeout exceptions of Objective Linda operations.

A timeout transition is based on a new kind of input arcs, called *non-deterministic input arcs*, and a new kind of output arcs, called *timeout output arcs*. Semantically, there is a distinction between transition *enabledness* and transition *ability to occur*. A timeout transition, with *timeout* as annotated value, is **enabled** if appropriate tokens (according to the corresponding arc expressions) are available from all but the "non-deterministic input places"; the transition **can occur** (normally) when it has been enabled for less than *timeout* time and if appropriate tokens are available from the "non-deterministic input places". In this case, no tokens are shifted towards the output places which are connected through timeout output arcs. A timeout transition is forced to occur if it has been enabled for *timeout* time without having occurred: tokens are withdrawn from all but the "non-deterministic input places" and tokens are shifted only through the timeout output arcs.

Graphically, non-deterministic input arcs and timeout output arcs are represented by an arc with open arrow heads; timeout output arcs are additionally annotated with a timeout value in square brackets. In general, it may be discussed whether such a feature might hamper the understandability of the intuitive dynamics of a net. However, the particular cases where we intend to use this feature are limited and do not cause confusion.

Consider the example in Figure 4.a. It shows a timeout transition with one normal ($P2$) and one non-deterministic input place ($P1$), and one normal ($P3$) and one timeout output place ($P4$). The marking M of this net has exactly one anonymous token in $P1$ and one in $P2$. The transition is enabled, since a token is available in $P2$. It can also occur because there is a token in $P1$, too. Occurring normally for this transition means retrieving tokens from all of its input places, and shifting a token towards the normal output arc. This results in the marking as shown in Figure 4.b. However, if the transition has been enabled for *timeout* time without having occurred, it is forced to occur. In this case, the token is retrieved from $P2$ only, and a token is shifted towards $P4$, the output-arc with the timeout annotation. This results in the marking as shown in Figure 4.c. Note that the same marking would be reached if $P1$ did not contain any token while Tr_1 was enabled.

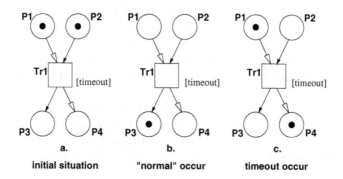

Fig. 4. A timeout transition and its dynamic behaviour.

Translation of timeout transitions to TPN transitions. A Petri net with timeout transitions can easily be translated into a TPN. TPNs associate with each transition t_i two times $t_{i,1}$ and $t_{i,2}$. A transition t_i can occur only if it has been enabled for at least $t_{i,1}$ time and it must occur before $t_{i,2}$ once it is enabled.

The translation of the transition from Figure 4.a into TPN components (without the markings of $P1$ and $P2$) is shown in Figure 5.

This translation serves as a definition of the semantics for timeout transitions. $Tr1$ is replaced by two TPN transitions, *Tr1-normal* and *Tr1-timeout*. *Tr1-normal* has the same input places as $Tr1$; *Tr1-timeout* is only adjacent to $Tr1$'s classical input places. Both share an additional, initially marked place *Pl-Tr1*. The time annotation of *Tr1-normal* is $[0, \infty]$, and is therefore left out. If the timeout transition $Tr1$ is enabled, *Tr1-timeout* is enabled. If there is also a

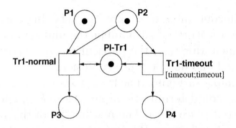

Fig. 5. The TPN equivalent of a timeout transition.

token in *P1*, *Tr1-normal* is enabled, too. When *Tr1-normal* occurs, tokens are shifted towards its output places, *P3* in the example. If *Tr1* has been enabled for *timeout* time without occurring, *Tr1-timeout* has been too. Therefore, *Tr1-timeout* will occur.

The additional place *Pl-Tr1* ensures that, if the timeout transition is multiply enabled (i.e. could occur more than once concurrently), the timeout counter of the *Tr1-timeout* transition is reset after the transition *Tr1-normal* has occurred. Consider the case in which *P1* initially contains one, and *P2* two tokens, and assume *timeout* to have a value indicating five seconds. Assume that after four seconds, the *Tr1-normal* occurs. If the place *Pl-Tr1* would not be considered, the transition *Tr1-timeout* would still be enabled for four seconds, and could fire at the fifth second. The place *Pl-Tr1* prohibits this by disabling *Tr1-timeout* for a very short time, and thus ensures that *Tr1-timeout* can only occur after being enabled for another five seconds.

TPNs as described above can be translated into time annotations for CPNs. This is however a rather technical matter with marginal contribution to the explanation of OPN. Moreover, because syntax and semantics of TPNs time annotation are simple and resemble timeout exceptions, this timeout mechanisms is used throughout this work.

Objects as Tokens. A characteristic of high-level nets is that tokens are not anonymous entities merely indicating state by their presence at a place, but they are structured and contain information. In OPN, tokens are passive objects, as described in Section 4.1, and places can contain tokens of any type.

Transitions in OPN can be annotated with a precondition and with a code segment. The precondition is a boolean expression that is a function of the tokens of its input places, as indicated by the input arc annotations. The code segment is a function of the tokens of its input places yielding objects which are stored in variables which are part of expressions that output arcs are annotated by.

Translation of functional (passive) objects in OPN to colours in CPN. The annotations containing objects and object types in OPN differ from colours and colour sets in CPN in mainly two ways.

objects vs. records First, CPN has no notion of objects as encapsulated entities of data and operations. However, CPN ML as the language for defining

colour sets supports complex data structures, including singletons, enumerated values, indexed values, tuples (colour set products), records, lists and multisets, as well as the most obvious functionality (record field selection, multiset operations). Below, we show how functional objects as defined in Section 4.1 translate to record structures and functions.

place colour set annotation Contrary to CPNs, where a particular colour set is associated with each place, denoting the type of tokens that places may contain, each place in OPN can in principle contain tokens of any type. Since object types are in subtype relation with at least the object type PassiveObject, and since a subtype relation denotes subset inclusion, we annotate each place with the colour set equivalent of the PassiveObject type.

Object types to records and functions. An object type for OPN objects consists of (1) a record structure for data items, and (2) a record structure of functional operations. A translation of object types into terms of CPN ML colour sets and functions results in two separate parts: a record structure for the data items of the object type, and a set of ("global") functions that take data record structures as parameters. E.g. the object type T as presented on page 46 translates to a record colour set, denoted $RecordType(T)$, and functions $oper_1, \ldots, oper_m$. The complete translation, however, of a functional specification of objects as in Section 4.1 into plain CPN ML record colour sets and functions, comes down to a small parser, the details of which are not relevant in this document. We therefore restrict ourselves to provide translation for the main aspects of object type definitions in order to convince the reader that such is possible.

- $RecordType(T) = $ **record**
$$typename : String(T) *$$
$$label_1 : RecordType(T_1) *$$
$$\ldots$$
$$label_n : RecordType(T_n);$$

  where $RecordType(T_i \in \mathcal{PT}) = T_i$
- **fun** $oper_1$ ($object$, $par_{1,1}$, $\ldots$, par_{1,n_1}) $= \ldots$
  $\ldots$
  fun $oper_m$ ($object$, $par_{m,1}$, $\ldots$, par_{m,n_m}) $= \ldots$

Each record type representing an object type includes the name of the object type as a string, along with a record field of the respective record type for each data item in the object type definition. The record type for primitive types is the primitive type itself, since these are supported by CPN ML.

The functions have the same number as parameters as the respective object operations. The function body contains a functional CPN ML expression which is the transaction from the operations' body expressions. An operation which is inherited through a *subtype* relation and hence has the same function signature is dealt with by one function. In that case, the function body starts with a case expression using the *typename* field of the record types of the function parameters as a selector.

An object variable **declaration** translates to a variable declaration, in particular of the *RecordType(PassiveObject)* type, which can refer to any value. An operation **invocation** comes down to a function call using the object variable as a first parameter. The case expression mentioned before deals with **dynamic binding**, and record type variables are **polymorphic** entities in CPN ML. Type checking is left over to the function body.

These observations suggest that each expression in terms of object types and variables of the functional object model presented in Section 4.1 can be provided with an equivalent in CPN ML in terms of colour sets and functions.

Multiset Annotations.   CPNs (among other formalisms) allow arcs to be annotated such that transition occurrences consume or produce multisets of tokens. We propose a syntax for such multiset expressions, suitable for our needs. The arc expressions we require allow a non-deterministic choice between several multisets by only demanding a minimal and a maximal number of tokens. Arc annotations look like $\{E\}_{min}^{max}$, where E is an expression yielding an object of a type $T \in \mathcal{T}$ (written $E : T$), indicating that the enabledness of the corresponding transition depends on the availability at the corresponding place of at least min tokens of type T which match the object yielded by E; when the transition occurs, at least min and at most max matching tokens of type T are withdrawn from the place. The expression can be as short as a variable, denoting that any (one) object of type T matches, or it can be a more complex functional expression that returns an object of type T. Multiset arc annotations allow us to provide an adequate model for the Objective Linda operations which use multisets of objects.

For simplicity reasons, we also allow an alternative syntax for denoting explicitly enumerated multisets, namely $\{e_1, e_2, \ldots, e_n\}$. As special cases, $\{e\}$ denotes a multiset containing exactly one object e, and $\{e\}_1^1$ denotes a multiset containing exactly one object matching e. Also, the expression E may be replaced by a type name, e.g. T, denoting that any objects of type T qualify.

An example is sketched in Figure 6. Assuming that $E : T$, transition *Tr1* consumes either three, four or five tokens of type T from place *P1*.

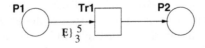

Fig. 6. A multiset arc annotation.

Translation of Objective Linda multiset arc annotations to CPN arc annotations. Because suitable multiset annotations are already part of the CPN definition, we need no explicit translation to a lower-level formalism, here. The translation of the OPN multiset arc annotations from Figure 6 into expressions known in CPN is presented in Figure 7. The translation consists of two components. First, the annotation is replaced by an ordinary multiset annotation m, which is declared to be a multiset of type T. Second, the transition precondition is extended by a predicate over this multiset m: the size of m should be at least min (3) and at most max (5), and all members of m must yield the value true when the function match is applied to them (the CPN ML function ext_col takes two parameters, a function and a multiset, and yields a multiset containing the results from the function calls to each member of the multiset ; the CPN ML function cf returns the number of appearances of a particular value in a multiset ; this predicate is true only when each member of multiset m yields the value true).

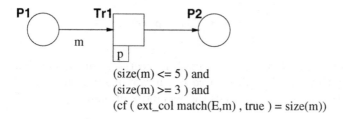

(size(m) <= 5) and
(size(m) >= 3) and
(cf (ext_col match(E,m) , true) = size(m))

var m : Multiset(T);

Fig. 7. Translation of multiset arc annotation into CPN terminology.

Named Places.   The last feature we introduce is *named places*. The idea is the following. A named place in an OPN is a place which is parametrized through a variable representing the places name. The content of this variable is the *identifier* of the *actual place* that the named place represents. As a result of transition occurrings, the content of these variables may be changed, which allows another actual place to be represented by the named place.

The intention is to model object space places as named places. Since within one agent an object space is represented by variables containing a "reference" to an actual object space, which can be changed at run time (e.g. by an attach operation), this kind of flexibility is necessary.

Graphically, named places are represented by a small oval, representing the container for the place name, within an oval representing the named place itself. Transitions that require to read or modify name variables of named places are adjacent to the inner ovals, as shown in Figure 8.a.

Named places allow a hidden form of dynamicity in the net structure. Changing the content of a named place variable means changing the actual input or output place for transitions that are adjacent to the named place. Hence, corresponding arcs are no longer a relation between places and transitions, but rather a relation between place variables and transitions.

Named places may seem to be a subversive feature, breaking the soundness of the OPN formalism. Yet again there is a simple high-level net equivalent. The

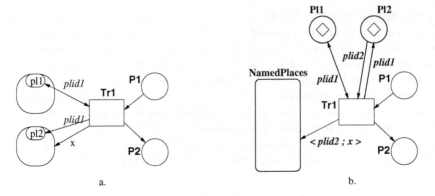

Fig. 8. Named places.

translation of named places to high-level net features can be performed by the following algorithm. This is illustrated by an example in Figure 8.

The net translation also defines the semantics of named places.

1. First, join all named places into one place (*NamedPlaces*), and provide a separate place per used named place variable (here: *Pl1* and *Pl2*). The type of the tokens of these new places is a special object type, PlaceId. Graphically, we represent PlaceId's by the ◇ sign.

   When a new object space reference is required, a new, unique PlaceId object is assigned to the place name. In the translation to CPNs, this new PlaceId object is withdrawn from a particular place, IdSpace, containing a set of unique PlaceId objects. The corresponding translation is shown in Figure 9.

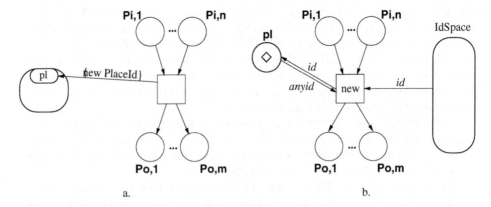

Fig. 9. Translation of object space reference creation to CPN.

The purpose of the introduction of these new places is two-fold. First, a token in such a place represents the variable and actually contains the reference to the named place as its value. Second, it is used to protect an agent from inconsistencies due to concurrent (re)assignments to the named place variables.

2. Then, each annotation of an arc that is connected to a named place is altered such that each expression denoting a token x is replaced by an expression denoting a token $<place\text{-}id;x>$, where *place-id* is an expression yielding an identifier of a place, e.g. the variable *pl1*. Finally, each transition that is adjacent to a named place and/or manipulates one or more place variables is connected through a bidirectional arc with the new places representing the place variables (such as *Pl1* and *Pl2* in the example).

Formal Definition of OPN.   In this section, we provide a formal definition of OPNs. We first introduce a notation we use for multisets and necessary type expressions for defining variable bindings, similar to the formal definitions of CPNs ([4]). Then we define OPNs through its structure and annotations. Finally, OPN dynamic behaviour is presented.

Multisets and expressions

Definition 5. *Multisets A* **multiset** *m over a non-empty and finite set S, is a function $m \in [S \to I\!N]$. The non-negative integer $m(s) \in I\!N$ denotes the number of appearances of the element s in the multiset m.*

By S_{MS} we denote the set of all multisets over S. The non-negative integers $\{m(s) : s \in S\}$ are called the **coefficients** *of the multiset m, and $m(s)$ is called the coefficient of s.*

We also subsume multiset operations such as summation, scalar multiplication, subtraction, comparison, multiplicity, and so on.

Similar to CPNs, we use the terms *variables* and *expressions* in the same way as in the typed lambda calculus and functional programming languages. This means that expressions do *not* have *side-effects* and variables are *bound* to values (rather than being assigned to).

We define the following:

- the *type of a variable* v is denoted by Type(v)
- the *type of an expression* E is denoted by Type(E)
- the *set of variables in an expression* E is denoted by Var(E). Var(E) however only includes free variables, i.e. those which are not bound e.g. by a local definition
- a *binding of a set of variables* $V = \{v_1, \ldots, v_n\}$ is denoted by $< v_1 = c_1, \ldots, v_n = c_n >$, where it is demanded that $Type(c_i) \preceq Type(v_i)$ for each v_i in V.
- the *value obtained by evaluating an expression* E *in a binding b* is denoted by $E < b >$. It is demanded that Var(E) is a subset of the variables of b, and the evaluation is performed by substituting each variable $v_i \in Var(E)$ with the value $c_i \in Type(v_i)$ determined by the binding b.

Definition of OPNs. OPNs are formally defined as follows. Remark that for each element (sets, mappings) of an OPN, we use the shorthand El instead of El_{OPN} if no confusion is possible (i.e. P instead of P_{OPN}, G instead of G_{OPN}, and so on).

Definition 6. *OPNs An* **OPN** *is an 11-tuple* $OPN = (P, T, F, NPF, \mathcal{T}, I, G, C, AE, NPAE, TO)$, *where*

(i) *P is a finite set of* **places**
  $P = OP \cup NP$ *where*
  – *OP is a set of (ordinary) places*

  – *NP is a set of named places,*
     where a mapping name $: NP \longrightarrow PlaceId$ *is defined*
(ii) *T is a finite set of* **transitions**, $P \cap T = \emptyset$
(iii) *F is the set of arcs (the* **flow relation***),*
  $F = DF \cup NF$ *where*
  – *DF is a set of (deterministic) arcs,*
     $DF \subseteq (P \times T) \cup (T \times P)$
  – *NF is a set of non-deterministic arcs,*
     $NF \subseteq (P \times T) \cup (T \times P)$
(iv) *NPF is a set of arcs which are adjacent to the place name of a named place,*
  $NPF \subseteq (NP \times T) \cup (T \times NP)$
(v) *$\mathcal{T}$ is a set of object types*
(vi) *I is an* **initialization** *function,*
  $I : P \longrightarrow \mathcal{E}$ *such that* $\forall\, p \in P : Var(I(p)) = \emptyset$
(vii) *G is a* **guard** *function,*
  $G : T \longrightarrow \mathcal{E}(Bool)$
(viii) *C is a* **code** *function,*
  $C : T \longrightarrow \mathcal{E}_{MS}$
(ix) *AE is an* **arc expression** *function,*
  $AE : F \longrightarrow Arc(\mathcal{E})$
(x) *$NPAE$ is a* **arc expression** *function for arcs* $\in NPF$,
  $NPAE : NPF \longrightarrow \mathcal{E}(PlaceId)$
(xi) *TO is a* **timeout** *function,*
  $TO : T \longrightarrow \mathbb{N}$

where

– *$\mathcal{E}$ denotes the set of all expressions, $\mathcal{E}(T)$ the set of all expressions $E \in \mathcal{E}$ such that $Type(E) = T$;*
– *$Arc(\mathcal{E})$ the set of multiset expressions as presented in Section 4.2.*

We also define a wrapper function $AAE : F \cup NPF \longrightarrow Arc(\mathcal{E})$, where $AAE(f \in F) = AE(f)$ and $AAE(f \in NPF) = NPF(f)$

(i) (ii) (iii) The **places**, **transitions** and **arcs** are described by three sets, P, T and F, which are demanded to be finite and pairwise disjoint. The set of arcs, F, is subdivided into "normal" arcs (DF) and non-deterministic arcs (NF).

(iv) NPF represents the set of arcs adjacent to named places and modelling place name related manipulations.

(v) $\mathcal{T}$ is the set of **object types**, i.e. $\mathcal{T} = \mathcal{PT} \cup \mathcal{OT}$, as previously defined in Section 4.1.

(vi) The **initialization** function I maps each place p into an expression which is a multiset of tokens. The expression is not allowed to contain any variables.

(vii) The **guard** function G maps each transition t into a predicate, i.e. an expression yielding a boolean.

(viii) The **code** function C associates an expression with each transition. The expression is a function representing the piece of code that is executed on transition occurrence.

(ix) (x) The **arc expression** functions associate an expression to each arc in an OPN.

(xi) The **timeout** function maps each transition into a natural number, denoting the timeout value for timeout exception.

Dynamic behaviour of OPNs. With the above definition of the static structure of OPNs, we can now formally define their dynamic behaviour. However, we do not consider time aspects in defining the dynamic behaviour since it adds substantial complexity to the definitions without a equally useful return, especially as we refrain from using timeout exception for any of the examples we propose further in this work. The work presented in [10] includes definitions of a time-based Petri net formalism which can be imitated for defining OPN dynamic behaviour with timeouts.

We first define an auxiliary notation:

- $\forall t \in T : Var(t) = \{v \mid v \in G(T) \text{ or } \exists \, a \in F(t) : v \in Var(E(a))\}$

Definition 7. *For a transition $t \in T$ with variables $Var(t) = \{v_1, \ldots, v_n\}$, we define the **binding type** of t as: $BT(t) = Type(v_1) \times \ldots \times Type(v_n)$, and we define the set of all **bindings** as $B(t) = \{(c_1, \ldots, c_n) \in BT(t) \mid G(t) < v_1 = c_1, \ldots, v_n = c_n >\}$.*

Definition 8. *A **token distribution** is a function M defined on P. We define the relations neq and leq as:*

- $M_1 \neq M_2 \iff \exists p \in P : M_1(p) \neq M_2(p)$
- $M_1 \leq M_2 \iff \forall p \in P : M_1(p) \leq M_2(p)$

The relations $<, <, \geq$ and $=$ are defined analogously to $\leq$.

Definition 9. *A **binding distribution** is a function Y defined on T such that $Y(t) \in B(t)_{MS}, \forall t \in T$. If $b \in Y(t)$, for $b \in B(t)$, we say that $(t, b) \in Y$.*

Definition 10. *A **marking** of an OPN is a token distribution; the **initial marking** is the marking obtained by evaluating the initialization expressions. A **step** is a non-empty binding distribution.*

Definition 11. *A step Y is enabled in a marking M iff:*

$$\forall p \in P : \sum_{(t,b) \in Y} AAE((p,t)) < b >\, \le M(p)$$

When a step is enabled, it may occur. When a step occurs, tokens are removed from the input places and added to the output places of the occurring transitions, based on the arc expressions, evaluated for the occurrence bindings.

Definition 12. *When a step Y is enabled in a marking M_1, it may **occur**, yielding a new marking M_2 such that:*

$$\forall p \in P : M_2(p) = (M_1(p) - \sum_{(t,b) \in Y} AAE(p,t) < b >$$

$$+ \sum_{(t,b) \in Y} C(t)(AAE(t,p) < b >))$$

where $C(t)(AAE(t,p) < b >)$ is the result of applying the transition code segment to the input tokens in the token binding b.

Similar to definitions for CPNs, we can define occurrence sequences, reachable markings, and so on.

4.3   A Formal Framework for Objective Linda

Thus far, we have presented Objective Linda for modelling concurrent applications, a formal functional model for passive sequential objects, and a high-level Petri net formalism with object-oriented extensions in Section 4.2. In the following, we employ these formal models for providing a formal basics for Objective Linda.

A Formal Computational Model. The Objective Linda computational model concerns the description of passive, sequential objects, and internally concurrent, autonomous active objects (agents). Agents can perform computational activities by creating and manipulating passive objects. Given the presented functional object model and the OPN formalism, providing a formal framework in which Objective Linda fits into is straightforward.

Passive objects. In most cases, it is very easy to transform an imperative piece of code that, at the end, yields one value (object), into a functional language construct that produces the same result. Therefore, in the context of a formal framework for Objective Linda, we demand each **passive object** class (type) specification to be a functional object type, without hampering the expressive freedom for application designers. This will result in a hierarchy of formal passive object types.

Default operations Passive objects in Objective Linda are required to provide at least the operations new, destroy and match. If these operations are not explicitly provided within a user definition, default versions are provided implicitly.

new The new operation is intended to establish initializations of object data items.

$$new : T \to T$$

destroy The destroy operation does not have memory de-allocating semantics, but rather allows to describe a number of actions that need to be performed when the object is actually destroyed.

$$destroy : T \to T$$

match This operation has a default implementation body, yielding the boolean value true. This operation can be redefined in order to specialize matching between objects.

$$match : T \to Boolean$$

Agents. **Agent classes** are formally represented by an OPN Petri net, called the agent net. Similar to the early (informal) introduction of a Petri net specification for internal agent behaviour, the Petri net describes internal concurrency, synchronization and causality between internal computations. The computational capacities of agents coincide with the expressiveness of functional expressions on passive objects.

Coordination Semantics. The Objective Linda coordination language introduces the concept of (multiple) object spaces, and a set of coordination operations on these object spaces. This section provides Petri net semantics for these Objective Linda coordination constructs.

Object spaces are represented in an agent net by means of named places (called object space places), in which the name serves as an object space reference. Coordination **operations** are modelled as agent net transitions which are adjacent to the object space places. The semantics of the operations are shown in Figure 10 through fragments of agent nets.

Figure 10.a shows a fragment of a OPN representing the in operation. It is a *timeout transition* with a *non-deterministic arc* with *multiset annotation*, and with a *named* input *place*, i.e. the object space place on which the in operation is performed. If the state of the agent (i.e. the marking of the agent net) is such that the transition is enabled, it can occur normally if appropriate objects are available in objsp within *timeout* time. In any case, if the transition did not occur within the time bound, it is forced to occur, retrieving tokens from the input places $P_{i,1} \ldots P_{i,n}$, and putting a token into the timeout output place P_{to}. Note that if appropriate objects are available in objsp within *timeout* time, the transition may, but is not forced to occur. This non-deterministic choice in a Petri net formalism reflects possible physical failures (such as network communication).

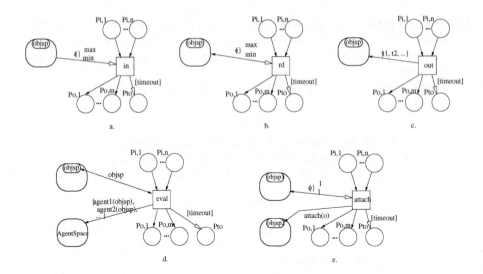

Fig. 10. OPN semantics for Objective Linda coordination operations.

Similarly, templates for rd and out are presented in Figure 10.b and 10.c. The rd transition is a *timeout transition* with a *non-deterministic arc* with *multiset annotation*, and with a *named* input *place*, the object space place on which the operation is invoked. If the transition is enabled, it can occur if appropriate objects can be retrieved from the object space. If it occurs, such objects are withdrawn from the object space place and replaced immediately.

The out transition is a *timeout transition* with an output arc with (explicitly enumerated) *multiset annotation*, and with a *named* output *place*, the object space place on which the operation is invoked. Since there is no non-deterministic input arc, the possibility to occur coincides with enabledness for this operation: if the transition is enabled, it either occurs normally, within *timeout* time, or the timeout exception mechanism makes it occur if it has not occurred after being enabled for *timeout* time.

The template for *eval* is presented in Fig. 10.d. If the transition occurs within the time bound, a multiset m of agents is shifted towards the *AgentSpace* place, representing the activation of new agents. In this case, every new agent $A(o) \in m$ is assigned o as its initial *context* object space. The bidirectional arc between the transition and the name variable of *objsp* denotes that the *PlaceId* token of the object space is used (withdrawn and replaced) for assigning a *context* object space to the newly created agents.

The *attach* operation (Fig. 10.e) is a *rd* operation with a multiset containing a single object o of a specific type, *OS_Logical*. Additionally, there is an output arc to the name variable of the object space to attach to where the value of *attach(o)* is stored. The function *attach* ∈ [OS_Logical → PlaceId], returns the PlaceId of the object space to which the OS_Logical matching o refers. Fig. 11 illustrates the expansion of the *attach* operation to lower level CPN/TPN components, analogous to Fig. 8.

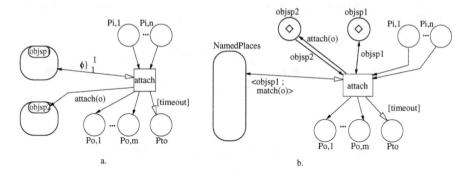

Fig. 11. The expansion of attach to CPN/TPN components.

Formal Verification of Objective Linda Applications. Open systems typically change configurations during execution. New, previously unknown kinds of agents and objects may be introduced, while others may disappear. Therefore, it is by definition impossible to construct one Petri net for the entire system.

However, this does **not** imply that one is not interested in reasoning on the behaviour of the system at a particular moment in time, or reasoning on the behaviour of a particular system configuration. For that purpose, one can take a snapshot of the system, consisting of a particular configuration and a particular set of agent definitions and object types. An outline of an algorithm for constructing a "snapshot overall net" is presented below. The resulting net is a CPN to which known analysis techniques as well as tools, such as Design/CPN [3], can be employed.

Constructing an Overall Net. We now provide an outline of an algorithm that takes an Objective Linda application with functional objects and agents with net specifications, and yields one CPN covering the entire application.

The algorithm is divided into two phases. The first phase translates each agent definition separately into a CPN. The second phase merges the CPNs for different classes of agents into one CPN. The global idea is that all agents of one class are represented by one CPN, where the state of each individual agent of a particular class is distinguished by tokens that include an identification of the agent of which they are part of the agent net. Separate CPNs then are merged through shared places (modelling the object spaces).

We illustrate the algorithm through an example. In particular, we consider a model of a Producer/Consumer setting, presented in Figure 12 and Figure 13. The model consists of two classes of agents: Producer, Consumer and StartUp. A StartUp agent initializes the system by evaling a producer and a consumer into

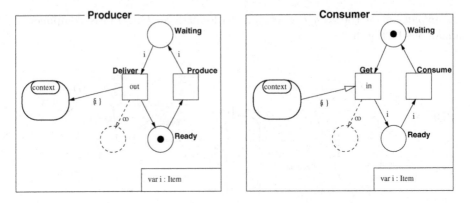

Fig. 12. Agent classes for Producers and Consumers.

an object space. In the model this is represented by the transition Startup which delivers two tokens to the AgentSpace when occurring.

The descriptions of the Producer and the Consumer agents yield no surprises. A Producer agent repeatedly produces new items (an internal action, represented by the transition produce) and outs them into its context object space. A Consumer agent repeatedly retrieves items from its context object space and consumes them (transition Consume). The time-out value of the Objective Linda operations is set to infinity (∞), and therefore the corresponding time-out annotated arcs can be left out.

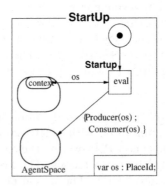

Fig. 13. An agent class for initializing Producer/Consumer setting.

Constructing one CPN from a Objective Linda application.

Phase 1: Translation of Agent Definitions. The goal of the first phase is to have a representation of all agent instances by one net per class of agents. The resulting net is a CPN. The first phase consists of four steps:

1. The first step is to translate all OPN **transitions** representing coordination and composition **operations** into CPN features, as explained in Section 4.2. At this time, per class of agents Agent, a new object type is defined, *CreationRequest*Agent. The output tokens of the eval transition are replaced by pairs < *CreationRequest*Agent ; *os*> where *os* refers to the object space the new agent is to be created in.

2. The second step is to translate the **named places**, i.e. the object space places of the agent nets. All object space places are replaced by one place (per agent class). Arcs that were adjacent to an object space place are replaced by an arc that is adjacent to the joint object space place. The arc annotation is changed and a new place per object space variable (of type PlaceId) is introduced, as explained in Section 4.2: the arc annotation is changed, such that, instead of identifying tokens by the annotation E, now tokens are pairs <*os_variable* ; *E*>, where *os_variable* equals the name of the object space which the transition originally was adjacent to; and new places containing a token of type *PlaceId* are introduced.

3. All **arcs** that are not adjacent to any object space place change **annotations**: each annotation, say E, denoting an object, is replaced by an annotation <*agent_id* ; *E* >. Arcs that have an implicit annotation, denoting anonymous tokens, are replaced by an annotation representing tokens as <*agent_id*>.

4. Finally, the **initial marking** of the agent is withdrawn. A new transition, called *InitNewAgent*Agent, is added. Its set of output places equals the set of places that are marked initially. The annotations of the respective output arcs correspond to the initialization expressions of the initially marked places.

   One occurrence of this transition creates a new agent identifier, and then forwards tokens, which correspond to the establishment of the initial marking of a new agent.

Figure 14 shows the result of this phase on the Producer agent specification.

Phase 2: Merge Agent Definitions. When all the agent definitions have been expanded separately, they can be merged into one overall net. This is achieved correctly by merging the object space places of the Petri nets from the agent class specifications resulting from the first phase into one overall object space place, and by merging all AgentSpace places into one common AgentSpace place. Furthermore, a new input arc is introduced for each transition *Init-NewAgent*Agent, connecting it with the AgentSpace place; it is annotated by <*CreationRequest*Agent ; *context*>. This ensures that the initialization transition of a new agent can occur if (and only if) a creation request for an agent of class Agent has been issued.

Figure 15 shows the overall net of the model of the Producer/Consumer setting.

And finally, passive object type definitions are translated to record structures and appropriate functions, as explained in Section 4.2.

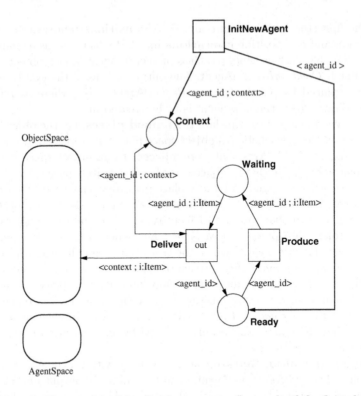

Fig. 14. The net describing Producer agents; the result of the first phase.

5 Case Study: The Hurried Philosophers

In this section we sketch a solution for the case study proposal on "the Hurried Philosophers" of C. Sibertin-Blanc. Although the proposal left quite some room for interpretation, it pinpoints several aspects that concern the coordination of the philosophers. It subsumes direct communication will be used. We feel that this fact hampers a thorough illustration of the strength of using generative communication, especially for joining or leaving philosophers. However, we intend to follow the case study as originally proposed, hence our model will to some extent simulate direct communication through generative communication.

5.1 Statement of the Case Study

The case is an extension of the well known dining philosophers problem. A number of philosophers are sitting around a table. Each philosopher has a plate and shares a fork with his right and one with his left neighbour. A philosopher can only eat when he is holding two forks, one in his left and one in his right hand.

Philosophers "talk" to each other. When a philosopher wants to eat, but he does not have a fork in his left hand, he can ask his left neighbour to give him

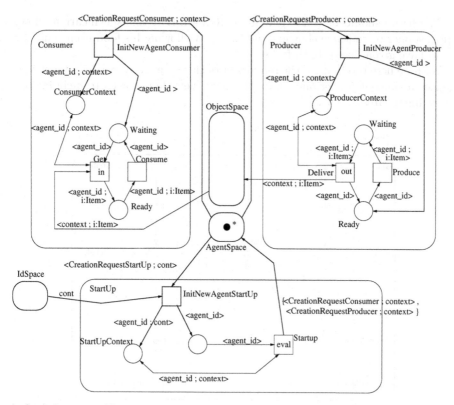

Fig. 15. The overall net of the Producer/Consumer model.

a fork. The same holds for his right-hand side. When a philosopher holds a fork e.g. at his left-hand side, and his left neighbour asks for a fork, he must give it. The same holds for his right-hand side.

Philosophers may introduce a new philosopher. A new philosopher brings a plate and has one fork in his left hand. A philosopher can only introduce another philosopher at his left-hand side.

A philosopher may leave the table, either based on an autonomous decision, or because one of his neighbours told him to do so. A philosopher can only leave when he holds a fork in his left hand.

5.2 Eating Philosophers

Obviously, there are two kinds of objects: philosophers and forks. Forks are mainly used as static data items, for which we do not provide a net. A first attempt to model the philosophers behaviour is presented in Figure 16. The marking of the net shows a philosophers initial state (i.e. its state at creation time). A philosopher holds one fork, and his mouth is empty.

The figure should be self-explanatory. Remark that this specification allows a philosopher to continue chewing and finally swallowing his food even if he/she would have given one or both forks.

We note here too that for the remainder of this case study, we do not consider the use of timeout exceptions, hence the agent nets do not need to include non-deterministic arcs either.

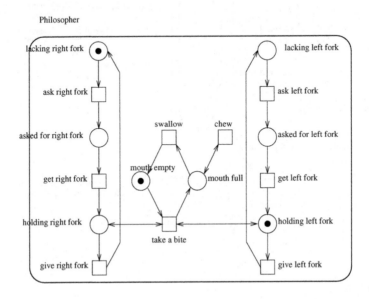

Fig. 16. Philosopher's internal behaviour.

Philosophers need to communicate: they need to ask each others forks, exchange forks, and so on. Hence, using the Objective Linda object model, we need to describe what object spaces are needed. Since object spaces are the forums for inter-object communication, and since each pair of neighbour philosophers communicate forks, we introduce one object space between two neighbours. This results in a "LeftObjectSpace", *OS-LN* , and a "RightObjectSpace", *OS-RN* (see Figure 17). Since in general, Objective Linda active objects know of two default object spaces, the *self* and the *context* object spaces, we note that, in this situation, *OS-RN* serves as the context, and *OS-LN* serves as the self object space.

To illustrate how the object spaces are used as common space for communication, Figure 18 shows a configuration of philosophers.

5.3   Arriving and Leaving Philosophers

Now, we add the more advanced features of the hurried philosophers problem: introducing a new philosopher, deciding to leave, and telling neighbour philosophers to leave.

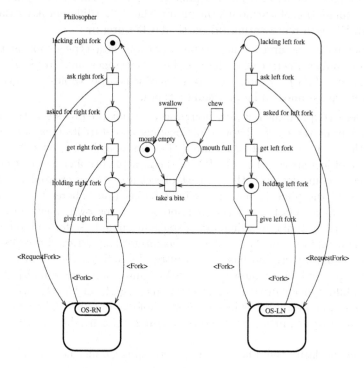

Fig. 17. Philosopher's behaviour taking communication with other philosophers into account.

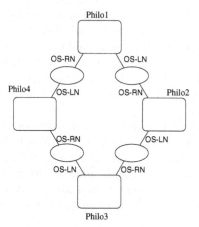

Fig. 18. A configuration of four philosophers with their respective object spaces.

The complete model of a hurried philosopher is presented in Figure 19. To assist the explanation of the model, the picture shows five different parts (labelled A through E), each representing (a) particular feature(s).

The net model assumes a new type: $Message$ is a type denoting the set of messages philosophers can exchange. These messages are: $ReqRightLeave$, $ReqLeftLeave$, I_Left, $FoundYou$. Besides of this type, tuples of objects and $OS\_Logical$ are assumed to be pre-defined types.

First, let us explain the initialization procedure, i.e. the procedure that agents follow for joining the table. Before a philosopher can start its task of eating, he must settle down at the table, meaning that his neighbours must "move up" and change object spaces. The $OS\text{-}RN$ of the new philosopher is set at creation time, since it coincides with the context object space it is created in, namely the $OS\text{-}LN$ of the philosopher who is introducing him. Thereafter, he inserts (by firing $init$) an OS-Logical (i.e. a reference) of his own, new $OS\text{-}LN$, for notifying his left neighbour which object space he should use for communicating with this new philosopher. When the left neighbour has retrieved the OS-logical of this new philosopher, and has thus adjusted his $OS\text{-}RN$, he notifies the new philosopher by sending the message $FoundYou$ into their common object space. The new philosopher can then retrieve this message from his $OS\text{-}LN$ (i.e. can fire $init\text{-}done$), and establishes his new state: holding only his right fork. At this time, he can start acquiring forks and responding to his neighbours who acquire forks.

Now, let us clarify the different parts of the philosopher net model:

A The first part deals with the eating behaviour of a philosopher, as it was introduced earlier. Notice that the initial marking has been changed for allowing a proper initialization as mentioned above.

B This part allows a philosopher to introduce a new philosopher. Since a philosopher is an active object, this action corresponds to an $eval$ operation in Objective Linda[5]. The precondition of this action is that the philosopher holds his left, but not his right fork, and that at least one fork is available at the fork heap. The fork heap is an additional common object space, and the corresponding place is initially marked with a set of tokens which are actually $Fork$ objects. When this transition fires, a new agent is created and shifted towards the $AgentSpace$. This new agent is given $OS\text{-}LN$ as its context object space.

C A philosopher may order his left or right neighbour to leave the table, but only when he holds his left fork and not his right fork himself. This is modelled by two transitions, $tell\_right\_to\_leave$ and $tell\_left\_to\_leave$ respectively. $tell\_right\_to\_leave$ puts a token with value $ReqRightLeave$ of type $Message$ into the $OS\text{-}RN$. $tell\_left\_to\_leave$ puts a token with value $ReqLeftLeave$ of type $Message$ into the $OS\text{-}LN$. How an agent deals with such a message can be found in part E.

D When a philosopher retrieves a token with an "I_Left" message, he knows that his right neighbour has left. The token then also contains an object space logical, called $logi$ in the picture, which identifies the object space to be used as off now for communicating with his "new" right neighbour. Therefore, the agent appropriately alters his OS-RN variable.

E Finally, philosophers can leave the table, either out of their own will, or because his left or right neighbour told him so. The latter case is represented by the transitions *right_told_to_leave* and *left_told_to_leave*. Let us consider *right_told_to_leave*. The *OS-RN* is an input place of this transition, which can only fire if a *ReqLeftLeave* message is available from that input place. This message will only be available there if the philosopher's right neighbour decided that this agent should leave. When this transition fires, it removes tokens as to prevent the philosopher to eat again, it inserts a message in the *OS-LN* for informing his left neighbour of his departure, and he puts his *Fork* back on the heap of forks.

A philosopher can leave (either because one of his neighbours told him so, or he decided to leave autonomously). The result is that a message is dropped into the object space shared by the philosopher and his left neighbour, telling him he is leaving, and sending along the name of the philosopher's right neighbour's object space (which is to become the right neighbour object space of his left neighbour).

A philosopher who notices that his right neighbour is leaving will modify his right neighbour object space as soon as this right neighbour's fork has been retrieved from the right neighbour's object space at this moment.

5.4   Evaluation

In his case study proposal, C. Sibertin-Blanc mentioned a few features to be tested on the formalism that is used for modelling the application. Here we formulate how the formalism we used conforms to these features.

- **local control**: One of the basic characteristics of the Objective Linda approach is the autonomicity of the agents. This is not harmed by introducing a Petri net as the specification of the agents behaviour and as the execution control engine. Each agent controls its own transitions, and hence retains the control over its actions and its decisions.
- **dynamic instantiation**: Agents may join and leave the system at run time. A restriction of this sort can only induced by the application, by imposing that only a finite set of forks be available to the table.
- **dynamic binding**, "the partner of communication is defined when the communication occurs, not by the model": Generative communication decouples communicating objects. This and the fact that object spaces, the forums of inter-object communication, can be passed on between objects allows dynamic partnership of communication.
- **inheritance and polymorphism**: This feature could not be demonstrated with the application. However, both passive and active objects are considered to have types, which can be related through inheritance or subtyping, and polymorphism is possible of course. Ongoing research focuses on types of

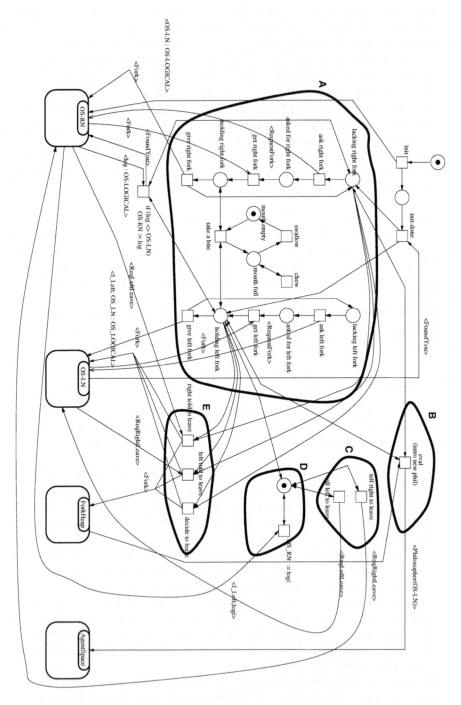

Fig. 19. Complete philosopher specification.

active objects in the Objective Linda. The Petri net behaviour representation plays a major role in defining types and subtypes.

Similar to the work of C. Sibertin-Blanc on Cooperative Nets [8], we have developed an algorithm for translating the application, modelled using Objective Linda agents with net-based behaviour specifications, into a very commonly used kind of nets, Coloured Petri nets (CPNs, [4]). The importance of this is obvious. It allows us to use analysis techniques and existing tools to reason upon the behaviour of agents and the behaviour of configurations of agents, and to deduce certain properties.

6   Conclusion

This paper presents ongoing research in combining concurrent object-orientation, coordination through generative communication and Petri nets. The advantages of (in particular, concurrent) object-orientation are known (reusability, flexibility, modular development, and so on). The coordination model of Objective Linda is well suited for coordination problems in open systems. Petri nets provide a sound basics not only for visually representing complex entities, but also for performing formal analysis on isolated objects, sets of communicating objects and entire systems.

The combination of agents, generative communication and Petri nets results in a high-level, visual, formal technique for designing, implementing and analyzing complex and open systems. The ability to translate system models onto well known net models is important. It allows to rephrase research results from the Petri net community in terms of agents and agent configurations, reuse analysis techniques and exploit existing tools.

In [2], the principles of PINOLI were introduced, however, without a thorough formal basics. In this paper, we provide a formal definition of a high-level Petri net formalism, called OPN which is used for providing semantics for Objective Linda coordination operations.

Introducing nets into active objects also yields the possibility to tackle certain type-theoretical issues. A formal definition of types of agents and a related formal notion of subtyping has been defined based on observable behaviour of agents in [2]. Based on the same notion, we can define types of particular components, namely configurations of cooperating agents. This allows us to define "plug-compatibility" of components, meaning that we can formally define when a component (a set of cooperating agents which become accessible to other components to one of more public object spaces) may be replaced by another component without affecting the behaviour of the overall system. This is an important step in dealing with component-oriented development for software in open systems.

References

1. D. Gelernter. Generative Communication in Linda. *ACM Transactions on Programming Languages and Systems*, 7(1):80–112, 1985.

2. T. Holvoet and T. Kielmann. Behavior Specification of Active Objects in Open Generative Communication Environments. In *Proc. HICSS30, Sw Track*, pages 349–358, Hawaii, 1997. IEEE Computer Society Press.

3. K. Jensen. *Coloured Petri Nets. Basic Concepts, Analysis Methods and Practical Use.* Springer, 1992.

4. Kurt Jensen. Coloured Petri Nets: A High Level Language for System Design and Analysis. In G. Rozenberg, editor, *Advances in Petri Nets*, number 483 in Lecture Notes in Computer Science, pages 342–416. Springer, 1990.

5. Thilo Kielmann. Designing a Coordination Model for Open Systems. In P. Ciancarini and C. Hankin, editors, *Coordination Languages and Models*, number 1061 in Lecture Notes in Computer Science, pages 267 – 284, Cesena, Italy, 1996. Springer. Proc. COORDINATION'96.

6. P. Merlin. *A Study of Recoverability of Computing Systems.* PhD dissertation, Dept. of Information and Computer Science, University of California, Irvine, California, 1974.

7. Y. Shoham. Agent-oriented Programming. *Artificial Intelligence*, pages 51–92, 1993.

8. C. Sibertin-Blanc. Cooperative Nets. In *Proceedings of the 15th International Conference on Application and Theory of Petri Nets, Zaragoza, Spain*, LNCS 815, June 1994.

9. SO/IEC JTC1/SC7/WG11. High-Level Petri Net Standard, August 1997. Working draft.

10. W. M. P. van der Aalst, K. M. van Hee, and P. A. C. Verkoulen. Interval Timed Coloured Petri Nets and their Analysis. In *Advances in Petri Nets '93*, number 691 in Lecture Notes in Computer Science, pages 453–472. Springer-Verlag, Berlin, 1993.

11. P. Wegner and S.B.Zdonik. Inheritance as an Incremental Modification Mechanism or What Like Is and Isn't Like. In *ECOOP'88 (European Conference on Object-Oriented Programming), Oslo, Norway*, LNCS 322, pages 55–77. Springer-Verlag, 1988.

Object-Oriented Nets with
Algebraic Specifications:
The CO-OPN/2 Formalism*

Olivier Biberstein, Didier Buchs, and Nicolas Guelfi

LGL-DI, Swiss Federal Institute of Technology, CH-1015 Lausanne, Switzerland
{Olivier.Biberstein|Didier.Buchs|Nicolas.Guelfi}@di.epfl.ch

Abstract. This paper presents and formally defines the CO-OPN/2 formalism (Concurrent Object-Oriented Petri Net) which is devised for
the specification of large concurrent systems. We introduce the basic
principles of the formalism, and describe how some aspects of object-
orientation – such as the notions of class/object, object reference, inheritance and subtyping – are taken into account. In CO-OPN/2, classes
(considered as templates) are described by means of algebraic nets in
which places play the role of attributes, and methods are external parameterized transitions. A semantic extension for the management of
the object references is defined. Inheritance and subtyping are clearly
distinguished. Interaction between objects consists of synchronizations.
Synchronization expressions are provided which allow the designer to
select interaction policies between the partners. We also provide a step
semantics which expresses the true concurrency of the object behaviors.
Finally, in order to illustrate the modeling capabilities of our formalism, we adopted a case study on groupware or, more specifically, on a
cooperative editor of hierarchical diagrams.

1 Introduction

The idea of modeling systems by combining Petri nets and data structures
emerged around 1985. Since then, several researches have introduced some notions peculiar to object-orientation.

In line with these research works, the CO-OPN (Concurrent Object-Oriented
Petri Nets) [8] approach and its tools have been devised so as to offer an adequate framework for the specification and design of large concurrent systems.
Recently, a new version of CO-OPN has been developed in order to enrich and
overcome some limitations of this object-based formalism. Notions of class, inheritance as well as subtyping have therefore been introduced. This version,
called CO-OPN/2 [4], is based upon two underlying formalisms: order-sorted

---

* This work has been sponsored partially by the Esprit Long Term Research Project
  20072 "Design for Validation" (DeVa) with the financial support of the OFES (Office Fédéral de l'Éducation et de la Science), and by the Swiss National Science
  Foundation project 20.47030.96 "Formal Methods for Concurrent Systems".

G. Agha et al. (Eds.): Concurrent OOP and PN, LNCS 2001, pp. 73–130, 2001.

algebras [10], for data structure matters, and the algebraic nets [16], for the operational and concurrent aspects.

In CO-OPN/2, classes are object templates described by a special kind of algebraic nets, while objects are net instances of these classes. A specific aspect of CO-OPN/2 is that it introduces subtyping and distinguishes it from inheritance. Interaction between objects is realized by using "synchronization expressions", a concept more general than the classical transition fusion of Petri nets. At the semantics level, transition systems are used to give a step semantics to CO-OPN/2 specifications which takes into account both inter- and intra-object concurrency. The transition systems are partially defined using SOS like inference rules and a semantic mechanism is introduced in order to manage the object identifiers.

Close to our work, let us mention [13,17,3], three approaches based on high level Petri nets. The reader interested in a finer description of these formalisms can either consult this book and the associated workshop proceedings or have a look at the study we propose in [11]. The latter presents a set of classification criteria and then situates each approach according to these criteria.

CO-OPN/2 is a specification formalism [7] designed for the development of complex systems from a software engineering point of view. Our approach aims at easing specification, design, implementation, and verification phases of system development, rather than focus on analysis aspects.

The organization of this paper is the following. Section 2 presents an informal introduction of the basic concepts of CO-OPN/2. Section 3 illustrates, by means of a simple example, various notions of the formalism such as the notions of object/class, object reference, inheritance, and subtyping. Then, Section 4 and 5 describe, respectively, the syntactic and the semantic aspects of the formalism. Finally, Section 6 deals with a common case study on groupware, namely a cooperative editor of hierarchical diagrams.

2   CO-OPN/2 Principles

In order to provide a progressive presentation of CO-OPN/2 and to allow the reader to easily compare CO-OPN/2 with other approaches, we briefly give, in this section, the most relevant characteristics of the formalism.

Object and Class. An object is considered as an independent entity composed of an internal state and which provides some services to the exterior. The only way to interact with an object is to ask one of its services; the internal state is then protected against uncontrolled accesses. Our point of view is that this protection mechanism, known as encapsulation, is an essential feature of object-orientation and there should be no way of violating it.

CO-OPN/2 defines an object as being an encapsulated algebraic net in which the places compose the internal state and the transitions model the concurrent events of the object. A place consists of a multi-set of algebraic values. The transitions are divided into two groups: the parameterized transitions, also called the methods, and the internal transitions. The former correspond to the services provided to the outside, while the latter compose the internal behaviors of an object. Contrary to the methods, the internal transitions are invisible to the exterior world and may be considered as being spontaneous events.

An important characteristic of the systems we want to consider is their potential dynamic evolution in terms of the number of objects they may include. In order to describe similar dynamic evolving systems, the objects are grouped into classes. A class describes all the components of a set of objects and is considered as an object template. Thus, all the objects of one class have the same structure.

Object Interaction. In our approach, the interaction with an object is synchronous, although asynchronous communications may be simulated. Thus, when an object requires a service, it asks to be synchronized with the method (parameterized transition) of the object provider. The synchronization policy is expressed by means of a synchronization expression, which may involve many partners joined by three synchronization operators (one for simultaneity, one for sequence, and one for alternative or non-determinism). For example, an object may simultaneously request two different services from two different partners, followed by a service request from a third object.

Concurrency. Intuitively, each object possesses its own behavior and concurrently evolves with the others. The Petri net model naturally introduces both inter-object and intra-object concurrency into CO-OPN/2 because the objects are not restricted to sequential processes.

The step semantics of CO-OPN/2 allows for the expression of true concurrency, which is not the case of interleaving semantics. A set of method calls can be concurrently performed on the same object. Nevertheless, the purpose of CO-OPN/2 consists of capturing the abstract concurrent behavior of each modeled entity, with the concurrency granularity not being found in the objects, but rather in the invocations of the methods. A set of method calls can be concurrently performed on the same object.

Object Identity. Within the CO-OPN/2 framework, each class instance has an identity, which is also called an object identifier, that may be used as a reference. Moreover, a type is explicitly associated with each class. Thus, each object identifier belongs to at least one type. An object identifier order-sorted algebra is constructed in order to reflect the subtyping relation that is established between the classes types, i.e. two carrier sets of object identifiers are related by inclusion if, and only if, the two corresponding types are related by subtyping.

Since object identifiers are algebraic values, it is possible to define data structures that are built upon object identifiers, e.g. a stack or a queue of object identifiers. Obviously, the places of algebraic nets may contain object identifiers.

Inheritance and Subtyping. We believe that inheritance and subtyping are two different notions used for two different purposes. Inheritance is considered as being a syntactic mechanism which frees the specifier from the necessity of developing classes from scratch and mainly serves in the re-use of parts of existing specifications. A class may inherit all the features of another and may also add some services, or change the description of some already defined services.

Our subtyping relationship is based upon the strong version of the substitutability principle [2,14]. This principle implies that, in any context, any class instance of a type may be substituted for a class instance of its supertype, while the behavior of the whole system remains unchanged. In other words, the instances of the subtype have a strong semantic conformance relationship with the supertype definition. This conformance relationship is based, in CO-OPN/2, upon the bisimulation between the semantics of supertype and the semantics the subtype restricted to the behavior of the supertype.

Both inheritance and subtyping relationships must be explicitly given, but the respective hierarchies generated by these relationships do not necessarily coincide. Identifying both inheritance and subtyping hierarchies leads to several limitations as stated by Snyder [18] and America [1].

3   Informal Specification Example

In this section we informally present the concrete CO-OPN/2 language by means of a simple example that encompasses the most important aspects of the formalism. The example makes a analogy with the production in a Swiss chocolate factory. The factory produces two kinds of chocolate packagings and a conveyor belt conveys the empty packagings to the packaging unit, which, in turn, fills in the empty packagings by means of two different kinds of chocolate.

Figure 1 shows a general view of the factory. The figure depicts the three main devices of the factory: the packaging producer, the conveyor belt, and the packaging unit.

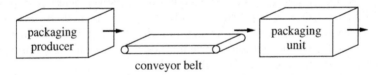

Fig. 1. General View of the Swiss Chocolate Factory.

Figure 2 depicts the two kinds of packagings produced by the factory. The regular packagings have a given number of square holes filled with pralines, while the deluxe packagings have some more round holes for truffles.

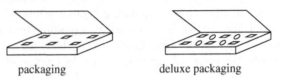

Fig. 2. Both Kinds of Packagings.

The decomposition of the example into abstract data types and classes is quite straightforward. In the CO-OPN/2 approach based on the object-oriented paradigm, it appears that the producer, the consumer, and the conveyor-belt can be naturally considered as objects and not as data structures. The same reasoning may be applied to the packagings. However, the chocolates are still considered as data structures, because it is not necessary to give them an identity, and, moreover, they do not have any operational behavior. Thus, four classes will be used to specify the four kinds of objects, while the chocolates will be specified by means of algebraic specifications.

Before introducing the data structures aspects and explaining the four objects involved in the example, let us have a look at Figure 3. This figure illustrates various entities of our example and the synchronizations between them. The shaded ovals represent the objects. The black and white rectangles are, respectively, the parameterized transitions and the internal transitions. The former corresponds to the methods, while the latter forms the internal behavior of the objects. The dashed arrows illustrate the synchronization requests.

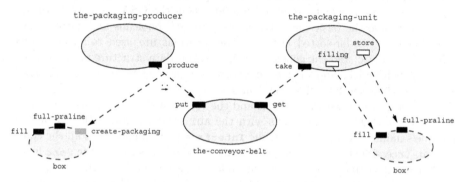

Fig. 3. Various Objects of the Swiss Chocolate Factory.

We observe that our example looks like the well-known producer-consumer example in which the packagings play the role of the messages. However, it is worthwhile to notice that:

1. The packagings are not algebraic values, but they are modeled by means of objects which are dynamically created. Actually, the elements conveyed to the packaging unit are identities of packagings, i.e. references represented by algebraic values. Since these objects are dynamically created, they are drawn as dotted ovals.

2. The nature of the conveyor-belt has a first-in-first-out structure. This data structure is modeled by means of order-sorted algebras.

3. The packaging unit has to fill the packagings conveyed by the conveyor-belt and store them when they are full. These internal operations are modeled by some internal transitions that ask to be synchronized with the boxes.

4. When many partners are involved in a synchronization, the synchronization is represented as a tree of dashed arrows where each node is labeled with a synchronization operator. Such kind of tree forms what is called a synchronization expression that allows to express the policy of the synchronizations between the partners involved in the synchronization. The orientation of the arrows represent the dependencies between the partners rather than the data-flow of the synchronization. For example such a synchronization expression appears in Figure 3. The synchronization involves

three methods: produce, create-packaging, and put. Informally the synchronization operator '..' means that the method produce calls in sequence create-packaging and put. In other words, the produce method behaves the same as the firing of create-packaging followed by put. Three synchronization operators are provided: one for simultaneity, one for sequence, and one for alternative or non-determinism.

3.1   Abstract Data Types

As previously mentioned, abstract data types (ADT) are described by order-sorted algebras (OSA). Subsorting, partial operations, polymorphism and overloading are notions which are encompassed within the OSA framework. In our example, beside the basic data structures (common booleans and naturals not provided here), order-sorted algebraic specifications are used to specify both kinds of chocolate, and the generic first-in-first-out data structure present in the conveyor-belt. Each kind of data structure is described by means of an ADT module as shown in Specifications 1 and 2.

One easily identifies the sections and the various fields of each section[1]. The header of an ADT module starts with the **ADT** reserved word followed by the module's name. On the one hand, the **Interface** section comprises all the components accessible from the outside. Usually a list of sorts follows the **Sort** field, while the subsorting relation is given under the **Subsort** field. Operations are defined under both **Generators** and **Operations** fields. These operations are coupled with their profile in which the underscore character '_' gives the position of the respective arguments. On the other hand, the operations properties expressed by means of equations or axioms lie under the **Axioms** field in the **Body** section. The equations are conditional positive and established as follows:

$$[\, Cond \Rightarrow]\ Expr_1 = Expr_2$$

Note that $Expr_1$ and $Expr_2$ may not be of the same sort. An optional condition *Cond* which determines the context in which an axiom holds true can be added. Of course, variables are available whenever defined under the **where** field. Since algebraic specifications are divided into modules, a module may need some components from another module. This kind of dependency is expressed by the **Use** field, followed by the list of all the required modules.

The ADT module in Specification 1 describes the two kinds of chocolate and illustrates a simple use of subsorting. The chocolates are arranged according three sorts in a subsorting relation. Thus, a value of the supersort chocolate represents either a praline or a truffle.

Genericity is another way of making specifications more concise without sacrificing their legibility. Generic abstract data types are data structures that depend on some formal parameter modules. The header of a generic data type must begin with the **Generic** keyword, followed by the module name and the list of its

---

[1] The Courier bold face words correspond to the reserved words of the language. These reserved words are not case sensitive and some words can be singular or plural.

```
ADT Chocolate;
Interface
  Sorts chocolate, praline, truffle;
  Subsort
    praline < chocolate,   truffle < chocolate;
  Generators
    P : praline;  T : truffle;
End Chocolate;
```

Spec. 1. Both the Kinds of Chocolate.

formal parameter modules. The generic first-in-first-out data type which is used by the conveyor belt is illustrated in Specification 2. Module `Elem` corresponds to the formal parameter module, which is replaced by an actual module during the instantiation process. It must be defined as a **Parameter** module.

One way of using partial operations, such as `extract` and `first`, consists of partitioning the domains and associating a sort `ne-fifo` for the non-empty fifos and a "larger" one `fifo` for the empty or non-empty fifos. In this way, both operations are partial across the entire domain, but total on their subdomains.

```
Generic ADT Fifo(Elem);              Parameter ADT Elem;
Interface                            Interface
  Use Natural;                         Sort elem;
  Sorts ne-fifo, fifo;               End Elem;
  Subsort ne-fifo < fifo;
  Generators
    [] : -> fifo;
    insert _ _ : elem fifo -> ne-fifo;
  Operations
    first    _ : ne-fifo -> elem;
    extract _ : ne-fifo -> fifo;
    size     _ : fifo -> natural;
Body
  Axioms
    first (insert e []) = e;
    first (insert e f)  = first f;

    extract (insert e []) = [];
    extract (insert e f)  = insert (e extract f);

    size [] = 0;
    size (insert e f) = 1 + (size f);

    where  e : elem;  f : ne-fifo;
End Fifo;
```

Spec. 2. A Generic First-In-First-Out Data Type and its Formal Parameter.

3.2   Classes

A class module is an encapsulated algebraic net used as a template from which objects are statically or dynamically instantiated. As for the ADT modules, the class modules have four sections that play the same role but contain different informations. The header starts with the **Class** reserved word also followed by the module's name. The **Interface** section comprises mainly the type of the class, the objects that are statically instantiated, and the methods provided to the outside. The state and the behavioral aspects of the class lie in the **Body** section.

The state of the class instances is represented by a collection of places that contains multi-set of algebraic values described by order-sorted algebraic specifications. The object state can be modified by means of two kinds of events: the methods (which are parameterized and correspond to the external object stimuli) and the internal transitions (which are concealed in the body and represent the spontaneous object reactions). Sometimes we call these two types of events: the *observable* events (methods) and the *invisible* events (internal transitions).

Cooperation between objects is accomplished by a synchronization mechanism. That is to say, any object event can request a synchronization with methods of one or a group of partners. Moreover, the synchronization policy can be chosen by means of a synchronization expression for which three synchronization operators are provided: '//' for simultaneity, '. .' for sequence, and '+' for alternative. The event's effects, data-flow and synchronization requests, are described by means of *behavioral axioms* which are established as follows:

$$Event \; [\textbf{with} \; Sync] \; :: \; [\, Cond \Rightarrow] \; Pre \text{->} Post$$

in which

- *Event* is an internal transition or a method name along with its parameters;
- *Sync* is an optional synchronization expression describing which methods of which partners are involved and how they interact with each other;
- *Cond* is an optional condition on the algebraic values of the formula which must be satisfied so that the event can occur;
- *Pre* and *Post* correspond, respectively, to the pre- and post-conditions of the common Petri nets, i.e. they represent what is consumed and what is produced at the various places within the net.

An event that requires a synchronization with some partners can occur if and only if the various methods involved in the synchronization can occur. In other words, a synchronization request abstracts the behavior of all the methods of its partners according to the synchronization operators. Note that a method is not a function (no value is returned) and can be regarded as a predicate. Nevertheless, formal and effective parameters are unified during a synchronization and informations can therefore be exchanged.

Regular Packagings and Dynamic Creation. The textual representation of the class module which describes the regular packagings is given in Specification 3. The methods along with their profile are grouped under the **Method** field and the **Type** field declares the type name of the instances. This name is used whenever an object identity has to be defined. This field has been introduced in order to avoid the confusion between the name of the class module and the type name of the instances. Both names are often very similar but address two different notions (class names are used for inheritance, while type names are used for subtyping). On the other side, the places and the behavioral axioms respectively given under the **Place** and **Axiom** fields remain concealed in the module's body.

```
Class Packaging;
Interface
  Use Chocolate;
  Type packaging;
  Methods
    fill _ : chocolate;
    full-praline;
  Creation
    create-packaging;
Body
  Use Natural, Capacity;
  Place
    #square-holes _ : natural;
  Axioms
    n > 0 => fill P :: #square-holes n -> #square-holes (n-1);

    full-praline :: #square-holes 0 -> #square-holes 0;

    create-packaging :: -> #square-holes praline-capacity;

    where n : natural;
End Packaging;
```

Spec. 3. The Modeling of the Packagings.

The regular packagings only contain a certain quantity of pralines (square chocolates). This quantity is stored in the place or attribute #square-holes, which is updated by the method fill. The method models the action of filling an empty square hole with a praline. The #square-holes place contains the number of empty holes in the packaging, and it is decreased by 1 if and only if the method's parameter is a praline. Note that the variable n has to be grater than zero; so, the method fill cannot occur when the state of the packaging has reached the value 0. In order to detect when a packaging is full, a method full-praline is provided. This method plays the role of a predicate because it can only occur when all the empty holes are filled; i.e. when value 0 is in #square-holes.

As to the dynamic creation, each class is equipped with a specific method **create**. When invoked, this method automatically determines a new object identifier and creates the new object. For example, the method call 'o.**create**' (o is variable of a given class type) assigns a new object identifier to o and creates the object according to the type of o.

It is often convenient to perform in one step the creation of an object as well as the initialization of its attributes according to some parameters. Thus, CO-OPN/2 provides an optional field **Creation** under which some creation/initialization methods can be declared. As usual, the properties of these methods lie under the **Axioms** field. It is necessary to mention that these methods do not have to deal with of the creation of the objects. Indeed, a creation/initialization method consists of an implicit synchronization with the **create** method followed by the creation/initialization method.

For example, in Specification 3, the create-packaging method under the **Creation** field first creates a new packaging object and then initializes its attribute with the praline-capacity constant.

The **create** and the creation/initialization methods are graphically represented by gray rectangles as shown in Figure 4 that depicts, on the left side, the

`Packaging` class. The places are drawn as circles and the data-flow corresponds to the labeled plain arrows. Explanations about the right side of the figure and the meaning of the arrows are given below.

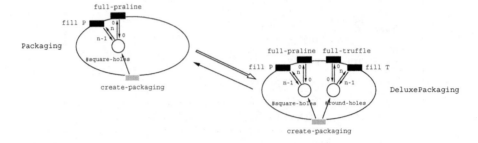

Fig. 4. Outline of the packagings classes.

Inheritance and Subtyping. The factory deals with two kinds of packagings, regular, as detailed above, and deluxe. Deluxe packagings are very similar to regular packagings, with the exception that they are bigger and contain both kinds of chocolate. Since the deluxe packaging shares some ingredients with the regular boxes, the new class associated with the deluxe packagings does not have to be developed from scratch and inheritance may be used.

Remember that, on the one hand, a class is considered as a template and, on the other hand, inheritance is viewed as a mechanism for design convenience. Moreover, inheritance which can be used for three purposes is clearly separated from the notion of subtyping.

CO-OPN/2 provides an **Inherit** section followed by the class modules that the current class inherits. In this section three fields can be specified:

- the **Rename** field allows us to rename some inherited identifiers;
- the **Redefine** field groups the components whose the properties are redefined in the inheriting class;
- the **Undefine** field groups the components that have to be eliminated in the inheriting class.

Specification 4 shows how the `DeluxePackaging` class module reuses the `Packaging` class. We observe that only enrichment inheritance is used; i.e. only new ingredients are added. Figure 4 illustrates clearly that the new attribute `#round-holes`, the `full-truffle` method, as well as the new behavior of the method `fill` are added. The thin and the thick arrows represent, respectively, the subtyping and the inheritance relations between the two kinds of packagings. Note that the creation-initialization methods are not inherited and, therefore, have to be defined whenever inheritance is used.

The explicit subtyping relation under the **Subtype** field specifies that the external behaviors of the subclass `DeluxePackaging` restricted to the external

```
Class DeluxePackaging;
Inherit Packaging;
   Rename packaging -> deluxe-packaging;
Interface
   Use Packaging;
   Subtype deluxe-packaging < packaging;
   Method
      full-truffle;
   Creation
      create-packaging;
Body
   Place
      #round-holes _  : natural;
   Axioms
      n > 0 => fill T :: #round-holes n -> #round-holes (n-1);

      full-truffle :: #round-holes 0-> #round-holes 0;

      create-packaging :: -> #square-holes praline-capacity,
                                #round-holes truffle-capacity;

      where n : natural;
End DeluxePackaging;
```

Spec. 4. The Deluxe Packagings Inherit from the Regular Packagings.

behaviors of the superclass Packaging are identical[2].This means that, in any context, an instance of the DeluxePackaging class can be substituted for an instance of the Packaging class. In other words, here, the subtype hierarchy and the inheritance hierarchy are the same.

Object Identity and Polymorphism. Subtyping naturally characterizes inclusion polymorphism, which is the ability for a function, a method, an attribute, etc. to deal uniformly with values of a given type as well as with values of any subtype.

Let us illustrate in our example where inclusion polymorphism occurs. Imagine a situation where, a few years ago, the factory did not have an automatic packaging producer. The packagings were merely arranged in a heap. Specification 5 describes such a heap of packagings.

Remember that, the deluxe-packaging type is a subtype of the packaging type. Thus, the place storage consists of a multi-set of object identifiers of the type packaging as well as of the type deluxe-packaging. In this sense, we can say that the attribute storage is polymorphic, since values of both types can be stored. Moreover, both methods, put and get, are polymorphic because different kinds of object references which belong to the packaging hierarchy can be handled by these methods.

In CO-OPN/2, object identifiers are algebraic values of an algebra called *object identifier algebra* which is organized in a very specific way. Actually, to each type (declared in a class module) corresponds a sort (and some operations) which defines a carrier set of object identifiers of that type (sort). The order-sorted algebra is constructed in such a way that the inclusion relation between

---

[2] At the semantic level the subtyping relation between two types is characterized by the bisimulation equivalence combined with the restriction of the subtype's behaviors with respect to the supertype's behaviors (c.f. Section 5.5).

```
Class Heap;
Interface
  Use  Packaging;
  Type heap;
  Methods put _ , get _ : packaging;
Body
  Place storage _ : packaging;
  Axioms
    put box :: -> storage box ;
    get box :: storage box -> ;
    where box : packaging;
End Heap;
```

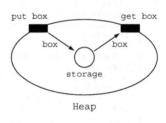

Spec. 5. The Modeling of a Heap of Packagings.

the carrier sets of the object identifier algebra reflects the subtyping relation between the types; i.e. two types are in a subtyping relation if and only if the corresponding carrier sets are related by inclusion. Thus, for example, saying that a variable has a given type, means that the variable represents an object identifier belonging to the carrier set (or any included carrier set) associated to the type of the variable (or to any subtype). More informations about the structure and the operations of the object identifier algebra are given in Section 5.1.

Similar Classes without Subtyping. When the factory acquired an automatic packaging producer, the `Heap` class has been replaced by a more sophisticated specification, the `Conveyor-Belt` class. In the example, the conveyor-belt is specified by means of the generic first-in-first-out data structure presented in Specification 2.

In order to obtain an actual ADT module from the generic one, it is necessary to invoke the instantiation mechanism as shown in Specification 6. The mechanism replaces the formal parameter module `Elem` by the actual parameter module `Packaging`, associates the sort `elem` to the type `packaging`, and performs some mandatory renamings. Thus, the `FifoPackaging` ADT module describes a first-in-first-out data structure of object identifiers of type `packaging` (and of course also of type `deluxe-packaging`).

```
ADT FifoPackaging As Fifo(Packaging);
Morphism elem -> packaging;
Rename
  fifo     -> fifo-packaging;
  ne-fifo -> ne-fifo-packaging;
End FifoPackaging;
```

Spec. 6. First-In-First-Out of Object Identifiers of Type `packaging`.

The modeling of the conveyor-belt based on the above data structure is given in Specification 7. We observe that the conveyor-belt device is very similar to the heap. However, now the `belt` place consists of a multi-set of fifos of packagings initialized with an empty fifo denoted '`[]`' (c.f. the **Initial** field). Consequently, it has been necessary to define both the `put` and `get` methods differently.

```
Class ConveyorBelt;
Interface
   Use   Packaging;
   Type conveyor-belt;
   Object   the-conveyor-belt;
   Methods put _ , get _ : packaging;
Body
   Use FifoPackaging;
   Place belt _ : fifo-packaging;
   Initial belt [];
   Axioms
      put box ::
         belt f -> belt (insert box f);
      get (first f') ::
         belt f' -> belt (extract f');
      where f  : fifo-packaging;
            f' : ne-fifo-packaging;
            box : packaging;
End ConveyorBelt;
```

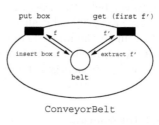

Spec. 7. The Conveyor-Belt Modeling.

We observe that the conveyor-belt type is not a subtype of the heap type. Indeed, on the one hand, both methods put and get cannot occur simultaneously in the conveyor-belt, as it was possible in the heap, because of the algebraic description of the fifo data structure. Furthermore, the sequence $s \xrightarrow{\text{put box1}} s' \xrightarrow{\text{put box2}} s'' \xrightarrow{\text{get box2}} s'''$, in which s, s', s'', and s''' denote four states, is valid for a heap of packagings, but cannot occur in a conveyor belt.

Inheritance without Subtyping. In order to illustrate that inheritance can be used independently of the notion of subtyping, we specify here a more realistic conveyor-belt than the previous one. From the ConveyorBelt class, we add a limit to the number of boxes that can be conveyed by the device. Beyond this limit, for any box put on the conveyor, the box at the other extremity will fall down. The introduction of this limit is modeled by means of a spontaneous reaction represented by the internal transition fall-down as shown in Specification 8.

```
Class LimitedConveyorBelt;
Inherit ConveyorBelt;
   Rename
      conveyor-belt -> limited-conveyor-belt;
Body
   Transition fall-down;
   Axiom
      fall-down ::
         (size f') >conveyor-capacity = true =>
         belt f' -> belt (extract f');
      where f' : ne-fifo-packaging;
End LimitedConveyorBelt;
```

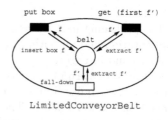

Spec. 8. The Modeling of the New Limited Conveyor-Belt.

Consequently, the two types have identical behaviors when the differences between the successive put and get operations are lower than the limit of the conveyor. However, although inheritance has been used to specify the new limited conveyor-belt, the limited-conveyor-belt type is not a subtype of the conveyor-belt type, because many behaviors of the previous conveyor-belt cannot occur in the limited conveyor-belt. This example shows the typical conflict between inheritance and subtyping when enrichment inheritance modifies the semantics of the original class.

Object Interaction. The two last classes of the example concern the packaging producer and the packaging unit. Specification 9 provides the modeling of the packaging producer. We observe that the axiom of the produce method in the PackagingProducer class expresses the synchronization request in sequence (synchronization operator '..') **with** two methods of two distinct objects.

The term box.create-packaging asks for the synchronization with the dynamic creation and the initialization of a packaging object identified by the free variable box, while the term the-conveyor-belt.put box asks for the synchronization with the method put of the static conveyor-belt object. As a result, a new packaging is created and initialized first, and then the packaging is "transmitted" to the conveyor-belt. Since the box variable is free, the axiom is satisfied for any value of box. This allows the dynamic creation of regular packagings as well as deluxe packagings.

```
Class PackagingProducer;
Interface
  Use Packaging;
  Type packaging-producer;
  Object the-packaging-producer;
  Method
    produce;
Body
  Axiom
    produce with
      box.create-packaging .. the-conveyor-belt.put box :: -> ;
    where box : packaging;
End PackagingProducer;
```

Spec. 9. Modeling of the Packaging Producer.

The packaging unit is the most complex device of our example. The textual representation and an outline of packaging unit's modeling are given in Specification 10 and in Figure 5.

In the PackagingUnit class, three events ask for a synchronization. The method take is quite simple: it removes a packaging from the conveyor-belt and stores it on the work-bench. The purpose of the filling transition is to fill the empty packagings which lies of the work-bench. The synchronization request of this internal transition consists of filling one corresponding empty hole of the current packaging. Note that the axiom which defines the filling event takes

```
Class PackagingUnit;
Interface
  Type packaging-unit;
  Method take;
Body
  Use Chocolate, ConveyorBelt, Packaging, DeluxePackaging;
  Transitions
    filling,
    store;
  Place
    work-bench _ : packaging;
  Axioms
    take with the-conveyor-belt.get box ::
      -> work-bench box;

    filling with box.fill choc;

    store with box.full-praline ::
      (sub-packaging-deluxe-packaging box) = box' =>
      work-bench box -> ;

    store with box'.full-praline // box'.full-truffle ::
      work-bench box' -> ;

    where  box : packaging;  box' : deluxe-packaging;
           choc : chocolate;
End PackagingUnit;
```

Spec. 10. Modeling of the Packaging Unit

advantage, on the one hand, of subsorting for the chocolates (choc) and, on the other hand, of subtyping for the packagings (box).

The store internal transition is a little bit more sophisticated. This event merely removes the packagings that are full from the work-bench. For this, it is mandatory to distinguish between the regular packagings and the deluxe ones because we modeled the fullness of the regular packagings by means of the full-praline method and the fullness of the deluxe packagings by means of the full-praline and full-truffle methods. In fact, the two axioms for the store event are mutually exclusive. However, the following condition requires some explanations:

$$(\texttt{sub-packaging-deluxe-packaging box}) = \texttt{box'}$$

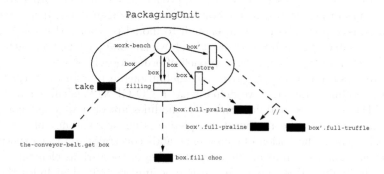

Fig. 5. Modeling the Packaging Unit.

As we mention earlier, all the object identifiers belong to a specific order-sorted algebra, the object identifier algebra. This algebra provides, for any two types related by subtyping, two additional operations which allow to determine, according to the subtype hierarchy, to which type a given object identifier belongs. Let us consider two types c and c' such that $c \leq c'$. Both the following operations sub-c-c' : $c \to c'$ and super-c'-c : $c' \to c$ are implicitly defined. The former takes an object identifier of type c and returns another object identifier of the subtype c', the latter operation has the reverse behavior. In this way, the above condition is satisfied if and only if the object identifier box is of type packaging but not of type deluxe-packaging. Note that this condition based on the "sub" operation could have been formulated by means of a "super" operation as follows:

```
box = (super-deluxe-packaging-packaging box')
```

The second axiom concerning the store event does not require such a condition because the use of a variable box' of type deluxe-packaging makes the selection of the packagings of type deluxe-packaging in a straightforward manner.

Finally, since both axioms of the store event are exclusive, as soon as the packaging which lies on the work-bench is full, the appropriate methods full-praline and full-truffle can occur and then the store internal transition is activated. Note the use of the simultaneous synchronization operator '//' within both the full-praline and the full-truffle methods when the packaging is of type deluxe-packaging. This implies that two chocolates are put at the same time in a deluxe packaging.

4   CO-OPN/2 Syntax

The purpose of this section is to describe the abstract syntax of the CO-OPN/2 formalism. Recall that a CO-OPN/2 specification is composed of two kinds of descriptions associated with two kinds of modules: *Adt modules* and *Class modules*. The Adt modules are used to describe the algebraic abstract data types involved in a CO-OPN/2 specification, while the Class modules correspond to the description of the objects that are obtained by instantiation. In this paper we use, without presenting it, the approach of hierarchical order-sorted algebraic specifications in which operations are total on subsorts. It has been shown that this approach is powerful enough to model partiality [10]. Nevertheless, a more flexible approach is presented in [4] which combines the approach of partial abstract data types with the approach of order-sorted algebras.

In this section, we first give some definitions in relation with the ADT and class modules, and then we define what a CO-OPN/2 specification is in terms of ADT and class modules. Since the modules are obviously not complete, we finally introduce the notion of completeness and consistency of a CO-OPN/2 specification. The notion of consistency adds constraints on the structure of the specification, as well as at the level of typing of the various elements.

Throughout this paper we consider a universe which includes the disjoint sets: **S**, **F**, **M**, **P**, **V**. These sets correspond, respectively, to the sets of all sort, function, method, place, and variable names. In the context of CO-OPN/2, the set of all sorts S is divided into two disjoint sets S^A and S^C, $S = S^A \cup S^C$ with $S^A \cap S^C = \varnothing$. The former is dedicated to the usual sort names involved in the algebraic description part, whereas the latter consists in all the type names of the classes. As already mentioned, each class defines a type, i.e. a sort of S^C, and every class instance has an identity which is an algebraic value of that sort.

The "S-sorted" notation facilitates the subsequent development. Let S be a set, a S-sorted set A is a family of sets indexed by S, we write $A = (A_s)_{s \in S}$. A S-sorted set of disjoint sets of variable is called a S-sorted variable set. Given two S-sorted sets A and B, a S-sorted function $\mu : A \to B$ is a family of functions indexed by S denoted $\mu = (\mu_s : A_s \to B_s)_{s \in S}$. We frequently use the extension of a partial order $\leq \subseteq S \times S$ to strings in S^* of equal length, $s_1 \cdots s_n \leq s'_1 \cdots s'_n$ iff $s_i \leq s'_i$ $(1 \leq i \leq n)$.

Let $\Sigma = \langle S, \leq, F \rangle$ be a signature and X be a S-sorted variable set. The S-sorted set $T_{\Sigma, X}$ is the set of all terms with variables.

In the context of CO-OPN/2 multi-sets are used for two purposes. The first one is the need for representing the values of the places and the second one is for the expression of concurrency in the semantics. The syntactic extension for a set of sort S consists of adding for every sort $s \in S$ the multiset sort $[s]$, the subsort relation related to the multi-sets and the operations *empty*, *singleton* and *union*. Similarly we semantically extend the models of A of an algebraic specification into $[A]$ by adding carrier set $[A]_s$ and the multiset functions.

4.1   Signature and Interface

As usual, an ADT module signature groups three elements of an algebraic abstract data type, i.e. a set of sorts, a subsort relation, and some operations. However, in the context of structured specifications, an ADT signature can intrinsically use elements not *locally* defined, i.e. defined outside the signature itself. For this reason, the profile of the operations as well as the subsort relation in the next definition are respectively defined over the set of *all* sorts names S and S^A, and not only over the set of sorts S^A defined in the module itself. When a signature only uses elements locally defined we say that the signature is *complete*.

Definition 1. *ADT module signature*
An *ADT module signature* (ADT signature for short) (over S and F) is a triple[3] $\Sigma^A = \langle S^A, \leq^A, F \rangle$, where

- S^A is a set of sort names of S^A;
- $\leq^A \subseteq (S^A \times S^A) \cup (S^A \times S^A)$ is a partial order (partial subsort relation);
- $F = (F_{w,s})_{w \in S^*, s \in S}$ is a $(S^* \times S)$-sorted set of function names of F.   ◇

---

[3] The A superscript indicates that the module and its components are in relation with the abstract data type dimension.

Observe that, since the profile of the operations is built over **S**, some elements of such profiles can obviously be of sort in $\mathbf{S^C}$. This suggests that ADT modules can describe data structures of object identifiers, stack or arrays of object identifiers for example.

Similarly to the notion of ADT module signature, the elements of a class module which can be used from the outside are grouped into a class module interface. The class module interface of a class module includes: the type of the class, a subtype relation with other classes, and the set of methods that corresponds to the services provided by the class.

Definition 2. *Class module interface*
A *class module interface* (class interface for short) (over **S**, **M**, and **O**) is a 4-tuple[4] $\Omega^C = \langle \{c\}, \leq^C, M, O \rangle$, where

- $c \in \mathbf{S^C}$ is the type[5] name of the class module;
- $\leq^C \subseteq (\{c\} \times \mathbf{S^C}) \cup (\mathbf{S^C} \times \{c\})$ is a partial order (partial subtype relation);
- $M = (M_{c,w})_{w \in \mathbf{S^*}}$ is a finite $(\{c\} \times \mathbf{S^*})$-sorted set of method names of **M**;
- $O = (O_c)_{c \in \mathbf{S^C}}$ is a finite $\mathbf{S^C}$-sorted set of static object names of **O**.   ◇

Remember that a method is not a function but a parameterized transition which may be regarded as a predicate. The set of methods M is $(\{c\} \times \mathbf{S^*})$-sorted, where c is the type of the class module and $\mathbf{S^*}$ corresponds to the sorts of the method's parameters. A method $m \in M_{c,s_1,\ldots,s_n}$ is often noted $m_c : s_1, \ldots, s_n$, while a method without any argument $m \in M_{c,\epsilon}$ is written m_c (ϵ denotes the empty string).

The transitive closure of the union of the subsort relations of a set of ADT signature's Σ and of the subtype relations of a set of class interfaces Ω builds a *global subsort/subtype relation* denoted $\leq_{\Sigma,\Omega}$.

For each class interface $\Omega^C = \langle \{c\}, \leq^C, M, O \rangle$, we induces an ADT signature $\Sigma^A_{\Omega^C} = \langle \{c\}, \leq^C, F_{\Omega^C} \rangle$ in which F_{Ω^C} contains the operations necessary to the management of the objects identifiers, as well as one constant for each static object. The semantics of these operations is detailed further in Section 5.1. These operations are defined on the global subsort/subtype relation as follows:

$$
\begin{aligned}
F_{\Omega^C} = \ &\{o_c :\to c \mid o : c \in O\} \ \cup \\
&\{\text{init}_c :\to c, \text{new}_c : c \to c\} \ \cup \\
&\{\text{sub}_{c,c'} : c \to c', \text{super}_{c,c''} : c \to c'' \mid c' \leq_{\Sigma,\Omega} c, c \leq_{\Sigma,\Omega} c''\}.
\end{aligned}
$$

On these basis, we define a global signature written $\Sigma_{\Sigma,\Omega}$ and a global interface written Ω_Ω . A global signature groups the sorts and types, the subsort and subtype relations, as well as the operations of ADT signatures Σ and of the ADT signatures derived from Ω. As for a global interface, it groups the types, the subtype relations, and the methods of the set of class interfaces Ω.

---

[4] Here the C superscript stresses the belonging to the class (algebraic net) dimension.
[5] In general, we use s symbols for sorts of the abstract data type dimension and c symbols for types (in fact sorts) of the classes,

Under completeness and monotonicity, $\Sigma_{\Sigma,\Omega}$ is an order-sorted signature.

In a similar way, a set of class interfaces must satisfy the contra-variance condition which guarantees, at the syntactic level, the substitutability principle of an object of type c' by any object of type c when c is a subtype of c'. Note that in the CO-OPN/2 context, a method is a parameterized synchronization rather than a function. Therefore, the usual co-variance of the function co-domain is not a relevant notion.

4.2   ADT Module

ADT modules describes abstract data types which may be used by other ADT or class modules. An ADT module consists of an ADT signature, a set of formulas also called axioms, and of course some variables. Remember that, in the context of structured specification, an ADT module may obviously use elements not locally defined, i.e. defined in other modules.

Definition 3. *ADT module*
Let Σ be a set of ADT signatures and Ω be a set of class interfaces such that the global signature $\Sigma_{\Sigma,\Omega} = \langle S, \leq, F \rangle$ is complete. An *ADT module* is a triple $Md^A_{\Sigma,\Omega} = \langle \Sigma^A, X, \Phi \rangle$, where

- Σ^A is an ADT signature;
- $X = (X_s)_{s \in S}$ is a S-disjointly-sorted set of variables of $\mathbf{V}$;
- Φ a set of formulas[6] (axioms) over $\Sigma_{\Sigma,\Omega}$ and X. $\Diamond$

4.3   Behavioral Formulas

Before defining a behavioral formula, let us recall that our formalism provides two different categories of events: the *invisible* events, and the *observable* events. Both of them can involve an optional *synchronization expression*. The invisible events describe the spontaneous reactions of an object to some stimuli. They correspond to the internal transitions which are denoted by τ and not by a specific name, as in the concrete CO-OPN/2 language. The observable events correspond to the methods that are accessible from the outside. A synchronization expression offers an object the means of choosing how to be synchronized with other partners (even itself). Three synchronization operators are provided: ' // ' for simultaneity, ' .. ' for sequence, and '$\oplus$' for alternative. In order to select a particular method of a given object, the usual dot notation has been adopted.

We write $\mathbf{E}_{A,M,O,C}$ for the set of all events over a set of parameter values A, a set of methods M, a set of object identifiers O, and a set of types of classes C. Because this set is used for various purposes, we give here a generic definition.

---

[6] A *formula* is either a pair (t, t'), denoted $t = t'$, or a set of pairs $\{(t_i, t'_i), (t, t') \mid 1 \leq i \leq n\}$, denoted $t_1 = t'_1, \ldots, t_n = t'_n \implies t = t'$, where t_i, t'_i ($1 \leq i \leq n$) and t, t' are terms of $T_{(\Sigma_{\Sigma,\Omega}),X}$.

Definition 4. *Set of all events*

Let $S = S^A \cup S^C$ be a set of sorts such that $S^A \in \mathbf{S^A}$ and $S^C \in \mathbf{S^C}$. Let us consider $A = (A_s)_{s \in S}$, $M = (M_{s,w})_{s \in S^C, w \in S^*}$, $O = (O_s)_{s \in S^C}$, and a set of types of classes $C \subseteq S^C$. The elements *Event* of $\mathbf{E}_{A,M,O,C}$ are syntactically built according to the following rules:

$$\begin{aligned}
Event &\to Inv \mid Inv \textbf{ with } Sync \mid Obs \mid Obs \textbf{ with } Sync \\
Inv &\to self.\tau \\
Obs &\to self.m(a_1, \ldots, a_n) \mid Obs \text{ // } Obs \mid Obs \text{ .. } Obs \mid Obs \oplus Obs \\
Sync &\to o.m(a_1, \ldots, a_n) \mid o.\text{create} \mid o.\text{destroy} \mid \\
&\quad Sync \text{ // } Sync \mid Sync \text{ .. } Sync \mid Sync \oplus Sync
\end{aligned}$$

where $s \in S^C$, $s_i, s_i' \in S$ $(1 \le i \le n)$, $a_1, \ldots, a_n \in A_{s_1} \times \cdots \times A_{s_n}$, $m \in M_{s, s_1' \cdots s_n'}$, $o \in O_s$, $c \in C$, and $self \in O_c$ and such that s_i, and s_i' $(1 \le i \le n)$ belong to the same connected component. $\diamond$

Note that the effective and the formal parameters of a method involved in a synchronization may not have the same sorts, or be comparable, but their sorts must be connected through the subsort relation.

Example 5.

Observable and invisible events build over Specification 9 and 10 using a term algebra for A are :

1. *self*.produce **with** box.create-packaging .. the-conveyor-belt.put(box)
2. τ.store **with** box'.full-praline // box'.full-truffle

We now give the definition of the behavioral formulas that are used to describe the properties of observable and invisible events of class modules. A behavioral formula consists of an event as established in the following definition, a condition expressed by means of a set of equations over algebraic values, and the usual Petri net pre/post-condition of the event. Both pre/post-conditions are sets of terms (of sort multi-set) indexed by the places of the net. A event can occur if and only if the condition on the algebraic values is satisfied, enough resources can be consumed/produced from/in the places of the module, and if the events involved in the synchronization can occur.

Definition 6. *Behavioral formula*

Let $\Sigma = \langle S, \le, F \rangle$ be an order-sorted signature such that $S = S^A \cup S^C$ ($S^A \subseteq \mathbf{S^A}$ and $S^C \subseteq \mathbf{S^C}$). For a given $(S^C \times S^*)$-sorted set of methods M, a S-disjointly-sorted set of places P, a set of types $C \subseteq S^C$, and a S-disjointly-sorted set of variables X. A *behavioral formula* is a 4-tuple $\langle Event, Cond, Pre, Post \rangle$, where

- $Event \in \mathbf{E}_{(T_{\Sigma,X}), M, (T_{\Sigma,X})_s, C}$ such that $s \in S^C$;
- $Cond$ is a set of equations over Σ and X;
- $Pre = (Pre_p)_{p \in P}$ and $Post = (Post_p)_{p \in P}$ are two families of terms over $[\Sigma], X$ indexed by P and of sort $[s]$ if p is of sort s.

We also denote a behavioral formula $\langle Event, Cond, Pre, Post \rangle$ by the expression

$$Event :: Cond \Rightarrow Pre \to Post \qquad \diamond$$

4.4   Class Module

The purpose of a Class module is to describe a collection of objects with the same structure by means of an encapsulated algebraic net. Actually, a class module is considered as a template from which objects are instantiated. A Class module consists of: a class interface, a set of places, some variables, the initial values of the places (also called the *initial state* of the module), and a set of behavioral formulas which describe the properties of the methods and of the internal transitions.

Note that the following definition establishes that class instances are able to store and exchange object identifiers because the sorts of the places, the variables, and the profile of the methods belong to the set of all sorts $\mathbf{S}$, therefore, these components can be either of sort $\mathbf{S^A}$ or $\mathbf{S^C}$.

Definition 7. *Class module*
Let Σ be a set of ADT signatures, Ω be a set of class interfaces such that the global signature $\Sigma_{\Sigma,\Omega} = \langle S, \leq, F \rangle$ is complete. A *Class module* is a 4-tuple $Md^C_{\Sigma,\Omega} = \langle \Omega^C, P, I, X, \Psi \rangle$, where

- $\Omega^C = \langle \{c\}, \leq^C, M \rangle$ is a class interface;
- $P = (P_s)_{s \in S}$ is a finite S-disjointly-sorted set of place names of $\mathbf{P}$;
- $I = (I_p)_{p \in P}$ is an initial marking, a family of terms indexed by P and of sort $[s]$ if p is of sort s.
- $X = (X_s)_{s \in S}$ is a S-disjointly-sorted set of variable of $\mathbf{V}$;
- Ψ is a set of behavioral formulas over the global signature $\Sigma_{\Sigma,\Omega}$, a set of methods composed of M and all the methods of Ω, the set of places P, the type of the class $\{c\}$, and X. $\Diamond$

4.5   CO-OPN/2 Specification

Finally, a CO-OPN/2 specification is a collection of ADT and Class modules.

Definition 8. *CO-OPN/2 specification*
Let Σ be a set of ADT signatures, Ω be a set of class interfaces such that $\Sigma_{\Sigma,\Omega}$ is complete and coherent[7], and such that Ω_Ω satisfies the contra-variance condition. A *CO-OPN/2 specification* consists of a set of ADT and class modules:

$$Spec_{\Sigma,\Omega} = \left\{ (Md^A_{\Sigma,\Omega})_i \mid 1 \leq i \leq n \right\} \cup \left\{ (Md^C_{\Sigma,\Omega})_j \mid 1 \leq j \leq m \right\}.$$

When Σ and Ω are, respectively, equal to the global signature and in the global interface of the specification, the specification is considered *complete*. We denote a CO-OPN/2 specification $Spec_{\Sigma,\Omega}$ by $Spec$ and the global subsort/subtype relation $\leq_{\Sigma,\Omega}$ by $\leq$. $\Diamond$

We consider in this paper, only acyclic specifications. That is, specifications, where it is not possible for two different modules (algebraic or class) to use themselves in their definition. Preliminary results can be found in [9] so as to allow cycles between class modules.

---

[7] The notion of coherence imposes conditions on the profile of the algebraic operations of a signature as defined in [4].

5   CO-OPN/2 Semantics

This section presents the semantic aspects of the CO-OPN/2 formalism which are based on two notions, the order-sorted algebras, and the transition systems.

First of all, we concentrate on order-sorted algebras as models of the data structures of a CO-OPN/2 specification. Then, we introduce an essential element of the CO-OPN/2 formalism, namely the order-sorted algebra of object identifiers, which is organized in a very specific way.

Afterwards, we present how the notion of transition system is used so as to describe a system composed of objects dynamically created. Then, we provide all the inference rules which allow us to construct the transition system of a CO-OPN/2 specification. Such a transition system is considered as the semantics of the specification.

5.1   Algebraic Models of a CO-OPN/2 Specification

Here we focus on the semantics of the algebraic dimension of a CO-OPN/2 specification. Remember that an ADT signature can be deduced from each class interface of the specification. It is composed of a type, of a subtype relation, and of some operations required for the management of the object identifiers. We now provide the definition of the ADT module induced by each class module of the specification. Such an ADT module is composed of the induced ADT interface and of the formulas which determine the intended semantics of the operations.

The ADT interface mentioned above includes, for syntactic consistency, a constant for each static object defined in the class interface. At the semantics level the role of those constants is just to abbreviate the object identifiers of the class instances statically created. Clearly, these abbreviations are not essential. Thus, without lost of generality and for the sake of simplicity, those constants are omitted in the following definition but we should create these objects in a preliminary phase of the semantics.

Definition 9. *ADT module induced by a class module*
Let *Spec* be a well-formed CO-OPN/2 specification and $\leq$ be its global subsort/subtype relation. Let $Md^C = \langle \Omega^C, P, I, V, \Psi \rangle$ in which $\Omega^C = \langle \{c\}, \leq^C, M, O \rangle$ be a class module of *Spec*. The ADT module induced by Md^C is denoted $Md^A_{\Omega^C} = \langle \Sigma^A_{\Omega^C}, V_{\Omega^C}, \Phi_{\Omega^C} \rangle$ in which $\Sigma^A_{\Omega^C} = \langle \{c\}, \leq^C, F_{\Omega^C} \rangle$, and where

- $F_{\Omega^C} = \{\text{init}_c : \to c, \text{ new}_c : c \to c\} \cup$
  $\{\text{sub}_{c,c'} : c \to c', \text{ super}_{c,c''} : c \to c'' \mid c' \leq c, c \leq c''\}$.
- $V_{\Omega^C} = \{o_c : c, o_{c'} : c' \mid c' \leq c\}$;
- $\Phi_{\Omega^C} = \{\text{sub}_{c,c'} \text{ init}_c = \text{init}_{c'},$
  $\text{sub}_{c,c'} (\text{new}_c \ o_c) = \text{new}_{c'} (\text{sub}_{c,c'} \ o_c),$
  $\text{super}_{c',c} \text{ init}_{c'} = \text{init}_c,$
  $\text{super}_{c',c} (\text{new}_{c'} \ o_{c'}) = \text{new}_c (\text{super}_{c',c} \ o_{c'}) \mid c' \leq c\}$

The variables of V_{Ω^C} are chosen in a way such that they do not interfere with other identifiers of the module signature. $\Diamond$

The presentation of a CO-OPN/2 specification consists of collapsing all the ADT modules of the specification and all the ADT modules which are induced by the class modules. Renamings are necessary to avoid name clashes between the various modules.

In the following parts, we often denote the set of all sorts, types, methods, and places of a specification $Spec$ by, Sorts($Spec$), Types($Spec$), Methods($Spec$), Places($Spec$), respectively. Note that Sorts($Spec$) represents the sorts defined in the ADT modules of the specification, while Types($Spec$) corresponds to the sorts (types) induced by the class modules.

The initial approach has been adopted throughout this work. The semantics of the algebraic dimension of a CO-OPN/2 specification $Spec$ is therefore defined as the semantics of the presentation of the specification. We denote this model $Sem(Pres(Spec))$.

The semantics of such a presentation is composed of two distinct parts. The first one consists of all the carrier sets defined by the ADT modules of the specification, i.e. the model of the algebraic dimension of the specification without considering the ADT modules induced by the class modules. The second part is what we previously called the *object identifier algebra*. This "sub-algebra" is constructed in a very specific way and plays an important role in our approach because it provides all the potential object identifiers as well as the operations required for their management.

Let $Sem(Pres(Spec)) = A$. The carriers set defined by the ADT modules of the specification are usually denoted $\ddot{A}$, while the object identifier algebra defined by the ADT modules induced by the class modules of the specification is $\widehat{A}$. Both $\ddot{A}$ and $\widehat{A}$ are disjoint and $A = \ddot{A} \cup \widehat{A}$.

Intuitively, the idea behind the object identifier algebra of a specification is to define a set of identifiers for each type of the specification and to provide some operations which return a new object identifier whenever a new object has to be created. Moreover, these sets of object identifiers are arranged according to the subtype relation over these types. It means that two sets of identifiers are related by inclusion if their respective types are related by subtyping. In other words, the inclusion relation reflects the subtype relation. Furthermore, two operations mapping subtypes and supertypes are provided. The structure of the object identifier algebra and the operations mapping subtypes and supertypes can be used to determine, for example, if an object identifier belongs to a given type as informally explained in Section 3.2.

Each class module defines a type and a subtype relation which are present in the ADT module induced by each class module (see Definition 9). On the one hand, each type (actually a sort) defines a carrier set which contains all the object identifiers of that type and, on the other hand, the global subtype relation imposes a specific structure over the carrier sets. Moreover, four operations are defined in each ADT module induced by each class module. These operations over the object identifiers are divided into two groups: the generators (the operations which build new values) and the regular operations. For each type c and c' of the specification these operations are as follows:

1. the generator $init_c$ corresponds to the first object identifier of type c;

2. the generator new_c returns a new object identifier of type c;
3. the operation $sub_{c,c'}$ maps the object identifiers of types c into the ones of type c', when $c' \leq c$;
4. the operation $super_{c',c}$ maps the object identifiers of types c' into the ones of type c, when $c' \leq c$;
5. as indicated by their names, $super_{c',c}$ is the inverse operation of $sub_{c,c'}$ (c.f. the next theorem).

Theorem 10.
Let *Spec* be a well-formed CO-OPN/2 specification and $\leq$ be its global relation. For any types c, c' such that $c' \leq c$ the following properties hold:

  i) $super_{c',c} (sub_{c,c'} \ o_c) = o_c$, where $o_c : c$;
 ii) $sub_{c,c'} (super_{c',c} \ o_{c'}) = o_{c'}$, where $o_{c'} : c'$.

Proof. See [4].

Example 11. Structure of the object identifier algebra
Let us assume that five class modules of a specification declare five types c_1, c_2, c_3, c_4, c_5, and the relation $\leq^C= \{(c_4, c_2), (c_4, c_3), (c_5, c_3), (c_2, c_1), (c_3, c_1)\}^*$.
    The object identifier algebra $\widehat{A}$ of the specification (defined by the five ADT module which are induced by the class modules) is depicted in Figure 6. As men-

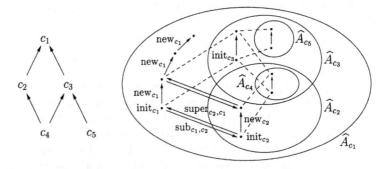

Fig. 6. Structure of $\widehat{A}$ according to the subtype relation.

tioned, the carrier sets $\widehat{A}_{c_i}$ $(1 \leq i \leq 5)$ (on the right side) are included according to the subtype relation (on the left side). The $init_{c_1}$ and new_{c_1} generators illustrate, for example, some object identifiers of type c_1, while the sub_{c_1,c_2} and $super_{c_2,c_1}$ operations show the mappings between the carrier sets $\widehat{A}_{c_1}$ and $\widehat{A}_{c_2}$. The dashed lines indicate some other sub and super mappings.

5.2   Management of Object Identifiers

Whenever a new class instance is created, a new object identifier must be assigned to it. This means that the system must know, for each class type and at any time, the last object identifier used, so as to be able to compute a new object identifier. Consequently, throughout its evolution, the system retains a function which returns the last object identifier used for a given class type. Moreover, another information has to be retained throughout the evolution of the system. This information consists of the objects that have been created and that are still alive, i.e. the object identifiers assigned to some class instances involved in the system at a given time. This second information is also retained by means of a function the role of which is to return, for every class type, a set of object identifiers which corresponds to the alive (or active) object identifiers.

For the subsequent development, let us consider a specification $Spec$, $A = Sem(Pres(Spec))$, and the set of all types of the specification $S^C = \text{Types}(Spec)$.

The partial function which returns, for each class, the last object identifier used, is a member of the set of partial functions:

$$Loid_{Spec,A} = \{l : S^C \rightarrow \widehat{A} \mid l(c) \in \widetilde{A}_c \text{ or is not defined}\}$$

in which $\widetilde{A}_c = \widehat{A}_c \setminus \cup_{c' \leq c c}\widehat{A}_{c'}$ represents the proper object identifiers of the class type c (excluding the ones of any subtype of c). Such functions either return, for each class type, the last object identifier that has been used for the creation of the objects, or is undefined when no object has been created yet.

For every class type c in S^C, the computation of a new last object identifier function starting with an old one is performed by the family of functions $\{newloid_c : Loid_{Spec,A} \rightarrow Loid_{Spec,A} \mid c \in S^C\}$ (new last object identifier) defined as:

$$(\forall c, c' \in S^C)(\forall l \in Loid_{Spec,A}) \ newloid_c(l) = l' \text{ such that}$$

$$l'(c') = \begin{cases} \text{init}_c^{\widehat{A}} & \text{if } l(c) \text{ is undefined and } c' = c, \\ \text{new}_c^{\widehat{A}}(l(c)) & \text{if } l(c) \text{ is defined and } c' = c, \\ l(c) & \text{otherwise.} \end{cases}$$

The second function retained by the system throughout the evolution of the system returns the set of the alive objects of a given class. It is member of the set of partial functions[8]:

$$Aoid_{Spec,A} = \{a : S^C \rightarrow C \mid C \subseteq \mathcal{P}(\widehat{A}), \ a(c) \in \mathcal{P}(\widetilde{A}_c)\}.$$

The creation of an object implies the storage of its identity and the computation of a new alive object identifiers function based on the old one. This is achieved by the family of functions $\{newaoid_c : Aoid_{Spec,A} \times \widehat{A} \rightarrow Aoid_{Spec,A} \mid c \in S^C\}$ (new alive object identifiers) defined as:

$$(\forall c, c' \in S^C)(\forall o \in \widetilde{A}_c)(\forall a \in Aoid_{Spec,A}) \ newaoid_c(a, o) = a' \text{ such that}$$

$$a'(c') = \begin{cases} a(c) \cup \{o\} & \text{if } c' = c, \\ a(c) & \text{otherwise.} \end{cases}$$

---

[8] The notation $\mathcal{P}(A)$ represents the *power set* of a set A.

Both of those families of functions $newloid_c$ and $newaoid_c$ are used in the inference rules concerning the creation of new instances, see Definition 14.

The set of functions $\{remaoid_c : Aoid_{Spec,A} \times \widehat{A} \to Aoid_{Spec,A} \mid c \in S^C\}$ is the dual version of the $newaoid_c$ family in the sense that, instead of adding an object identifier, they remove a given object identifier and compute the new alive object identifiers function as follows:

$$(\forall c, c' \in S^C)(\forall o \in \widetilde{A}_c)(\forall a \in Aoid_{Spec,A}) \quad remaoid_c(a, o) = a' \text{ such that}$$

$$a'(c') = \begin{cases} a(c) \setminus \{o\} & \text{if } c' = c, \\ a(c) & \text{otherwise.} \end{cases}$$

This family of functions is necessary when the destruction of class instances is considered, see Definition 14.

Here are three operators and a predicate in relation with the last object identifier used and the alive object identifiers functions. These operators and this predicate are used in the inference rules of Definition 17; they have been developed in order to allow simultaneous creation and destruction of objects. We now only provide their formal definition; the explanations related to their meaning and their use is postponed until the informal description of the inferences rules in which they are involved. The first two operators are ternary operators which handle an original last object identifiers function and two other functions. The third binary operator and the predicate handle alive object identifiers functions.

$\triangle: Loid_{Spec,A} \times Loid_{Spec,A} \times Loid_{Spec,A} \to Loid_{Spec,A}$ such that

$$(\forall c \in S^C) \, (l' \triangle_l l'')(c) = \begin{cases} l'(c) & \text{if } l'(c) \neq l(c) \wedge l''(c) = l(c), \\ l''(c) & \text{if } l'(c) = l(c) \wedge l''(c) \neq l(c), \\ l(c) & \text{otherwise.} \end{cases}$$

$\stackrel{\triangle}{=} \, : Loid_{Spec,A} \times Loid_{Spec,A} \times Loid_{Spec,A}$ such that

$$(\forall c \in S^C) \, (l' \stackrel{\triangle}{=}_l l'')(c) = ((l(c) = l'(c) = l''(c)) \vee (l'(c) \neq l''(c)))$$

$\cup : Aoid_{Spec,A} \times Aoid_{Spec,A} \to Aoid_{Spec,A}$ such that

$$(\forall c \in S^C) \, (a \cup a')(c) = a(c) \cup a'(c)$$

$P : Aoid_{Spec,A} \times Aoid_{Spec,A} \times Aoid_{Spec,A} \times Aoid_{Spec,A}$ such that

$$P(a_1, a_1', a_2, a_2') \iff$$

$$(\forall c \in S^C) \, (((a_1(c) \cap ((a_2(c) \setminus a_2'(c)) \cup (a_2'(c) \setminus a_2(c)))) = \varnothing) \wedge$$
$$((a_1'(c) \cap ((a_2(c) \setminus a_2'(c)) \cup (a_2'(c) \setminus a_2(c)))) = \varnothing) \wedge$$
$$((a_2(c) \cap ((a_1(c) \setminus a_1'(c)) \cup (a_1'(c) \setminus a_1(c)))) = \varnothing) \wedge$$
$$((a_2'(c) \cap ((a_1(c) \setminus a_1'(c)) \cup (a_1'(c) \setminus a_1(c)))) = \varnothing))$$

5.3   State Space

In the algebraic nets community, the state of a system corresponds to the notion of marking, that is to say a mapping which returns, for each place of the net, a multi-sets of algebraic values. However, this current notion of marking is not suitable in the CO-OPN/2 context. Remember that CO-OPN/2 is a structured formalism which allows for the description of a system by means of a collection of entities. Moreover, this collection can dynamically increase or decrease in terms of number of entities. This implies that the system has to retain two additional informations as explained above. In that case, the state of a system consists of three elements. The first two ones manage the object identifiers, i.e. a partial function to memorize the last identifiers used, and a second function to memorize which identifiers are created and alive. The third element consists in a *partial* function that associates a multi-set of algebraic values to an object identifier and a place. Such a partial function is undefined when the object identifier is not yet assigned to a created object. This is a more sophisticated notion of marking than the one presented in usual algebraic nets. This new notion of marking is necessary in the CO-OPN/2 context because, here, a net does not describe a single instance but a class of objects which can be dynamically created and destroyed.

Definition 12. *Marking, definition domain, state*
Let *Spec* be a specification and $A = Sem(Pres(Spec))$. Let $S = \text{Sorts}(Spec)$ and $P = \text{Places}(Spec)$, a *marking* is a partial function $m : \widehat{A} \times P \to [A]$ such that if $o \in \widehat{A}$ and $p \in P_s$ with $s \in S$ then $m(o, p) \in [A]_s$. We denote the set of all markings over *Spec* and A by $Mark_{Spec,A}$. The *definition domain* of a marking $m \in Mark_{Spec,A}$ is defined as

$$Dom_{Spec,A}(m) = \{(o, p) \mid m(o, p) \text{ is defined}, p \in P, o \in \widehat{A}\}.$$

A marking m is denoted $\perp$ when $Dom_{Spec,A}(m) = \varnothing$. The *state* of a system over *Spec* and A is a triple $(l, a, m) \in Loid_{Spec,A} \times Aoid_{Spec,A} \times Mark_{Spec,A}$. We denote the *state space*, i.e. the set of all states, by $State_{Spec,A}$.   ◊

The notion of transition system is an essential element of the semantics of a CO-OPN/2 specification. We define a transition system as a graph in which the arcs are labelled by a multi-set of transition names, in order to allow the simultaneous firing of transitions. Although CO-OPN/2 is based on such a step semantics, the events of a system described by a CO-OPN/2 specification are not restricted to transition names, but are much more sophisticated. The introduction of the distinction between invisible and observable events, the synchronizations between the objects and then the parameterized transitions (methods), as well as the three operators ' // ', ' .. ', and '⊕', lead us to adopt a different notion of transition system. In this new notion for the transition systems, the state space is as defined above, and each transition is labelled by an element of the set of all events as established in Definition 4.

Definition 13. *Transition system*

Let *Spec* be a specification and $A = Sem(Pres(Spec))$. Let $S^C = \text{Types}(Spec)$ and $M = \text{Methods}(Spec)$. A *transition system* over *Spec* and A is a set of triples

$$TS_{Spec,A} \subseteq State_{Spec,A} \times \mathbf{E}_{A,M,\widehat{A},S^C} \times State_{Spec,A}.$$

The set of all transitions systems over *Spec* and A is denoted $\mathbf{TS}_{Spec,A}$. A triple $\langle st, e, st' \rangle$ is called a *transition*; such an event e between two states st and st' is commonly written either $st \xrightarrow{e} st'$ or $st \xRightarrow{e} st'$.   ◊

Before introducing the set of inference rules designed for the construction of the transition system associated to a given CO-OPN/2 specification, we now introduce informally some basic operators on markings and for the management of the object identifiers. These operators are intensively used in those inference rules.

Informally, the *sum* of markings '+' adds the multi-set values of two markings and takes into account the fact that markings are partial functions. The *common markings* predicate '⋈' determines if two markings are equal for their common places. As for the *fusion* of markings '$m_1 \unlhd m_2$', it returns a marking whose the values are those of m_1 and those of m_2 which do not appear in m_1.

5.4   Semantics and Inference Rules

In order to construct the semantics of a CO-OPN/2 specification which consists mainly of a transition system, we provide here a set of inference rules expressed as *Structural Operational Semantics*, a widely used formalism for describing the computational meaning of systems.

The idea behind the construction of the semantics of a specification composed of several class modules, is to build the semantics of each individual class modules first, and compose them subsequently by means of the synchronizations. Such a semantics of an individual class module is called a *partial semantics* in the sense that it is not yet composed with other partial semantics (by the synchronizations), and it still contains some invisible events.

The distinction between the observable events (composed of methods) and the ones that are invisible (in relation with the internal transitions τ) implies a *stabilization process*. This process is necessary so that the observable events are performed only when all executable invisible events have occurred. A system in which no more invisible event can occur is said to be in a *stable* state.

Another operation called the *closure operation* is necessary to take into account the three operators (sequence, simultaneity, alternative) as well as the synchronization requests. Such a closure operation determines all the sequential, concurrent, and non-deterministic behaviors of a given semantics and composes the different parts of the semantics by means of the synchronizations.

The successive composition of both the stabilization process and the closure operation on all the class modules of the specification will finally provide a transition system in which

- all the sequential, simultaneous, and non-deterministic behaviors will have been inferred;
- all the synchronization requests will have been solved;
- all the invisible or spontaneous events will have been eliminated; in other words every state of the transition system is stable.

Such a transition system will be considered as the semantics of a CO-OPN/2 specification.

As we will see, the inference rules introduced further for the construction of the semantics of a specification, generate two kinds of transitions. The transitions that involve both invisible and observable events are denoted by a single arrow $st \xrightarrow{e} st'$, while the ones that involve only observable events are denoted by a double arrow $st \overset{e}{\Longrightarrow} st'$. A transition system can then include two kinds of transitions which must be distinguished during the construction of the semantics. Thus, in order to identify these two kinds of transitions, any transition system is $\{\rightarrow, \Rightarrow\}$-disjointly-sorted. This means that any transition system is divided into two disjoint sub-transition systems: the sub-transition system which contains only $\rightarrow$-transitions and the one which is composed of $\Rightarrow$-transitions.

The inference rules are arranged into three categories and realize the following tasks:

- the rules CLASS and MONO build, for a given class, its partial transition system according to its methods, places, and behavioral formulas; CREATE and DESTROY take charge of the dynamic creation and destruction of class instances;
- SEQ, SIM, ALT-1, and ALT-2 generate all deductible sequential, simultaneous, and non-deterministic behaviors; SYNC composes the various partial semantics by means of the synchronization requests between the transition systems;
- STAB-1 and STAB-2, involved in the stabilization process, "eliminates" all invisible or spontaneous events which correspond to internal transitions of the classes.

Partial Semantics of a Class

We now develop the partial semantics of a given class module of a specification. First of all, we give some auxiliary definitions used in the subsequent construction of the partial semantics. Let us consider a specification $Spec$, $A = Sem(Pres(Spec)))$, and a class module $Md^C = \langle \Omega^C, P, I, X, \Psi \rangle$ of type c. Let $S^A = \text{Sorts}(Spec)$, $S^C = \text{Types}(Spec)$, $M = \text{Methods}(Spec)$, and Σ be the global signature of $Spec$.

The evaluation of a set of terms of $T_{[\Sigma],X}$ indexed by P for a given assignment σ and a given class instance o into the set of markings $Mark_{Spec,A}$ is defined as:

$$[\![(t_p)_{p \in P}]\!]^\sigma_o = m \text{ such that } (\forall p \in \text{Places}(Spec))(\forall o' \in \widehat{A})$$

$$m(o',p) = \begin{cases} [\![t_p]\!]^\sigma & \text{if } o' = o \text{ and } p \in P, \\ \text{undefined} & \text{otherwise.} \end{cases}$$

Such terms form, for example, a pre/post condition of a behavioral formula or an initial marking. This kind of evaluation is used in the inference rules as shown in Definition 14.

Another kind of evaluation required by the inference rules is the evaluation of an event which consists of the evaluation of all the arguments of the methods, but also the evaluation of the objects identifiers terms. The event evaluation $[\![\ ]\!]^\sigma : \mathbf{E}_{(T_\Sigma,X),M,(T_\Sigma,X)_s,\{c\}} \to \mathbf{E}_{A,M,\widehat{A},\{c\}}$ with $s \in S^C$ naturally follows from Definition 4 and is inductively defined as:

$$[\![t.\tau]\!]^\sigma = [\![t]\!]^\sigma.\tau$$
$$[\![t.m(a_1,\ldots,a_n)]\!]^\sigma = [\![t]\!]^\sigma.m([\![a_1]\!]^\sigma,\ldots,[\![a_n]\!]^\sigma)$$
$$[\![t.\text{create}]\!]^\sigma = [\![t]\!]^\sigma.\text{create}$$
$$[\![t.\text{destroy}]\!]^\sigma = [\![t]\!]^\sigma.\text{destroy}$$
$$[\![Event' \text{ with } Event'']\!]^\sigma = [\![Event']\!]^\sigma \text{ with } [\![Event'']\!]^\sigma$$
$$[\![Event' \text{ op } Event'']\!]^\sigma = [\![Event']\!]^\sigma \text{ op } [\![Event'']\!]^\sigma$$

for all $Event, Event', Event'' \in \mathbf{E}_{(T_\Sigma,X),M,(T_\Sigma,X)_s,\{c\}}$ with $s \in S^C$, for all $t \in (T_\Sigma,X)_s$ and for all methods $m \in M_{s,s_1,\ldots,s_n}$ with $s \in S^C$ and $s_i \in S^A$, and for all synchronization operators $op \in \{//,\ldots,\oplus\}$. Note that the evaluation of any term t of $(T_\Sigma,X)_s$ with $s \in S^C$ belong to $\widehat{A}$ and then represents an object identifier. The evaluation of such terms, in the previous definition, is essential when data structures of object identifiers are considered.

Finally, the satisfaction of a condition of a behavioral formula is defined as:

$$A,\sigma \models Cond \iff (Cond = \varnothing) \vee (\forall (t = t') \in Cond,\ A,\sigma \models (t = t')).$$

Definition 14. *Partial semantics of a class module*
Let *Spec* be a specification and $A = Sem(Pres(Spec))$. Let $Md^C = \langle \Omega^C, P, I, X, \Psi \rangle$ be a class module of *Spec*, where $\Omega^C = \langle \{c\}, \leq^C, M, O \rangle$. The *partial semantics* of Md^C is the transition system denoted $PSem_{Spec,A}(Md^C)$ which is the least fixed point resulting from the application of the inference rules: CLASS, MONO, CREATE, and DESTROY given in Table 1. ◇

The inference rules introduced in Table 1 can be informally formulated as follows:

- The CLASS rule generates the basic observable – as well as invisible – transitions that follow from the behavioral formulas of a class. For all the object identifiers of the class, for all last object identifier function l, and for all alive object identifier function a, a *firable* (or *enabled*) transition is produced provided that:
  1. there is a behavioral formula $Event :: Cond \Rightarrow Pre \to Post$ in the class;
  2. there exists an assignment $\sigma : X \to A$;
  3. all the equations of the global condition are satisfied ($A,\sigma \models Cond$);

Table 1. Inference Rules for the Partial Semantics Construction.

$$\text{CLASS} \quad \frac{\begin{array}{l} Event :: Cond \Rightarrow Pre \rightarrow Post \in \Psi, \ \exists \sigma : X \rightarrow A, \\ A, \sigma \models Cond, \ o \in a(c) \end{array}}{\langle l, a, [\![Pre]\!]_o^\sigma \rangle \xrightarrow{\ [\![Event]\!]^\sigma\ } \langle l, a, [\![Post]\!]_o^\sigma \rangle}$$

$$\text{CREATE} \quad \frac{\begin{array}{l} \exists \sigma : X \rightarrow A, \\ l' = newloid_c(l), \ a' = newaoid_c(a, o), \ o = l'(c), \ o \notin a(c) \end{array}}{\langle l, a, \bot \rangle \xrightarrow{\ o.\text{create}\ } \langle l', a', [\![I]\!]_o^\sigma \rangle}$$

$$\text{DESTROY} \quad \frac{o \in a(c), \ a' = remaoid_c(a, o)}{\langle l, a, \bot \rangle \xrightarrow{\ o.\text{destroy}\ } \langle l, a', \bot \rangle}$$

$$\text{MONO} \quad \frac{\langle l, a, m \rangle \xrightarrow{\ e\ } \langle l', a', m' \rangle}{\langle l, a, m + m'' \rangle \xrightarrow{\ e\ } \langle l', a', m' + m'' \rangle}$$

for all l, l' in $Loid_{Spec,A}$, for all a, a' in $Aoid_{Spec,A}$, for all m, m', m'' in $Mark_{Spec,A}$, for all o in $\widetilde{A}_c$, and for all e in $\mathbf{E}_{A,M,\widehat{A},\{c\}}$.

4. the object o has already been created and is still alive, i.e. it belongs to the set of alive objects of the class ($o \in a(c)$).

The transition generated by the rule guarantees that there are enough values in the respective places of the object. The firing of the transition consumes and produces the values as established in the pre-set and post-set of the behavioral formula.

- The CREATE rule generates the transitions aimed at the dynamic creation of new objects provided that:
  1. for any last object identifier function l and any alive object identifier function a;
  2. a new last object identifier function is computed ($l' = newloid_c(l)$);
  3. a new object identifier o is determined for the class ($o = l'(c)$);
  4. this new object identifier must not correspond to any active object ($o \notin a(c)$).

The new state of the transition generated by the rule is composed of the new last object identifier function l', an updated function a' in which the new object identifier has been added to the set of created objects of the class, and the initial marking $[\![I]\!]_o^\sigma$.

- The DESTROY rule, aimed at the destruction of objects, is similar to the CREATE rule. The DESTROY rule merely takes an object identifier out of the set of created objects, provided that the object is alive.
- The MONO rule (for monotonicity) generates all the firable transitions from the transitions already generated.

Example 15.

Figure 7 illustrates the partial semantics of the 'ConveyorBelt' class of the Swiss chocolate factory example given in Specification 7. For the sake of simplicity, only few relevant events are depicted. Moreover, each state is represented by a triple $\langle o, O, m \rangle$, where o is the last object identifier used for the creation of new instances of the class, O is the set of the active objects of the class, and m is the multi-set values of the 'belt' place. Brackets '[_]' denote multi-set values, while vertical lines '|_|' denote fifo data structures of packagings. The abbreviation 'tcb' stands for 'the-conveyor-belt' object identifier; x, y are object identifiers of type 'packaging' or 'deluxe-packaging'. For instance the state $\langle \text{tcb}, \{\text{tcb}\}, [|x, x|] \rangle$ establishes that the 'tcb' object is alive and its 'belt' place is composed of one fifo which contains twice the same packaging object identifier x.

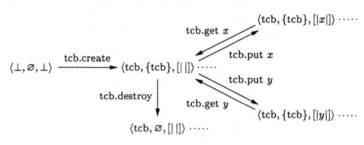

Fig. 7. Partial Semantics of the Conveyor-Belt.

Semantics of a CO-OPN/2 Specification

The purpose of this section is to define the closure operation and the stabilization process, and show how their composition is used to build the whole semantics of a specification from the various partial semantics of the class modules.

The construction of the whole semantics of a CO-OPN/2 specification composed of several class modules consists in considering each partial semantics and combine them by means of the successive composition of the stabilization process and a closure operation. This cannot be done in random order because observable events (methods) can be performed only when invisible events have occurred.

In order to build the whole semantics of a specification $Spec$, we introduce a total order over the class modules of $Spec$ which depends on the partial order induced by the client-ship relation D^C_{Spec}. This total order is used to construct the semantics; it is denoted $\sqsubseteq$ and defined such that $D^C_{Spec} \subseteq \sqsubseteq$.

Given Md^C_0 the least module of the total order and the fact that $Md^C_i \sqsubseteq Md^C_{i+1}$ $(0 \leq i < n)$, we introduce the partial semantics of all the modules Md^C_i $(0 \leq i \leq n)$ of a specification from the bottom to the top.

Stabilization Process. The purpose of the stabilization process is to provide a transition system in which all the invisible events (internal transitions) have been taken into account. More precisely, the stabilization process consists in merging all the observable events and the invisible ones into one step.

The stabilization process proceeds in two stages. The first stage is the application of two inference rules on a given transition system to produce the merged transitions. This step is called the *pre-stabilization*. The second step produces the intended transition system which contains only the relevant transitions, i.e. all the transitions except the transitions which do not lead to a stable state. We observe that the STAB-1 and STAB-2 involve a new kind of transitions denoted with a double arrow ($\Rightarrow$-transitions). This kind of transitions is introduced in order to distinguish between a transition system composed of stable states and the one in which some invisible events have to be taken into account. Now, we understand a little bit better why we defined a transition system as $\{\rightarrow, \Rightarrow\}$-disjointly-sorted, that is to say divided into two disjoint sub-transition systems.

Definition 16. *Stabilization process*
Let *Spec* be a specification and $A = Sem(Pres(Spec))$. The *stabilization process* consists of the function $Stab : \mathbf{TS}_{Spec,A} \to \mathbf{TS}_{Spec,A}$ defined as follows:

$$Stab(TS) = \{m \xrightarrow{\ e\ } m' \in TS\} \cup$$
$$\{m \xRightarrow{\ e\ } m' \in PreStab(TS) \mid \nexists\, m' \xrightarrow{\ o.\tau\ } m'' \in PreStab(TS)\}$$

in which $PreStab : \mathbf{TS}_{Spec,A} \to \mathbf{TS}_{Spec,A}$ is a function such that $PreStab(TS)$ is the least fixed point which results from the application on TS of the inference rules[9] STAB-1 and STAB-2 given in Table 2. $\Diamond$

Table 2. Inference Rules of the Stabilization Process.

$$\text{STAB-1} \quad \frac{e \neq o.\tau,\ e \neq o.\tau \text{ with } e',\ \langle l,a,m \rangle \xrightarrow{\ e\ } \langle l',a',m' \rangle}{\langle l,a,m \rangle \xRightarrow{\ e\ } \langle l',a',m' \rangle}$$

$$\text{STAB-2} \quad \frac{m_1' \bowtie m_2,\ \langle l,a,m_1 \rangle \xRightarrow{\ e\ } \langle l',a',m_1' \rangle,\ \langle l',a',m_2 \rangle \xrightarrow{\ o.\tau\ } \langle l'',a'',m_2' \rangle}{\langle l,a,m_1 \trianglelefteq m_2 \rangle \xRightarrow{\ e\ } \langle l'',a'',m_2' \trianglelefteq m_1' \rangle}$$

for all $m, m', m_1, m_1', m_2, m_2'$ in $Mark_{Spec,A}$, for all l, l', l'' in $Loid_{Spec,A}$, for all a, a', a'' in $Aoid_{Spec,A}$, for all o in $\widehat{A}$, and for all e, e' in $\mathbf{E}_{A,M,\widehat{A},S^C}$.

The inference rules introduced in Table 2 can be informally formulated as follows:

- The STAB-1 rule generates all the observable events which will be merged with invisible events if they lead to an unstable state; note that neither the pure internal transitions nor the internal transitions asking to be synchronized with some partners are considered by this rule;
- The STAB-2 rule merges an event leading to a non-stable state and the invisible event which can occur "in sequence". This rule is very similar the SEQ introduced further when the closure operation is presented. Thus, the same comments regarding its functioning and the meaning of the operators involved in the rule hold.

---

[9] The application of the inference rules on TS obviously includes TS itself.

It is worthwhile to note that when infinite sequences of transitions are encountered, the stabilization process does not retain any collapsed transition. From an operational point of view, such infinite sequence of internal transitions can be considered as a program that loops. However, in a distributed software setting, when an object (or a group of objects) loops, it does not mean that the whole system loops; it simply means that such an object is not able to give services any more and, therefore, it can be ignored.

Closure Operation. The closure operation consists of adding to a given transition system all the sequential, simultaneous, alternative behaviors, and to perform the synchronization requests. A set of inference rules are provided for these aims.

Definition 17. *Closure operation*
Let *Spec* be a specification and $A = Sem(Pres(Spec))$. The *closure operation* consists of the function $Closure : \mathbf{TS}_{Spec,A} \to \mathbf{TS}_{Spec,A}$ such that $Closure(TS)$ is the least fixed point which results from the application on TS of the inference rules SEQ, SIM, ALT-1, ALT-2, and SYNC given in Table 3.   ◊

Table 3. Inference Rules of the Closure Operation.

$$\text{SEQ} \quad \frac{m'_1 \bowtie m_2, \ \langle l,a_1,m_1\rangle \xrightarrow{e_1} \langle l',a'_1,m'_1\rangle, \ \langle l',a'_2,m_2\rangle \xrightarrow{e_2} \langle l'',a'_2,m'_2\rangle}{\langle l,a,m_1 \trianglelefteq m_2\rangle \xrightarrow{e_1 \ \cdot\cdot \ e_2} \langle l'',a'_2,m'_2 \trianglelefteq m'_1\rangle}$$

$$\text{SIM} \quad \frac{l' \stackrel{\triangle}{=}_l l'', \ P(a_1,a'_1,a_2,a'_2), \ \langle l,a_1,m_1\rangle \xrightarrow{e_1} \langle l',a'_1,m'_1\rangle, \ \langle l,a_2,m_2\rangle \xrightarrow{e_2} \langle l'',a'_2,m'_2\rangle}{\langle l,a_1 \cup a_2,m_1+m_2\rangle \xrightarrow{e_1 \ // \ e_2} \langle l' \, \triangle_l \, l'',a'_1 \cup a'_2,m'_1+m'_2\rangle}$$

$$\text{ALT-1} \quad \frac{\langle l,a,m\rangle \xrightarrow{e_1} \langle l',a',m'\rangle}{\langle l,a,m\rangle \xrightarrow{e_1 \oplus e_2} \langle l',a',m'\rangle} \qquad \text{ALT-2} \quad \frac{\langle l,a,m\rangle \xrightarrow{e_1} \langle l',a',m'\rangle}{\langle l,a,m\rangle \xrightarrow{e_2 \oplus e_1} \langle l',a',m'\rangle}$$

$$\text{SYNC} \quad \frac{l' \stackrel{\triangle}{=}_l l'', \ P(a_1,a'_1,a_2,a'_2), \langle l,a_1,m_1\rangle \xrightarrow{e_3 \ \text{with} \ e_2} \langle l',a'_1,m'_1\rangle, \langle l,a_2,m_2\rangle \xrightarrow{e_2} \langle l'',a'_2,m'_2\rangle}{\langle l,a_1 \cup a_2,m_1+m_2\rangle \xrightarrow{e_3} \langle l' \, \triangle_l \, l'',a'_1 \cup a'_2,m'_1+m'_2\rangle}$$

for all m_1,m'_1,m_2,m'_2 in $Mark_{Spec,A}$, for all l,l',l'' in $Loid_{Spec,A}$, and for all a,a',a_1,a'_1,a_2,a'_2 in $Aoid_{Spec,A}$, for all e_1,e_2 in $\mathbf{E}_{A,M,\widehat{A},S^C}$ which are not equal to $o.\tau$ or to $o.\tau$ with e' and for all e_3 in $\mathbf{E}_{A,M,\widehat{A},S^C}$.

The inference rules of Table 3 can be informally formulated as follows:

- The SEQ rule infers the sequence of two transitions provided that the markings shared between m'_1 and m_2 are equal. The double arrow under the e_1 event forces that e_1 leads to a stable state. This guarantees that all the invisible events are taken into account before inferring the sequential behaviors.

- The SIM rule infers the simultaneity of two transitions, provided that some constraints on the l and a functions are satisfied. The purposes of these constraints are:
    1. to avoid that an event can use a given object being created by the other event (i.e. which does not already exist);
    2. to avoid that an event can use a given object being destroyed by the other event (i.e. which does not exit any more).

  Informally, the operators defined in Section 8 are used to:
    1. $l' \overset{\triangle}{=}_l l''$ avoids the conflicts when simultaneous creation is considered. Remember that the assignment of a new object identifier is handled by a different function for each type. Consequently, this characteristic does not permit the simultaneous creation of two objects of the same type.
    2. $l' \triangle_l l''$ combines the last object identifier functions according to the creations involved in e_1 and e_2;
    3. $a \cup a'$ makes merely the union of the a and a' for each type;
    4. the predicate $P(a_1, a'_1, a_2, a'_2)$ guarantees that the objects created or destroyed by the events e_1 do not appear in the upper tree related to the event e_2 and vice versa; more precisely, for each type c the active objects of $a_1(c)$ (and $a'_1(c)$) and the "difference" between $a_2(c)$ and $a'_2(c)$ have to be disjoint, as well as the active objects of $a_2(c)$ (and $a'_2(c)$) and the "difference" between $a_1(c)$ and $a'_1(c)$.

- The ALT-1 and ALT-2 rules provides all the alternative behaviors. Two rules are necessary for representing both choices of the alternative operator $\oplus$.

- The SYNC "solves" the synchronization requests. It generates the event which behaves the same way as the event 'e_3 with e_2' asking to be synchronized with the event e_2. The double arrow under the event e_2 guarantees that the synchronizations are performed with events leading to stable states. Note that e_3 can be an invisible event because internal transitions may ask for a synchronization.

  The similarities between the SIM and SYNC are not surprising because of the synchronous nature of CO-OPN/2.

In consequence, several intuitive but important intended events can never occur in a system that is built by means of such formal system. These are :

1. the use of an object followed by the creation of this object;
2. the destruction of an object followed by the use of this object;
3. the creation (or destruction) of an object and the simultaneous use of this object;
4. the creation (or destruction) of an object and the simultaneous creation (or destruction) of another object of the same type;
5. the synchronization of the use of an object with the creation (or destruction) of this object;
6. the multiple creation of the same object;
7. the multiple destruction of the same object;
8. the destruction followed by the creation of the same object;

The semantics expressed by the following definition is calculated starting from the partial semantics of the least object (for a given total order), and repeatedly adding the partial semantics of a new object. For each new object added to the system, we observe that the stabilization process is obviously performed before the closure operation.

Definition 18. *Semantics of a specification for a given total order*
Let *Spec* be a specification composed of a set of class modules $\{Md_j^C \mid 0 \le j \le m\}$ and $A = Sem(Pres(Spec))$. Let $\sqsubseteq$ be a total order over the class modules such that $D_{Spec}^C \subseteq \sqsubseteq$. The *semantics of Spec for* $\sqsubseteq$ is denoted $Sem_A^{\sqsubseteq}(Spec)$ and inductively defined as:

$$Sem_A^{\sqsubseteq}(\varnothing) = \varnothing$$

$$Sem_A^{\sqsubseteq}(\{Md_0^C\}) = \lim_{n \to \infty} (Closure \circ Stab)^n (PSem(Md_0^C))$$

$$Sem_A^{\sqsubseteq}(\cup_{0 \le j \le k}\{Md_j^C\}) =$$
$$\lim_{n \to \infty} (Closure \circ Stab)^n (Sem_A^{\sqsubseteq}(\cup_{0 \le j \le k-1}\{Md_j^C\}) \cup PSem_A(Md_k^C))$$

for $1 \le k \le m$. $\diamond$

The above definition of the semantics is not independent of the total order as demonstrated further by Example 21. Thus, we define the semantics of a CO-OPN/2 specification only when it does not depend of such a total order.

Definition 19. *Semantics of a specification*
Let *Spec* be a specification, $A = Sem(Pres(Spec))$, $\sqsubseteq$ be a total order over the class modules D_{Spec}^C. The *semantics of Spec* denoted $Sem_A(Spec)$ is defined as the $Sem_A(Spec) = Sem_A^{\sqsubseteq}(Spec)$ iff it is independent of the total order $\sqsubseteq$ over the class modules of *Spec* (i.e. the semantics are equal for different total orders), otherwise it is undefined. $\diamond$

This definition is implied by the fact that the stabilization needs a precise hierarchy between modules in order to be deterministic for its behavior. Alternative definitions could consist in imposing, from the specifier, to fix the total order of stabilization; or to reduce the stabilization domain to specific objects.

Finally, we define the *step semantics* of a CO-OPN/2 specification from the above semantics in which we only retain the $\Rightarrow$-transitions whose the events are atomic or simultaneous. Moreover, we only consider the transitions from states that are reachable from the initial state.

Definition 20. *Step Semantics of a specification*
Let *Spec* be a specification and $A = Sem(Pres(Spec))$. The *step semantics of Spec*, denoted $SSem_A(Spec)$, is defined as the greatest set in $\mathbf{TS}_{Spec,A}$ such that $SSem_A(Spec) \subseteq Sem_A(Spec)$ and for any transition $st \xRightarrow{e} st'$ in $SSem_A(Spec)$ the following properties holds[10]:

---

[10] The symbol $\Vdash^*$ corresponds to the reflexive transitive closure of the reachability relation, however, $\Vdash$ is defined for the $\Rightarrow$-transitions. The initial state is denoted $\langle \perp, \varnothing, \perp \rangle$.

i) $e = e_1 /\!/ e_2 /\!/ \cdots /\!/ e_n$, where $e_i = o_i.m_i(a_{1i}, \ldots, a_{ki})$ $(1 \leq i \leq n)$;
ii) $\langle \bot, \varnothing, \bot \rangle \Vdash^* st$;

where $e, e_i \in \mathbf{E}_{A,M,\widehat{A},S^c}$ $(1 \leq i \leq n)$. $\Diamond$

Example 21.
Let us consider the specification *Spec* composed of the three class modules A,
B, and C given in Specification 11, and $A = Sem(Pres(Spec))$. The graphical
representation of the three static objects defined in the classes is depicted on the
right side of the specification.

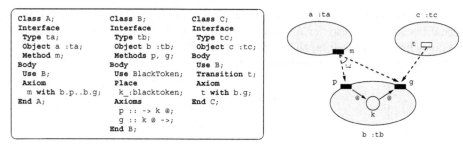

Spec. 11. Three Classes, Each of Them Defines a Static Object.

These three classes can be organized according to two total orders:

$$B \sqsubset_1 A \sqsubset_1 C \quad \text{and} \quad B \sqsubset_2 C \sqsubset_2 A.$$

Thus, we demonstrate that these two total orders lead to different semantics;
in other words, the semantics of such a system is not independent of the total
order, $Sem_A^{\sqsubset_1} \neq Sem_A^{\sqsubset_2}$.

Since this example depicts several transitions systems, we adopt, for the sake
of clarity, the following conventions:

1. we represent the state of the objects by the content of the places; natural
   numbers denote the number of tokens, while $\bot$ is used for object without
   any place;
2. the events in relation with the creation of the objects are not represented;
3. black and gray arrows denote, respectively, $\rightarrow$- and $\Rightarrow$-transitions;
4. only the synchronization events of the closure operation are depicted, the
   numerous other events are omitted.
5. both the semantics $Sem_A^{\sqsubset_i}$ $(i = 1, 2)$ have an observational flavor; i.e. only
   the $\Rightarrow$-transitions from states which are reachable from the initial state are
   drawn and the $\rightarrow$-transitions are omitted.

First, we provide the partial semantics of the three individual classes A, B,
and C in Figure 8. Note that, for the sake of clarity, only the relevant transitions
are shown in these figures, especially those in relation with the static objects.

Figure 9 depicts the incremental construction of the semantics of the specification according the first total order $\sqsubset_1$. As a result, the third transition system

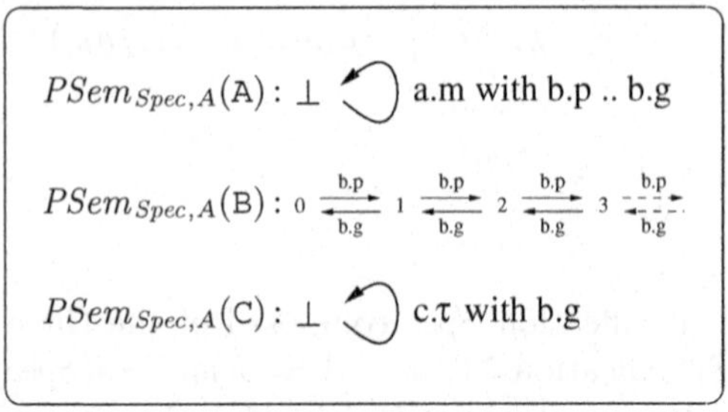

Fig. 8. Partial Semantics of the Three Classes A, B, and C.

is due to the fact that the internal transition τ (which corresponds to the internal transition t) asks for a synchronization with the method g as soon as it becomes firable. Therefore, the system cannot offer the service g to the outside world once class C has been incorporated.

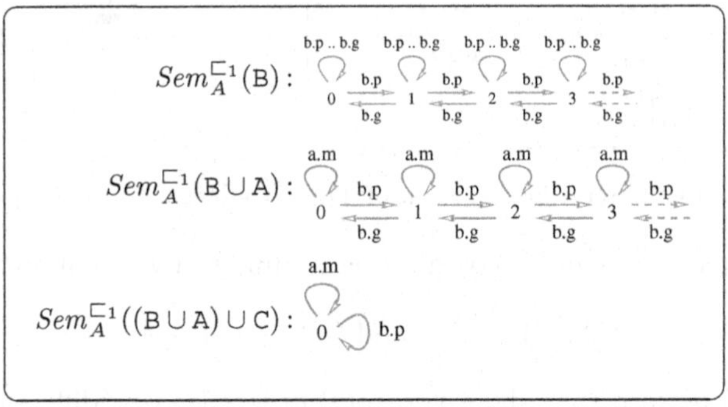

Fig. 9. Semantics of the Specification According to $\sqsubseteq_1$.

The semantics of the specification according to the second total order $\sqsubseteq_2$ is given in Figure 10. We observe, as above, that the incorporation of class C makes the service g unavailable. Consequently, adding class A is useless, because of the prevailing internal transition which breaks the atomicity of the sequence $b.p \mathinner{..} b.g$ in method m.

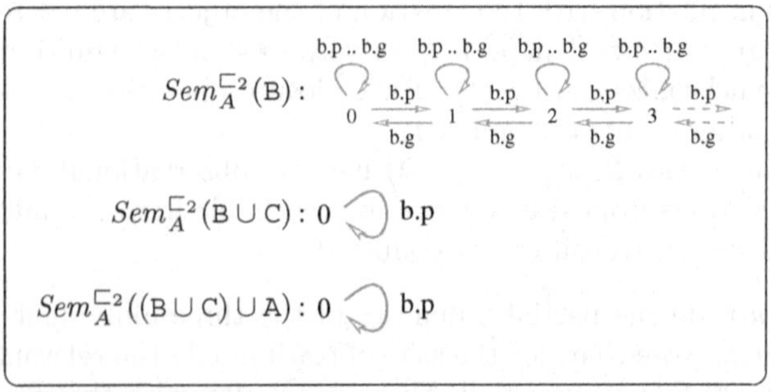

Fig. 10. Semantics of the Specification According to $\sqsubseteq_2$.

5.5   Subtyping

As mentioned earlier, we believe, as some other researchers [2,14], that the sub-typing relationship is related to the behavior of the objects and addresses a semantic conformance between a subtype and its supertypes. Moreover, we are convinced that, at the specification level, the strong form of the substitutability rather than the weak form should be adopted.

The subtype relationship we are going to state is based on the strong bisim-ulation equivalence between the semantics of the supertype and the semantics of the subtype restricted to the behaviors of the supertype; i.e. between two transition systems. Bisimulation, introduced in [15] is adequate in our context because it is weaker than isomorphic equivalence and stronger than trace equiv-alence. On the one hand, bisimulation disregards the internal state of the objects and, on the other hand, takes into account the non-determinism of the transition systems. The bisimulation equivalence establishes a relation between the states of two transition systems: two states are related if and only if each transition in the first system corresponds to another transition in the second one. As we will only consider transition systems as model of a CO-OPN/2 specification, then all the states will be reachable from a specific state, the initial state. Unicity of the bisimulation relation is imposed by the relation of initial states. We denote, for a transition system TS, an initial state by st^{init}.

In the CO-OPN/2 context, the usual definition of bisimulation is extended for the subtyping. Thus, the subtyping definition we establish below compares two semantics which clearly involves object identifiers of two different types (in subtyping relation). These object identifiers are then obviously not equal. The following definition takes into account the fact that the object identifier belong to two types in subtyping relation by means of the operations between the carriers of the object identifier algebra. Thus, it is necessary to define the natural extension of a subtype relation $\leq$ on the object identifiers sorts according to the event structure of Definition 4.

Definition 22. *Event substitution relation*
Let $S = S^A \cup S^C$ be a set of sorts such that $S^A \in \mathbf{S^A}$ and $S^C \in \mathbf{S^C}$. Let us consider $A = (A_s)_{s \in S}$, $M = (M_{s,w})_{s \in S^C, w \in S^*}$, $O = (O_s)_{s \in S^C}$, a set of types of classes $C \subseteq S^C$, let, then the event substitution relation $\leq \in \mathbf{E}_{A,M,O,C} \times \mathbf{E}_{A,M,O,C}$ is defined inductively as follows:

$$o'.m(a_1, \ldots, a_n) \leq o.m(a_1, \ldots, a_n) \text{ iff } o' = \mathrm{sub}_{c,c'}\ o$$
$$(e_1 \mathrel{..} e_2) \leq (e_1' \mathrel{..} e_2') \qquad \text{iff } e_1 \leq e_1' \text{ and } e_2 \leq e_2'$$
$$(e_1 \mathbin{/\!/} e_2) \leq (e_1' \mathbin{/\!/} e_2') \qquad \text{iff } e_1 \leq e_1' \text{ and } e_2 \leq e_2'$$
$$(e_1 \oplus e_2) \leq (e_1' \oplus e_2') \qquad \text{iff } e_1 \leq e_1' \text{ and } e_2 \leq e_2'$$

where $e_1, e_1', e_2, e_2' \in \mathbf{E}_{A,M,O,C}$, $s \in S^C$, $s_i, s_i' \in S$ $(1 \leq i \leq n)$, $a_1, \ldots, a_n \in A_{s_1} \times \cdots \times A_{s_n}$, $m \in M_{s,s_1' \cdots s_n'}$, $o \in O_s$, $c \in C$, and such that s_i, and s_i' $(1 \leq i \leq n)$ belong to the same connected component. $\Diamond$

Definition 23. *Strong subtype bisimulation*
Let *Spec* be a specification, $A = Sem(Pres(Spec))$, and $\leq$ be the event substitution relation. A *strong subtype bisimulation* between two transition systems $TS_1, TS_2 \in \mathbf{TS}_{Spec,A}$ is the relation $R \subseteq State(TS_1) \times State(TS_2)$ such that

1. if $st_1 \ R \ st_2$ and $st_1 \overset{e_1}{\Longrightarrow} st'_1 \in TS_1$ then there is $st_2 \overset{e_2}{\Longrightarrow} st'_2 \in TS_2$ such that $st'_1 \ R \ st'_2$ and $e_1 \leq e_2$;

2. if $st_2 \ R \ st_1$ and $st_2 \overset{e_2}{\Longrightarrow} st'_2 \in TS_2$ then there is $st_1 \overset{e_1}{\Longrightarrow} st'_1 \in TS_1$ such that $st'_1 \ R \ st'_2$ and $e_1 \leq e_2$;

3. $st_1^{\text{init}} \ R \ st_2^{\text{init}}$;

where $e_1, e_2 \in \mathbf{E}_{A,M,\widehat{A},S^C}$. We say that TS_1 and TS_2 are *strongly subtype bisimilar*, for short if there exists such a non-empty relation R, and we denote this by $TS_1 \leftrightarrow_{\leq} TS_2$. $\Diamond$

We say that a type c' is a subtype of another type c when the step semantics of the class $Md_{c'}^C$, in which c' is declared, along with its dependencies restricted to the behaviors of the class Md_c^C, in which c is declared, is strongly subtype bisimilar to the step semantics of Md_c^C and its dependencies. Formally we have:

Definition 24. *Validity of subtype relation*
Let *Spec* be a specification which includes two class modules $Md_{c'}^C$ and Md_c^C such that $c' \leq c$ in *Spec*, we say that $c' \leq c$ is valid iff

$$SSem_A(Dep(Md_{c'}^C) \cup (Md_{c'}^C))|_{Md_c^C} \leftrightarrow_{\leq} SSem_A(Dep(Md_c^C) \cup (Md_c^C))$$

where $Dep(Md^C)$ denotes the set of all the class modules used by Md^C. The notation $SSem_A(Spec)|_{Md_c^C}$ means that $SSem_A(Spec)$ is limited to the services of the class module Md_c^C. $\Diamond$

For example, let us consider both the deluxe-packaging type defined in Specification 4, and the packaging type defined in Specifications 3. It is clear that deluxe-packaging is a subtype of packaging because after the appropriate restriction both the semantics are isomorphic. Figure 11 illustrates the step

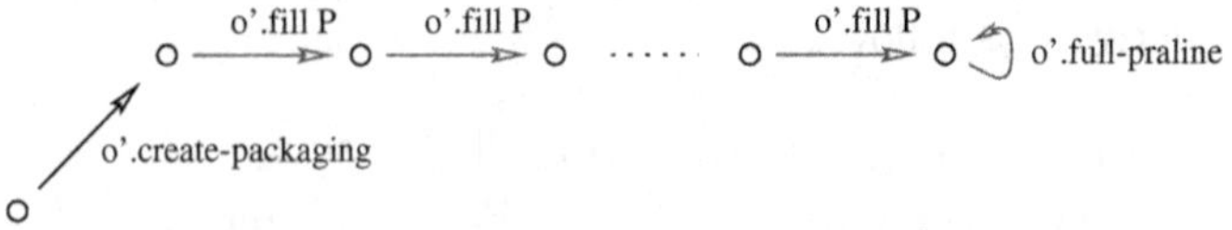

Fig. 11. Step Semantics of the Packagings

semantics of the packagings, while Figure 12 depicts the step semantics of the deluxe packagings. In Figure 13 shows the step semantics of the deluxe packagings restricted to the packagings services, i.e. all the events added by the deluxe packagings are ignored, moreover, only the states reachable from the initial state are considered.

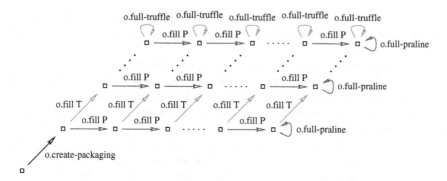

Fig. 12. Step Semantics of the Deluxe Packagings

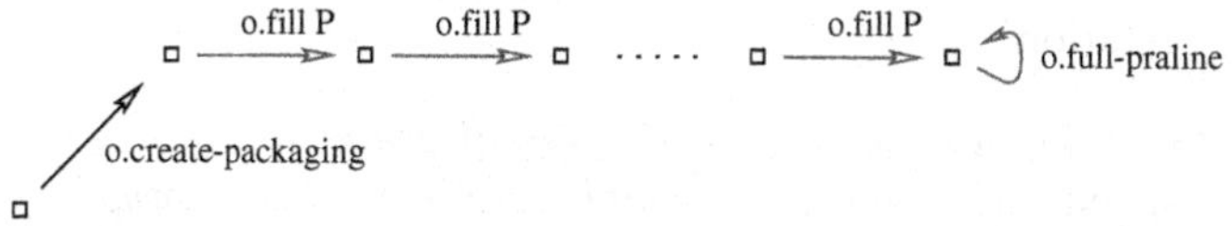

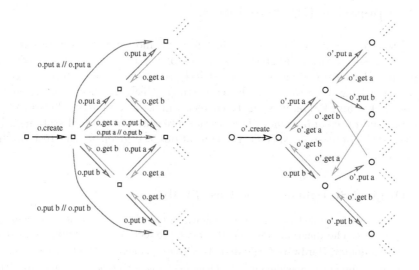

Fig. 13. Step Semantics of the Deluxe Packagings Restricted to the Packagings

The following counter-example which also involves two classes of the Swiss chocolate factory example is less obvious. Let us consider both the heap and the conveyor-belt types defined, respectively, in Specification 5 and 7. Figure 14 depicts both these semantics. On the one hand, we observe, on the left side, the step semantics of the heap, while the right side shows the step semantics of the conveyor belt.

Fig. 14. Step Semantics of the Heap and the Conveyor Belt

We observe easily that both these transition systems are not strongly subtype bisimilar. Thus the heap type is not a subtype of the conveyor-belt type and vice versa.

The conveyor-belt type is not a subtype of the heap type, because the conveyor-belt type does not allow any concurrent event, while the heap type does. Moreover, the nature of the conveyor-belt (first-in-first-out) forces some sequences of events. Consequently, the conveyor-belt type cannot give all the services of the heap type.

On the other side, the heap type is not a subtype of the conveyor-belt type because the heap type possesses some concurrent events. It means that if we substitute an instance of type heap for an instance of type conveyor-belt in a context that possibly uses some concurrent events of the heap type, the behavior of the whole system will be probably not guaranteed.

Finally we are able to introduce the semantics of a specification which satisfies the subtyping relation.

Definition 25. *Subtyping semantics of a specification*
Let *Spec* be a specification and $A = Sem(Pres(Spec))$. The *subtyping semantics of Spec*, denoted $TSem_A(Spec)$, is defined as the transition system $SSem(Spec)$ which validates all the subtype relations between classes of *Spec*.   ◇

We have defined in this section a subtype relation for the classes of a specification as a semantic relation between behaviors. This approach follows the principle of substitutivity and clearly distinguishes the subtype relation from the inheritance relation. A last property which should be mentioned is that our semantics of subtyping based on bisimulation ensures the substitutivity principle.

6   Cooperative Diagram Editor

The objective of this section is to illustrate the modeling capabilities of the CO-OPN/2 specification language by means of a medium-size case study on groupware. We have chosen to adhere strictly to the original statement of the Cooperative Diagram Editor (CDE) case study which has been suggested by Bastide, Lakos and Palanque so as to correctly make a comparison between our modeling and other approaches. This section presents the original statement of the case study and some specific interpretations are given in the rest of the description.

6.1   Original Statement of the Case Study

The system to be studied is a software allowing for cooperative editing of hierarchical diagrams. The diagrams may be the work products of some Object-Oriented Design methodology, Hardware Logic designs, *Petri Net diagrams*[1], etc. (Note that if the diagrams happen to coincide with *the formalism you are proposing*[2], be careful to distinguish clearly between the two.)

One key aspect of this problem is that the editor should cater for several users, working at different workstations, and cooperating in the construction of the one diagram. In the Computer Supported Cooperative Work (CSCW) vocabulary, such a tool could be ranked amongst synchronous groupware (each user is informed in real time of the actions of the others) allowing for relaxed WYSIWIS (What You See Is What I See) : each user may have his own customized view of the diagram under design, viewing different parts of the drawing, or examining it at a different level of detail.

A *second key aspect*[3] of this problem is that the editor should cater for hierarchical diagrams, i.e. components of the diagram can be exploded to reveal subcomponents.

A simple *coordination protocol*[4] is proposed to control the interactions between the various users:

1. (a) Users may join or leave the editing session at will, and may join with different levels of editing privileges. For example, a user may join the session merely to view the diagram, or perhaps to edit it as well (see below).
   (b) The current members of the editing session ought to be visible to all, together with their editing privileges.
2. (a) Graphical elements may be free or owned by a user.
   (b) Different levels of ownership should be supported, including ownership for deletion, encapsulation, modification, and inspection.
   (c) The ownership must be compatible with the user's editing privileges.
3. Ownership for deletion requires that no one else has any ownership of the component – not even for inspection.
4. Ownership for encapsulation requires that only the owner can view the internal details of the component – all other users can only view the top level or *interface*[5] to the component.
5. Ownership for modification allows the user to modify attributes, but not to delete the component.
6. Ownership for inspection only allows the user to view the *attributes*[6].
7. Only ownership for encapsulation can persist between editing sessions. (Note that this ownership is tied to a particular user, not to a particular workstation.) All other ownerships must be surrendered between sessions.
8. Ownership for inspection is achieved simply by selecting the component.
9. Other forms of ownership (and release) are achieved by an appropriate command, having first selected the component.
10. The level of ownership of a component is visible to all other users, as is the identity of the owner.
11. The creator of an element owns it for deletion, until it is explicitly released.

6.2 Specific Interpretations

As it has been suggested in the statement of the case study, several kinds of cooperative editors may be considered and the designers may exercise a certain degree of freedom when making their interpretations. In order to present the informal specification with as much precision as possible, the following explains in details our specific interpretations of each highlighted word that appears in the previous sub-section.

[1],[2] We have decided to consider one kind of CDE: hierarchical Petri nets. The hierarchical Petri nets editor is used as the basic case study and

a complete modeling of this editor is presented, SADT and CO-OPN/2 diagrams editors are briefly presented as variants.

③ We have made the choice of representing a component by means of an object. In order to describe a hierarchical structure, two kinds of components have been introduced. The former correspond to the nodes while the latter represent the leaves of the hierarchical structure.

④ The coordination protocol is only used to coordinate the diagram accesses of the users and must be distinguished from the synchronization protocol which, in turn, must fulfill the cooperative requirements that allow all users to act upon the same document. Moreover, it must ensure the integrity of the document.

⑤ With regard to the notion of interface, we have introduced the notion of anchor, the notion of link along with the notion of interface. An anchor of a component is a location to which another component may be attached. A link of a component is used to link two anchors which belong to components These two notions form the concept of the interface of a component corresponding to the set of sub-components which may be linked to any anchor of the component itself. For example, the interface of the transition of the hierarchical Petri nets that are under consideration consists of the input and output places.

⑥ The notion of attribute is present but we do not provide the specification associated with it. For example, the date of creation as well as the name of the creator of a component could be regarded as being attributes. Note that position and rotation are not considered as being attributes because they are integral parts of the component.

6.3   Example of an Editing Session

Here we give an example of a hierarchical Petri net which could be edited by means of our cooperative diagram editor. Figure 15 depicts a two level Petri net involving some transitions, some places, some arcs, and one token. The figure shows that both the transitions and the places have been placed in a hierarchical relationship (a transition is associated with a sub-net and a place may contain some tokens). For instance, hierarchical transition t (level 0) is associated with a sub-net (level 1) composed of the component labeled p_i^t, t_i^t, a_j^t for $i = 1, 2$ and $j = 1, \ldots, 4$ and the place p_1 (level 0) contains one token labeled r (level 1) which is also displayed at level 0. The anchors are represented by small black squares placed at the border of each component.

The three names 'Nicolas', 'Olivier' and 'Didier' represent the three users involved in the editing session. 'Nicolas' and 'Olivier' edit the Level 0 while 'Didier' acts upon the Level 1.

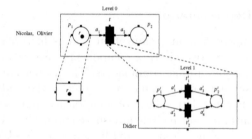

Fig. 15. An Example of Editing Session.

6.4   Modeling Structure of a CDE

Among the kinds of architectures (centralized and replicated), the centralized architecture has been chosen because it encompasses all the requirements of the case study. Moreover, the design of applications seems to be easier with respect to centralized architectures in contrast to those that are replicated.

We use a layered abstract model of a cooperative software upon which we base our specification of a CDE. This model is a simplification of the one that has been proposed by Karsenty [12] and has already been presented for the elaboration of a general cooperative system modeling which uses CO-OPN/2 [6].

We consider a Graphical Interface Layer (the viewports), a Centralized Synchronization Layer (the server) and an Abstract Document Representation Layer (the diagram document). This structure and its elements are described in this section and depicted in Figure 16.

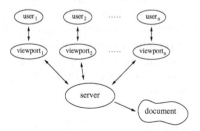

Fig. 16. Overview of a CDE.

All the users collaborate in the edition of a part of a hierarchical diagram by means of a viewport which provides some basic functions that allows aspects of the document to be modified. Viewports serve to collect the actions of its logged-in user, to transmit these actions to the server, and to display the document which pertains to the information sent by the server. The role of the server is to collect and handle the actions of the users, to update the document in accordance to the required coordination protocol, to inform every concerned viewport of the actions of the users and to allow for a simultaneous access to the document.

6.5   Document Representation

As required, a document has an organization that is hierarchical. In order to represent a tree structure, two kinds of components have to be introduced: hierarchical components and atomic components. The former kind of components may include some sub-components and correspond to the nodes of the hierarchical document, while the latter kind of components represent the leaves of the structure and thus may not include any sub-components. A document is, in itself, a hierarchical component represented by means of an object.

6.6   Three Levels of Entities

In the interest of establishing generalizations and progressively refining our discussion, our modeling is organized in three main entity levels as is shown in Figure 17. Each level is, in fact, the result of a further step in the development of the specification.

   We observe the four entities which compose an editor presented horizontally, i.e. User, Viewport, Server, and Component. Vertically, the top level corresponds to classic editors, which does not permit any cooperation between many users. Thus, only one user at a time may edit a diagram. At the second level (C-User, C-Server, C-Viewport, C-Component), the notion of cooperation as well as the notions of attribute, ownership and interface are introduced. This second level, in fact, complies with all the features of the proposed case study. The third level, depicted in Figure 17 shows an example of a concrete cooperative hierarchical Petri nets editor. It is composed of a PN-C-Server, PN-C-Viewport and its PN-C-Components. The effective components of a Petri nets, i.e. the arcs, the places, and the transitions, are not represented in this figure. Since they compose the document, they should be connected with the PN-C-Component entity.

   This structure makes it possible to build various cooperative editors. At the third level of the Figure 17 we illustrate a cooperative Petri net editor but other kinds of CDE such as SADT diagrams or even CO-OPN/2 diagrams editor, could equally have been derived.

6.7   Structure of the Classes

The classes that compose our modeling of a CDE, as well as the relationship between the classes, clearly arise from Figure 17. Three kinds of relationships between classes are relevant here: clientship, inheritance, and subtyping.

Clientship. In Figure 17, for example, one may see, at the top level, that the User class uses (as a client) the Viewport class because a user simply submits some requests to the viewport, e.g. to select a component or add a new component to the diagram, and no information is sent back from the viewport to the user. A similar argument holds for the two other levels. Regarding the other entities (viewports, servers and components), the clientship relationship is symmetric. In the case of a cooperative editor, if we consider, for example, the user's

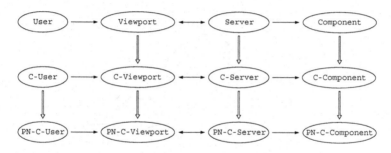

Fig. 17. The Three Levels of Entities.

action of adding a new sub-component to a given component, it will generate the following interaction between the other entities. First the viewport transmits the request to the server which asks for the creation of the new component and summons the concerned component to incorporate the new component. Then, the component involved communicates its new aspect to the server that must subsequently compute and send the information which has to be transmitted to each viewport. This natural cycle can be easily eliminated by adding, at each level, a class which interacts with the viewport and the server classes.

Inheritance. We believe that inheritance and subtyping are two different notions which are used for two different purposes. Inheritance is mainly considered to be a syntactic mechanism for reusing a part of an existing class while subtyping pertains to the behavior of the instances and is a semantics concern.

In Figure 17, it clearly appears that the thick vertical arrows represent inheritance relationships. The classes User, C-User, Server, Viewport, and Component have been built from scratch while the C-Server, C-Viewport and C-Component classes reuse or inherit from the initial classes and add what is needed in order to allow for cooperative edition. In a similar manner, the classes located at the lower level inherit from the higher level and add services required by a cooperative editor of hierarchical Petri nets, e.g services which are mainly related to the creation of new places, tokens, arcs or transitions.

Subtyping. Recall that, in CO-OPN/2, subtyping is based on the strong version of the substitutability principle.

With respect to our modeling of a CDE, we have shown that inheritance could be used without necessary implying a subtyping relationship (c.f. Figure 17). Now, we introduce, in Figure 18, the classes that are related by subtyping and by inheritance. The classes at the lowest level represent the effective components involved in a cooperative hierarchical Petri net editor (i.e. transitions, places, arcs, and tokens), while the classes at the top and the second levels are used to classify and to make the distinction between the hierarchical components (i.e. the transitions and the places) and the atomic ones (i.e. the arcs and the tokens).

It is not surprising that the depicted classes in Figure 18 alone are related by subtyping. In fact, we will see in the next section that the contra-variant rule

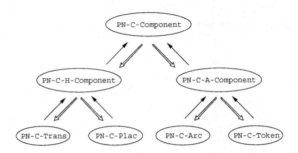

Fig. 18. The Classes Related by Subtyping.

is violated between Component, C-Component, and PN-Component. A similar argument will be given for the viewport classes as well as for the server and user classes.

6.8   CO-OPN/2 Design of the CDE Entities

Unfortunately, due to the size of most modules, we do not provide a complete specification of these but rather a specification of the relevant parts. Nevertheless, the complete specification of the case study can be found in [5]. Now, we describe the design of the various entities presented in Figure 17.

Users. The users involved in a CDE are modeled by means of the C-User class described in Specification 12 and in Figure 19. This is a small and simple class, built from scratch.

Missing algebraic modules may be found in [5]. Nevertheless, we mention that the Coord ADT module defines the sort coord, which is a pair of naturals and that the Privilege module defines both the view and edit user privileges as required. The four ownerships required by the proposal are defined in the ADT module Ownership. As for the Unique module, it only defines the generator '@' of sort unique, which plays the role of a black token.

Every name within the **Methods** field models an action that a user would like to accomplish by means of the viewport. Among these methods we may mention the join and leave methods which correspond to the will of a user to join or to leave an editing session. The select method is used to select a component of the diagram at a given location, the go-down allows a hierarchical component, previously selected, to be visited, and the go-back rises again within in the hierarchy, delete removes a selected component. The move, resize, and rotate methods are used to modify aspects of selected components. The own-for method represents the user's will to own a component for a given ownership.

The instance variables (**Places** field) are involved in the behavioral axioms which describe the properties of the methods. For instance, the behavioral axiom

    join pr with v.(self join pr) :: idle @ -> active @, cur-vp v

```
Abstract Class C-User;
Interface
  Use Coord, Size, Angle, Privilege, Ownership;
  Type c-user;
  Methods
    join _ : privilege; leave;
    select   _ : coord; unselect;
    go-down; go-back;
    delete;   move     _ : coord;
    resize _ : size;  rotate _ : angle;
    own-for _ : ownership;
Body
  Use Unique, C-Viewport;
  Places
    idle     _ , active _ : unique;
    cur-vp _ : c-viewport;
  Initial
    idle @;
  Axioms
    join pr with v.(self join pr) :: idle @ -> active @,cur-vp v;
    leave with v.leave :: active @,cur-vp v -> idle @;
    select p with v.select p :: active @,cur-vp v -> active @,cur-vp v;
    move p with v.move p :: active @,cur-vp v -> active @,cur-vp v;
    own-for os with v.own-for os:: active @,cur-vp v -> active @,cur-vp v;
    where p : coord; v : c-viewport; pr : privilege;  os : ownership;
  ;; ... and more axioms ...
End C-User;
```

Spec. 12. The Cooperative Users.

indicates that the `join` method is asking to be synchronized with the `join` method for the privilege `pr` of an available existing viewport `v`. The user has to be idle and will become active. Moreover, the free variable `v` will be unified with an available existing viewport and memorized by the current viewport attribute (`cur-vp v`). The self-reference **self** informs the viewport of the identity of the user who wants to be connected to it. The viewport will store this object identifier in order to know what connection has been established.

The role of the method `select` is the transmission of the user's will to select an existing component at a given position `p` to the current viewport `v`. The same remark holds true for the other methods.

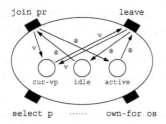

Fig. 19. Outline of the C-User Class.

Figure 19 provides a partial graphic view of the `C-User` class. Note that for the sake of clarity not all the methods have been represented.

In order to obtain an effective class of users which may be involved in a hierarchical Petri net editor we make use of inheritance. In Specification 13 one may see the inheriting class PN-C-User which inherits from the class C-User as declared under the **Inherit** field. In this situation some renamings, declared under the **Rename** field, are necessary. As expected, the profiles and the properties of the new services provided by the inheriting class have been introduced within the **Methods** and **Axioms** fields, respectively. The new methods introduced in this class model the user's will to create the effective components involved in the hierarchical Petri net diagrams at a given location, i.e. places, transitions, arcs, and tokens. Moreover, one observes that three static objects have been declared under the **Objects** field.

```
Class PN-C-User;
Inherit C-User;
   Rename c-user -> pn-c-user; C-Viewport -> PN-C-Viewport;
          c-viewport -> pn-c-viewport;
Interface
   Objects Didier, Nicolas, Olivier : c-user;
   Methods
      new-trans;  new-plac;
      new-token;  new-arc;
Body
   Axioms
      new-trans with v.new-trans :: cur-vp v -> cur-vp v;
   ;; ... and more axioms ...
      where v : pn-c-viewport;
End PN-C-User;
```

Spec. 13. Cooperative Petri Net Users.

Viewport. A viewport has two main responsibilities. First a viewport has to transmit the user's actions to the server. Its second role is to receive the information of the server and to redisplay the document whenever it is necessary.

In Specification 14 we give a partial textual form of the C-Viewport class. One may see that the majority of the services concerns the requests of the users. For example, the behavioral axiom of the _ join _ method informs the server that the user u wishes to be logged-in with the privilege pr. The select _ method asks the server to select a component which would be in location p. The display method (abstractly modeled here) activate the server when the picture has to be displayed on the screen.

We deliberately have not provided the specification of the effective class of the viewports involved in a hierarchical Petri net editor.

Components. A document is composed of components organized into a hierarchy. Some of these components are said to be hierarchical in the sense that they may contain sub-components, and some are said to be atomic because they represent the leaves of the tree structure of the document and, consequently, cannot contain any sub-components. In fact, a document is a hierarchical component in itself.

```
Abstract Class C-Viewport;
Interface
  Use Coord, Size, Angle, Picture, Privilege, Ownership, C-User;
  Type c-viewport;
  Methods
    _ join _ : c-user privilege;  leave;
    select _ : coord;  unselect;
    go-down; go-back;
    delete; move      _ : coord;
    resize  _ : size;        rotate  _ : angle;
    own-for _ : ownership;
    display _ : picture;   clear-screen;
Body
  Use C-Server, C-User;
  Places
    priv _ : privilege;      cur-user _ : c-user;
  Axioms
    u join pr with the-server.(u login self)::-> priv pr,cur-user u;
    leave with the-server.(u logout) :: cur-user u, priv pr -> ;
    select p with the-server.(u select p wth pr) ::
      cur-user u, priv pr -> cur-user u, priv pr;
    move p with the-server.(u move p wth pr) ::
      cur-user u, priv pr -> cur-user u, priv pr;
    own-for os with the-server.(u own-for os) ::
      cur-user u -> cur-user u;
    v.display pic with broadcast v pic :: -> ;
  ;; ... and more axioms ...
    where p:coord; pr:privilege; os:ownership; u:c-user; pic:picture;
          v: c-viewport;
End C-Viewport;
```

Spec. 14. General Class of Viewports.

The Component class given in Specification 15 models the simplest components of our class structure. This kind of components does not include any notion of cooperation specific to the case study. The component state consists of six places, four of them concern the graphic aspects of the component itself, i.e. position, dimension, rotation and label. The anchors and links places contain, respectively, the set of the locations at which another component may be attached, and which components are linked to (a link is a triple of two anchors and one component).

The notions of cooperation between several users required by the case study are introduced in the C-Component class which inherits from the previous class as illustrated in Specification 16. The particular authorized-for place associates a set of users with each ownership. The associations user-ownership are surrendered by the surrender service as required. The authorized-for? method is used by the server to determine if a given user owns the component for a given ownership, while the own-for method allows a user to modify its ownerships. The behavioral axiom of the own-for event ensures that nobody owns the component for deletion and adds the user u to the users who already own the component for modification.

The PN-C-Component class models the components in relation with a Petri net editor. This class inherits from the C-Component with some necessary renamings.

```
Abstract Class Component;
Interface
  Use Coord, Size, Angle, String, Link,
      Set-Of-Links, Anchor, Set-Of-Anchors;
  Type component;
  Methods
    move _ : coord; resize _ : size; rotate _ : angle;
    get-pos  _ : coord; put-label _ : string;
    add-link  _ , del-link  _ : link;
    get-links _ : set-of-links;
    add-anchor  _ , del-anchor  _ : anchor;
    get-anchors _ : set-of-anchors;
Body
  Places
    position _ : coord;     dimension _ : size;
    rotation _ : angle;    label     _ : string;
    links   _ : set-of-links; anchors _ : set-of-anchors;
  Axioms
    move p     :: position p' -> position p;
    resize s   :: size s'     -> size s;
    rotate r   :: rotation r' -> rotation r;
    get-pos p :: position p  -> position p;
    put-label lb  :: label lb' -> label lb;
    add-link l :: links lks -> links lks + l;
    del-link l :: links lks -> links lks-1;
    get-links lks :: links lks -> links lks;
;; ... and more axioms ..
    where p, p' : coord;    s, s' : size;  r, r' : angle;
          lb, lb' : string;  l : link;      lks : set-of-links;
End Component;
```

Spec. 15. General Class of Components.

Server. The server is a crucial element which has to :

- manage the identification of the users,
- save and restore a document,
- provide all the services required by the users,
- cope with the accesses to the shared document in accordance with the coordination protocol,
- allow the users to simultaneously access the document,
- send to the viewports the relevant information to be displayed.

A partial textual form of the C-Server class which allows several users to edit simultaneously the same document is given in Specification 17.

In a centralized architecture only one server is present. Thus, we define in the C-Server interface within the **Object** field a static instance of this class the-server. In order to give an insight of the behavior of such a server, we explain in detail the _move_wth_ method which may also be observed in all the classes. Moreover, in Figure 20, we have provided a graphic outline of the C-Server class stressing the move method behavior.

```
Abstract Class C-Component;
Inherit Component;
  Rename component -> c-component; Link -> C-Link; link -> c-link;
Interface
  Use Attributes,Ownership,Picture,C-User;
  Methods
    get-attrib _, put-attrib _ : attributes;
    _ own-for _ , _ authorized-for? _ : c-user ownership;
    surrender; component-pic _ : picture;
Body
  Use Set-Of-C-Users;
  Places
    attrib _ : attributes; _ authorized-for _ : set-of-c-users ownership;
  Initial
    [] authorized-for deletion;      [] authorized-for inspection;
    [] authorized-for modification;[] authorized-for encapsulation;
  Axioms
    u own-for deletion ::
      (((users1=[]) or (users1=[]+u)) and ((users2=[]) or (users2=[]+u)) and
      ((users3=[]) or (users3=[]+u)) and ((users4=[]) or (users4=[]+u))) =>
      users1 authorized-for deletion, users2 authorized-for inspection,
      users3 authorized-for modification, users4 authorized-for encapsulation
      ->
      ([]+u) authorized-for deletion, users2 authorized-for inspection,
      users3 authorized-for modification, users4 authorized-for encapsulation;
    u own-for os :: not (os = deletion) =>
      [] authorized-for deletion, users authorized-for os ->
      [] authorized-for deletion, (users + u) authorized-for os;
    u authorized-for? os ::
      (users+u) authorized-for os -> (users+u) authorized-for os;
    where os : ownership;  u : c-user;
          users, users1, users2, users3, users4 : set-of-c-users;
End C-component;
```

Spec. 16. Cooperative Components.

The set of triple-uvc contained in the instance variable assoc associates each logged-in user with the viewport he is connected to, and the current component he is currently acting upon.

The behavioral axiom of the move method indicates that the user u wants to move the selected component to location p according to its privilege pr. To accomplish this task the server has to determine the already selected component and be synchronized, sequentially, with it. An additional necessary verification is to make sure that the user possesses the edit privilege. The synchronization expression first ensures that the user u has the authorization for the modification of the component c and then asks the component to move itself to the position p. The production of a black token @ in the place broadcast-flag triggers off the transition start-broadcast to be fired.

As one can see in Figure 20, the start-broadcast internal transition copies the assoc place into the broadcast-to which will be able to send the information to be displayed to each viewport v. A synchronization request gets the graphic aspect of the expanded current component. This process is ended when all triples of the broadcast-to place have been removed and only the empty set resource is left. Finally, the finish-broadcast transition consumes the empty set. For the outside world the events which compose the move method seem to be performed simultaneously, whereas they are actually serialized.

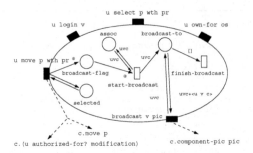

Fig. 20. Outline of the Cooperative Server.

```
Abstract Class C-Server;
Interface
  Use Coord, Size, Angle, Privilege, Ownership, C-User, C-Viewport;
  Type c-server;
  Object the-server : c-server;
  Methods
    _ login _ : c-user c-viewport;
    _ select _ wth _ , _ move _ wth _ : c-user coord privilege;
    _ own-for _ : c-user ownership;
    save; restore; new-document;
    broadcast _ _ : c-viewport picture;
  ;; ... and more methods ...
Body
  Use Unique, C-Component,
      Set-Of-C-Components, C-Triple-uvc, Set-Of-C-Triple-uvc, Picture;
  Places
    document _ : c-component;
    broadcast-flag _ : unique;    broadcast-to   _ : set-of-c-triple-uvc;
    assoc _ : set-of-c-triple-uvc;
    _ selected _ : c-user set-of-c-components;
  Initial
    broadcast-to []; assoc [];
  Transitions
    start-broadcast; finish-broadcast;
  Axioms
    u login v :: assoc uvc, document doc ->
                  assoc (uvc + <u v doc>), u selected [], document doc;
    u select p wth pr with c.get-child c' at p ::
      (pr = edit) or (pr = view) =>
      assoc (uvc + <u v c>), u selected compnts ->
      assoc (uvc + <u v c>), u selected (compnts + c'), broadcast-flag @;
    u move p wth pr with c'.(u authorized-for? modification) .. c'.move p ::
      pr = edit =>
      u selected ([] + c') -> u selected [], broadcast-flag @;
    u own-for os with c'.(u own-for os) ::
      u selected ([] + c') -> u selected [];
    start-broadcast ::
      broadcast-flag @, assoc uvc -> assoc uvc, broadcast-to uvc;
    broadcast v pic with c.component-pic pic ::
      broadcast-to (uvc + <u v c>) -> broadcast-to uvc;
    finish-broadcast :: broadcast-to [] -> ;
    where p : coord;  pr : privilege; os : ownerskip;  pic : picture;
          uvc : set-of-c-triple-uvc;  c, c', doc : c-component;
          compnts : set-of-c-components; u : c-user; v : c-viewport;
  ;; ... and more axioms ...
End C-Server;
```

Spec. 17. Cooperative Server Class.

Finally, from the abstract C-Server class we derive the PN-C-Server which describes the effective server involved in the hierarchical Petri net editor. The PN-C-Server class is presented in [5] introduces the missing services related to the creation of the effective Petri net components.

6.9   Behavioral Properties of the Classes

In Figure 21 we have provided a snapshot of a system of objects taking part in the development of a hierarchical Petri net, in accordance with the example given in Section 6.3 and illustrated in Figure 15. Two kinds of relationships are represented, the solid arrows correspond to the clientship, and the dotted arrows represent the link between the connected components.

The server object is a key element of the whole system. All the modifications concerning any component are transmitted to the server that dispatches the result of the modifications to the appropriate viewports. In CO-OPN/2, the concurrency is naturally managed by the places which are multi-sets of algebraic values and build states concurrently accessible by the object methods. Despite this modeling power, it is sometimes difficult to model concurrent accesses when operations on global multi-set states have to be considered. It is, for instance, the case for the children place of the PN-C-H-Component class which is a set of references modeled by means of an algebraic abstract data type and not a multi-set. This modeling choice prevents from accessing simultaneously some of the sub-components of a node.

The server component allows intra-concurrency. Indeed, most of the services (not login and logout) of the server may be simultaneously activated by two different users for two different components, for instance the _move_wth_ method. Moreover, some services may be requested by two different users for the same component, for instance the rotation and the shifting of a component. We did not optimize the broadcast performed after a modification, for example a simultaneous modification of two components will produce two simultaneous request for the display service of the viewports (synchronously performed after the component modification).

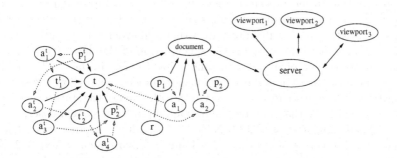

Fig. 21. Class Instances of the Example.

6.10   Modeling of Other Editors

We give here one direction for the evolution of our modeling of the hierarchical Petri net editor, in order to obtain other kinds of cooperative editors.

The class structure presented in this paper is flexible enough to take into account other cooperative editors, such as an editor of SADT diagrams, as well as an editor of CO-OPN/2 diagrams. The mandatory modifications concern only the classes derived from second level of Figure 17. Thus, the lowest level of Figure 17 as well as the component hierarchy must be redefined according to the type of the diagram. Here follow two suggestions concerning the subtyping relationship between the components of SADT diagrams and CO-OPN/2 diagrams.

SADT diagrams are quite simple, they consist of two kinds of components: the boxes which may be arbitrarily nested, and the arrows.

A CO-OPN/2 specification is a set of interconnected classes which consist of an encapsulated Petri net equipped with some methods or parameterized transitions. Thus, CO-OPN/2 formalism defines only two levels of hierarchy with regard to the component nesting: the net level and the class level. At the net level, the CO-OPN/2 components are the places, the internal transitions, the methods and the arrows that represent the control flow. At the class level, the components are the classes and the arrows which express the synchronization requests.

7   Conclusion

The main benefit of this article is to provide a presentation of the latest evolution of CO-OPN, in such a way that the abstract syntax and formal semantics are complete and coherent, and that the specification capabilities of the formalism can be compared with other approaches in the field of high level Petri nets.

The abstract syntax covers the full concrete syntax of our specification formalism which is illustrated in an introductory example dedicated to an intuitive understanding of the CO-OPN/2 semantics. The main idea is that a CO-OPN/2 specification defines a set of modules which can be either algebraic Petri net classes or abstract data types modules. Algebraic Petri nets are Petri nets with parameterized transitions (methods) and internal spontaneous transitions. A new synchronization mechanism is introduced for the composition of algebraic Petri net classes that generalize mechanisms found in other Petri net approaches. The abstract data type specifications are structured order-sorted algebraic specifications and are used not only for the specification of the data structures, but also for the description of the class instances' identifiers. On the one hand, algebraic specifications as well as the synchronization mechanism allow the specifier to describe a system in an abstract way. On the other hand, the state/event based modeling provided by the Petri nets allows an operational system description.

The formal semantics of CO-OPN/2 will be adapted to an operational semantics for a further version of our development environment SANDS/2. The semantics given here is a step semantics expressing the full inter and intra concurrent behaviors of the objects of a specification and it is built using SOS inference rules. The sub-typing relation needs quite a complex treatment at the semantics level through observational semantics and is sketched here.

Lastly, a medium-sized case study is summarized to illustrate the specification phase with CO-OPN/2 and to provide a way for evaluating our formalism by comparison with the same exercise with other object-oriented Petri net formalisms.

Future work will be mainly devoted to provide a complete methodology for the development of distributed systems. For this purpose we will include a coordination layer [9], and introduce intermediate levels of sub-typing to provide the specifier with more flexible specialization constraints. Refinements techniques are also under study in order to have a framework for the development of concrete application from abstract specifications.

Acknowledgments

We wish to thank Julie Vachon and Jarle Hulaas for their numerous and valuable comments on this paper. We also thank anonymous referees for their comments on an earlier version of this paper. We are also grateful to all members of the ConForM group who contributed to this work.

References

1. Pierre America. Inheritance and subtyping in a parallel object-oriented language. In J. Bézivin, J.-M. Hullot, P. Cointe, and H. Lieberman, editors, *ECOOP'87: European conference on object-oriented programming: proceedings*, volume 276 of *Lecture Notes in Computer Science*, pages 234–242, Paris, France, June 1987. Springer-Verlag.
2. Pierre America. A behavioural approach to subtyping in object-oriented programming languages. Technical Report 443, Philips Research Laboratories, Nederlandse Philips Bedrijven B. V., April 1989. Revised from the January 1989 version.
3. E. Battiston, A. Chizzoni, and F. De Cindio. Inheritance and concurrency in CLOWN. In *Proceedings of the "Application and Theory of Petri Nets 1995" workshop on "Object-Oriented Programming and Models of Concurrency"*, Torino, Italy, June 1995.
4. Olivier Biberstein. *CO-OPN/2: An Object-Oriented Formalism for the Specification of Concurrent Systems*. PhD thesis, University of Geneva, July 1997.
5. Olivier Biberstein, Didier Buchs, and Nicolas Guelfi. CO-OPN/2 applied to the modeling of cooperative structured editors. Tech. Report 96/184, Swiss Federal Institute of Technology (EPFL), Software Engineering Laboratory, Lausanne, Switzerland, 1996.
6. Olivier Biberstein, Didier Buchs, and Nicolas Guelfi. Using the CO-OPN/2 formal method for groupware applications engineering. In *Proceedings of the IMACS-IEEE-SMC conference on Computational Engineering in Systems Application (CESA'96)*, Lille, France, July 1996. Also available as Tech. Report (EPFL-DI-LGL No 96/187).
7. Olivier Biberstein, Didier Buchs, and Nicolas Guelfi. CO-OPN/2: A concurrent object-oriented formalism. In *Proc. Second IFIP Conf. on Formal Methods for Open Object-Based Distributed Systems (FMOODS)*, pages 57–72, Canterbury, UK, March 1997. Chapman and Hall, London.

8. Didier Buchs and Nicolas Guelfi. CO-OPN: A concurrent object-oriented Petri nets approach for system specification. In M. Silva, editor, *12th International Conference on Application and Theory of Petri Nets*, pages 432–454, Aahrus, Denmark, June 1991.

9. Mathieu Buffo. *Contextual Coordination: a Coordination Model for Distributed Object Systems*. PhD thesis, University of Geneva, 1997.

10. Joseph A. Goguen and José Meseguer. Order-sorted algebra I: Equational deduction for multiple inheritance, overloading, exceptions, and partial operations. *TCS: Theoretical Computer Science*, 105(2):217–273, 1992. (Also in technical report SRI-CSL-89-10 (1989), SRI International, Computer Science Lab).

11. N. Guelfi, O. Biberstein, D. Buchs, E. Canver, M-C. Gaudel, F. von Henke, and D. Schwier. Comparison of object-oriented formal methods. Technical Report Technical Report of the Esprit Long Term Research Project 20072 "Design For Validation", University of Newcastle Upon Tyne, Department of Computing Science, 1997.

12. Alain Karsenty. *GroupDesign : un collectitiel synchrone pour l'édition partagée de documents*. PhD thesis, Université Paris XI Orsay, 1994. Also in Computing System: "GroupDesign: Shared Editing in a Heterogeneous Environment", vol. 6, no. 2, pp. 167–192, 1993.

13. Charles Lakos. The consistent use of names and polymorphism in the definition of object Petri nets. In *Proceedings of the "Application and Theory of Petri Nets 1996"*, volume 1091 of *Lecture Notes in Computer Science*, pages 380–399, Osaka, Japan, June 1996. Springer.

14. Barbara Liskov and Jeanette M. Wing. A behavioral notion of subtyping. *ACM Transaction on Programming Languages and Systems*, 16(6):1811–1841, November 1994.

15. David Park. Concurrency and automata on infinite sequences. In P. Deussen, editor, *Theoretical Computer Science: 5th GI-Conference, Karlsruhe*, volume 104 of *Lecture Notes in Computer Science*, pages 167–183. Springer-Verlag, March 1981.

16. Wolfgang Reisig. Petri nets and algebraic specifications. In *Theoretical Computer Science*, volume 80, pages 1–34. Elsevier, 1991.

17. C. Sibertin-Blanc. Cooperative nets. In Robert Valette, editor, *Application and Theory of Petri Nets 1994*, volume 815 of *Lecture Notes in Computer Science*, pages 471–490, 15th International Conference, Zaragoza, Spain, June 1994. Springer-Verlag.

18. Alan Snyder. Encapsulation and inheritance in object-oriented programming languages. In *Proceedings OOPSLA '86*, volume 21, 11 of *ACM SIGPLAN Notices*, pages 38–45, November 1986.

CLOWN as a Testbed
for Concurrent Object-Oriented Concepts

Eugenio Battiston, Alfredo Chizzoni, and Fiorella De Cindio

Università degli Studi di Milano - Computer Science Dept.
decindio@dsi.unimi.it, chizzoni@tin.it

Abstract. Petri nets have been considered among the variety of formalisms and theories taken into account for giving a sound semantic basis to (concurrent) object-oriented systems and languages. Pursuing this research direction, we have defined a formalism called CLOWN (CLass Orientation With Nets) and discussed on it many object-oriented concepts, with special regards to a net-based notion of inheritance.

This paper presents the insights about the principles and the open problems of object-orientation that we derived by modelling in CLOWN a number of small to medium case studies.

The current results suggest, as promising future efforts, the tighter integration with object-oriented analysis and design methodologies, especially aimed at the Smalltalk language.

1. Introduction

The strong demand of robust distributed and network-based applications calls for specification and programming languages provided with abstraction features to manage their inherent complexity. Many efforts are aimed at the extensions of the object-oriented paradigm, already employed in the development of a significant share of business and industrial applications, with distributed and concurrent features. Relevant research topics investigated in this thread includes:

- object-oriented principles, such as data specification, encapsulation, inheritance, subtyping and polymorphism;
- concurrency and distribution issues, such as abstraction tools (e.g., modularity, compositionality, observability), synchronous vs. asynchronous communication, degree of concurrency (multithreading), qualitative and quantitative analysis techniques for proving behavioural properties (e.g. liveness and safeness);
- and their interplay which is often source of difficulties, as the well known "inheritance anomaly" problem shows [AWY93, McH94].

Our approach is based on Petri nets, considered also by the other research teams working on these topics [Sib94, BB95, HV95, Lak96b, BP96, HD96, Lil96, Val96] for giving a sound semantic basis to (concurrent) object-oriented systems and languages. Pursuing this research direction, we have defined a formalism called CLOWN (CLass Orientation With Nets) [Chi94] and formalised in it many object-oriented concepts, with special regards to a net-based notion of inheritance. CLOWN

G. Agha et al. (Eds.): Concurrent OOP and PN, LNCS 2001, pp. 131–163, 2001.

principles and syntax are briefly presented in Sect. 3. Section 4 discusses how CLOWN faces with some of the above recalled major topics from the experimental perspective obtained testing the formalism against several case studies. This emphasis on practical experience is justified by our will to employ CLOWN as a testbed for semantically sound object-oriented specifications. The conclusion summarises the achievements, the major issues for further research and some promising development toward a tighter integration of CLOWN with object-oriented analysis and design methodologies, especially aimed at the Smalltalk language.

2.   Foundations of CLOWN

The CLOWN main sources of inspiration are summarised in Fig. 1., where arrows stand for an inheritance-like relation. In the following we sketch the insights that brought us to conceive CLOWN, introducing concepts which are helpful in the sequel of the paper.

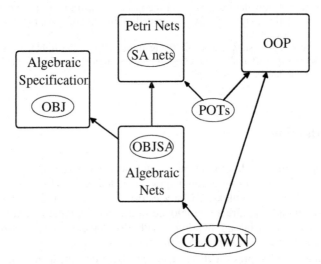

Fig. 1. CLOWN ancestors

Among the formalisms considered in [BRR91] for giving sound foundations to (concurrent) object-oriented systems and languages, [ELR91] introduces Parallel Object-based Transition systems (POTs), a low level Petri net-based formalism for object-based systems. Object-based systems are at the lowest level of the object-oriented system hierarchy, according to Wegner's taxonomy [Weg90], because there is no notion of class and hence inheritance. In the POTs view, a system is a set of sequential objects synchronously interacting. POTs can be given a finite high-level representation by Parallel Object-based Programs (POPs). POPs are object-based, as is their low level counterpart. POTs and POPs have been applied in [ELR91] to model SIMPOOL, a rather simple subset of POOL [Ame89, Ame92], and some features of actor systems [Agh86].

In [ELR91] the authors point out that POTs essentially coincide with (1-safe) Superposed Automata nets [DDPS82]; moreover they identify a strong similarity and some differences between POPs and OBJSA nets [BDM88, BDM96], a class of algebraic high-level nets, in the sense of [Rei91], enhanced with modularity features. Their name reflects that they integrate Superposed Automata (SA) nets (see also [BD92]) and the algebraic specification language OBJ [GW88]. OBJSA nets are modular as they can be obtained by combining OBJSA (elementary) components by transition fusion (we refer to [BDM96] for further details).

Starting from this intuition, first we realised that as a high-level formalism, OBJSA nets have the same abstraction power as POPs, but in contrast to them, they provide a natural representation of classes of objects. This turns out to be a good consequence of the restrictions on the structure of the underlying net and of the tokens, which characterise OBJSA nets among the other classes of algebraic nets. In fact these features match with the constraints characterising POTs and also allow an easy representation of an object's identity and data structure. In this way, the main idea behind POTs and POPs is preserved and enriched with the possibility of modelling classes of objects.

Once realised that OBJSA nets are well-suited to represent class-based systems, the natural further step, according to the Wegner's classification, is to deal with inheritance, exploiting net-based notions for dealing with behavioural aspects. The outcome of this effort [BD93, Chi94, BCD95, BCD96] is a formal notation called CLOWN, which includes a net-based notion of (multiple) inheritance, whose semantics is given by an associated OBJSA net.

3.   CLOWN Basics

In this section we first give a brief overview of the main clauses of CLOWN, together with hints about its semantics in terms of OBJSA nets, and then we illustrate them in more detail making use of the well known Dining Philosophers example. The discussion on the main design choices and their consequences is postponed to Sect. 4.

The main building block of a CLOWN specification is the *elementary class* (just class in the following). The template of a class consists of:
- a set of textual clauses that define the inheritance links, the instance data structure and methods and the class interface (plus an additional clause concerning class invariant properties);
- a labelled net that describes the causal relationships between methods execution; it can be seen as the body (in the sense of Simula67 [BOMN79] and POOL) of the class, i.e. its main method which rules the execution of the others on the basis of the instance state.

The semantics of a class is given by a corresponding OBJSA elementary component. Every *object*, instance of a class, is represented by a structured token flowing in the associated OBJSA elementary component.

The *communication* between objects isn't managed by the traditional message passing mechanism of the object-oriented programming languages, but by the mutually synchronous execution of corresponding methods. Each object may ignore the identity and/or the features of the communication partner(s), but imposes some synchronisation constraints, which are specified in the class *interface*, to guarantee that object synchronisation can perform correctly. The actual *synchronisation* is specified by a different template: the compound class, whose semantics is an OBJSA (non-elementary) component. An application consists of a compound class together with the specification of a set of objects that initially exist, while other objects are created dynamically.

Notice that each elementary class can thus be included in a variety of different applications.

3.1.   Elementary Class

Each elementary class is specified by the following clauses (bracketed clauses are optional):

> *class*
> The elementary class name.
>
> *[inherits]*
> The inheritance clause describes the relation between the class and all its parents, and the adjustments needed to avoid inheritance conflicts. When omitted it is assumed that the class inherits directly from the top class ROOT (Fig. 2.). This implies that *every* CLOWN class inherits ROOT.
>
> *[const]*
> Constants are typed entities, whose type is defined by means of sorts in an algebraic specification (ADT); their value is fixed at instance creation and never changes later on. There is a special constant, called ID, whose value is the object name.
>
> *[var]*
> Variables are typed entities, whose type is defined by means of sorts in an algebraic specification (ADT); their value can be modified at method execution (transition occurrence).
>
> *[interface]*
> Each interface specifies a list of typed formal parameters detailing the incoming data from partner objects and, optionally, a predicate *pre*, specifying further conditions over the interacting partners and their formal parameter.
>
> *[places]*
> By means of this clause a predicate, i.e. a correctness condition, can be specified over objects (tokens) when residing in the specified place.
>
> *[method]*
> The methods specify the actions that an object can execute. Their specification can include:
>
>> *[with]*
>> This is the list of the interacting partners. When omitted the method cannot read values from the interacting objects, but can be read from them.
>>
>> *[pre]*
>> A boolean guard whose arguments can be the local variables or the interface parameters. Whem omitted it's assumed the default value *true*.

[post]
The modification of the object variables value. When omitted the object variables value doesn't change by method execution.

net

The class net is a SA net. Each transition is labelled with a method name, which is executed at transition occurrence, and, for readability purpose, with the interfaces and the multiplicity of the interacting objects. Places with underlined names indicate that objects (tokens) occurring in that place satisfy the associated predicate. Since every CLOWN class inherits ROOT (Fig. 2.), every class net represents the objects life-cycle (exploiting an intuition of POTs) beginning from the *unborn* place, becoming *alive* by means of the transition *create* and later disappearing by moving to *dead* through occurrence of the transition *leave*.

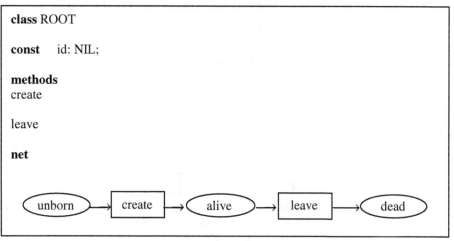

class ROOT

const   id: NIL;

methods
create

leave

net

Fig. 2. The ROOT class

To illustrate these clauses, we now present the specification of the well known Dining Philosophers problem with the slight variation of a meal counter attached to each philosopher to show the data handling primitives.

Let us explain some features of the specification of Fig. 3. The algebraic specification associated to this class, that we omit for space reason, consists of four sorts: NAT are the standard natural numbers; PHNAME and FKNAME are countable sets of names for philosophers and forks, whose generic element is denoted, respectively, by p_i and f_i; FORKSNAME represents couples of forks and employs some ad hoc algebraic operators, such as the straightforward *left* and *right*.

The interface FORK requires a parameter *fid* of sort FKNAME, i.e., the name of the interacting fork. FORK appears in the *with* clauses as *FORK (2)* because each philosopher interacts at the same time with two different forks: *FORK(1)* and *FORK(2)* in the *pre* clauses identify the two forks.

class PHIL

const   id: PHNAME;

var   myforks: FORKSNAME;
                 meals: NAT;

interface
         FORK (fid: FKNAME);

methods
get forks
          with   FORK (2);
          pre   left(myforks) = FORK(1).fid and right(myforks) = FORK(2).fid;

release
          with   FORK (2);
          pre   left(myforks) = FORK(1).fid and   right(myforks) = FORK(2).fid;
          post   meals <- meals + 1;

wake up

net

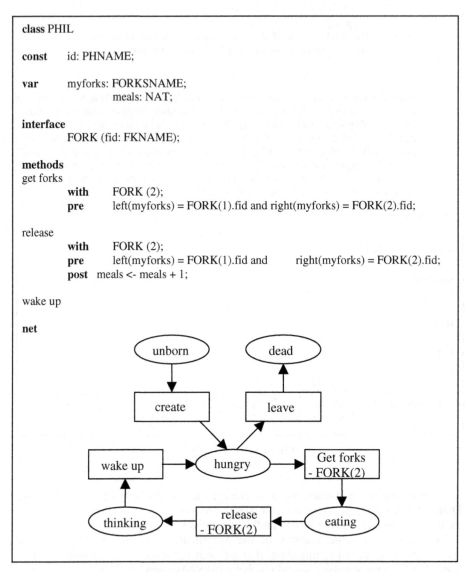

Fig. 3. The dining philosopher class

Note that not every method employs all the three subclauses; it's also possible to have a method without any clause, as e.g. *wake up*, that's anyway significant in the class specification to mark a step of the objects life-cycle.

The semantics of the above class is the OBJSA open elementary component of Fig. 4. It can be read as a standard SPEC-inscribed net [Rei91]. Note that output arc labelling is omitted when it coincides with the labelling of the corresponding input arc.

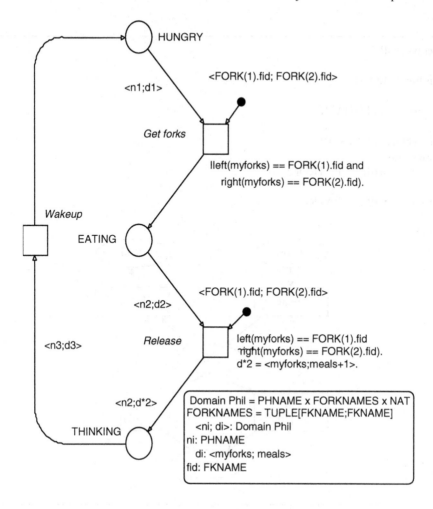

Fig. 4. The dining philosopher OBJSA net

3.2.   Compound Class

Before introducing the compound class let us specify in Fig. 5. the partner of the former philophers, i.e. the forks.

The desired Dining Philosophers behaviour comes from the synchronization between a set of philosophers, instances of the elementary class PHIL, and a corresponding set of forks, instances of the elementary class FORK.

The compound class clause specifies how a set of (elementary) classes mutually interact. In particular it has to fix, for each formal parameter in the class *interface*, its corresponding actual parameter. This is done by a *view,* strongly inspired by OBJ views, as the *interface* is inspired by OBJ *theories* (the interested reader finds a presentation of these OBJ features in [Gog84]).

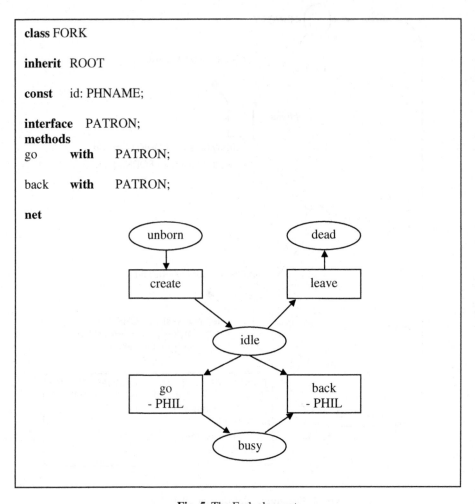

class FORK

inherit ROOT

const   id: PHNAME;

interface PATRON;
methods
go   **with**   PATRON;

back   **with**   PATRON;

net

Fig. 5. The Fork class net

> *compound class*
> The compound class name.
>
> *component*
> The list of the elementary classes building the compound class.
>
> *view*
> The view is the set of all the possible instances of the interfaces formal parameters.
>
> *actions.*
> The actions define the method composition and list the methods of the elementary classes involved.

The specification of the compound class DINING PHIL is given in Fig. 6. Let us notice that in the *view* clause the left sides stand for a formal paremeter, while the right sides are the corresponding actual parameters.

The semantics of the compound class DINING PHIL is a corresponding OBJSA net obtained by composing the OBJSA nets corresponding to the two shown elementary classes. It is shown in Fig. 7.: bold labelled transitions result from method synchronisations.

```
compound class DINING PHIL

component
        PHIL;
        FORK;

view
        PHIL: FORK.fid = FORK.id;
        FORK: PATRON = PHIL;

action
        pick_forks is (PHIL.get_forks, FORK.go);
        release_forks is (PHIL.release, FORK.back);
```

Fig. 6. The Dining Philosopher compound class

Fig. 7. The Philosophers overall system semantics

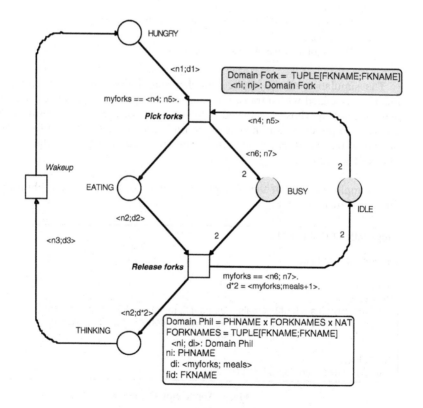

In less trivial examples, the specification of the compound class can be more complex, essentially because the CLOWN inter-object communication scheme is a multi-object synchronisation instead of a simple message call. To handle this feature and preserve a fairly readable compound class clause, we rely upon interactive syntax-driven commands in the CLOWN support environment [Con96], developed on the top of the OBJSA Net Environment [BCCD96]. [DeR97] gives a comprehensive mapping of the communication pattern behind the compound class synchronization clause into Smalltalk code (see also section 5.4 below).

3.3. System

CLOWN applications consist of a class together with the specification of a set of initial objects, while other objects are created dynamically. This is equivalent to specify an initial marking for the corresponding OBJSA component where some objects are in *alive* places, while all the other initially reside in the *unborn* place, that acts as an object collector, from where they are removed to an alive state by the *create* method execution.

> *system*
> The name of the system (application).
>
> *model*
> The class (elementary or compound) to be instanced.
>
> *marking*
> The list of all alive objects (tokens) of a system.

The simulation of the application therefore follows the net occurrence rule: a method associated with a transition is executable when all the input places contain at least one token whose associated data structure (i.e., the object instance variables) satisfies the pre clause; at the execution time, each object is moved from the input to the output place and its data structure is modified according to the algebraic rewriting rules of the methods.

A sample instance of the philophers system, consisting of five philosophers and forks, is shown in Fig. 8.

```
application CONFERENCE

model DINING PHIL

marking
PHIL.hungry:
   (id=p1; myforks = [f1,f2]; meals = 0); (id=p2; myforks = [f2,f3]; meals = 0);
   (id=p3; myforks = [f3,f4]; meals = 0); (id=p4; myforks = [f4,f5]; meals = 0);
   (id=p5; myforks = [f5,f1]; meals = 0)

FORK.idle:
   (id=f1); (id=f2); (id=f3); (id=f4); (id=f5)
```

Fig. 8. Application Conference

Let us note that the system illustrated in this example generates objects only by means of the initial marking. It's also possible to generate new instances (and remove active ones) dynamically at simulation time simply customising the *create* and *leave* methods, specified in every class.

4.   Theoretical Issues

In this following we present the most relevant outcomes and insights arising from our experimentation with CLOWN, focusing on some general   issues of relevance for concurrent object-oriented systems and languages mentioned in the Introduction.

We begin with some remarks about relationships and differences between abstract data types and objects in concurrent system specifications; then a major section about the inheritance follows, including a discussion of the so-called inheritance anomaly problem; finally, inter-object vs. intra-object concurrency is discussed.

4.1.   ADTs vs. Objects

As a way to specify data types, such as natural numbers, ADT are recognised to be sound, effective and, by definition, independent from the implementation. Moreover, the specific parameterization facilities offered by OBJ, namely the notion of *theory* which supports the CLOWN *interface* clause, let the designer employ a very good level of encapsulation: a component doesn't have to know the data structure of the partners, but just to ask for a term of a given sort.

When the system under specification includes a significant data structure, then in CLOWN, as in other approaches (e.g. [BB95]) a choice arises to the designer: whether to represent it as an ADT or as an object of a class. Think for example to a stack object, to be used in a few instances in a system specification: the designer can just reuse the elegant OBJ specification of a stack (cf. e.g. [Gog84]) instead of using a more complex CLOWN class template enriched by a not too meaningful net. The choice concerns the deep distinction between (abstract) data types and objects ([Coo91] is a good source of examples and discussions).

Our experience with CLOWN leads us to believe that if the entity to be specified - the stack - behaves mostly as a private data of another object, then draw it as an ADT of the owner; otherwise, i.e. if it's a common and persistent component interacting with various partners, the best choice is to build the class and instance it out. While ADTs do not have an identifier, objects do have. This confirm the above guideline: a variable name is sufficient to access a private/local data, while a persistent identifier (a name) is necessary in communications, for instance for a Receiver to be sure that gets an item from the same stack in which the Sender had stored it before.

4.2.   Inheritance

Inheritance is one of the fundamental concepts of the object-oriented paradigm and is the main mechanism for classes organisation. Despite its popularity, it's well known that a heavy employment of inheritance as a code reuse technique collides with the correctness of the subtyping hierarchy built among classes (cf. e.g. [Ame92, Mey93]).

Solutions to this problem have been already proposed both in the theoretical field, by means of the creation of two different hierarchies, one for code reuse and one for subtyping (examples are POOL-I [Ame92] and Sather [Omo93]) and in the industrial framework, with the development of in-house mandatory reuse policies that help designers to build consistent hierarchies. Developing CLOWN we have spent a major effort to solve this dichotomy between conceptual rigour (strict subtyping) and practical usefulness (code reuse).

The principle inspiring CLOWN inheritance is that each class can extend parents' specifications and specialise them in a restricted domain. It is therefore very close to the *substitutability* principle introduced and discussed in [WZ88, LW93a, LW93b]. Let A and B two objects, instances of homonymous class, being the latter subclass of the former. Any object C, interacting with A by means of a sequence of messages m, obtains the same answers and causes analogous internal modifications, sending the messages m to B.

We have therefore thoroughly looked for a notion capable of supporting substitutability when dealing with the behaviour of the CLOWN classes, represented by the class net. Moreover we wanted to support multiple inheritance to allow complete and straightforward representation of systems. The well-known technical difficulties related to multiple inheritance (repeated inheritance, conflicting hierarchies) are eased by the high level of abstraction of CLOWN specification and, if necessary, can be unambiguously solved by the designer with a few directives.

4.2.1.   Net Inheritance

The formalisation of the inheritance relation adopted in CLOWN to verify that the descendant's class behaviour includes and extends the parent's one relies upon the observability notion bound to the ST-preorder relation, introduced in [PS91]. After the seminal work of R. Milner [Mil80], a lot of equivalence notions for comparing concurrent systems based on *action* observation have been imported into or defined for net based models (see [PRS92] for a survey). ST-equivalence (also surveyed in [PRS92]) is instead based on *state* observation, and can therefore be seen as a dual notion. By identifying a set of *observable places*, the designer declares to be interested to compare states (i.e., markings) which involve those places. Adopting this state-oriented perspective, the capabilities of comparing net systems are improved by introducing the notion of ST-preorder ($\cdot_{ST}$), which takes into account the possibility of behaviour extension.

The net systems Σ_1 and Σ_2 in Fig. 9., where shaded places are the observable ones, are ST-equivalent, under a mapping of observable places which is represented in the figure by using the same name for corresponding places. Analogously, the systems Σ_4 and Σ_5 are equivalent.

Between the net systems Σ_2 and Σ_3 in Fig. 9. the $\cdot_{ST}$ relation holds. In fact Σ_3 performs the analogous state-transformation from p_1 to p_2 as Σ_2 does, and, moreover, performs a further state transformation from p_2 to p_3. Analogously, $\Sigma_3 \cdot_{ST} \Sigma_4$ as Σ_4 extends the state transformations of Σ_3 by adding a branch.

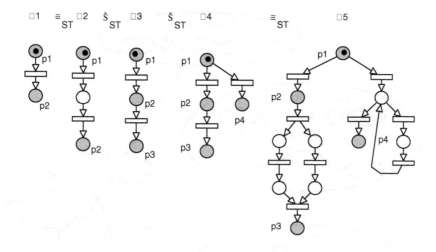

Fig. 9. ST-equivalence and ST-preorder

Therefore ST-preorder ($\bullet_{ST}$) let us represent both the behaviour preservation ($\cong_{ST}$), necessary for supporting the substitutability principle and the behaviour extension ($<_{ST}$), necessary to deal with new methods, added in the descendant class.

Fig. 10. illustrates an inheritance hierarchy concerning a family of printers. The ST-preorder allows the designer to verify that the behaviour specified in each descendant class net includes the parent's net behaviour.

E.g. class PRINTER can behave like its parent MONO_PRINTER (formally: $\Sigma_{MONO\_PRINTER} \bullet_{ST} \Sigma_{PRINTER}$), because it can print just one document. The opposite is not possible, as a MONO_PRINTER instance can't print more than one copy. This relation applies to multiple inheritance too. The behaviour of the class LOCK_PRINTER, as modelled by its class net, shown in the figure, includes both the behaviour of the (generic) class PRINTER and the behaviour of a generic LOCKER.

The formal verification of the above ST-preorder relations relies upon a suitable choice of the nets observable places (a heuristic based on a simple marking scheme is illustrated in [Chi94]) and upon the identification of an injective morphism among observable places (according to Def. 5. of [PS91], note that, for technical reasons, in order to prove the ST relation, it is also necessary to remove the unpureness of the net by splitting the side conditions into two transitions).

4.2.2. Class Inheritance

CLOWN inheritance notion consists of the condition over the class net illustrated in the former section, plus some conditions over the textual clauses, which assure that a parent's client can employ the descendant's (specialised) method without any additional requirement w.r.t. the parent's version.

The textual conditions are:
   (a) each typed entity of the parent class (constant, variable, interface parameter) can be redefined to a sort that algebraically extends its domain;

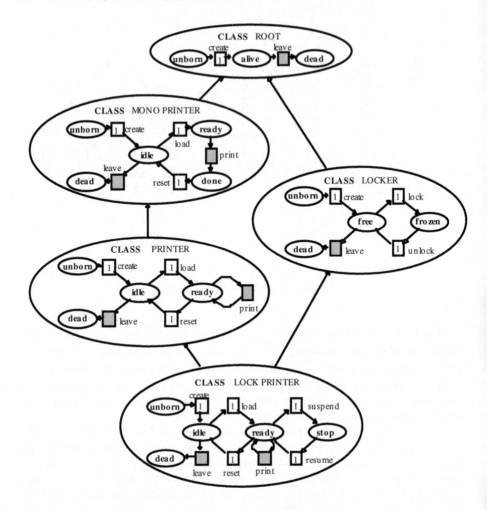

Fig. 10. The ST-preorder relation between printers

(b) parent's methods can be redefined by its descendants according to the
    following constraints:
    (b1) all parent's method interfaces listed in the with clauses must be
          included in the corresponding descendant's ones;
    (b2) descendant's precondition (pre clause) can be extended w.r.t. the
          parent's version, but only by adding conditions over the newly
          defined attributes.
New methods, attributes, interfaces and properties can be added at will.

Conditions over the class net and over the textual clauses are integrated as follows:

Let $A_1...A_n$ and B n+1 CLOWN classes and let $N_{A_1}...N_{A_n}$ and N_B their corresponding nets. Let $\Sigma_{A_1}...\Sigma_{A_n}$ and Σ_B the net systems obtained by marking $N_{A_1}...N_{A_n}$ and N_B with a token in the place unborn. Let O_{A_i} and O_B the set of observble places of each class net, consisting of the places belonging to the pre-set of an open transition.

Class B is descendant of classes $A_1...A_n$ only if:

(i) N_B extends the behaviours of $N_{A_1}...N_{A_n}$ for each A_i, that is it exists an injective mapping from O_{A_i} to O_B such that it holds:

$\Sigma_{A_i} <_{ST} \Sigma_B$.

(ii)   New methods, attributes, interfaces and properties can be added.

(iii) B's typed entities (interface parameters and variables) inherited from A_i are either the same type or an extension of their parents.

(iv)   B's inherited methods include parent's method interfaces.

Fig. 11. Inheritance guidelines

However, this notion isn't enough to completely solve the problems bound to the semantics of inheritance, i.e., the above conditions are necessary but not sufficient. For instance, condition (iv) does not exclude that an user defines a subclass whose behaviour isn't compatible with its parent.

To illustrate this unpleasant possibility, let's consider classes MonoPrinter and Printer in Fig.10. The textual clauses of class Printer have to include an interface attribute to get from the user the (natural) number of required copies to be printed: a value, given by the user, is assigned to such attribute *counter* in the post-condition of the load method. *Count* is then used in the method *print,* namely in the predicate controlling the enabling of the corresponding transition, to produce the correct number of copies. If the designer has defined *count* to be a natural number and the user types "zero", no print is produced, but this would violate the ST-preorder relation between the behaviours of classes MonoPrinter (which requires that one copy is printed) and Printer. This odd behaviour can be easily avoided declaring *count* as a positive natural number (the implementation will check that this constraint is satisfied).

The example shows that the problems arises because of the interplay between the class net and the textual clauses; the OBJSA net, which gives the semantics of the class, integrates them, and therefore might be the right field for giving a sound definition of inheritance between CLOWN classes. Unfortunately, neither ST-equivalence and preorder, nor any other of the equivalence notions surveyed in [PRS91] have been formulated for high-level nets.

A possible solution is to unfold the net (with some associated heuristics for dealing with infinite domain, as discussed in [Rac94]) and define inheritance either over the resulting Elementary Net system or over the associated Labelled Transition system (as in [BB95]). This would formally solve the problem, but is practically unusable. Another alternative, e.g. undertaken in [Lak96b], is to support inheritance, without any constraints for guaranteeing behaviour compatibility, but this choice looks too weak in a concurrent object-oriented framework.

We therefore adopt an intermediate solution, that is to look at the inheritance principle formalised in Fig.11. as a *strong guideline* to be followed by the designer instead of a *formal definition*. While we don't abandon the research to find out a more satisfactory solution, we believe, in the same direction of [HV95], that guidelines are helpful to support the design of sound hierarchies of classes. Our experience over several case studies, teaches us that the above guidelines are effective even in non toy examples and can be easily supported in a specification development environment.

4.2.3.   Inheritance Anomaly

As said in the Introduction, the so called inheritance anomaly problem is a well-know example of the difficulties which can arise because of the interplay between concurrency and object-orientation.

Nearly all the analysis of inheritance anomaly [MWY90, Neu91, GW91, BI92, MY93, Ber94] claim that it is caused by the conflict between synchronisation protocols and the inheritance mechanism. This hypothesis is also supported by a similar problem, discussed in [ABSB94], the conflict between inheritance and real-time constraints. According to these analysis, two opposite kinds of solutions have been proposed: (i) the development of more expressive synchronisation mechanisms [MY93] and (ii) the removal of all the synchronisation primitives (Maude [Mes93a]). Neither solution has closed the debate, but both partially reduce the harmful effects of the anomaly.

A different interpretation has been suggested in [McH94], where the author claims that many of the difficulties encountered in tackling the inheritance anomaly are due to a misleading intuition, since the anomaly origin wouldn't lie in the synchronisation vs. inheritance conflict, but in the need of different rules to handle inheritance of the synchronisation code and inheritance of the sequential code. McHale proposes a tool, the "ISVIS Matrix", able to support designers of concurrent object-oriented languages in checking the presence of inheritance anomaly through a set of case studies.

Although developed independently, CLOWN obeys McHale's methodological approach as it separates the sequential code (embedded in the *class* clauses, but for the *interface* directives) from the synchronisation code (collected in the *interface* clause of each class and in the *compound* class code). In particular, the different role played by the net and by the algebraic specification goes in the same direction of separating sequential and synchronization features. In fact (method) synchronization is charged over the net of the elementary classes and over the *action* clause of the compound class, while method sequential semantics is given in a non-procedural way through the method *pre* and *post* clauses.

This peculiar mix of notions supporting CLOWN inheritance (ADT, Petri nets, semantic constraints) looks therefeore effective to almost prevent the harmful effect of inheritance anomaly.

4.3.   Inter-object Communication

Concurrent object-oriented languages show a variety of approaches to the concurrency issue. The term *inter-object* concurrency is usually used when it is obtained by the communication among active sequential objects, while *intra-object* concurrency is used when concurrency comes from the concurrent execution of internal methods. By means of a well-known example, we intend to show that, though inter-object concurrency is native in CLOWN, while intra-object concurrency is not, this latter can be obtained in a straightforward way through the compound class clause.

Fig. 12. shows the CLOWN class specifying a bounded buffer, where items can be inserted at the bottom and read and removed from the top. Items are provided by a *SERVER* interface (parameter *input*) and consumed by a generic *READER*. Methods executability is constrained by predicates on the buffer size. The operators *size, max-size, put_in, tail*, used to give method semantics, are defined in the associated algebraic specification. (Regarding method *read*, recall that when the post condition is omitted, the object variable value does not change by method execution).

This buffer specification is intrinsically sequential since in CLOWN each elementary class corresponds to a finite state machine. Let's compare it with the net specification of a fully concurrent bounded buffer (Fig. 13.) drawn with a CLOWN-like notation. The main difference is that the method *insert* can be executed concurrently with methods *remove* and *read* since the former operates on the buffer tail, and the other ones concerns just the buffer head.

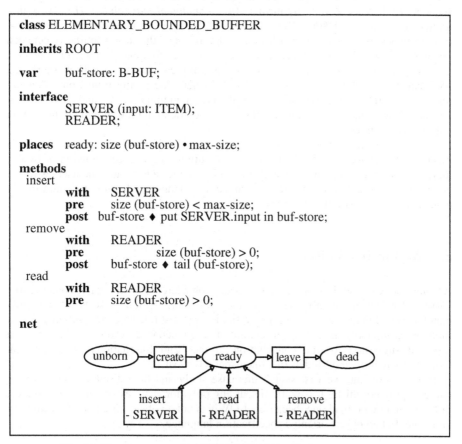

class ELEMENTARY_BOUNDED_BUFFER

inherits ROOT

var   buf-store: B-BUF;

interface
         SERVER (input: ITEM);
         READER;

places   ready: size (buf-store) • max-size;

methods
 insert
              with   SERVER
              pre   size (buf-store) < max-size;
              post buf-store ♦ put SERVER.input in buf-store;
 remove
              with   READER
              pre   size (buf-store) > 0;
              post   buf-store ♦ tail (buf-store);
 read
              with   READER
              pre   size (buf-store) > 0;

net

Fig. 12. The elementary class ELEMENTARY_BOUNDED_BUFFER

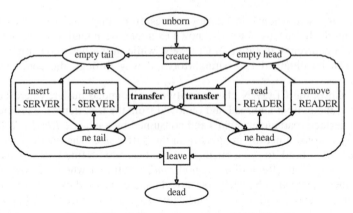

Fig. 13. The fully concurrent bounded buffer net

To capture in CLOWN this *internal* concurrency, the designer has to specify two classes: Fig. 14. shows the TAIL of the buffer, receiving the new items, and Fig. 15. presents the buffer HEAD, to read and remove items. These two (elementary, thus sequential) classes are joint, by means of the *compound class* clause (Fig. 16.), in a unique one, the COMPOUND_BOUNDED_BUFFER that exhibits the same concurrency capabilities of Fig. 14. In fact, let us notice the strong similarity between the net of Fig. 14. and the net associated with the compound class (cf. Fig.17.), generated from the two class nets on the basis of the *action* clause. They are indeed identical but for the shaded places of the former, due to the transition balancing required in OBJSA nets. For sake of simplicity, the methods *create* and *leave* of both nets are closed (internal), while in a complete specification they should be interaction with the environment.

More in general, when a class of objects exhibits intra-object concurrency, this means that it contains multiple threads of execution, i.e. it indeed consists of two or more (sequential) components which mutually interact and relates themselves to the environment as a whole. Therefore, an ad hoc language feature for intra-object concurrency is not strictly necessary.

5.   Applicative Issues

Among the case studies which have been considered for mutually compare the various formalisms combining nets and object-orientation - sorts of common benchmarks of the teams working in the field - the most relevant one, both in size and inherent complexity is the specification of a cooperative editing environment, a proposal suggested for the contribution to the "2nd International Workshop on Object-Orientation and Models of Concurrency [OOMC95].

In the following, we first sketch this case study and then discuss a major issue, namely polymorphism. It is worth discussing polymorphism as an applicative topics, rather than a theoretical one, because CLOWN is not an object-oriented programming language, but an object-oriented specification language.

class TAIL
inherits
       ROOT
              redefine var   id: BB-NAME;
var   rear: B-BUF;
interface
       SERVER (input: ITEM);
       CLIENT;
place
       empty_tail, ne_tail: size (rear) • max-size - 1;
methods
 insert
       with   SERVER
       pre   size (rear) < max-size;
       post   rear <- put SERVER.input in rear;
 put
       with   CLIENT
       post   rear <- tail (rear);
net

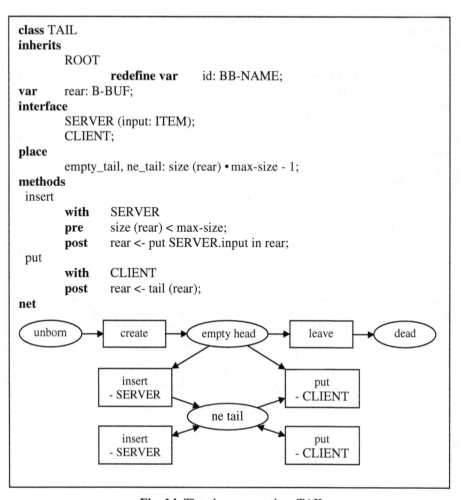

Fig. 14. The elementary class TAIL

5.1.   Cooperative Editor Case Study

The proposed case study [BLP96] concerns the architecture of a tool for synchronous groupware, whose main functionality is the possibility to manage a hierarchical graphical specifications, that can be accessed at the same time by many users with different privileges.

    Our solution specialises the case study proposal, quite generic in essence for sake of flexibility, with the principles of a CASE tool already available for ONE [BCCD96], the development environment for OBJSA nets. The existing tool that we have considered is ONESyst [Ber95], the ONE module supporting the preliminary phase of a system specification with OBJSA nets. It employs a hierarchical notation inspired both by SADT diagrams [MM88] and by some object-oriented design methodologies [CY91, DLF93].

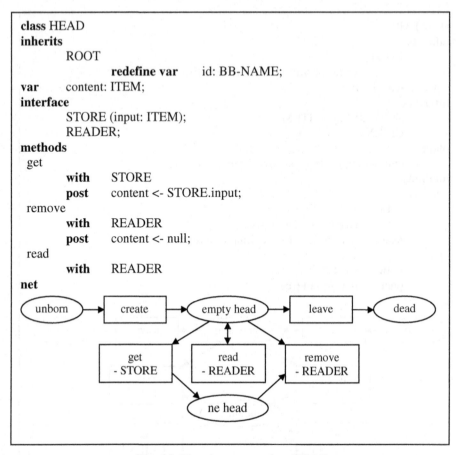

class HEAD
inherits
  ROOT
     redefine var id: BB-NAME;
var content: ITEM;
interface
  STORE (input: ITEM);
  READER;
methods
 get
   with STORE
   post content <- STORE.input;
 remove
   with READER
   post content <- null;
 read
   with READER
net

Fig. 15. The elementary class HEAD

compound class COMPOUND_BOUNDED_BUFFER
component
 HEAD
 TAIL

action
 create **is** {HEAD.create, TAIL.create};
 transfer1 **is** {TAIL.put, HEAD.get} **synch** size (TAIL.rear) == 1;
 transfer2 **is** {TAIL.put, HEAD.get} **synch** size (TAIL.rear) • 2;
 release **is** {HEAD.leave, TAIL.leave};
binding
 HEAD:
  STORE **is** TAIL **with** input = head (buf-store);
 TAIL:
  CLIENT **is** HEAD;
end

Fig. 16. The compound class COMPOUND_BOUNDED_BUFFER

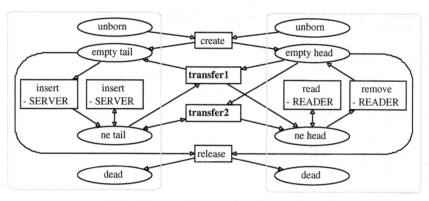

Fig. 17. The COMPOUND_BOUNDED_BUFFER net

The CLOWN specification that we have developed models ONESyst[++], which inherits the main functionalities of ONESyst, extending it to support more users cooperating on the same specification. It is required that the tool supports more than one active ONESyst diagram (i.e. specs presently in use by one or more registered users). The CLOWN specification obeys to almost all the general requirements of the case study, but for some minor changes [suggested by the notation employed]. The complete CLOWN specification is in [BCD95]; here in §5.2 we present some non trivial classes related each other thorugh inheritance and in §5.3 we show how CLOWN, as a concurrent object-oriented specification formalism, deals with polymorphism.

Before doing this, it's worth recalling that the client side of ONESyst[++] consists of USERS that interact with a Graphical User Interface (GUI) module. The clients communicate with the server, whose main task is the coordination of the users commands by means of an USERS COORDINATOR module. [Additional server modules are: the REPOSITORY, where all the data needed to maintain the consistency of the cooperative environment administration are stored, and the PERMISSION module, that checks the correctness of users privileges].

5.2.   The Main Classes

ONESyst employs three kinds of graphical objects (to avoid confusion, said "shape" or "items" in the following): boxes, components and pages. Boxes are abstract items of which just some data and actions are known; they correspond to subsystems which the designer is not interested to fully specify. Components are characterised by an algebraic data structure completely specified; they are the candidates for a corresponding OBJSA elementary components. Pages can be opened and filled with further items, building the hierarchical structure of the diagram. Every item is connected by one or more links to the other items of the diagram.

As boxes are the most generic items, the corresponding BOX class, presented in Fig.18, describes the standard behaviour and the basic attributes shared by all the ONESyst items.

```
class BOX

const    id: BOX-NAME;

var
  appearence: GX;
  actions: LIST-OF-ACTION;
  data: LIST-OF-DATA;
  links: LIST-OF-LINK;
  owner: USER-NAME
  delete: BOOL;

interface
  IG (iid:ITEM-NAME; app: GX; acts: LIST-OF-ACTIONS;
    data: LIST-OF-DATA; own:USER-NAME);
  GUI (app: GX; acts: LIST-OF-ACTIONS; data: LIST-OF-DATA;
       links: LIST-OF-LINK);
  SM (uid: USER-NAME);

methods
create
  with   IG;
  post   id <- IG.iid;  appearence <- IG.app; actions <- IG.acts;
         data <- IG.data;  links <- nil;  owner <- IG.own;  delete <- true;
add
  with   IG;
cancel
  with   IG;
delete
  with   SM;
lock
  with   SM;
  post   owner <- SM.uid;
unlock
  with   SM;
  post   owner <- none;
                   delete <- false;
link
  with   SM;
modify
  with   SM;
  post   appearence <- GUI.app;
         actions <- GUI.acts;
         data <- GUI.data;
         links <- GUI.links;
info
  with   SM;
```

Fig. 18. The BOX class

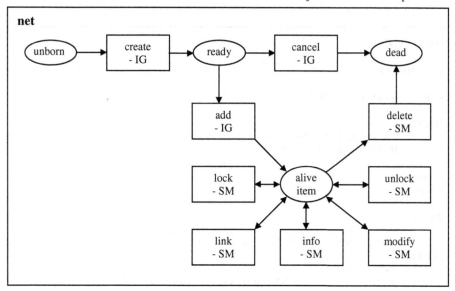

Fig. 18. (continued)

Pages are similar to boxes, but for the following features:
- pages can contain inner items;
- all page actions and data must be enclosed in one of the inner items;
- a page can be opened for inspection and modification according to the permission associated to the current user.

Therefore, the PAGE class, presented in Fig. 19, inherits from the BOX class, adding the inner attribute and the open, put item and get item methods.

Components are almost identical to boxes, but for the property that it's data structure is complete, so that the (redefined) method add can produce the algebraic domain of the associated OBJSA component. The domain is [automatically] defined by the algebraic operator choose, that browses the sorts associated to the component. Note that the COMPONENT class, shown in Fig. 20, does not include the net, as it is identical to the net of the parent class.

5.3. Polymorphism

Polymorphism is the mechanism that let objects, instanced by different classes in inheritance relation, answer the same message by executing the most appropriate method. It's a key concept for object-orientation, whose handling is tightly bound to the language run time support.

When object-oriented design languages are considered, polymorphism looses this role of allowing the dynamic binding between a procedure call (a message) and the best corresponding code (a method). But it is still important to avoid code duplication and enhance its reuse.

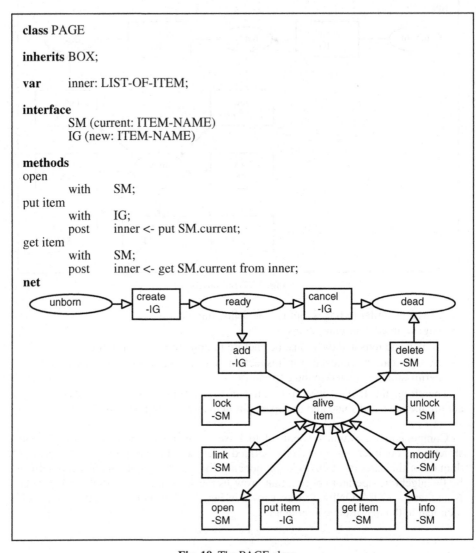

class PAGE

inherits BOX;

var   inner: LIST-OF-ITEM;

interface
        SM (current: ITEM-NAME)
        IG (new: ITEM-NAME)

methods
open
        with   SM;
put item
        with   IG;
        post   inner <- put SM.current;
get item
        with   SM;
        post   inner <- get SM.current from inner;
net

Fig. 19. The PAGE class

CLOWN provides a simple notion of polymorphism that takes advantage of the interface clause and of their semantics in terms of OBJ theories. An example is shown in the following.

The class ITEM GENERATOR (Fig. 21.). takes care of creating a graphic shape of the type chosen by the user (box, component or page).

Let us focus on the structure of the class net, namely on its two conflicts. The first one corresponds to the user's choice among one of the possible items: a box item (BIT), a component item (CIT) and a page item (PIT). The user is then asked to confirm or decline her/his choice.

```
class COMPONENT

inherits     BOX;

var domain: LIST-OF-DATA;

methods
add
  with IG;
  post domain <- choose (data)
```

Fig. 20. The COMPONENT class

```
class ITEM GENERATOR

const          id: IG-NAME;

var
  appearence: GX;
  actions: LIST-OF-ACTION;
  data: LIST-OF-DATA;
  token: DOMAIN
  owner: USER-NAME;

interface
  SM (box: ITEM-NAME);
  GUI (app:GX; acts:LIST-OF-ACTION; data:LIST-OF-DATA; uid:USER-NAME);
  OPAGE (outer: ITEM-NAME) pre OPAGE.outer == SM.box;
  ITEM; BIT; PIT; CIT;

methods
create item
  with          GUI;
fill header
  with          GUI;
  post          iappearence <- GUI.app;  actions <- GUI.acts;
                data <- GUI.data;  owner <- GUI.uid;
box item
  with          BIT;
page item
  with          PIT;
component item
  with          CIT;
add item
  with          ITEM, GUI, OPAGE;
cancel
  with          ITEM, GUI;
```

Fig. 21. The ITEM GENERATOR class

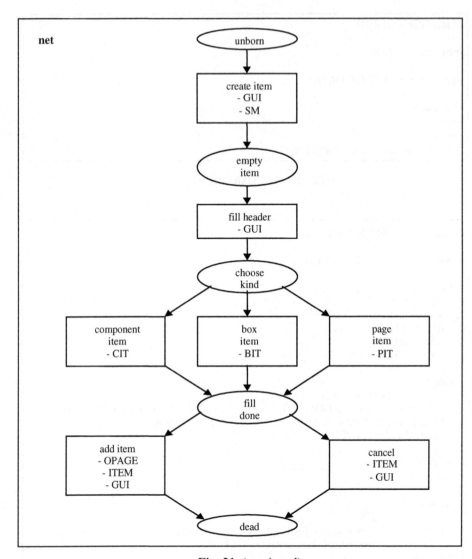

Fig. 21. (continued)

The straightforward specification of this choice would consist (cf. Fig. 22) of a couple of add+cancel methods for each possible shape (Boxes, Pages and Components), each one with the appropriate interface specification: BIT, CIT, PIT, respectively. However, a more effective solution can be adopted in CLOWN, and is indeed employed in the class net of Fig.21. It consists in using a single couple of add+cancel methods, whose interface is a generic ITEM that matches with all the classes representing the graphical shapes (BOXES, PAGES and COMPONENTgiven in Fig.19, 20 and 21). Through the specification of method composition in the *action* clause of the compound class USERS_COORDINATOR, the required pattern of Fig. 22 is automatically produced by the environment, by merging the corresponding transitions.

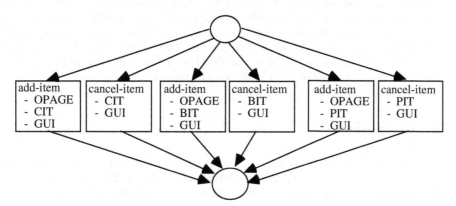

Fig. 22. Polymorphic item generation

5.4. Code Generation

An issue that we recently tackled with CLOWN is code generation (cf. also [EK94]), i.e. the automated production of object-oriented code starting from CLOWN specifications. This topics introduces both theoretical and engineering difficulties. On the theoretical side one has to deal with the translation of inherent concurrent specification in sequential ones, while on the engineering side the effort is to produce good code which can then be manipulated independently from the CLOWN specification it comes from.

The target language that we chose is Smalltalk [GR89], the forefather of pure object-oriented languages, now by a mature development environment. The whole translation process of a CLOWN specification in Smalltalk code is given in [DeR97].

The first step is the translation of the net associated to each CLOWN class into a textual format, called TCLOWN. The generation of a Smalltalk class corresponding to a TCLOWN elementary class is then straightforward, but for the resolution of the inheritance legacy clause whose technique is shown in [Chi94].

Compound classes are harder to translate for the reason why Smalltalk virtual machine implements message passing by means of remote procedure calls and doesn't support real multitasking. To produce Smalltalk code reusable in real applications, we did the choice to exclude the net simulation engine, underlying each CLOWN specification. Instead, each class has the capability to check the firing rules of its own superposed transitions. The events handling mechanism, native in the Smalltalk kernel, has shown particularly apt to fulfill this task.

Smalltalk choice looks quite appropriate for handling algebraic specifications too. In fact, most of the OBJ specification associated to CLOWN classes in the case studies we developed, can be ported to Smalltalk with a little effort. This occurs for the extreme dinamicity of the language and because Smalltalk is a pure object-oriented language, i.e. all of its computational entities are homogeneously treated as objects. For handling the (few) remaining cases, [DeR97] gives the specifications for emulating the OBJ interpreter in the Smalltalk environment.

A definite advantage in generating Smalltalk code starting from CLOWN specification is the automatic production of the preconditions associated to the net places and pre- clauses, allowing the production of a safe Smalltalk code from the start. For safe Smalltalk code we mean an application whose method calling usually depends upon the state of the receiver object. Smalltalk programmers are often forced to use ad hoc attributes (instance variables) to include these constraints, while Smalltalk code generated from a CLOWN spec's includes them by free.

6.   Conclusions

Pursuing the previous development of a specification environment for distributed systems based on a class of modular algebraic nets, namely OBJSA nets, we engaged in combining Petri nets and object-orientation. As pointed out at the very beginning of the project [BD93], our aim was twofold: on the one hand, to integrate the effectiveness of the object-oriented approach with the advantages of an executable specification language; on the other hand, to take profit of existing notions of equivalence and preorder developed within net theory to give sound foundations to inheritance in concurrent object-oriented (specification) languages.

This paper reports about CLOWN, the resulting object-oriented concurrent specification language, and illustrates some of the lessons learned through its use for modelling several case studies. The outcome of designing and experimenting with CLOWN includes difficulties, successes and failures, that we have tried to realistically describe. Neither of the two initial goals has been fully achieved, but both suggest topics to be faced further on.

First of all, inheritance deserve further investigation to understand if and how the net based notion of state-observability, that we employ in CLOWN, can be merged with action-observability, as suggested, e.g., in [Nie93, BB95, HV95]. Along this thread, a worth suggestion comes from [Ame89], where the author claims: "we propose ... to make a clear distinction between inheritance and subtyping, where inheritance deals with the internal structure of the objects, while subtyping, on the other hand, deals with their externally observable behaviour" (i.e. the offered services). State-observability looks appropriate for dealing with objects internal structure, while action-observability naturally faces with the externally observable behaviour.

A second major topics concerns modelling of dynamic systems, a challenging field for formalism combining nets and object-orientation. Preliminary examples, e.g. the specification of a "dynamic dining philosophers system" worked out in various formalisms including POPs and CLOWN (even though not shown in this paper for space reasons), indicate that CLOWN is adequate to describe systems whose components are dynamicaly created and destroyed and whose topologies evolve over time. Moreover, operations on OBJSA nets (presented in [BBCD95] and fully formalised in [Bar96]) which fold a component into a concurrent algebraic specification (in the line of [Mes93b]), which can recursively become the data structure of another net (tackling problems similar to those treated in [Lak96a, Val96]), open toward the modelling of reflective systems [Agh86, HA93, CELM96]. These two improvements would remove the major limitations introduced in [MAS92] for modelling actors systems by means of Petri nets.

Finally, our interest for non toy applications requires to complete and improve the CLOWN support environment. The recent and very satisfactory implementation of a code generator [DeR97] looks very promising and opens new opportunities in the field of object-oriented analysis and design techniques, since CLOWN templates seems to provide a guideline for producing well-formed classes, and look close to the principles of the Smalltalk main role-driven design technique [WWW90]. This suggests to reimplement in Smalltalk the tool supporting CLOWN classes definition, which was previously impelmented on the top of the ONE tool [BCCD96].

Acknowledgements

This research has been carried on with the financial support of the Italian MURST, 40% on "Modelli della computazione e dei linguaggi di programmazione". Moreover the authors wish to thank Claudia Balzarotti, Stefano Beretta, Fabio Conca, Lorenzo Capra and Piercarlo Della Rossa for the friendly collaboration and Olivier Biberstein, Didier Buchs and Charles Lakos for the helpful discussions during all the CLOWN development.

References

[ABSB94] M. Aksit, J. Bosch, W. van der Sterren, L. Bergmans, *Real-time specification inheritance anomalies and real-time filters,* in ECOOP '94 Proc., M. Tokoro, R. Pareschi (eds.), LNCS 821, 1994.

[AD95]   G. Agha and F. De Cindio (eds.), *Proc. of the 1st Workshop on Object-Oriented Programming and Models of Concurrency,* held within 16th Int. Conf on Appl. and Theory of Petri Nets, Torino (I), 1995.

[AD96]   G. Agha and F. De Cindio (eds.), *Proc. of 2nd Workshop on Object-oriented Programming and Models of Concurrency,* held within 17th Int. Conf on Appl. and Theory of Petri Nets, Osaka (JP), 1996.

[Agh86]   G. Agha, Actors: *A Model of Concurrent Computation in Distributed Systems.* MIT Press, 1986.

[Ame89]   P. America, *Issues in the design of a parallel object-oriented language,* in "Formal Aspects of Computing", Volume 1, pp.366-411, 1989.

[Ame92]   P. America, *Designing an object-oriented programming language with behavioural subtyping,* in [BRR91].

[AWY93]   G. Agha, P. Wegner, A. Yonezawa (eds.), *Research Directions in Concurrent Object-Oriented Programming,* MIT Press, 1993.

[Bar96]   G. Baratta: *Operazioni su componenti OBJSA* Master degree thesis, University of Milan, , 1996 (In italian).

[BB95]   O. Biberstein and D. Buchs: *Structured Algebraic Nets with Object-orientation;* in [AD95].

[BBCD95] E. Battiston, O. Botti, E. Crivelli and F. De Cindio: *An incremental specification of a Hidroelectric Power Plant Control System using a class of modular algebraic nets.* in G. De Michelis and M. Diaz (eds.), Appl. and Theory of Petri Nets 1995, LNCS 935, Springer, 1995.

[BCCD96] E. Battiston, L. Capra, A. Chizzoni and F. De Cindio: *The OBJSA Net Environment ONE.* Tool presentation at the 17th ICATPN, Osaka, J, 1996.

[BCD95]   E. Battiston, A. Chizzoni and F. De Cindio: *Inheritance and Concurrency in CLOWN.;* in [AD95].

[BD92]   L. Bernardinello, F. De Cindio, *A survey of basic net models and modular net classes,* in G. Rozenberg (ed.), Advances of Petri Nets 1992, LNCS 609, Springer, 1992.

[BD93]   E. Battiston, F. De Cindio, *Class Orientation and Inheritance in Modular Algebraic Nets,* in Proc. "IEEE 1993 Intern. Conf. on Systems, Man and Cybernetics, Le Touquet, F, October 17-20 1993", Vol. II, pp.717-723, 1993.

[BDH92]   E. Best, R. Devillers and J. Hall, *The Petri Box Calculus: a new causal algebra with multilabel communication,* in G. Rozenberg (ed.), Advances of Petri Nets 1992, LNCS 609, Springer, 1992.

[BDM88]   E. Battiston, F. De Cindio, G. Mauri, *OBJSA Nets: a class of high level nets having objects as domains,* in Advances in Petri Nets 88, G. Rozenberg (ed.), LNCS 340, Springer Verlag, 1988

[BDM96] E. Battiston, F. De Cindio and G. Mauri, *Modular Algebraic Nets to Specify Concurrent Systems,* IEEE Trans.on Software Engineering, 22(10), 1996.

[Ber94]   L. Bergmans, *Composing concurrent objects: applying composition filters for the development and reuse of concurrent object-oriented programs,* PhD Thesis, TRESE, Univ. of Twente, The Netherlands, 1994.

[Ber95]   S. Beretta: *Uno strumento per la fase iniziale della specifica di un sistema con le Reti OBJSA.,* Master degree thesis, Univ. of Milan (I) , 1995 (In italian).

[BI92]   M. Benveniste, V. Issarny, *Concurrent programming notations in the object-oriented language Arche,* techical report 1882, IRISA, France, 1992.

[BLP96]   R. Bastide, C. Lakos, P. Palanque, *A Cooperative Petri Net Editor,* Case Study proposal for the 2nd Workshop on Object-Oriented Programming and Models of Concurrency, available at

http://wrcm.dsi.unimi.it/PetriLab/ws96/case.html

[BOMN79] G.M. Birtwistle, OJ. Dahl, B. Myhrhaug and K. Nygaard, *Simula Begin,* Studentlitteratur, Lund, Sweden, 1979.

[BP96]   R. Bastide and P. Palanque, *Modeling a groupware editing tool with Cooperative Objects,* in [AD96].

[BRR91]   J.W. de Bakker, W.P. de Roever, G. Rozenberg (eds.), *Proc. of "School/Workshop in Foundations of Object-Oriented Languages",* LNCS 489, Springer, 1991.

[CDF93] G. Chiola, C. Dutheillet, G. Franceschinis and S. Haddad, *Stochastic well-formed coloured nets for symmetric modelling applications*, IEEE Trans. on Computers, 42(11), 1993.

[CELM96] M.Clavel, S. Eker, P. Lincoln and J. Meseguer, *Principles of Maude*, in Proc. of 1st Workshop on Rewriting Logic and its Applications, Pacific Grove, CA, pp. 65-89, 1996.

[Chi94]   A. Chizzoni: *CLOWN: CLass Orientation With Nets*. Master degree thesis, Univ. of Milan (I) , 1994 (In italian).

[Con96]   F. Conca: *Implementazione di CLOWN nell'ambiente ONE*. Master degree thesis, Univ. of Milan (I) , 1996 (In italian).

[Coo91] W.R. Cook, *Object-oriented programming versus abstract data types*, in [BRR91].

[CY91]   P. Coad and E. Yourdon: *Object-Oriented Analysis*. 2nd edition, Yourdon Press 1991.

[DDPS82] F. De Cindio, G. De Michelis, L. Pomello and C. Simone, *Superposed Automata nets*, in C. Girault and W. Reisig (eds.), Application and Theory of Petri Nets, IFB 52, , Springer , 1982.

[DeR97]   P. Della Rossa: *Ttranslation of CLOWN classes in Smalltalk classes*, Master degree thesis, Univ. of Milan (I) , 1997 (In italian).

[DLF93]   D. De Champeaux, D. Lea and P. Faure: *Object-Oriented software Development*. Addison Wesley 1993.

[ELR91]   J. Engelfriet, G. Leih and G. Rozenberg, *Net based description of parallel object-based systems, or POTs and POPs*, in [BRR91].

[EK94]   W. El Kaim and F. Kordon, *An integrated Framework for Rapid System Prototyping and Automatic Code Distribution*, in Procs. of 5th International Workshop or Rapid System Prototyping, N. Kanopoulos (ed.), IEEE Comp. Soc. Press, Grenoble, June, 1994.

[Gog84]   J. Goguen, *Parametrized programming*, in IEEE Trans. on Software Engineering", 10(5), pp. 528–543, 1984.

[GR89]   A. Goldberg and D. Robson, *Smalltalk-80: The Language*, Addison-Wesley, Reading, MA, 1989.

[GW88]   J. Goguen, T. Winkler, *Introducing OBJ3*, Technical Report SRI-CSL-88-9, SRI International, Computer Science Lab, 1988.

[GW91]   W. Gerteis, W. Wiraz, *Synchronising objects by conditional path expressions*, in TOOLS Pacific '91, pp. 193-201, 1991.

[HA93]   C. Houck and G. Agha, HAL: *A High-level language and its distributed implementation*, in [AWY93], 1993.

[HD96]   X. He and Y. Ding, *Object-Oriented Specification Using Hierarchical Predicate Transition Nets*, in in [AD96].

[HJS90]   P. Heber, K. Jensen and R.M. Shapiro, *Hierarchies in Coloured petri Nets*, in G. Rozenberg (ed.), Advances in Petri Nets 1990, LNCS 383, Springer, 1990.

[HV95]   T. Holvoet and P. Verbaeten: *PN-TOX: a Paradigm and Development Environment for Object Concurrency specifications*. in [AD95].

[JR91]   K. Jensen and G. Rozenberg (eds.): *High-Level Nets*. Springer, 1991.

[Lak96a]   C. Lakos, *The Consistent use of Names and Polymorphism in the Definition of Object Petri Nets*, in J. Billigton and W. Reisig (eds.), Appl. and Theory of Petri Nets 1996, LNCS 1091, Springer, 1996.

[Lak96b]   C. Lakos, *A Cooperative Editor for Hierarchical Diagrams: An Object Petri Net Model*, in in [AD96].

[Lil96]   J. Lilius, *OOB(PN)2: An Object Oriented Petri Net Programming Notation (A status report)*, in in [AD96].

[LW93a]   B. Liskov, J. M. Wing, *Specifications and their use in defining subtypes*, in OOPSLA '93 Proc., ACM SIGPLAN Notices, Vol. 28, No. 10, pp. 16-28.

[LW93b]   B. Liskov, J. M. Wing, *A new definition of the subtype relation*, in ECOOP '93 Proc., LNCS 707, Springer, 1993.

[Omo93]   S. Omohundro, *The Sather Language: Efficient, Interactive, Object-Oriented Programming*, Dr. Dobbs Journal, 1993

[OOMC95] http://wrcm.dsi.unimi.it/PetriLab/ws95/home.html

[MAS92]   S. Miriyala, G. Agha and Y. Sami, *Visualizing Actor Programs using Predicate Transition Nets*, in Journal of Visual Languages and Computing, 1992.

[Mes93a]   J. Meseguer, *Solving the inheritance anomaly in concurrent object-oriented programming*, in ECOOP '93 Proc., Springer, 1993.

[Mes93b]   J. Meseguer, *A logical theory of concurrent objects and its realization in the Maude language*, in [AWY93], pp. 314-390, 1993.

[Mey93]   B. Meyer, *Object-Oriented Software Construction, 2nd Ed.*, Prentice-Hall, Englewood Cliffs, NJ, 1993.

[Mil80]   R. Milner, *A calculus for communicating systems*, LNCS 92, Springer, 1980.

[MM88]   D. Marca and C.L. McGowan: *SADT. Structured Analysis and Design Technique*. McGraw Hill 1988.

[MWY90]   S. Matsuoka, K. Wakita, A. Yonezawa, *Synchronisation constraints with inheritance: What is not possible - So what is?*, technical report 90-010, Dept. of Information Science, University of Tokyo, Japan, 1990.

[MY93]   S. Matsuoka, A. Yonezawa, *Analysis of Inheritance Anomaly in Object-Oriented Concurrent Programming Languages*, in [AWY93], pp.107-150.

[Neu91]   C. Neusius, *Synchronising actions*, in ECOOP '91 Proc., P. America (ed.), LNCS 512, Springer, 1991.

[Nie93]   O. Nierstrasz, *Regular Types for Active Objects*, in Proceedings of the 8th OOPSLA'93, ACM Sigplan Notices, 28(10), pp. 1-15, 1993.

[PRS92] L. Pomello, G. Rozenberg, C. Simone, *A Survey of Equivalence Notions for Net based systems*, in "Advances of Petri Nets 92", G.Rozenberg(ed.), LNCS 609, Springer, 1992.

[PS91] L. Pomello and C. Simone, *A state transformation preorder over a class of EN systems*, in "Advances of Petri Nets 90", G. Rozenberg (ed.), LNCS 483, Springer, 1991.

[Rac94]   G. Raccagni: *Sviluppo e integrazione di strumenti su supporto alla ricerca di invarianti su reti modulari di alto livello,* Master degree thesis, Univ. of Milan (I) , 1994 (In italian).

[Rei91]   W. Reisig: *Petri Nets and Algebraic Specifications.* TCS 80, pp. 1-34, North-Holland, 1991. Also in [JR91].

[RTS96]   L. Recalde, E. Teruel and M. Silva, *{SC}*ECS: A Class of Modular and Hierarchical Cooperating Systems,* in Appl. and Theory of Petri Nets 1996, J. Billigton and W. Reisig (eds.), LNCS 1091, Springer, 1996.

[Sib94]   C. Sibertin-Blanc, *Cooperative Nets,* in Appl. and Theory of Petri Nets 1994, R. Valette (ed.), LNCS 815, pp. 471-490, Springer, 1994.

[Val96]   R. Valk, *On Processes of Object Petri Nets,* TR FBI-HH-B-185/96, Hamburg University, 1996.

[Weg90] P. Wegner, *Concepts and Paradigms of Object-Oriented Programming,* OOPS Messenger, Volume 1, Number 1, August 1990.

[WWW90] R. Wirfs-Brock, B. Wilkerson and L. Wiener: *Designing object-oriented software,* Prentice Hall, Englewood Cliffs, New Jersey, 1990.

[WZ88]   P. Wegner, S. B. Zdonik, *Inheritance as an Incremental Modification Mechanism or What Like Is and Isn't Like,* in ECOOP '88 Proc., LNCS 322, Springer, 1988.

Concurrency in Communicating Object Petri Nets

Rüdiger Valk

Fachbereich Informatik, Universität Hamburg
Vogt-Kölln-Str. 30, D-22527 Hamburg, Germany
valk@informatik.uni-hamburg.de

Abstract. Objects are studied as higher-level net tokens having an individual dynamical behaviour. In the context of Petri net research it is quite natural to also model such tokens by Petri nets. To distinguish them from the system net, they are called object nets. Object nets behave like tokens, i.e., they are lying in places and are moved by transitions. In contrast to ordinary tokens, however, they may change their state (i.e. their marking) when lying in a place or when being moved by a transition. By this approach an interesting and challenging two-level system modelling technique is introduced. Similar to the object-oriented approach, complex systems are modelled close to their real appearance in a natural way to promote clear and reliable concepts. Applications in fields like workflow, agent-oriented approaches (mobile agents and/or intelligent agents as in AI research) or open system networks are feasible. This paper gives a precise definition of the basic model together with a suitable process semantics. The focus is set more on basic concepts and their fundamental study than on high modelling capability.

1 Introduction

With the emergence of object systems and object-oriented programming also a number of papers have been published combining this modelling technique with Petri net models [2], [6], [7], [11]. This appears to be quite natural since both, object-oriented modelling as well as modelling by Petri nets, intend to support software development by abstraction of objects from the real world and then using the model to build a language-independent design organized around these objects. Both approaches promote better understanding of requirements, clearer designs, and more maintainable systems.

Object-oriented modelling means that software is designed as the interaction of discrete objects, incorporating both data structure and behaviour [10]. However, in most contributions, if formal techniques for describing the behaviour of objects in an object-oriented model are provided at all, these are usually equivalent to finite automata. In particular if concurrent behaviour is important, system modellers have to fall back on rather intuitive and informal methods. Here are the advantages of system modelling by Petri nets. They combine intuitive approaches with a formal tratment of systems and behavioural description. In addition they provide a deep and fundamental theory of concurrency.

G. Agha et al. (Eds.): Concurrent OOP and PN, LNCS 2001, pp. 164-195, 2001.

From a Petri net view objects appear in the form of tokens. During the last decade tokens have been considered as more and more complex data objects. In this paper we continue our previous work [13] by adding dynamical aspects to such token-objects. To integrate this approach into the systematics of Petri net modelling, it is quite natural to consider dynamical tokens as Petri nets themselves.

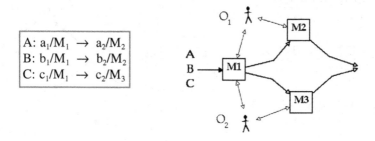

Fig. 1.1. Three machines with 3 tasks

Before giving an introduction to the formalism used and an overview on the structure of the paper we motivate the approach by some less formal examples. In the first example there are three tasks A, B and C to be processed on three machines M_1, M_2 and M_3 (Fig. 1.1.). There are limited resources for the machines of the following kind. Machines M_1 and M_2 are operated by an operator O_1. He can only operate one of the machines M_1, or M_2 at a given time. The same holds with operator O_2 with respect to M_1 and M_3. Machine M_1 can work, in mutual exclusion, only in a mode 1 with O_1 or in a mode 2 with O_2. Each of the tasks is divided in two subtasks, e.g. a_1 and a_2 in the case of A. The subtasks have to be executed by particular machines, as specified on the left-hand of Fig. 1.1. In the case of task A the second subtask a_2 must be executed on M_2 after the execution of a_1 on M_1. We take an „object-oriented"

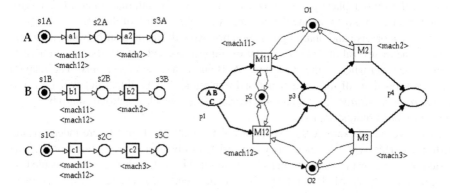

Fig. 1.2. Three machines example as object system

approach in the sense that the task is to be modelled as an object that enters machine M_1 and leaves it after execution to be then transferred to machine M_2. Attached with the object there is an „execution plan" specifying the machines to be used and the order for doing so. Also the current „status" of the execution is noted.

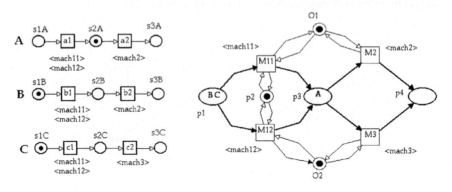

Fig. 1.3. Follower marking of the object system from Fig. 1.2.

This situation is formalized by the Petri nets of Fig. 1.2., where the two modes of machine M_1 are modelled by two different transitions M_{11} and M_{12}. Mutual exclusion is obtained by the places p_2, O_1 and O_2. Initially, all three tasks A, B and C are in the place p_1 (in the net on the right-hand side). By the „object -oriented" approach they are not represented by an unstructured token, but by their entire execution plan, as given on the left-hand side, also as Petri nets. Note that by the marking in the nets A, B and C also their „status" description is given.

Labels at a subtask of the form <mach11> indicate that this subtask can be executed by any machine with the same label, i.e. by M_{11} in this case. Following this convention, transition M_{11} of the „machine net" on the right-hand side of Fig. 1.2. can occur with respect to A in its input place p_1. The whole „task net" is then moved as a token to the output place p_3, as shown in Fig. 1.3. The internal token of A is also moved from s_{1A} to s_{2A} to update the current execution „status" of the task.

To have a precise and unambiguous notation we will distinguish between a *system net* and one or several *object nets*. In the example presented before, the „machine net" is the *system net* whereas the „task nets" A, B and C are the *object nets*. The relation between transitions of the system net and transitions of the object net will be called the *interaction relation*. This relation is represented by labels (enclosed sharp brackets) in the graphical representation.

In the preceding example the executions of the tasks A, B and C are independently performed. This type of concurrency is restricted only by resource limitations of the system net. For instance, in the marking given in Fig. 1.3. task A can be executed on machine M_2 while task B (or task C) is concurrently executable on M_{12}. A more realistic application to flexible manufacturing systems is given in [19].

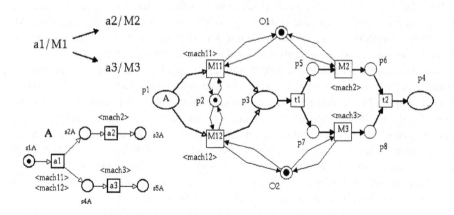

Fig. 1.4. Machine example with concurrent execution

Another type of concurrency is exemplified by the object system of Fig. 1.4. In its object net, after termination of the subtask a_1 (on machine M_1), there are two successor tasks a_2 and a_3. They can be executed independently if there are suitable functional units (machines) being able to perform the execution concurrently. The system net in Fig. 1.4. (on the right-hand side) offers such a suitable system architecture: after the occurrence of transition t_1 two identical descriptions of the object net (both with its current marking) are generated and placed in p_5 and p_7. With respect to the instance of the object in p_5 the subtask a_2 is executable on M_2, while concurrently the subtask a_3 of the object net instance in p_7 is executed by M_3. The „results" of the partial executions are then „brought together" by transition t_2, which outputs the combined and final result to the place p_4. Hence, from an intuitive point of view, M_2 and M_3 produce partial results that are independent from each other. They are „put together" into a single task description by transition t_2. Here the precise semantics of this action in terms of Petri nets will be left open. This will be one of the results to be described in section 3. We will refer to this kind of concurrency as *intra-object concurrency*, as opposed to concurrent behaviour of two different objects.

Using standard definitions of Petri net theory the preceding examples can be encoded either as a „flat" net by identifying corresponding transitions or as a coloured net with appropriate data type definitions in the colour sets (compare with [8]). In this paper we follow a different approach: object and system nets are defined (as simple as possible) as Elementary Net Systems ("EN systems", formerly condition/event-systems), including the occurrence rule of this net type. Using the individual occurrence rules of the component nets, by combination of the individual EN-systems, *Object Systems* will be introduced, which allow to model real systems directly in the style of object oriented modelling. In doing this, we are led by the experience that has been made by the development of higher Petri nets (Pr/T-nets, coloured nets) from "lower" Petri nets (C/E-nets, P/T-nets), namely, that new features should be introduced in accordance with basic principles of concurrency theory, as formulated by C.A. Petri.

For the first time, Petri nets as dynamical objects have been considered for describing the execution of task systems in systems of functional units [4],[13],[14]. In [16] the formalism is extended in such a way that the objects are allowed to be general EN systems not necessarily restricted to (non-cyclic) causal nets.

Object-oriented modelling and programming, as appearing in current literature, is characterized by a specific object notion together with many features like generation and deletion of objects, defining classes and subclasses, inheriting attributes and many more. We do not incorporate all these features in our model as we are concentrated on such properties that can be expressed on the level of EN systems. This is done to master the complexity of the new approach. It is easier to define new models than to derive formal results. However, working on formal theories gives important hints for a suitable design of the model. Where the system net and all object nets are EN systems object systems will be called *Elementary Object Systems*.

Section 2 is concentrated on the study of object systems having only a single object net (*Unary Elementary Object Systems*). This class is introduced to investigate the behaviour of concurrent task execution. Different notions of markings are introduced. Finally a suitable formalism for the modelling of „fork/join"-concurrency structures is proposed. Also a process-oriented semantics for unary elementary object systems is given (section 3). One of the main formal results of this paper is a theorem characterizing processes of unary elementary object systems by classical processes of EN systems.

In section 4 *Unary Elementary Object Systems* are studied that allow for more than a single object net. This class is restricted to system nets that essentially are state machines. Therefore duplication of objects is not possible as well as intra-object concurrency. This restriction is not necessary. It has only been made to allow for a simpler occurrence rule and simpler semantical descriptions. On the other hand, a new feature is introduced with this model: the direct interaction of different object nets. Further complex examples using such an „inter-object communication" are presented, showing advantages of the object-oriented approach.

The main results of this contribution may be summarized as follows:

- A simple and clear notion of object Petri net is introduced such that most principles of the elementary net theory are respected.
- A formal semantics of the behaviour is given for this net class.
- It was discovered that from the different choices for the definition of markings and occurrence rules, not all of them allow a meaningful and consistent modelling of object-oriented concurrency.
- Processes of Elementary Net systems are extended to the model in a natural way. A low-level characterization of such higher-level processes is given and the equivalence is formally proven.
- Unary object systems are consistently extended to multiple objects.

At the end of this introduction we now give a less artificial example of an unary elementary object system with intra-object concurrency. In the example a workflow of the Dutch Justice Department is modelled. It has been used for demonstration of modelling and analysis of workflow applications using Petri nets [1].

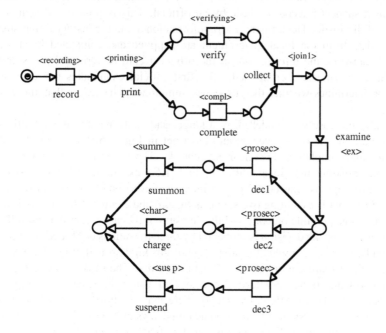

Fig. 1.5. Object net of the work flow example

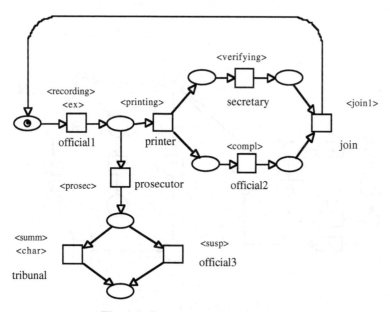

Fig. 1.6. System net of the work flow example

170   R. Valk

The example was introduced as follows. When a criminal offence happened and the police has a suspect a record is made by an official. This is printed and sent to the secretary of the Justice Department. Extra information about the history of the suspect and some data from the local government are supplied and completed by a second official. Meanwhile the information on the official record is verified by a secretary. When these activities are completed, the first official examinates the case and a prosecutor determines whether the suspect is summoned, charged or that the case is suspended.

Originally the case was modelled by a single and „flat" net for the workflow. A slightly modified version is given as an object net in Fig. 1.5. Observe that, indeed, verification and completion are concurrent subtasks. The labels in sharp brackets refer to the corresponding functional units (in Fig. 1.6.) executing these subtasks. For instance „printing" is executed by a printer and „verifying" is executed by the secretary. Official1 is executing two subtasks (record and examine) for this object net. As there are three possible outcomes of the decision of the prosecutor that are followed by different actions, the decision is modelled by three transitions dec1, dec2 and dec3.

Though being more complex the advantage of this kind of modelling by object nets lies in the direct representation of the functional units. The system net in Fig. 1.6. reflects the organisational structure of the system while the object net (Fig. 1.5.) represents a particular workflow. Obviously there may be different workflows (object nets) for the same system of functional units (system net). The simultaneous simulation of different such executions can be used to determine bottlenecks and execution times.

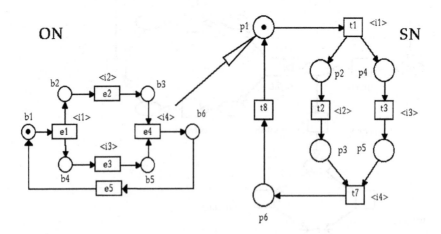

Fig. 2.1. Elementary object system "con-tasks"

2 Unary Elementary Object Systems

In this section Unary Elementary Object Systems are introduced, consisting of a *system net* SN and an *object net* ON, both being elementary net systems. These are used in their standard form as given in [12].

An *Elementary Net System* (EN system) $N = (B,E,F,C)$ is defined by a finite set of *places* (or conditions) B, a finite set of *transitions* (or events) E, disjoint from B, a flow relation $F \subseteq (B{\times}E) \cup (E{\times}B)$, and an *initial marking* (or initial case) $C \subseteq B$. The occurrence relation for markings C_1, C_2 and a transition t is written as $C_1[t > C_2$ or $C_1 \to_t C_2$. If t is enabled in C_1 we write $C_1[t >$ or $C_1 \to_t$. These notions are extended to words $w \in E^*$ as usual and written as $C_1[w > C_2$ (or $C_1 \to_w C_2$) and $C_1[w >$ (or $C_1 \to_w$), respectively. N is called a *structural state machine* if each transition $t \in T$ has exactly one input place ($|{\bullet}t| = 1$) and exactly one output place ($|t{\bullet}| = 1$). N is said to be a *state machine* if it is a structural state machine and C contains exactly one token ($|C| = 1$). $FS(N) := \{ w \in E^* \mid C\ [w > \}$ is the set of *firing* or *occurrence sequences* of N , and $R(N) := \{C_1 \mid \exists\, w : C[w > C_1\}$ is the set of *reachable markings* (or cases), also called the *reachability set* of N (cf. [9]). Processes of EN systems will be defined in the appendix .

Definition 2.1
An *unary elementary object system* is a tuple EOS = (SN,ON,ρ) where
SN= (P,T,W,M_o) is an EN system with $|M_o| = 1$, called *system net* of EOS,
ON = (B,E,F,m_o) is an EN system, called *object net* of EOS, and
$\rho \subseteq T \times E$ is the *interaction relation*.
An elementary object system is called *simple* if its system net SN is a state machine.

Fig. 2.1. gives an example of an elementary object system with the components of an object net ON on the left-hand and a system net SN on the right-hand side. The interaction relation ρ is given by labels $<i_n>$ at t and e iff tρe ("i_n" stands for interaction number n). A similar object net is used in Fig. 2.3. (i_1 is removed to illustrate autonomous transitions), but with a different system net. By this system net the "parallel" transitions e_2 and e_3 perform in a serial way. Since the system net is a state machine, the object system is simple.

Before coming to formalization we describe the intuition behind the occurrence rule to be defined afterwards. The object net ON of Fig. 2.1. should be thought of lying in place p_1 of the system net SN. It is represented by a token in that place. The occurrence of transition t_1 of the system net SN should coincide with e_1 in the object net ON by the interaction i_1. The object net ON is then removed from p_1 and added to p_2 and p_4 in two copies, both of them being in the marking $\{b_2,b_4\}$. Then we observe some concurrent behaviour ending with a kind of „join" operation by the interaction i_4 of t_7 and e_4. Furthermore, there are transitions without interaction like the so-called "autonomous" occurrences of t_8 or e_5

In the definitions of the occurrence rule we will use the following well-known notions for a binary relation ρ. For $t \in T$ and $e \in E$ let $t\rho := \{e \in E \mid (t,e) \in \rho \}$ and

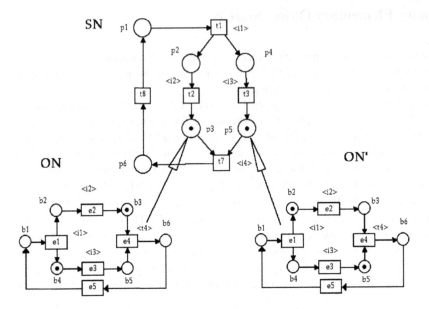

Fig. 2.2. Successor marking of Fig. 2.1.

$\rho e: = \{t \in T \mid (t,e) \in \rho\}$. Then $t\rho = \emptyset$ means that there is no element in the interaction relation with t.

Definition 2.2

A *bi-marking* of an unary elementary object system EOS = (SN,ON,ρ) is a pair (M,m) where M is a marking of the system net SN and m is a marking of the object net ON.

a) A transition $t \in T$ is *enabled* in a bi-marking (M,m) of EOS if $t\rho = \emptyset$ and t is enabled in M. Then the *successor bi-marking* (M',m') is defined by $M \to_t M'$ (w.r.t. SN) and m'=m. We write $(M,m) \to_{[t,\lambda]} (M',m')$ in this case.

b) A pair $[t,e] \in T \times E$ is *enabled* in a bi-marking (M,m) of EOS if $(t,e) \in \rho$ and t and e are enabled in M and m, respectively. Then the *successor bi-marking* (M',m') is defined by $M \to_t M'$ (w.r.t. SN) and $m \to_e m'$ (w.r.t. ON).
We write $(M,m) \to_{[t,e]}(M',m')$ in this case.

c) A transition $e \in E$ is *enabled* in a bi-marking (M,m) of a EOS if $\rho e = \emptyset$ and e is enabled in m. Then the *successor bi-marking* (M',m') is defined by $m \to_e m'$ (w.r.t. ON) and $M' = M$. We write $(M,m) \to_{[\lambda,e]} (M',m')$ in this case.

In transition occurrences of type b) both the system and the object participate in the same event. Such an occurrence will be called an *interaction*. By an occurrence of type c), however, the object net changes its state without moving to another place of the system net. It is therefore called *object-autonomous* or *autonomous* for short. The symmetric case in a) is called *system-autonomous* or *transport*, since the object net is transported to a different place without performing an action.

By extending this notion to occurrence sequences for the EOS of 2.3., for example, we obtain the following sequence: $[\lambda,e_1]$, $[t_1,\lambda]$, $[t_4,e_3]$, $[t_5,e_2]$, $[t_6,\lambda]$, $[t_7,e_4]$, $[\lambda,e_5]$. After this sequence, the initial bi-marking is reached again. We call this the *occurrence sequence semantics*. It is possible to characterize the set of all such occurrence sequences of simple EOS by some kind of intersection of the individual occurrence

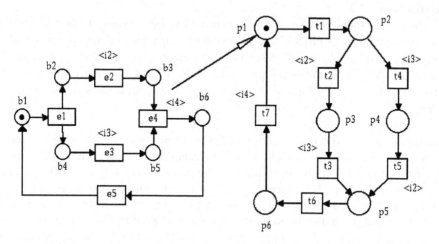

Fig. 2.3. Elementary object system "ser-task"

sequences of SN and ON. As simple object systems appear quite frequently in applications, this definition of a bi-marking and transition occurrence semantics is useful. However, the question must be asked whether it is also adequate for general EOS.

To discuss the problem consider the EOS of Fig. 2.1. again in a bi-marking (M,m) = $(\{p3,p5\},\{b3,b5\})$ that is reached after the occurrence sequence $[t_1,e_1]$, $[t_2,e_2]$, $[t_3,e_3]$. Apparently this notion of a bi-marking is not adequate since the distributed character of this state is not represented, namely it is not visible that b_3 and b_5 hold in *different* copies of the object net, as graphically visualized in Fig. 2.2. In the next transition occurrence the tokens b_3 and b_5 should be used, since these tokens represent those parts of the object net processes which are the „most advanced". It might be possible to modify the object system in such a way that the tokens b_2 and b_4 are used by a transition occurrence. This would be contraintuitive since b_2 represents a part of the process of one copy of the object net which is „less advanced", but where the other copy was more progressive. In the same way one could argue for b_4 .

A solution different from bi-markings is a marking where the pairs $(ON,\{b_3,b_4\})$ and $(ON,\{b_2,b_5\})$ are assigned to p_3 and p_5, respectively (cf. Fig. 2.2.). As shown by a counter example in [18], [19] also such a modelling is not adequate for non-simple unary elementary object systems. By this observation we are lead to follow a different approach for representing markings of object systems. Instead of object net markings the corresponding processes will be used. In a bi-marking of an elementary object system, a place may be empty or contain the object net ON, being in some specific state of its execution. Such a state is now described by a process of the object net ON. Hence a marking associates a process proc ∈ PROC(ON) to every place p ∈ P. In order to distinguish this form of a marking from the one in the previous section, we call it a "process-marking" or in short a "p-marking".

Definition   2.3

A *process-marking (p-marking)* $\underline{M}$ of an elementary object system EOS = (SN,ON,ρ) is a mapping $\underline{M}$: P → PROC(ON), associating to each place of the system net SN a process proc of the object net ON (including the empty process). If in a p-marking M of an EOS $\underline{M}$(p) = Ø (the empty process), we say the place is *empty*, else *occupied*. The set CM := {p ∈ P | $\underline{M}$(p) ≠ Ø } of occupied places defines a *case* or *marking* of EOS.

In this definition PROC(ON) denotes the set of all processes of ON. (For definitions concerning processes see appendix.) For introducing the new occurrence rule, consider the EOS of Fig. 2.1. In the initial marking the process consisting of b_1 is in the place p_1 of the system net. Now consider a follower state after the occurrence of $[t_1,e_1]$, $[t_2,e_2]$ and $[t_3,e_3]$. For the next step t_7 is enabled since all its input places are non-empty and $(t_7,e_4) ∈ ρ$. But in addition e_4 should be enabled in ON. The preconditions of e_4 are satisfied if *all* copies of the object net processes lying in the input places in p_3 and p_5 of t_7 are taken into consideration (see 2.4.). The joint information is obtained by the least upper bound „lub" of these processes.

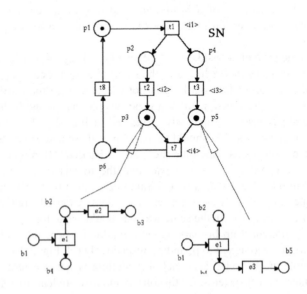

Fig. 2.4. Elementary object system "con-task" with p-marking

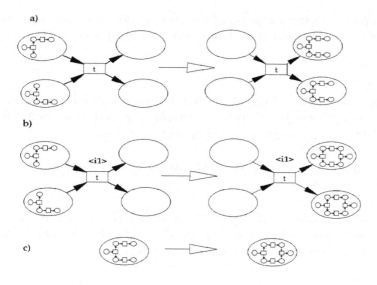

Fig. 2.5. Object system occurrence rule: a) transport, b) interaction, c) autonomous

Definition 2.4

Let $\underline{M}$ be a p-marking of an unary elementary object system EOS, $t \in T$ a transition of the system net SN and $e \in E$ a transition of the object net ON.

To enable t, it is necessary in any of the following cases a) and b) that all input places $p \in {}^{\bullet}t$ are occupied, the process LUB $:= \text{lub}(\{\underline{M}(p) \mid p \in {}^{\bullet}t \})$ exists and all output places $p \in t^{\bullet}$ are empty in $\underline{M}$.

In case a), where $t\rho=\emptyset$, transition t is enabled and we write $\underline{M} \rightarrow_{[t,\lambda]}$.

In case b), where $t\rho e$, the pair [t,e] is enabled (denoted $\underline{M} \rightarrow_{[t,e]}$) if e is enabled for LUB (cf. appendix).

For both cases a) and b) the follower p-marking $\underline{M}'$ is defined by $\underline{M}'(p) = \emptyset$ if $p \in {}^{\bullet}t$, and for $p \in t^{\bullet}$ we define $\underline{M}'(p) = $ LUB in case a) and $\underline{M}'(p) = $ LUB$^{\circ}$e in case b) and $\underline{M}'(p) = \underline{M}(p)$ otherwise.

(Case a) is called a *system-autonomous* or a *transport occurrence* and is denoted by $\underline{M} \rightarrow_{[t,\lambda]} \underline{M}'$ whereas case b) is called an *interaction* and is denoted by $\underline{M} \rightarrow_{[t,e]} \underline{M}'$).

case c): If in some place $p \in P$ an object net transition $e \in E$ with $\rho e=\emptyset$ is enabled in proc $\in \underline{M}(p)$, we write $\underline{M} \rightarrow_{[\lambda,e]}$ and define a follower marking $\underline{M}'$ by

$$\underline{M}'(p) = \text{proc}^{\circ}e \text{ and}$$
$$\underline{M}'(p') = \underline{M}(p) \text{ for } p' \neq p.$$

(Case c) is called an *object-autonomous* or an *autonomous* occurrence and is denoted by $\underline{M} \rightarrow_{[\lambda,e]} \underline{M}'$).

In Fig. 2.5. all cases a), b) and c) of the occurrence rule are represented symbolically. In this figure process inscriptions are not given. In general, however, it may depend on these inscriptions whether the process LUB exists or does not exist. Next we extend the definition to sequences of the form $[\lambda,e_1]$ $[t_1,\lambda]$ $[t_2,e_2]$ $[t_3,e_3] \in ((T \cup \{\lambda\}) \times (E \cup \{\lambda\}))^*$.

Definition 2.5
For an elementary object system EOS = (SN,ON,ρ) we consider occurrence sequences $w \in Q^*$ where $Q := \underline{T_l} \times \underline{E_l}$ and $\underline{T_l} := T \cup \{\lambda\}$, $\underline{E_l} := E \cup \{\lambda\}$. For such sequences and p-markings $\underline{M}$ and $\underline{M}'$ the relation $\underline{M} \to_w \underline{M}'$ is inductively defined by :
1. $\underline{M} \to_w \underline{M}$ if $w = [\lambda,\lambda]$
2. $\underline{M} \to_{wq} \underline{M}'$ if $w \in Q^*$, $q \in Q$ and $\underline{M} \to_w \underline{M}''$, $\underline{M}'' \to_q \underline{M}'$ for a p-marking $\underline{M}''$.

Definition 2.6
The *initial p-marking* of EOS is defined using the initial markings M_0 and m_0 of SN and ON, respectively: $\underline{M}_0(p) :=$ if $p \in M_0$ then m_0 else $\varnothing$. (Note that in this context m_0 means the initial process of EOS, as defined in the appendix).

Definition 2.7
Given an elementary object system EOS = (SN,ON,ρ), then FS(EOS) := { $w \in Q^*$ | $\exists \underline{M}$: $\underline{M}_0 \to_w \underline{M}$} is the *set of occurrence sequences* of EOS, and
R(EOS) := {$\underline{M}$ | $\exists$ w : $\underline{M}_0 \to_w \underline{M}$ } is the set of *reachable p-markings*, also called the *reachability set* of EOS.

3   Processes of Unary Elementary Object Systems

The definition of processes of unary elementary object systems is quite obvious if autonomous occurrences are not considered. To give an example, in Fig. 3.2. a process of the elementary object system "con-tasks" EOS = (SN,ON,ρ) from Fig. 2.1. is constructed as follows. A process of the system net SN is extended in such a way that the places contain the object net process in the corresponding p-marking. Autonomous transitions are represented in black. Throughout this section all elementary object systems are assumed to be unary.

Concurrency of transitions is a fundamental topic of Elementary Net Systems. Transitions may occur concurrently if their input and output places are disjoint. A more interesting situation occurs if these transitions occur in markings where the object net process takes part in the system net transition occurrence. This is formally treated in lemma 3.1. and graphically represented in Fig. 3.1. a) and b) for the case n=2.

Lemma 3.1
Let EOS = (SN,ON) be an unary elementary object system and „t" a system-autonomous transition of SN and „e" an object-autonomous transition of ON.

Suppose •t= {p_1} and |t•| = n ≥ 1 and $\underline{M}_1(p_1)$ = proc°e for some p-marking $\underline{M}_1$ (i.e. p_1 contains a process where „e" is „at the end" (see appendix)). Then there are p-markings $\underline{M}$ and $\underline{M}'$ such that

$$\underline{M} \to_w \underline{M}' \quad \text{where } w=[\lambda,e][t,\lambda] \text{ and}$$
$$\underline{M} \to_v \underline{M}' \quad \text{where } v=[t,\lambda][\lambda,e]^n.$$

Proof: Starting from $\underline{M}_1$ p-markings $\underline{M}$ and $\underline{M}'$ are constructed as follows in a), b):
a) Since $\underline{M}_1(p_1)$ has the form proc°e there is a predecessor marking $\underline{M}$ such that
$\underline{M} \to_{[\lambda,e]} \underline{M}_1$ where $\underline{M}(p_1)$ = proc and $\underline{M}(p) = \underline{M}_1(p)$ for p ≠ p_1
b) Since t is enabled in $\underline{M}_1$ there is a follower marking $\underline{M}'$ of $\underline{M}_1$ (i.e. $\underline{M}_1 \to_{[t,\lambda]} \underline{M}'$)
where proc°e is contained in all n output places of t. By a) and b) we have $\underline{M} \to_{[\lambda,e]} \underline{M}_1$
$\to_{[t,\lambda]} \underline{M}'$.

To prove the second part of the lemma, we observe that [t,λ] is also enabled in $\underline{M}$, since the unique input place p_1 of t contains the process proc i.e. $\underline{M} \to_{[t,\lambda]} \underline{M}_2$ for some p-marking $\underline{M}_2$. In $\underline{M}_2$ all n output places of t contain the process proc in each of which the transition e is enabled. This leads to $\underline{M} \to_{[t,\lambda]} \underline{M}_2 \to_u \underline{M}'$ with u = $[\lambda,e]^n$.
q. e. d.

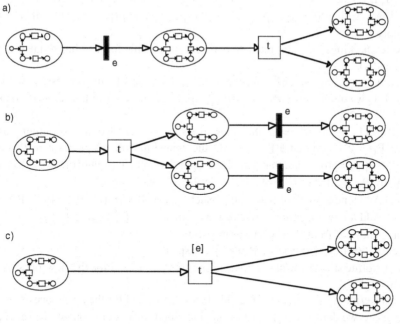

Fig. 3.1. Concurrent autonomous transitions

While object autonomous transition occurrences are not part of SN processes, they should be visible in the object system process. For distinction we denote them by small and solid (black) transitions as in Fig. 3.1. and 3.2. Furthermore it is straight forward to introduce an equivalent step containing both transitions as in Fig. 3.1. c). We will call this a reduced process equivalent.

Next we inductively define a process of an elementary object system, having a set of places $\underline{P}_\pi$ and a set of transitions $\underline{T}_\pi$, together with mappings ϕ and μ. For a place x $\in \underline{P}_\pi$ or a transition y $\in \underline{T}_\pi$ of this process $\phi(x)$ and $\phi(y)$ give the corresponding place or transition of the system net, respectively. $\mu(x)$ will be the process of the object net ON associated to a place x. Furthermore for each place x $\in \underline{P}_\pi$ the interaction relation $\rho \subseteq T \times E$ is extended to $\rho_x \subseteq T_\pi \times E_\pi$, where E_π is the set of places of the process $\mu(x)$. These mappings are given by inscriptions in the example of Fig. 3.2. as explained after the following definition.

Definition   3.2

For a given firing sequence w $\in$ FS(EOS) of an unary elementary object system EOS = (SN,ON,ρ), where SN = (P,T,W,M_0), ON = (B,E,F,m_0), a *process* proc(w) = $(\underline{P}_\pi, \underline{T}_\pi, \underline{E}_\pi, \phi, \mu)$ is a structure consisting of a causal net $(\underline{P}_\pi, \underline{T}_\pi, \underline{E}_\pi)$ and mappings

$$\phi : \underline{P}_\pi \cup \underline{T}_\pi \to P \cup T \cup E \quad \text{and} \quad \mu: \underline{P}_\pi \to \text{PROC(ON)}.$$

proc(w) is defined by induction over Q*. Furthermore for each object net process $proc_2$ = $(B_\pi, E_\pi, F_{2\pi}, \phi_2) = \mu(x)$, x $\in \underline{P}_\pi$ an *extended interaction relation* $\rho_x \subseteq T_\pi \times E_\pi$ is defined.

I. If w = λ, then $\underline{P}_\pi = \{p_\pi \mid p \in M_0\}$ with $\phi(p_\pi) = p$ and $\mu(p_\pi) = m_0$ for all $p_\pi \in P$. ρ_x is empty.

(Note: markings are interpreted here as processes in the form of an initial process (see appendix).

II. Let $\underline{M}_0 \to_w \underline{M} \to_{[u,v]}\underline{M}'$ and proc(w) = $(\underline{P}_\pi, \underline{T}_\pi, \underline{E}_\pi, \phi, \mu)$ be the process of w. Then for [u.v]$\in$ Q we define proc(w[u,v]) = $(\underline{P}'_\pi, \underline{T}_\pi, \underline{E}'_\pi, \phi', \mu')$ for each of the cases a), b) and c) of definition 2.4:

a) If [u,v] = [t,λ] and tρ=$\varnothing$, then there is a subset $P_1 \subseteq \underline{P}_\pi$ having no output transitions (i.e. $\underline{P}_1 \bullet = \varnothing$) such that $\phi(\underline{P}_1) = \bullet t$. By the enabling rule all places x in $\underline{P}_1$ contain a process $\mu(x)=proc_1$ such that their least upper bound LUB:=lub$\{\mu(x)|x\in\underline{P}_1\}$ exists. To obtain $(\underline{P}_\pi', \underline{T}_\pi', \underline{E}_\pi', \phi', \mu')$ we have to do the following steps :

a_1) Add a new set P_2 of places to $\underline{P}_\pi$ such that $\phi'(P_2)=t\bullet$, (i.e. $\underline{P}_\pi' = \underline{P}_\pi \cup P_2$).

a_2) Add a new transition y with $\phi'(y) = t$ to $\underline{T}_\pi$ (i.e. $\underline{T}_\pi' := \underline{T}_\pi \cup \{y\}$).

a_3) Add arcs from P_1 to y and from y to P_2 (i.e.: $\underline{E}_\pi' := \underline{E}_\pi \cup \{ (x,y) \mid x \in P_1 \} \cup \{(t',p)|p\in P_2\}$).

a_4) Define $\phi' = \phi$ for all old places and transitions and for the new ones as defined in a_1) and a_2).

a_5) Define $\mu'(x) = \mu(x)$ for the old places x $\in \underline{P}_\pi$ and for the new places $x_2 \in P_2$ with $\phi(x_2) \in t\bullet$ we define $\mu'(x_2) := LUB$ (i.e. the output places of t contain the same process LUB. ρ_x for x $\in P_2$ remains as in the old places.

b) If [u,v] = [t,e] and tρe, then P_1 exists as in case a). The steps b_1) to b_4) are defined as a_1) to a_4), respectively.

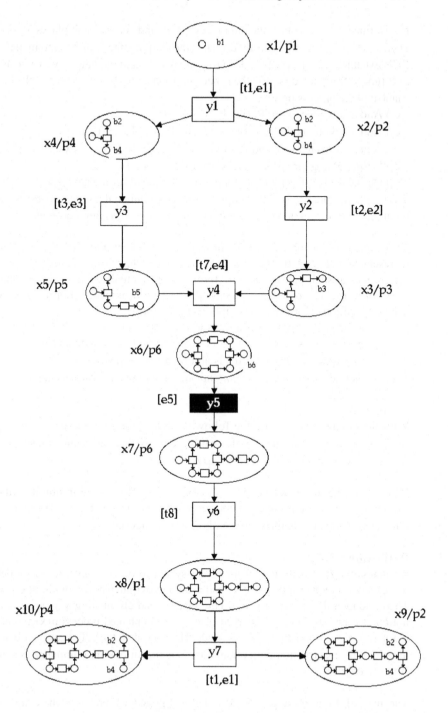

Fig. 3.2.: A process of the object system „Con-Task" from Fig. 2.1.

b_5). Define $\mu'(x) = \mu(x)$ for the old places $x \in P_\pi$ and for the new places $x_2 \in P_2$ with $\phi(x_2) \in t\bullet$ we define $\mu'(x_2) := LUB°e$ (i.e. the output places of t contain the process LUB extended by e). (y,e_2) is added to ρ_x for $x \in P_2$ and the new e_2 with $\phi_2(e_2) = e$.

c) If $[u,v] = [\lambda,e]$ and $\rho e = \varnothing$, then there is a place $x \in P_\pi$ with $x\bullet = \varnothing$ such that "e" is enabled in the process $proc_1 = \mu(x)$.

c_1) Add a new place x_2 to P_π (i.e. $\underline{P_\pi}' = \underline{P_\pi} \cup \{x_2\}$).

c_2) Add a new transition y with $\phi(t) = e$ to $\underline{T_\pi}$ (i.e. $\underline{T_\pi}' := \underline{T_\pi} \cup \{y\}$

c_3) Add arcs from x to y and from y to x_2 (i.e.: $\underline{E_\pi}' := \underline{E_\pi} \cup \{ (x,y), (y,x_2)\}$).

c_4) Define $\phi' = \phi$ for all old places and $\phi'(x_2) = \phi(x_1)$.

c_5) Define $\mu'(x) = \mu(x)$ for the old places $x \in \underline{P_\pi}$ and for the new place $x_2 \in P_2$ we define $\mu'(x_2) := proc_1°e$ (i.e. the output places of y contain the same processes as the input places but extended by a new e_2 with $\phi_2(e_2) = e$). ρ_x is not modified.

An example of an object system process is shown in Fig. 3.2. In the graphical representation ϕ and μ are given as follows. Places are named x_1, x_2,... and inscriptions at such places have the form $x_i/\phi(x_i)$ (actually, due to the graphical tool used: xi/ϕ(xi)). The object net process $\mu(x)$ is drawn into the ellipse of x. $y_1,y_2,...$ denote transitions. They have inscriptions of the form

a) $[\phi(y_i)]$ if $\phi(y_i) = t_i$ is a transport, i.e. $t_i\rho = \varnothing$,

b) $[\phi(y_i),e] \in T \times E$ if $\phi(y_i) = t_i$ interacts with e, i.e. $t_i\rho e$ and

c) $[\phi(y_i)]$, $\phi(y_i) = e \in E$ if e is autonomous i.e. $\rho e = \varnothing$.

Transitions y of case c) are called *autonomous* and drawn as black rectangles.

Lemma 3.3

With the notation of def. 3.2 the following holds: for each place $x \in \underline{P_\pi}$, $proc_x = (B_\pi,E_\pi,F_{x\pi},\phi_{x\pi}) = \mu(x)$, and $e \in E_\pi$ with $\rho_\pi e \neq \varnothing$ there is some transition y with $y <_{proc(w)} x$ (i.e. y "before" x) such that $y\rho_\pi e$.

Proof: e is either introduced to $proc_x$ in $\mu(x)$ in step II b) of definition 3.2 (then $y\rho_\pi e$ for some $y \in \bullet x$) or e is created with a copy of $proc_2$ from some $x \in \bullet y$ in one of the other steps (in that case the statement holds by induction).   q.e.d.

Definition 3.4

An autonomous transition y of an object system process as introduced in definition 3.2 c) has a unique input place x_1 and a unique output place x_2 with $\phi(x_1) = \phi(x_2)$. Therefore identifying x_1 and x_2 to a new place x and eliminating y gives a consistent notion of a process. For the merged place x the contained process $\mu(x)$ is defined by $\mu(x_1)$ if $x_2\bullet \neq \varnothing$ and $\mu(x_2)$ if $x_2\bullet = \varnothing$. The causal net $(\underline{P_\pi},\underline{T_\pi},\underline{E_\pi},\phi,\mu)$ obtained by iterating this construction until all autonomous transitions are eliminated is said to be "*in reduced form*".

The reduced form of a process can be interpreted as process where autonomous transition occurrences are hidden.

It is a general observation in net theory that behavioural effects appear in a similar way in high level nets, e.g. Coloured Petri Nets, as in low level nets, for instance in Elementary Net Systems. It is often easier to study these effects in the low level form to profit from the gained experience for use with high level nets. This was our major motivation to search for a low level equivalent of the high level process notion of object systems. In fact, proving the theorem of this section has much influenced our efforts in finding a consistent formalisation of the object system semantics. Furthermore the theorem provides general insight into the nature of distributed computing.

A natural approach for representing processes of elementary object systems is to construct the two processes of the system and the object net side by side as co-operating processes. This is done in Fig. 3.3. for the EOS "con-tasks" (Fig. 2.1.) and its process (Fig. 3.2.). This figure contains at the top a process of the object net ON from the elementary object system EOS = (SN,ON,ρ). Below a process of the system net SN is drawn. Interacting transitions t and e with tρe are connected. A formal definition follows the induction principle of def. 3.2 and is not given here in full detail.

Definition 3.5

Let EOS = (SN,ON,ρ) be an elementary object system with SN= (P,T,W,M_o) and ON = (B,E,F,m_o). Given a process proc = ($\underline{P}_\pi,\underline{T}_\pi,\underline{E}_\pi,\phi,\mu$) of EOS (def. 3.2) having a latest place x_ω (see appendix for a definition) a *cop-process* (process in co-operating process form) of EOS is defined as a triple

$$\Theta = (proc_1,proc_2,\rho_\pi) \text{ where}$$
$$proc_1 = (P_\pi,T_\pi,F_{1\pi},\phi_1) \in PROC(SN),$$
$$proc_2 = (B_\pi,E_\pi,F_{2\pi},\phi_2) \in PROC(ON) \text{ and}$$
$$\rho_\pi \subseteq T_\pi \times E_\pi.$$

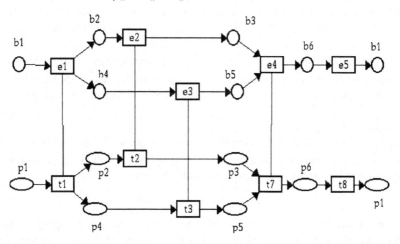

Fig. 3.3. Cooperating process representation of a subprocess from Fig. 3.2.

where $proc_1 = (\underline{P}_\pi,\underline{T}_\pi,\underline{E}_\pi,\phi)$ (i.e proc without μ), $proc_2 = \mu(x_\omega)$ and ρ_π as defined in def. 3.2 w.r.t. $proc_2$ and x_ω i.e. $\rho_\pi := \rho_{x_\omega}$.

For an example consider the process of Fig. 3.2. from the beginning x_1 up to the place x_7 (and delete y_6, x_8, y_7, x_9 and x_{10}). With respect to this shorter process we construct the reduced form (def. 3.4) by merging x_6 and x_7 to a new place x with $\mu(x)$ = $\mu(x_7)$ (since $x_7{}^\bullet = \varnothing$) and $\phi_1(x) = \phi_1(x_6) = \phi_1(x_7) = p_6$. x is a latest place x_ω. Fig. 3.3. gives the result of the construction: $proc_1$ is drawn in the lower part and $proc_2 = \mu(x_7)$ in the upper one. The definition can be extended to cover the case of the whole process of Fig. 3.2. as well. Then the object net processes in the terminal cut (cf. appendix) of proc(w) should have a least upper bound LUB. Note that a cop-process representation of EOS-processes leads to a more consistent notion of concurrency. While Lemma 3.1 is describing concurrency properties int he style of interleaving semantics, cop-processes represent independent actions by unrelated transition. To give an example, in the cop-process form of a EOS-process in Fig. 3.3., the object autonomous transition e_5 and the system autonomous transition t_8 are represented without causal dependence .

Remark: Given a cop-process $\Theta = (proc_1,proc_2,\rho_\pi)$ of an EOS as defined above, then a corresponding process of the EOS can be recovered. For $proc_1 = (P_\pi,T_\pi,F_\pi,\phi)$ and each $p \in P_\pi$ a suitable process $\mu(p)$ of ON has to be defined. This can be done by first constructing the set $T_1 := \{e \in E \mid \exists\, t \in T_\pi : t < p \wedge e\rho_\pi t \}$, where "<" is the causal order of $proc_1$. Then $\mu(p)$ is the subprocess $past_{proc1}(T_1)$ (see appendix).

Lemma 3.6
  a) Given a cop-process $\Theta = (proc_1,proc_2,\rho_\pi)$ of an EOS = (SN,ON,ρ) then
     $\forall\, y_1 \in T_\pi\ \forall\, e_1, e_2 \in E_\pi : y_1\rho_\pi e_1 \wedge y_1\rho_\pi e_2 \Rightarrow e_1 = e_2$ holds.
  b) There is a cop-process $\Theta = (proc_1,proc_2,\rho_\pi)$ of an EOS = (SN,ON,ρ) such that
     $\forall\, y_1,y_2 \in T_\pi\ \forall\, e_1 \in E_\pi : y_1\rho_\pi e_1 \wedge y_2\rho_\pi e_1 \Rightarrow y_1 = y_2$ is *not* true in general.

Proof: In the construction of ρ_π each transition $y \in T_\pi$ appears only once, whereas $e \in E_\pi$ may appear in different copies. In Fig. 3.5. a cop-process of the EOS from Fig. 3.4. is shown together with the relation ρ_π. The cop-process fails to have property b).

                                                                              q.e.d.

Definition 3.7
Let be $T_{int} := \{y \in T_\pi \mid y\rho_\pi \neq \varnothing \}$ and $E_{int} := \{e \in E_\pi \mid \rho_\pi e \neq \varnothing \}$ the set of interactive transitions of $proc_1$ and $proc_2$, respectively. To simplify the following definitions and proofs from now on we exclude object autonomous transitions, i.e. we assume $E_{int}=E_\pi$.

Then (by lemma 3.6 a)) $\varphi : T_{int} \to E_\pi$ with $(\varphi(y) = e \Leftrightarrow y\rho_\pi e)$ is a mapping. φ may be non-injective (by lemma 3.6b) but is surjective, however. Hence, a *cop-process* $\Theta = (proc_1,proc_2,\rho_\pi)$ can be represented by $\Theta = (proc_1,proc_2,\varphi)$. Using this notation lemma 3.3 can be rewritten as follows.

Lemma 3.8

Given a *cop-process* $\Theta = (proc_1, proc_2, \varphi)$, then $e_1 <_{proc2} \varphi(y)$ implies

$$\exists y_1 : y_1 <_{proc1} y \wedge \varphi(y_1) = e_1 .$$

Proof: By definition $e = \varphi(y)$ iff $y\rho_\pi e$ in the corresponding EOS-process "proc". By induction on the construction of proc transitions e_1 and e are in $\mu(p)$ for any $p \in y^\bullet$. By lemma 3.3 there is a transition $y_1 <_{proc1} p$ with $y_1 \rho_\pi e$. By lemma 3.6 (since $y\rho_\pi e$) $y_1 \neq y$, hence $y_1 <_{proc} y <_{proc} p$, and also $y_1 <_{proc1} y <_{proc1} p$.   q.e.d.

This lemma motivates a property, called *extended process morphism* property (EMP), that generalizes the notion of process morphism.

Definition 3.9

Given an elementary object system EOS = (SN, ON, ρ) and processes
$proc_1 = (P_\pi, T_\pi, F_{1\pi}, \phi_1) \in PROC(SN)$, $proc_2 = (B_\pi, E_\pi, F_{2\pi}, \phi_2) \in PROC(ON)$ and a
mapping $\varphi : T_{int} \rightarrow E_\pi$.
φ is called interaction true or true if

    a) φ is surjective,
    b) $\forall y \in T_{int} \; \forall e \in E_\pi : \varphi(y) = e \Leftrightarrow \phi_1(y)\rho\phi_2(e)$
    c) $\forall y_1, y_2 \in T_\pi : y_1 <_{proc1} y_2 \Rightarrow \varphi(y_1) \neq \varphi(y_2)$

The triple $\Theta = (proc_1, proc_2, \varphi)$ has the *extended process morphism property (EMP)* iff:

$$e_1 <^\bullet_{proc2} e_2 \wedge y_2 \in \varphi^{-1}(e_2) \Rightarrow \exists y_1 <_{proc1} y_2 : \varphi(y_1) = e_1$$

$(<^\bullet_{proc2} \subseteq <_{proc2}$ denotes the immediate successor relation of $<_{proc2}$ restricted to transitions.)

By a) the whole object net process is reached by φ. b) relates the interaction relation of the EOS to a corresponding relation on the processes. By c) causally dependent actions are excluded to execute the same task. If φ is an injection, then $\psi : E_\pi \rightarrow T_{int}$ where $\psi := \varphi^{-1}$ is a T-morphism (cf. appendix). There is a convincing interpretation of the extended morphism property. Consider object net transitions as tasks being executed by functional units, given here in the form of system net transitions. Then two sequential tasks e_1 and e_2 with $e_1 < e_2$ cannot be executed by concurrent system net transitions (formally: $e_1 < e_2$ implies $\psi(e_1) < \psi(e_2)$), as for the execution of the second task e_2 the "result" of the execution of e_1 is required. Hence concurrent object net transitions may be executed sequentially but not vice versa.

Theorem 3.10

Let be given an elementary object system EOS = (SN, ON, ρ) and a triple $\Theta = (proc_1, proc_2, \varphi)$, where $proc_1 \in PROC(SN)$, having a latest place x_ω, and $proc_2 \in PROC(ON)$ are processes and $\varphi : T_{int} \rightarrow E_\pi$ is an interaction true mapping. Then Θ is a cop-process of EOS if and only if φ has the extended morphism property.

Proof: The necessity of the condition follows from lemma 3.8.

To prove that the condition is also sufficient, assume that $\Theta = (proc_1, proc_2, \varphi)$, where $proc_1 = (P_\pi, T_\pi, F_\pi, \phi_1) \in PROC(SN)$, $proc_2 = (B_\pi, E_\pi, F_{2\pi}, \phi_2) \in PROC(ON)$ are processes and $\varphi: T_\pi \to E_\pi$ is a true mapping satisfying the EMP.

First we have to find a mapping $\mu: P_\pi \to PROC(ON)$ such that $proc = (P_\pi, T_\pi, F_\pi, \phi_1, \mu)$ is an EOS-process. This is done by defining:

$$\mu(x) := past_{proc2}(\{\varphi(y_1)| \ y_1 <_{proc1} x\}) \cup init(proc_2)$$

(As defined in the appendix, $past_{proc2}(A)$ is the subprocess „generated" by the set A.)

Next it must be shown that proc is an EOS-process, i.e. that μ is consistent with the occurrence of definition 3.2. This is done by induction on P_π w.r.t. the partial order $<_{proc}$ of proc.

a) If for $x \in P_\pi$ the condition $(\exists \ y_1 \in T_{int}: y_1 <_{proc1} x)$ does not hold then $\mu(x) = init(proc_2)$ by definition.

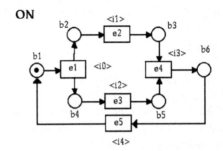

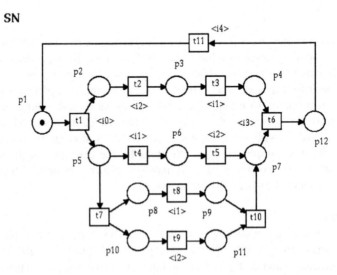

Fig. 3.4. A more complex unary EOS

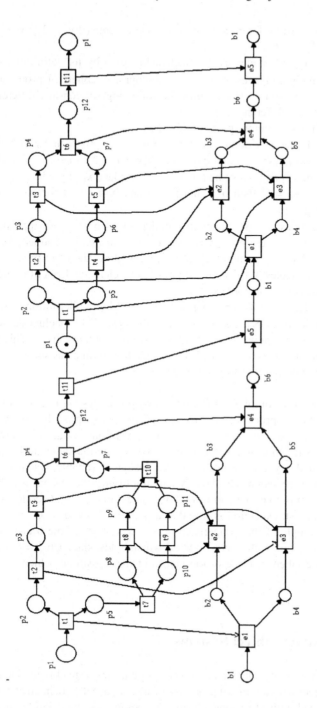

Fig. 3.5. Cop-process of the more complex unary EOS from Fig. 3.4.

b) (induction step) Different to case a) we may assume $(\exists\ y_1 \in T_{int}\colon y_1 <_{proc1} x)$, i.e. $\bullet x \neq \emptyset$, $\{y\} := \bullet x$.

case b_1): $y \in T_{int}$. Then $e: = \varphi(y)$ is included in $\mu(x)$ by its definition and it must be shown that the input places of y contain appropriate subnets of $proc_2$, such that their lub enables transition e. Therefore in the following we consider different subnets of $proc_2$ contained in different $\mu(x)$.

Now let be $b \in \bullet e$ an input place of e in $\mu(x)$.

subcase b_{11}) If there is some input transition $e_1 \in \bullet b$ then by (EMP) there is a transition y_1 with $y_1 <_{proc1} y$ and $\varphi(y_1) = e_1$. By def. 3.9 c) $\varphi(y_1) = e_1$ and $\varphi(y) = e$ are different. Hence e_1 and b also belong to an input place x_1 of y with $y_1 <_{proc1} x_1 <_{proc1} y$.

subcase b_{12}) If $\bullet b = \emptyset$ then $b \in init(proc_2)$ and $b \in \mu(x_1)$ for all $x_1 \in \bullet y$ by the definition of μ.

All the input places x_1 of y contain initial parts of $proc_2$. (see appendix for „initial part"). Hence the process $proc_{lub}$, defined as their lub, exists and as proved before the terminal cut of $proc_{lub}$ contains all input places of e. Therefore $proc_{lub}$ enables e (see appendix for „enables"). By similar arguments all elements from $\mu(x_1)$ are also in $\mu(x)$. This concludes the proof for case b_1).

case b_2): $y \notin T_{int}$. Then $\mu(x) = lub\{\mu(x_1) \mid x_1 \in \bullet y\}$ by the definition of μ. Thus the occurrence rule for EOS is also respected in this case. This concludes case b).

Finally it has to be shown that the latest place x_ω contains $proc_2$. This follows from the definition of $\mu(x_\omega)$ since by def. 3.9 a) each transition e has some $y \in \varphi^{-1}(e)$ and $y <_{proc1} x_\omega$ by the definition of the latest place.   q.e.d.

In Fig. 3.4. a more complex unary elementary object system is given to illustrate the theorem by its cop-process as in Fig. 3.5. The mapping φ is obviously not injective. Moreover there are system autonomous transitions (e.g. t_7). Two concurrent transitions, as t_2 and t_9 with $\varphi(t_2) = \varphi(t_9)$ may execute the same „task" $\varphi(t_2) = \varphi(t_9) = e_3$. This redundancy can be useful in the design of reliable systems. The extended morphism property can be checked. When simplifying the system net SN by deleting the subnet from p_5 to p_7 the corresponding process in Fig. 3.5. becomes sequential and no concurrent task execution is possible any more. Then $\psi := \varphi^{-1}$ is a T-morphism. To see an example of this property the first occurrences of the transitions labelled e_3 and e_5. Then $e_3 < e_5$ implies $\psi(e_3) < \psi(e_5)$ which holds, since $\psi(e_3) = t_2$ and $\psi(e_5) = t_{11}$ (the labels are taken in place of the names of the transitions which are not drawn in the figure).

4 Elementary Object Systems

In this section unary elementary object systems are extended in such a way that different object nets are moving around in a system net and interact with both, the system net and with other object nets. As before, the model is kept as simple as possible in order to have a clear formalism.

Definition   4.1

An *elementary object system* is a tuple EOS = (SN,<u>ON</u>,Rho,type,<u>M</u>) where

- SN= (P,T,W) is a net (i.e. an EN system without initial marking), called *system net* of EOS,
- <u>ON</u> = {ON$_1$,...,ON$_n$} (n≥1) is a finite set of EN systems, called *object nets* of EOS, denoted by ON$_i$ = (B$_i$,E$_i$,F$_i$,m$_{oi}$),
- Rho = (ρ,σ) is the *interaction relation*, consisting of a system/object interaction relation ρ ⊆ T×**E** where **E** := ∪{E$_i$|1≤i≤n} and a symmetric object/object interaction relation σ ⊆ (**E**×**E**)\id$_E$,
- type : W → $2^{\{1,...,n\}}$∪**N** is the *arc type* function, and
- <u>M</u> is a marking as defined in definition 4.2.

Fig. 4.1. gives a graphical representation of an elementary object system with a system net SN and three object nets ON$_i$ (1≤i≤3). The value of type(p$_1$,t$_1$) = {1,2,3} is given by a corresponding arc inscription (1)+(2)+(3). Intuitively, an object net ON$_i$ can be moved along an arc (x,y) if i ∈ Type(x,y). Arcs of type type(x,y) = k ∈ **N** are labelled by k ∈ **N**. They are used as in the case of P/T-nets. xρy holds iff x and y are marked by the same label of the form <i$_1$> (e.g. t$_1$ρe$_{1a}$) and xσy is given by a label of the form [r] (e.g. e$_{2a}$e$_{2b}$). On the right-hand side the relation ρ∪σ is represented as an undirected digraph. Next, a marking will be defined as an assignment of a subset of the object nets together with a current marking to the places. It is also possible to assign a number k of tokens.

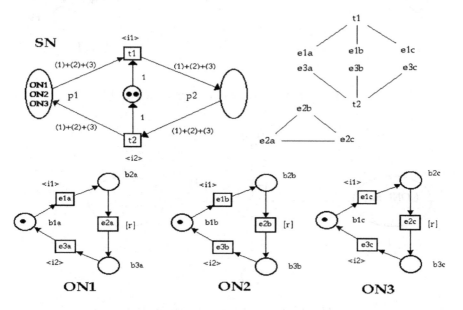

Fig. 4.1. A simple Elementary Object System with 3 object nets

Definition 4.2

The set **Obj** := {$(ON_i, m_i) \mid 1 \leq i \leq n, m_i \in R(ON_i)$} is the set of objects of the EOS. An *object-marking* (O-marking) is a mapping $\underline{M}$: P → $2^{Obj} \cup \mathbb{N}$ such that $\underline{M}(p) \cap$ **Obj** $\neq \varnothing \Rightarrow \underline{M}(p) \cap \mathbb{N} = \varnothing$

The (initial) O-marking of the EOS in Fig. 4.1. is obvious. By restriction to a particular object type from EOS we obtain a unary EOS (i-component, $1 \leq i \leq n$). The 0-component (zero-component) describes the part working like an ordinary P/T-net. This will be used to define simple elementary object systems.

Definition 4.3

Let EOS = (SN,$\underline{ON}$,Rho,type,$\underline{M}$) be an elementary object system as given in def. 4.1 but in some arbitrary marking $\underline{M}$. Rho = (ρ,σ) is said to be *separated*, if $i\sigma j \Rightarrow \rho i = \varnothing = \rho j$. The *i-component* ($1 \leq i \leq n$) of EOS is the EN system SN(i) = $(P,T,W(i),M_{0i})$ defined by $W(i) = \{(x,y) \mid i \in type(x,y)\}$ and $M_{0i}(p) = 1$ iff $(On_i, m_i) \in \underline{M}(p)$. The

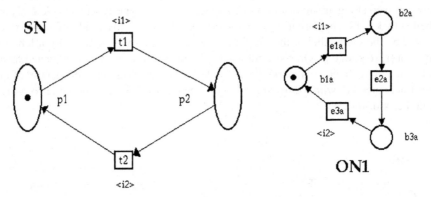

Fig. 4.2. The 1-component EOS(1) of Fig. 4.1.

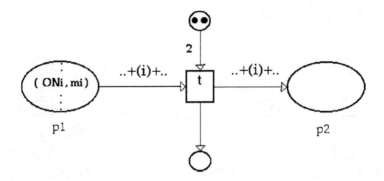

Fig. 4.3. Occurrence rule for simple EOS

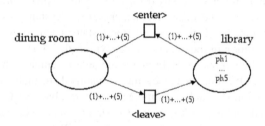

Fig. 4.4. Five Philosophers I: system net SN

0-component (zero-component) is the P/T-net $SN(0) = (P,T,W(0),M_{00})$ with the arc weight function $W(0)(x,y) = k$ if $type(x,y) = k \in \mathbb{N}$ and $M_{00}(p) = k \in \mathbb{N}$ iff $k \in \underline{M}(p)$. The subnet $SN(1..n) = (P,T,W(1..n),M_{1.n})$, where $W(1..n) = \cup\{W(i)|1 \leq i \leq n\}$ and $M_{1.n}(p) = \underline{M}(p) \cap \mathbf{Obj}$ is said to be the *object-component*.

EOS is said to be a *simple elementary object system* if $SN(1..n)$ is a structural state machine, all i-components of SN are state machines and Rho is separated.

Remark: For each $i \in \{1,...,n\}$ the *i-component* $EOS(i) := (SN(i),ON_i,\rho(i))$ is an unary EOS, where $\rho(i) := \rho \cap (T \times E_i)$.

The EOS from Fig. 4.1 is simple since each $SN(i)$ $(1 \leq i \leq 3)$ is a state machine and

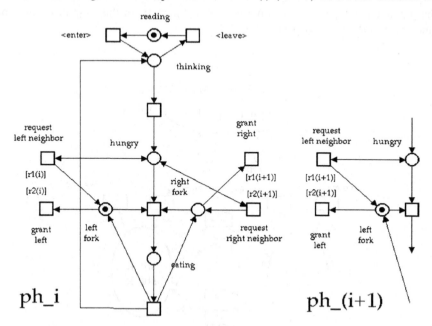

Fig. 4.5. Five Philosophers I: system net SN

Rho is separated. The latter property is easily deduced from the depicted graph of $\rho \cup \sigma$. The 1-component is a simple and unary elementary object system (see Fig. 4.2). Dropping the condition that SN(1..n) is a structural state machine would lead to inconsistencies in the definition of the dynamical behaviour (def. 4.4).

By the introduction of i-components of EOS we are able to connect the models of unary EOS to general EOS. For instance, the semantical formalization of the behaviour of the more complex model of a simple elementary object system can profit from the results obtained earlier in this paper for simple unary elementary object systems. The property of separated interaction relation Rho allows to separate system/object interactions from the new concept of object/object interaction. The latter form of interaction is restricted to the case where the i-components perform autonomous transitions in the same place of the system net. Therefore in the following definition of transition occurrence of simple EOS system/object interactions are defined using case b) of def. 2.2 whereas object/object interactions are associated with case c) of this definition.

Definition 4.4

Let be EOS = (SN,$\underline{ON}$,Rho,type,$\underline{M}$) an elementary object system as in def. 4.1 and $\underline{M}$: P $\rightarrow 2^{Obj} \cup \mathbb{N}$ an O-marking (def. 4.2) and t $\in$ T, $e_i \in E_i$, $e_j \in E_j$, $i \neq j$.

a) Transition t $\in$ T is *enabled* in $\underline{M}$ (denoted $\underline{M} \rightarrow_t$) if $t\rho = \emptyset$ and the following holds:

a_1) t is enabled in the zero-component of SN (def. 4.3) (i.e. in the P/T-net part)

a_2) By the state machine property there is at most one type $i \in \{1,..,n\}$ such that $i \in$ type(p_1,t) and $i \in$ type(t,p_2) for some $p_1 \in$ •t and $p_2 \in$ t•. In this case there must be some object (ON_i,m_i) $\in \underline{M}(p_1)$.(cf. Fig. 4.3.)

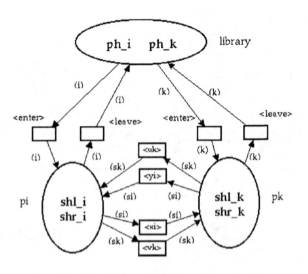

Fig. 4.6. Five philosophers II: system net SN

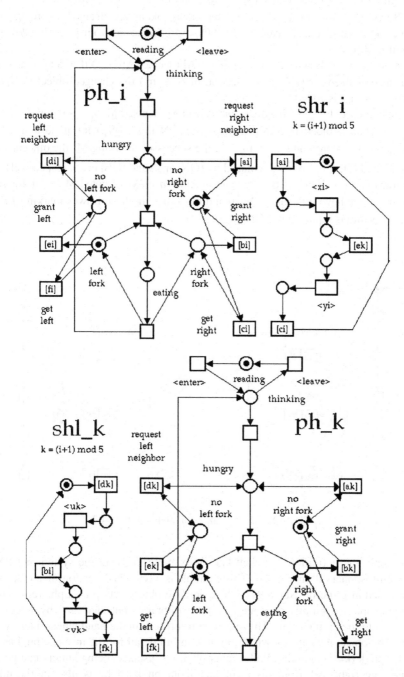

Fig. 4.7. Five Philosophers II: four object nets ph_i, shr_i, ph_k and shl_k (k = (i+1) mod 5)

If t is enabled, then t may occur ($\underline{M} \to_t \underline{M}'$) and the *successor marking* $\underline{M}'$ is defined as follows: with respect to the zero-components tokens are changed according to the P/T-net occurrence rule. In case of a_2) (ON_i, m_i) is removed from p_1 and added to p_2 (only if $p_1 \neq p_2$).

b) A pair $[t,e] \in T \times E_i$ with $t\rho e$ is *enabled* in $\underline{M}$ (denoted $\underline{M} \to_{[t,e]}$) if in addition to case a) e is also enabled for ON_i in m_i. Instead of (ON_i, m_i) the changed object (ON_i, m_{i+1}) where $m_i \to_e m_{i+1}$ is added.

c) A pair $[e_i, e_j] \in E_i \times E_j$ with $e_i \sigma e_j$ is *enabled* in $\underline{M}$ (denoted $\underline{M} \to_{[e_i, e_j]}$) if for some place $p \in P$ two objects $(ON_i, M_i) \in \underline{M}(p)$ *and* $(ON_j, m_j) \in \underline{M}(p)$ are in the *same* place p and $m_i \to_{e_i} m_{i+1}$ and $m_j \to_{e_j} m_{j+1}$. In the successor marking $\underline{M}'$ the objects (ON_i, m_i) and (ON_j, m_j) in p are replaced by (ON_i, m_{i+1}) and (ON_j, m_{j+1}), respectively.

d) A transition $e \in E_i$ with $e\sigma = \sigma e = \varnothing$ is *enabled* in $\underline{M}$ (denoted $\underline{M} \to_e$) if for some place $p \in P$ we have $(ON_i, m_i) \in \underline{M}(p)$ and $m_i \to_e m_{i+1}$. In the follower marking $\underline{M}'$ the object (ON_i, m_i) is replaced by (ON_i, m_{i+1}).

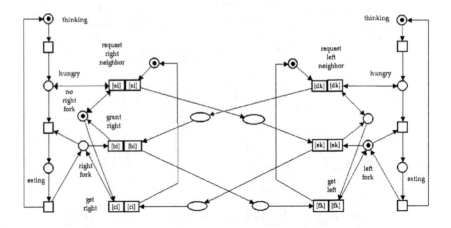

Fig. 4.8. EN system equivalent of Fig. 4.7.

To apply the definition to a well-known example, consider the system net SN of „Five Philosophers I" in Fig. 4.4. This example is a restricted version of a case study as proposed in [Sibertin-Blanc 94]. There are five object nets $ph_1, ..., ph_5$ representing the philosophers. Initially they are in a place „library", but can „go" by interaction <enter> into the dining room. They have their left fork in the hand when entering this room. The object nets are shown in Fig. 4.5. in full detail for the instance ph_i and in part for ph_k (k=(i+1)mod5). Philosopher ph_i, for instance, being hungry can borrow the missing right fork from his right neighbour ph_k (if he is also in the dining room) by interaction of the transitions labelled [r2(k)]. All neighbouring philosophers can exchange the shared fork in the same way, under the condition that they are in the dining room. If both ph_i and ph_k are hungry at the same time, it can happen, that they permanently exchange the fork without ever managing to eat. Such „after you - after you" effects are well known from ordinary Petri nets and can be excluded by similar methods. Solutions for this problem are out of the scope of this paper. It should be observed, however, that side conditions are used, which are not allowed by standard EN systems. By simple constructions they can be eliminated, resulting in a more complex net, however.

In a truly distributed environment the philosophers can only communicate by sending messages. This is assumed for the system of „Five Philosophers II". An extract of the system net is given in Fig. 4.6. Each philosopher ph_i can enter his own place p_i by an arc of type (i). In p_i he finds a „fork shuttle right" shr_i, that can be used to send a request to his right neighbour ph_k by the interaction [ai] (see Fig. 4.7.).

The shuttle then moves to p_k using interaction $<x_i>$ to take the fork of ph_i using interaction [ek], provided philosopher ph_k is now at his place and the fork is free. Then it goes back, delivering the fork to ph_i by [ci]. The type of this object net is (si). In a symmetrical way ph_k uses shuttle shl_k („fork shuttle left") to obtain the fork back. Since the partners for communication are fixed in this example by merging communicating transitions, an ordinary net (see Fig. 4.8.) can be constructed, representing the behaviour of shuttle exchange. This net can be seen as a communica - tion protocol for distributed mutual exclusion, being similar to the method of [15] and [5].

Many different settings of the distributed philosophers problem could be realized, as well. For instance, a fork shuttle could move around and distribute forks to arbitrary participants. Also, different approaches for handling forks on leave of the dining room could be realized (e.g.: a philosopher leaves with „his" left fork, as he came in, or he leaves without forks granting the resource to a neighbour.) Such variants of specifications are out of the scope of this paper.

The semantical description techniques discussed in sections 2 and 3 can be extended to the model of general EOS. In particular, the description of processes by (an extended version of) co-operating processes has been applied to this case and has been proved to be very useful [18]. Further research is necessary, in order to well understand the behaviour of non-simple (general) Elementary Object Systems.

5   Conclusion

An intuitive notion of object system is introduced and then formalized. *Unary object systems* are restricted to contain only a single object net, but allow for „intra-concurrency" to model concurrent task execution. Using net processes a suitable definition of marking was found. Processes of unary object systems have been defined and were represented as *cop-processes*. This representation was characterized by the necessary and sufficient *extended morphism property*.

In the second part *simple elementary object systems* have been considered, where intra-concurrency is excluded, but concurrent behaviour and interaction of different object nets is possible. Such object systems are used to model two instances of Philosophers, showing the usefulness of the approach for a simple and direct way to model in the object paradigm on the level of classical Petri nets.

6 Appendix: Processes of EN Systems

The non-sequential behaviour of EN systems is given by causal nets (occurrence nets (cf [9])). A process of an EN system $N = (B,E,F,C)$ is defined by a node-labelled causal net $N_\pi = (B_\pi, E_\pi, F_\pi, \phi)$ such that $\phi : B_\pi \cup E_\pi \to B \cup E$ satisfies

    a) $\phi(B_\pi) \subseteq B$ and $\phi(E_\pi) \subseteq E$ and $\forall\, t \in E_\pi$: $[\phi(\bullet t) = \bullet\phi(t)$ and $\phi(t\bullet) = \phi(t)\bullet]$

    b) ϕ is injective on every B_π-cut of N_π

    c) $\forall\, b \in B_\pi$: $\bullet b = \varnothing \Leftrightarrow \phi(b) \in C$

The initial process of N_π $\mathrm{init}(N_\pi) = (B_C, E_C, F_C, \phi_C)$ consists just of the initial case C, i.e. $B_C = C$, $E_C = F_C = \varnothing$ and $\phi_C(b) = b$. The set of places $\mathrm{term}(N_\pi) := \{b \in B_\pi | b\bullet = \varnothing\}$ is called a *terminal cut*. (It is assumed that all transitions $e \in E_\pi$ have an output place: $e\bullet \neq \varnothing$). Only finite processes are considered in this paper.

We use PROC(N) to denote all processes of N together with the "empty process" $\varnothing$. By $<_{N\pi} := (F_\pi)^+$ we denote the partial order "before". A place b_ω is called *latest place*, if all other places are before b_ω, i.e.: $\forall\, b \in B_\pi \backslash b_\omega : b <_{N\pi} b_\omega$. Given N_π and a subset $A \subseteq E_\pi$ of process transitions, then $\mathrm{past}_{N\pi}(A)$ is the subnet generated by A. The process $\mathrm{past}_{N\pi}(A) = (B'_\pi, E'_\pi, F'_\pi, \phi')$ is defined by all transitions "before or in A", i.e. $E'_\pi := \{t \in E_\pi | t \in A \vee \exists t_1 \in A: t <_{N\pi} t_1\}$ and all input or output places of E'_π, i.e.: $B'_\pi := \{b \in B_\pi | \exists b \in E'_\pi : b \in \bullet t \cup t\bullet\}$. ϕ' is the restriction of ϕ to $B'_\pi \cup E'_\pi$. Note that $A = \varnothing$ implies $B'_\pi = E'_\pi = \varnothing$.

Given two processes $N_{\pi 1}$ and $N_{\pi 2}$ then a *T-morphism* from $N_{\pi 1}$ to $N_{\pi 2}$ is a mapping $\alpha : E_{\pi 1} \to E_{\pi 2}$ such that $\forall\, x,y \in E_{\pi 1}$: $x <_{N\pi 1} y \Rightarrow \alpha(x) <_{N\pi 2} \alpha(y)$. Every firing sequence $w = e_1 \ldots e_n \in \mathrm{FS}(N)$ uniquely determinates a process $\mathrm{proc}(w) = (B_\pi, E_\pi, F_\pi, \phi)$ such that $\phi(e_i) <_{\mathrm{proc}(w)} \phi(e_j) \Rightarrow i < j$. On the set PROC(N) of all processes there is a partial order $\leq_\pi$ "initial part": $\mathrm{proc}_1 \leq_\pi \mathrm{proc}_2$ if $\exists\, w_1, w_2 \in \mathrm{FS}(N)$: $\mathrm{proc}_1 = \mathrm{proc}(w_1)$ and $\mathrm{proc}_2 = \mathrm{proc}(w_2)$ and $\exists\, v \in E^* : w_1 v = w_2$. If in this definition $v = e \in E$, then we say that e is enabled by proc_1 (denoted $\mathrm{proc}_1 \to_e$) and $\mathrm{proc}_1{}^\circ e := \mathrm{proc}_2$ (i.e. $\mathrm{proc}_1{}^\circ e$ is the prolongation of proc_1 by a transition e_2 with $\phi(e_2) = e$ and the output places of e_2). If for two processes proc_i ($i \in \{1,2\}$) there is a process proc_3 such that $\mathrm{proc}_i \leq_\pi \mathrm{proc}_3$ ($1 \leq i \leq 2$), then there is a least upper bound of proc_1 and proc_2, which we denote $\mathrm{lub}(\mathrm{proc}_1, \mathrm{proc}_2)$. This definition is extended to finite subsets $Q \subseteq \mathrm{PROC}(N)$ and denoted by $\mathrm{lub}(Q)$.

References

1. v. d. Aalst, W.: private communication, (1997)
2. Becker, U., Moldt, D.: Object-Oriented Concepts for Coloured Petri Nets, in Proc. IEEE Int. Conf. on Systems, Man and Cybernetics, (1993) 279-286

3. Brauer, W., Reisig, W., Rozenberg. G. (eds.): Petri Nets: Central Models and their Properties, Lecture Notes in Computer Science No 254, 255, Springer, Berlin (1987)

4. Jessen, E., Valk, R.: Rechensysteme, Springer, Berlin (1987)

5. Kindler, E. Walter, R.: Message Passing Mutex, in J. Desel (Ed.): Structures in Concurrency Theory, Proceedings, Workshops in Computing, Springer, Berlin (1995)

6. Lakos, C.A.: Object Petri Nets, Technical Report TR94-3, Computer Science Department, University of Tasmania (1994)

7. Lakos, C.A.: From Coloured Petri Nets to Object Petri Nets, in G. De Michelis and M. Diaz (Eds): Application and Theory of Petri Nets 1995, LNCS No. 935, Springer, Berlin (1995) 278-297

8. Moldt, D., Wienberg, F.: Multi-Agent-Systems Based on Coloured Petri Nets, in P. Azema, G. Balbo (Eds): Application and Theory of Petri Nets 1997, LNCS Vol. 1248, Springer, Berlin (1997) 82-101

9. Rozenberg, G.: Behaviour of Elementary Net Systems, in [3], part I, pp 60-94

10. Rumbaugh, J. et al.: Object-Oriented Modeling and Design, Prentice-Hall, London (1991)

11. Sibertin-Blanc, C.: Cooperative Nets, in Valette, R (Ed): Application and Theory of Petri Nets 1994, LNCS Vol. 815, Springer, Berlin (1994) 471-490

12. Thiagarajan, P.S.: Elementary Net Systems, in [3], part I, pp 26-59

13. Valk, R.: Nets in Computer Organisation, in [3], part II, pp 218-233.

14. Valk, R.: Modeling of Task Flow in Systems of Functional Units, report FBI-HH-B-124/87, University Hamburg (1987)

15. Valk, R.: On Theory and Practice: an Exercise in Fairness, in: Petri Net Newsletter No. 26, pp. 4-11. Bonn, Germany: Gesellschaft für Informatik (GI), Special Interest Group on Petri Nets and Related System Models, April (1987)

16. Valk, R.: Modeling Concurrency by Task/Flow EN Systems, Proceedings 3rd Workshop on Concurrency and Compositionality, GMD-Studien Nr. 19, Gesellschaft f. Mathematik und Datenverarbeitung, St. Augustin, Bonn (1991)

17. Valk, R.: Petri Nets as Dynamical Objects, Proc. Workshop on Object-Oriented Programming and Models of Concurrency, Torino, June (1995)

18. Valk, R.: On Processes of Object Petri Nets, Report 185/96, Fachbereich Informatik, University Hamburg (1996)

19. Valk, R.: Petri Nets as Token Objects - An Introduction to Elementary Object Nets, in J. Desel and M. Silva (Eds.): Application and Theory of Petri Nets 1998, LNCS No. 1420, Springer-Verlag, Berlin (1998) 1-25

Object Orientation in Hierarchical Predicate Transition Nets

Xudong He

Department of Computer Science
North Dakota State University
Fargo, ND 58105, U.S.A.

Yingjia Ding

Great Plains Software, Inc.
Fargo, ND 58103, U.S.A.

Abstract. In this paper, an approach of using hierarchical predicate transition nets (HPrTNs in the sequel) for object-oriented specification is proposed. The realization of various object-oriented features (including encapsulation, inheritance, and polymorphism) in HPrTNs is presented and is demonstrated with examples. We believe that the approach can achieve the benefits of the object-oriented methods while maintaining the analyzability of HPrTNs.

1 Introduction

Recently there have been many research efforts in integrating object-oriented (OO in the sequel) methods with Petri nets, which can be classified into two main paradigms: (1) use OO or special algebraic specifications to define class relationships and use Petri nets to define the dynamic semantics of objects and object interactions ([2], [3], [4], [20]), and (2) make Petri nets OO by adapting OO concepts and features ([21], [22], [23]).

In [2], Cooperative Objects were introduced, in which high-level Petri nets were used to model the object behavior and communications (called service invocations) and OO approach was used to define class structure and relationships. In [4], CLOWN (Class Orientation With Nets) was presented, where class relationships were defined using structured templates and high-level Petri nets were used to define object behavior and synchronized object interactions. In [3], CO-OPN/2 (Concurrent Object-Oriented Petri Nets) was proposed, in which algebraic specifications were used to define class structure and relationships, and algebraic nets were used to define object behavior and interactions. Although high-level Petri nets were used in all of the above research, they do not have well-defined structural mechanism and thus are not capable to deal with more general class relationships such as containment and inheritance.

In [23], some special structural mechanism was informally defined on high-level Petri nets and thus class containment relationships and class use relationships could be expressed. However no discussion of inheritance relationship and dynamic object creation was given. In [21] and [22], OPNs (Object Petri Nets) were presented, which extend hierarchical colored Petri nets [18] with various OO concepts (including encapsulation, inheritance, polymorphism, and dynamic binding). Classes were defined as subnets so that containment relationships could be easily defined. Furthermore, token types could be classes as well, which means tokens could be subnets and be instantiated. Polymorphism was realized based on token types (classes).

This paper presents an approach of using HPrTNs [14] for OO specification without adding any new features. Specifically, we show how to realize various OO features such as classes (data encapsulation), class instances (objects) and their creation and destruction, class relations, class hierarchy and inheritance (subclasses), and

G. Agha et al. (Eds.): Concurrent OOP and PN, LNCS 2001, pp. 196-215, 2001.
© Springer-Verlag Berlin Heidelberg 2001

polymorphism (operation overriding) in HPrTNs. Many results are complete new or major improvements over those given in [12]. A simple C++-like class schema [27] is used to aid the understanding of a resulting OO HPrTN specification; however the class schema is not a part of OO HPrTN specification since the class schema does not add any additional information to the given OO HPrTN. We believe our approach can help users write OO specifications in HPrTNs and maintain the analyzability of HPrTNs. Furthermore, we expect the resulting OO HPrTN specifications are easier to understand and to reuse than behavioral equivalent non-OO HPrTN specifications.

2 Hierarchical Predicate Transition Nets

The development of HPrTNs was motivated by the need to construct specifications for large systems using Petri nets [25] and inspired by the development of modern high-level programming languages and other hierarchical and graphical notations such as data flow diagram [28] and statechart [11]. With the introduction of hierarchical structures into predicate transition nets, the resulting net specifications are more understandable and the specification construction process becomes more manageable. HPrTNs were used in specifying several systems including an elevator system [15], a library system [16], and a hurried dining philosophers system [12]. HPrTNs can be analyzed directly by using a hybrid technique combining structural, behavioral, and logical reasoning [13], and can be translated into program skeletons in a concurrent and parallel object-oriented programming language CC++ [17] (an introduction to CC++ can be found in [7]). A complete formal definition of HPrTNs was given in [14]. In the following sections, basic concepts and notation of HPrTNs are briefly introduced.

2.1 The Syntax and Static Semantics of HPrTNs

An HPrTN N consists of (1) a finite hierarchical net structure (P, T, F, ρ), (2) an algebraic specification $SPEC$, and (3) a net inscription (φ, L, R, M_0).

(P, T, F) is the essential net structure, where $P \cup T$ is the set of nodes satisfying the condition $P \cap T = \varnothing$. P is called the set of *predicates* and T is called the set of *transitions*. There are two kinds of nodes for both predicates and transitions - *elementary nodes* (represented by solid circles or boxes) and *super nodes* (represented by dotted circles or boxes). Elementary nodes have the traditional meaning in flat Petri net models. Super nodes are introduced to abstract and refine data and processing in HPrTNs [15]. In particular, we identify two subsets $IN \subseteq P \cup T$ and $OUT \subseteq P \cup T$ such that IN contains the heads of all incoming *non-terminating* arcs (an arc inside a super node is a non-terminating arc if one of its end is connected to the boundary of the super node) and OUT contains the tails of all outgoing non-terminating arcs. Nodes in $IN \cup OUT$ are called *interface* nodes. We use •IN to denote the set of the pre-sets of all elements in IN, i.e. •$IN = \{•n \mid n \in IN\}$; and OUT• to denote the set of the post-sets of all elements in OUT. F is the set of arcs and is called the *flow relation* satisfying the conditions: $P \cap F = \varnothing$, $F \cap T = \varnothing$, and $F \subseteq (•IN \times IN \cup P \times T \cup T \times P \cup OUT \times OUT•)$. An arc f can be uniquely identified by a pair of nodes (n1,n2) denoting its source and sink, in which n1 (n2) may denote the pre-set (post-set) of n2 (n1) when f is a non-terminating arc. An arc in an HPrTN may represent a cluster of flows due to the use of super nodes, and individual component flows are defined by the arc label to be discussed below.

$\rho: P \cup T \rightarrow \wp(P \cup T)$ is a hierarchical mapping that defines the hierarchical relationships among the nodes in P and $T;$ and satisfies the constraint that the interface nodes $\in IN \cup OUT$ be all predicates if their parent node is a predicate or all transitions if their parent node is a transition. For any node n, $\rho(n)$ defines the immediate descendant nodes of n. The ancestor and descendants of any node can be easily expressed by using well-known relations such as transitive closure on ρ. A node in an HPrTN is local to its parent, and can be uniquely identified by prefixing its ancesters' names separated with periods to its own name; however often its own name is referred whenever there is no name clash occurs.

The underlying specification $SPEC = (S, OP, Eq)$ consists of a signature $S = (S, OP)$ and a set Eq of S-equations. Signature $S = (S, OP)$ includes a set of sorts S and a family $OP = (OP_{s1,...,sn, s})$ of sorted operations for $s1, ..., sn, s \in S$. For each $s \in S$, we use CON_S to denote $OP_{,s}$ (the 0-ary operation of sort s), i.e. the set of constant symbols of sort s. The S-equations in Eq define the meanings and properties of operations in OP. We often simply use familiar operations and their properties without explicitly listing the relevant equations. $SPEC$ is a meta-language to define the tokens, labels, and constraints of an HPrTN. As a matter of fact, only one general $SPEC$ is needed for HPrTNs.

Tokens of an HPrTN are essentially constant symbols of the family OP. The tokens of sort s are elements in CON_S.

The set of labels, $Label_S(X)$ (X is the set of sorted variables disjoint with OP), can be simple labels (a simple label is a tuple with an identifer and a multi-set flow expression), or compound labels involving label constructor + (non-deterministic flow relation, in which not all component data are consumed or generated in a transition firing) and $\times$ (concurrent flow relation, in which all component data are consumed or generated in a transition firing). Label identifiers are used to connect non-terminating arcs and normal arcs at different levels, and flow expressions define the flow capacity.

Constraints of an HPrTN are a subset of first order logic formulas (where the domains of quantifiers are finite and any free variable in a constraint appears in the label of some connecting arc of the transition), and thus are essentially propositional logic formulas. The subset of first order logical formulas contains the S-terms of sort $bool$ over X, denoted as $Term_{OP,bool}(X)$.

The net inscription (φ, L, R, M_0) associates each graphical symbol of the net structure (P, T, F, φ) with an entity in the underlying $SPEC$, and thus defines the static semantics of an HPrTN. Thus different HPrTNs have different net inscriptions.

Each predicate in an HPrTN is a data structure and a component of the overall system state. The sort of each predicate defines its valid values, i.e. proper tokens. The sorts of elementary predicates are members of S in $SPEC$. The sort of a super predicate is defined as the union of sorts of its interface child predicates. Therefore, we associate each predicate p in P with a subset of sorts in S, and give the following sort assignment: $\varphi : P \rightarrow \wp(S)$.

$L: F \rightarrow Label_S(X)$ is a sort-respecting labeling of N. All simple labels of a compound label must have distinct identifiers, and all simple labels of arcs connected to the same node must have distinct identifiers. Since compound labels define data flows as well as control flows. The following basic control flow patterns [15] must be correctly labeled: (1) data flows into and out of an elementary transition must take place concurrently, and (2) data flows into and out of an elementary predicate can occur at different times. Furthermore, data flows between different levels of hierarchies must be balanced, i.e. a simple label occurs in a non-terminating arc if and only if it also appears in an arc with the same direction connected to the enclosing super node.

$R: T \rightarrow Term_{OP,bool}(X)$ is a well-defined constraining mapping of N, which associates each transition t in T with a first order logic formula defined in the underlying algebraic specification. Furthermore, the constraint of an elementary transition defines the meaning of the transition, and in general contains two parts - the pre-condition part involving only label variables in incoming arcs and the post-condition part specifying the relationships between the variables of the incoming arcs and label variables of the outgoing arcs. The pre-condition specifies the required tokens and the post-condition defines the values of generated token in terms of the selected tokens. Therefore the pre-condition is essentially the guard of the functionality (processing) defined by the post-condition. Therefore the canonical form of the constraint $R(t)$ of an elementary transition t can be written as $\text{Pre}(t) \wedge \text{Post}(t)$. A super transition is an abstraction of low-level actions and its meaning is thus completely defined by the low-level refinement. Therefore the constraint of a super transition is true by default (it is conceivable that a non-trivial constraint for a super transition might be useful; however in general it is very difficult to define such a constraint and also very difficult to interpret the constraint with regard to the operational (dynamic) semantics of the super transition).

$M_0: P \rightarrow MCONS$ is a sort-respecting initial marking of N, which assigns a multi-set of tokens to each predicate p in P. The tokens of a super predicate is a sorted union of the tokens of its interface child predicates since only those tokens are externally accessible.

2.2 Dynamic Semantics of HPrTNs

A marking M of an HPrTN is a mapping $P \rightarrow MCONS$ from the set of predicates to multi-sets of tokens. An elementary transition is enabled if its pre-set contains enough tokens and its constraint is satisfied with an occurrence mode. The firing of an enabled elementary transition consumes the tokens in the pre-set and produces tokens in the post-set. A super transition is enabled if at least one of its interface child transition in IN is enabled and its firing is defined by an execution sequence of its child transitions, and thus its behavior is fully defined by its child transitions. The firing rule of a transition is formally defined in [14]. Two transitions (including the same transition with two different occurrence modes) can fire concurrently if they are not in conflict (the firing of one of them disables the other). Conflicts are resolved non-deterministically. The firing of an elementary transition is atomic, and the firing of a super transition implies the firing of some elementary transition and may not be atomic. We define the behavior of an HPrTN to be the set of all possible maximal execution sequences containing only elementary transitions. Each execution sequence represents consecutively reachable markings from the initial marking, in which a successor marking is obtained through a step (firing of some enabled transitions) from the predecessor marking.

2.3 An HPrTN Example

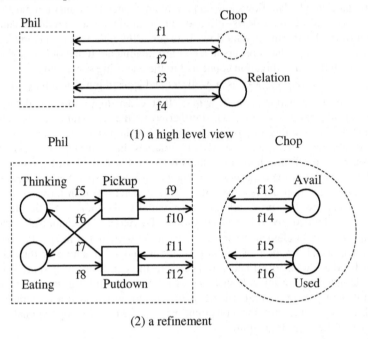

φ(Thinking) = φ(Eating) = ℘(PHIL),   φ(Avail) = φ(Used) = ℘(CHOP),
φ(Chop) = φ(Avail) ∪ φ(Used) ,   φ(Relation) = ℘(PHIL × CHOP × CHOP),
$L(f3) = <1,re>$, $L(f4) = <2,re>$, $L(f5) = <3,ph>$,
$L(f6) = <4,ph>$, $L(f7) = <5,ph>$, $L(f8) = <6,ph>$,
$L(f13) = <7,\{ch1,ch2\}>$, $L(f14) = <8,\{ch1,ch2\}>$,
$L(f15) = <9,\{ch1,ch2\}>$, $L(f16) = <10,\{ch1,ch2\}>$,
$L(f9) = L(f3) \times L(f13)$, $L(f10) = L(f4) \times L(f16)$,
$L(f11) = L(f3) \times L(f15)$, $L(f12) = L(f4) \times L(f14)$,
$L(f1) = L(f13) \times L(f15)$, $L(f2) = L(f14) \times L(f16)$,
$R(\text{Pickup}) = (ph = re[1]) \wedge (ch1 = re[2]) \wedge (ch2 = re[3])$,
$R(\text{Putdown}) = (ph = re[1]) \wedge (ch1 = re[2]) \wedge (ch2 = re[3])$, $R(\text{Phil}) = True$,
$M_0(\text{Thinking}) = \{1, 2, ..., k\}$, $M_0(\text{Eating}) = \{ \}$,
$M_0(\text{Avail}) = \{1, 2, ..., k\}$, $M_0(\text{Used}) = \{ \}$,
$M_0(\text{Chop}) = M_0(\text{Avail}) \cup M_0(\text{Used}) = \{1, 2, ..., k\}$,
$M_0(\text{Relation}) = \{(1,1,2), (2,2,3),...,(k,k,1)\}$

Fig. 1 - An HPrTN Specification of Dining Philosophers Problem

The above HPrTN specifies the well-known dining philosophers problem. The high-level net structure abstraction Fig.1 (1) shows an elementary predicate Relation (defining the association among each philosopher and his left and right chopsticks), and two super nodes, super transition Phil and super predicate Chop, which are connected through two arcs with labels f1 and f2 respectively. The low-level refinement Fig.1(2) shows the internal structure of Phil with two states denoted by predicates Thinking and

Eating respectively and two transitions Pickup and Putdown, and the internal structure of Chop with two predicates Avail and Used denoting the states of chopsticks.

In the underlying specification $SPEC = (S, OP, Eq)$,

(1) S includes elementary sorts such as Integer and Boolean, and also sorts PHIL, CHOP, and ORDER derived from Integer. S also includes structured sorts such as set and tuple obtained from the Cartesian product of the elementary sorts;

(2) OP includes standard arithmetic and relational operations on Integer, logical connectives on Boolean, set operations, and selection operation on tuples (we use A[i] to denote the ith component of tuple A); and

(3) Eq includes known properties of the above operators.

The above specification allows concurrent executions such as multiple non-conflicting (non-neighboring) philosophers picking up chopsticks simultaneously, and some philosophers picking up chopsticks while others putting down chopsticks. The constraints associated with transitions Pickup and Putdown also ensure that a philosopher can only use two designated chopsticks defined by the tokens in predicate Relation.

From this example, it can be seen that a hierarchical specification provides different levels of abstraction, supports information encapsulation, and facilitates specification composition and modification.

3 Object-Oriented Features in Hierarchical Predicate Transition Nets

In the following sections, we show how to realize various OO concepts and features in HPrTNs. To simplify the discussion, only necessary net components and thus incomplete HPrTNs are used to illustrate relevant OO concepts and features. Furthermore arcs are extended over super nodes boundaries to show particular connections, which are instead defined through label expressions in the formal definition of an HPrTN.

3.1 Classes

One of the central ideas of OO paradigm is data encapsulation captured by the class concept. A class is essentially an abstract data type with a name, a set of related data fields (or attributes), and a set of operations on the data fields. It is straightforward to use a predicate to denote a data field (structure) and a transition to represent an operation in Petri nets. The current value of a data field is determined by the tokens of the denoting predicate under the current marking. The meaning or definition of an operation is specified by the constraint associated with the denoting transition.

HPrTNs were originally developed for structured analysis, which provide separate mechanisms for data abstraction and processing abstraction through super predicates and super transitions respectively. Therefore we can use a super predicate and super transition pair in an HPrTN to capture the notion of a class; although it is adequate to define a class by using a super predicate when there is no externally visible operation or using a super transition when there is no externally visible attribute. This view of class is a major improvement over that in [12], where a class was represented by a super predicate only. In this view, the interface of the class is defined by the super predicate and the super transition. The super predicate defines data and internal operations of the class while the super transition mainly defines the externally visible operations of the class. The corresponding subnets further define the internal structures of the data and the operations and the net inscription defines the meanings of net components through predicate types, token values, and transition constraints. When the resulting HPrTN is simple enough, there is no need to separate the super nodes from their subnets, i.e. the subnets are directly embedded inside the super nodes. An attribute or operation is externally visible if the corresponding denoting predicate or transition is an interface node (i.e. connected with a non-terminating arc). It should be noted however that not every super predicate or transition needs to be considered as a class. A super predicate or transition may simply denote a data abstraction or operation abstraction as originally intended; for example a super predicate can be used to hide the internal states of an attribute that is defined by several related predicates, and a super transition can be used to define alternative implementations of an operation to realize operation overloading or overriding. Thus our approach supports the co-existence of various modeling paradigms.

Based on the above analysis, we use the following C++-like class schema to document a class defined by the super node(s) in an HPrTN (it is worth noting that the class schema is only used for understanding purpose, and does not add functionality to the given HPrTN):

class Name [:superclass(es)]
{   public:
          predicates and transitions
     [private:
          predicates and transitions]
}

where brackets [...] denote optional items. Predicates and transitions listed in both public and private sections are those contained in the super node(s). The name(s) of the super node(s) are used to form the class name.

In the HPrTN shown in Fig. 1, both super transition Phil and super predicate Chop can be viewed as classes. Thus the following class definitions can be obtained:

class Chop
{   public:
          Avail, Used
},
class Phil
{   public:
          Pickup, Putdown
     private:
          Thinking, Eating
}.

3.2 Class Instances - Objects

An instance or object of a class has its own copy of data while sharing operations with other objects of the same class. To distinguish an object from other objects, a unique identifier is needed for each object.

In an HPrTN, an object is essentially defined by a set of tokens related through the same identifier, and thus the sort of any predicate p needs to contain a component sort of relevant identifiers, i.e. $\varphi(p) = \wp(ID \times ...)$. Different objects of a class share the same class data structure, i.e. tokens with different identifiers can reside in a predicate at the same time in an HPrTN, however in general objects of the same class cannot interact with each other directly. The above problem is easily solved by defining a subexpression comparing token identifiers in the constraint of each transition such that only proper tokens are used in the transition firing. Movements of tokens and/or changes of token values while maintaining the object identifier indicate state changes of the object.

In the HPrTN shown in Fig. 1, there are k philosopher objects with identifiers of sort PHIL; and there are k chopstick objects with identifiers of sort CHOP.

3.3 Object Creation / Destruction

Objects of a class may be created statically (the resources for the objects are allocated before the execution of the underlying system) or dynamically (the resources for the objects are allocated during the execution of the underlying system.

HPrTNs support both static and dynamic object creation. Static object creation in an HPrTN is achieved through the initial marking M_0. Dynamic object creation in an HPrTN is realized through (constructor) transition firings that produce new objects with unique identifiers. Object constructors are system dependent and often create multiple tokens belonging to the same object at once (i.e. such transitions have multiple output arcs connected to the internal predicates denoting distinct class attributes). Unique object identifiers can be defined through the initial marking M_0 and/or are hidden in an internal predicate within a constructor transition as shown in Fig.2. Often a function with a prefix new_ is used in the constraint of a constructor transition to denote a new unique identifier generated for each firing of the transition without explicitly defining the net structure shown in Fig.2. Similarly, an object destructor can be represented by a transition that consumes the last token belonging to an object. Multiple constructors and / or destructors may exist in a class defined by an HPrTN.

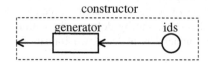

Fig. 2 - An HPrTN for generating unique identifiers

The objects in the HPrTN shown in Fig. 1 are created statically through the initial marking M_0.

3.4 Class Reference Relation

Classes work together to fulfill the functionality of the underlying system. A class can use the operations and / or data provided by other classes.

In HPrTNs, the interface of a transition includes a box with a name and the labels of relevant arcs (the label identifiers determining the calling context and the flow expressions specifying parameters). The meaning of an elementary transition is defined by its constraint and the meaning of a super transition is defined through its corresponding subnet.

It is easy to model a class reference by adding some arc when a class needs to access some public attribute of another class. For example, Fig.3(1) illustrates simple class reference relationships where some operation in class C1 (p1 & t1) uses some public attributes defined in class C2 (p2 & t2) and some operation in class C2 uses some public attributes defined in class C1. Fig.1 contains simple reference relationships between classes Phil and Chop.

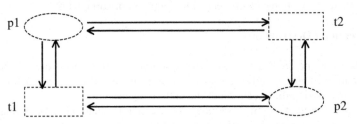

Fig.3(1) - Simple public attributes access

To define an operation in one class using another operation in a different class, we cannot simply add an arc since Petri nets do not allow direct connections between transitions. As discussed in the introduction section, there are two main ways to handle class reference relationships in the existing research works: (1) to fuse the two operations in two classes into one such that only synchronized communication is allowed ([3], [4], [22]), and (2) to create some places inside one class to hold parameters to simulate message passing and function calls ([2], [22]), which supports asynchronous communication. HPrTNs support both synchronous and asynchronous communications through *reference predicates* to model different communication protocols. These reference predicates do not belong to any class, which can be viewed as connectors in software architecture languages [26]. It is quite easy to model a function call through two message passings by using one reference predicate to hold the input values and another to hold output results, and by defining the calling operation (function) as a super transition whose subnet has at least two transitions to handle sending and receiving values.

Fig.3 (2) shows the general pattern of a function call from class C1 containing t1 to class C2 containing t2, in which p1 and p2 are reference predicates. The above pattern defines a one way synchronized communication, i.e. the caller must wait for the callee to continue its execution. While a simple message passing from an operation in one class to an operation in another class in general defines an asynchronous communication.

Fig.3(3) defines a general synchronization pattern such that two operations in class C1 containing t1 and C2 containing t2 must execute mutual exclusively, where p is a reference predicate with an initial dummy token.

It is quite natural and easy to define class reference relationships by using the decomposition and synthesis techniques of HPrTNs discussed in [15].

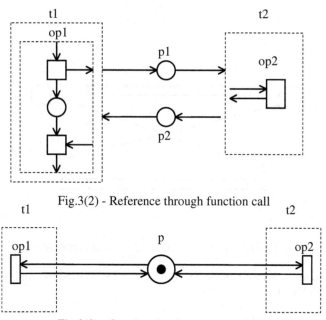

Fig.3(2) - Reference through function call

Fig.3(3) - Synchronized communication

3.5 Class Containment Relation

Many classes are conceptually related through a "whole-part" relationship [8] (or aggregation relationship in [5]), which is modeled through a class containment relationship, i.e. one class being a logical part of another class.

In HPrTNs, a class containment relationship can be easily defined through nested super nodes denoting containing and contained classes. The super nodes representing the contained class (i.e. the interface of the contained class) become data and /or operations of the containing class, however the internal data and operations of the contained class do not belong to the containing class. The containing and the contained classes usually have different internal data and operations, which may be accessed by both.

The general pattern of a containment relationship is shown in Fig.4. The following class definitions are obtained from Fig.4:

```
class p1&t1
{    public:
          p2, p3, t2, t3, ...
     private:
          ...
};
class p3&t3
{    public:
          ...
     private:
```

}.

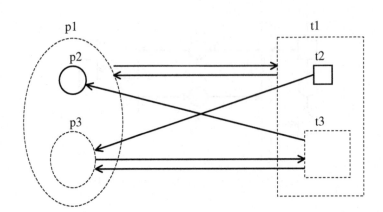

Fig.4 - The general pattern of class containment relation

Class hierarchy structure can be explored using the abstraction and refinement techniques of HPrTNs discussed in [15].

3.6 Class Inheritance Relation

Another major feature of OO paradigm is class inheritance relation that captures the "generalization-specialization" relationships in the real world [8]. A class inheritance relationship exists between a superclass and a subclass such that the subclass inherits data structures as well as operation definitions from the superclass without the need to define them again. Thus class inheritance relation supports a flexible and managed way to reuse existing data structures and operations.

A class inheritance relation is realized in HPrTNs through the reuse of the net structures of inherited super nodes; and the net inscription of inherited elementary nodes (the sorts of predicates, the label expressions of relevant arcs, and the constraints of transitions) defined in an existing HPrTN denoting a class. However the inherited predicates and transitions are explicitly represented or embedded in the subclass to clearly define its role (the same convention was used in [22]). An inherited element in a subclass has a name of the form: super_node.element_name, where super_node is the partial name of the superclass and element_name is the internal name of the element within the superclass. Renaming of relevant arcs are also necessary to reflect the current context and to ensure flow balance. It is clear that inheritance does not reduce the size of an HPrTN specification since inherited elements are embedded (an alternative way to embedding is through delegation [1]), however the advantages are obvious since the meaning or structure (the most difficult part in writing an HPrTN specification) of an inherited element is already available and is obtained without any additional effort; and furthermore many known properties of the inherited element might be maintained through inheritance (structural properties are surely kept, but behavioral properties may need additional validation). It is worth noting that (1) only public components of a superclass can be inherited; (2) inheritance from multiple superclasses are supported and an element can be inherited by multiple subclasses since no ambiguity will occur due to the naming convention; and (3) a re-defined (overriding) operation is considered as a new operation in a subclass and is distinguished from an inherited operation such that an overriding operation in a subclass has the same name as the overriden operation in the superclass (this distinction between inheritance and overriding was also made in [1]). The above realization of class inheritance relation in HPrTNs is complete new and is more general and useful than using nested super nodes in realizing inheritance proposed in [12].

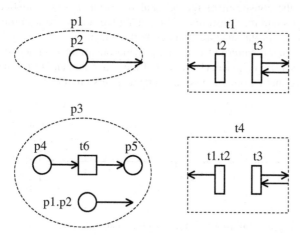

Fig.5 - An example of class inheritance relation

Fig.5 shows a class inheritance relationship defined as follows:
class p1&t1
{   public:
        p2, t2, t3
};
class p3&t3 : p1&t1
{   public:
        t3, p1.p2, t1.t2
     private:
        p4, p5, t6
}.

3.7 Polymorphism

OO paradigm also supports polymorphism such that an operation's name (with possibly different signatures) may have different meanings or behaviors (implementations) through inheritance or overriding.

Polymorphism can be achieved in HPrTNs in two different yet related ways. First, polymorphism is a major feature of the underlying algebraic specification *SPEC* of an HPrTN (detailed discussions of algebraic specifications and polymorphism can be found in [9]). The same operation symbol in *SPEC* is used for many derived sorts. A simple example is the overloaded equality "=" operator when an algebraic specification *SPEC* contains two elementary data types (or classes) INT and CHAR with a single parameterized definition of "=". Second, polymorphism can be accomplished through net structure and inscription. An operation provides overriding capability if its constraint distinguishes a superclass object and a subclass object (or two objects from two different subclasses with the same superclass) and processes them differently. To realize polymorphism in an HPrTN, a shared predicate can be used to hold tokens of a superclass as well as tokens of subclasses and the shared predicate is connected to the transition defined in the superclass and its inherited (or overriding) versions in the subclasses such that the constraint of the original transition is only satisfied by the tokens of the superclass and the constraint of each inherited (or overriding) transition is only satisfied with tokens of the subclass containing the transition.

Fig.6 shows the general pattern of realizing polymorphism through net structure in HPrTNs, in which p is a shared place, t1 is a part of the superclass and t2 is a part of the subclass, and op1* is either an inherited or an overriding version of op1.

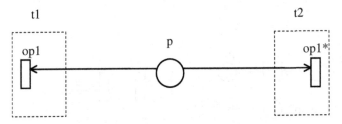

Fig.6 - Polymorphism through choice structure

Furthermore, operation overriding can be achieved through a super transition in an HPrTN such that the firing of a particular component transition is determined partly by the (dynamic) instantiation of an object identifier (and thus its sort). The use of the above case-like net structure in realizing polymorphism can be avoided in the implementation of an HPrTN.

3.8 Concurrency

HPrTNs maintain the distributed and concurrent features of elementary Petri nets. Concurrency takes place not only at the object level but also at the finer object component level (thus multiple threads can exist within an object's behavior). Furthermore, no inheritance anomaly discussed in [24] will occur in our approach for the following main reasons: (1) HPrTNs are a specification method (not a concurrent OO programming language that may cause inheritance anomaly), (2) the dynamic semantics of HPrTNs ensure object access synchronization without using any explicit synchronization mechanisms, and (3) HPrTNs naturally support the main solution proposals to the anomaly problem [24] through transition constraints to achieve method guards and necessary internal (hidden) predicates to realize history-sensitive synchronization.

4 An OO HPrTN Example

In this section, an example of using HPrTNs to realize class inheritance relationship and polymorphism is given.

4.1 The Hurried Philosophers Problem

There are initially m ≥ 2 philosophers sitting around a table:
(1) A philosopher at the table is either thinking or eating,
(2) A philosopher can eat only when he has two chopsticks (left chopstick shared with his left neighbor and right chopstick shared with his right neighbor),
(3) When a philosopher is thinking, he does not hold any chopstick,
(4) A philosopher can leave the table with his right chopstick (his own chopstick), and
(5) A new philosopher can arrive at the table with his own chopstick (right chopstick).

Requirements (1) to (3) constitute the traditional dining philisophers problem, which has many different versions of Petri net specifications, including the HPrTN shown in Fig.1. Requirements (4) and (5) make the system dynamic such that the number of philosophers and the association between neighboring philosophers (and thus the left chopstick of a philosopher) change from time to time during system execution.

4.2 An HPrTN Specification Using Class Inheritance Relation

Two classes are identified from the problem description, a class of hurried philosophers with two distinct states: thinking and eating, and two operations: pickup and putdown; and a class of chopsticks with two distinct states: available and being used. Thus classes Phil and Chop in Fig.1 can be used as superclasses from which two subclasses HPhil and HChop defining the hurried philosophers problem are derived. To fulfill requirements (4) and (5), two new transitions Arrive (denoting the constructor) and Leave (denoting the destructor) are added to HPhil class; and two new transitions Bring (denoting the constructor) and Take (denoting the destructor) are added to HChop class. Since these constructors and destructors are externally accessible to dynamically change philosophers and chopsticks, HChop class now needs to be defined by using a pair of super nodes - super predicate HChop_S for chopstick status and super transition HChop_O for chopstick operations (the constructor and the destructor). Furthermore a new arriving philosopher brings in a new chopstick (his right one) and changes the neighborhood of two existing thinking philosophers. Similarly a leaving philosopher takes away his right chopstick and changes the neighborhood of his left and right philosophers. To define the above changes, a super transition Rush containing two elementary transitions Add and Delete, and two reference predicates Pid and Cid are used. Fig.7 shows an HPrTN specification of the hurried philosophers problem.

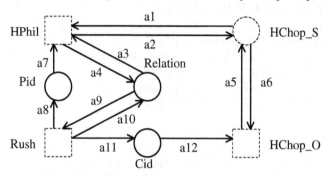

Fig.7(1) - The overall system structure

HPhil

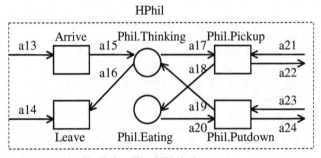

Fig.7(2) - The HPhil class

HChop_S   HChop_O

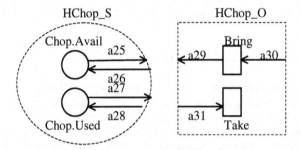

Fig.7(3) - The HChop Class

Rush

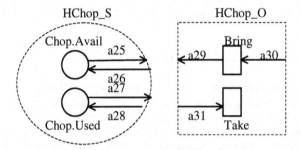

Fig.7(4) - The refinement of Rush

The net inscription (φ, L, R, M_0) is as follows:

(1) Sorts of predicates:

φ(Phil.Thinking) = φ(Phil.Eating) = $\wp$(HPHIL),

φ(Pid) = φ(Cid) = $\wp$(INT × CHAR), φ(Id) = INT,

φ(Chop.Avail) = φ(Chop.Used) = φ(HCHOP),

φ(HChop_S) = φ(Chop.Avail) ∪ φ(Chop.Used) ,

φ(Relation) = $\wp$(HPHIL × HCHOP × HCHOP),

(2) Arc definitions:

L(a1) = <1,{ch1,ch2}> + <2,{ch1,ch2}>,

L(a2) = <3,{ch1,ch2}> + <4,{ch1,ch2}>,

L(a3) = <5,re> + <6,re>, L(a4) = <7,re> + <8,re>,

$L(a5) = <9,ch>$, $L(a6) = <10,ch>$,

$L(a7) = <11,id> + <12,id>$, $L(a8) = <13,pid> + <14,pid>$,

$L(a9) = <15,\{re1,re2\}> + <16, \{re1,re2,re3\}>$,

$L(a10) = <17, \{re1',re2',re3\}> + <18, \{re1',re3'\}>$,

$L(a11) = <19, cid> + <20, cid>$, $L(a12) = <21, id> + <22,id>$,

$L(a13) = <11,id>$, $L(a14) = <12,id>$,

$L(a15) = <23,ph>$, $L(a16) = <24,ph>$,

$L(a17) = <25,ph>$, $L(a18) = <26,ph>$,

$L(a19) = <27,ph>$, $L(a20) = <28,ph>$,

$L(a21) = <1,\{ch1,ch2\}> \times <5,re>$, $L(a22) = <3,\{ch1,ch2\}> \times <7,re>$,

$L(a23) = <2,\{ch1,ch2\}> \times <6,re>$, $L(a24) = <4,\{ch1,ch2\}> \times <8,re>$,

$L(a25) = <1,\{ch1,ch2\}> + <29,ch>$, $L(a26) = <4,\{ch1,ch2\}> + <30,ch>$,

$L(a27) = <2,\{ch1,ch2\}>$, $L(a28) = <3,\{ch1,ch2\}>$,

$L(a29) = <30,ch>$, $L(a30) = <21, id>$,

$L(a31) = <22, id> \times <29,ch>$, $L(a32) = <15,\{re1,re2\}>$,

$L(a33) = <13,pid> \times <19, cid> \times <17, \{re1',re2',re3\}>$,

$L(a34) = <31, x>$, $L(a35) = <32, x'>$,

$L(a36) = <16, \{re1,re2,re3\}>$,

$L(a37) = <14,pid> \times <20, cid> \times <18, \{re1', re3'\}>$,

(3) Constraints of transitions:

$R(Arrive) = (id[2] = \text{'A'}) \wedge (ph = id[1])$,

$R(Leave) = (id[2] = \text{'D'}) \wedge (ph = id[1])$,

$R(Phil.Pickup) = (ph = re[1]) \wedge (ch1 = re[2]) \wedge (ch2 = re[3])$,

$R(Phil.Putdown) = (ph = re[1]) \wedge (ch1 = re[2]) \wedge (ch2 = re[3])$,

$R(HPhil) = True$,

$R(Bring) = (id[2] = \text{'A'}) \wedge (ch = id[1])$,

$R(Take) = (id[2] = \text{'D'}) \wedge (ch = id[1])$,

$R(HChop\_O) = True$,

$R(Add) = (pid[1] = x) \wedge (pid[2] = \text{'A'})$

$\qquad \wedge (cid[1] = x) \wedge (cid[2] = \text{'A'})$

$\qquad \wedge (x' = x + 1)$

$\qquad \wedge (re1[3] = re2[2]) \wedge (re1[1] \geq k+1)$

$\qquad \wedge (re1'[1] = re1[1]) \wedge (re1'[2] = re1[2]) \wedge (re1'[3] = x) \wedge (re2' = re2)$

$\qquad \wedge (re3[1] = x) \wedge (re3[2] = x) \wedge (re3[3] = re2[2])$,

$R(Delete) = (re1[3] = re2[2]) \wedge (re2[3] = re3[2]) \wedge (re1[1] \geq k+1)$

$\qquad \wedge (pid[1] = re2[1]) \wedge (pid[2] = \text{'D'})$

$\qquad \wedge (cid[1] = re2[2]) \wedge (cid[2] = \text{'D'})$

$\qquad \wedge (re1'[1] = re1[1]) \wedge (re1'[2] = re1[2]) \wedge (re1'[3] = re3[2]) \wedge$

$\qquad \wedge (re3' = re3)$,

$R(Rush) = True$,

(4) The initial marking:

$M_0(Phil.Thinking) = \{k+1, k+2, ..., k+m\}$,

$M_0(Phil.Eating) = \{ \}$,

$M_0(Chop.Avail) = \{k+1, k+ 2, ..., k+m\}$,

$M_0(Chop.Used) = \{ \}$,

$M_0(HChop\_S) = M_0(Chop.Avail) \cup M_0(Chop.Used) = \{k+1, k+2, ..., k+m\}$,

$M_0(Relation) = \{(k+1,k+1,k+2), (k+2,k+2,k+3),...,(k+m,k+m,k+1)\}$,
$M_0(Pid) = \{\ \}$, $M_0(Cid) = \{\ \}$,
$M_0(Id) = \{k+m+1\}$.

Remarks:
- HPHIL is the sort of hurried philosopher identifiers and is a subtype of integer INT, and HCHOP is the sort of chopstick identifiers and is also a subtype of INT;
- The label expressions can be understood by following the same label identifier and basic Petri net semantics. For example, the label expression of arc a1 indicates two data flows from HChop_S to HPhil, specifically one from Chop.Avail to Phil.Pickup with label identifier 1, and another from Chop.Used to Phil.Putdown with label identifier 2;
- The constraints of transitions Phil.Pickup and Phil.Putdown are exactly the same as those defined in the superclass shown in Fig.1;
- In the constraint of transition Add, the 1st line defines an identifier for an arriving philosopher, the 2nd line uses the same identifier for the new chopstick, the 3rd line selects two adjacent hurried philosophers (the object idenitifier of each hurried philosopher is greater than k) hurried philosophersuch that the new philosopher sits between them specified by the new neighborhood (the 4th and 5th lines);
- The subexpression (re1[1] $\geq$ k+1) in the constraints of transitions Add and Delete can be omitted by restricting the domain of sort HPHIL to {k+1, k+2, ...};
  Internal predicate Id holds a single token with an initial value k+m+1, which is used to generate unique object identifiers.

The class schemas of HPhil and HChop are as follows:
class HPhil: Phil
{   public:
          Arrive, Leave, Phil.Pickup, Phil.Putdown
      private:
          Phil.Thinking, Phil.Eating
},
class HChop: Chop
{   public:
          Bring, Take, Chop.Avail, Chop.Used
}.

The above HPrTN specification of the hurried philosophers problem has benefitted significantly from inheriting partial net structure as well as relevant net inscription from the HPrTN specification of the traditional dining philosophers problem given in Fig.1. Furthermore, an HPrTN specification of a more general dining philosophers problem combining the traditional dining philosophers problem and the hurried philosophers problem can be easily obtained by merging the predicate Relation in both Fig.1 and Fig.7. Polymorphism is realized through the net structure and inscription since the firing of which Pickup transition (either in Phil or in HPhil) is determined by a specific token relevant to either Phil or HPhil in predicate Relation.

5 Discussion and Conclusion

In this paper, an approach to write OO HPrTNs is presented. The approach is powerful enough to realize most known OO features in HPrTNs without introducing any new notation. The OO HPrTNs obtained from the approach can be analyzed by using existing analysis techniques for HPrTNs, and may also be easier to understand and reuse than equivalent HPrTNs obtained using an ad hoc approach. An additional advantage of OO HPrTNs is that they can be readily implemented in concurrent OO programming languages such as CC++ [17]. We are also adapting the translation rules in [17] to derive Java program skeletons from given HPrTN specifications.

One of the open research problems with regard to our approach is to develop some useful and reusable specification patterns similar to those proposed in object-oriented design community [10]. Some case studies will be carried out.

Compared with popular informal OO methods such as UML [6], OO HPrTNs are relatively more difficult to understand and use due to the following reasons: (1) HPrTNs are a formal method supporting formal semantics definition and analysis, and thus are intrinsically more complicated than informal methods, (2) HPrTNs define all system aspects (functionality, data, structure, and behavior) in one notation while UML offers different notations such as Class diagrams, Interaction diagrams, State diagrams, and Activity diagrams for different system aspects, and (3) HPrTNs are not developed particularly for OO modeling and analysis, and thus lack high-level mechanisms to explicitly and directly represent various class relationships. (1) and (2) are not drawbacks but rather strengths of HPrTNs since they are important for preventing and revealing human errors in early stages of software development. To facilitate the application of HPrTNs, we are doing research to establish some relationships between HPrTNs and UML with the goal to develop some derivation technique from UML specifications to HPrTN specifications so that the strengths of informal methods (object oriented methods) - understandability, and formal methods (Petri nets) - analyzability can be combined. This research topic is based on our previous work of deriving algebraic Petri nets from structured analysis [19]. To overcome the problem in (3), new notations such as different types of arcs are needed; however we must ensure that the newly introduced notations are well-defined and are representable by using some patterns of conventional Petri net symbols so that the resulting specifications are still analyzable.

Acknowledgements

We thank one anonymous referee for careful reading of the first draft of this paper and for making helpful suggestions to improve the quality of the paper. The research of Xudong He was supported in part by the National Science Foundation (NSF) of the USA under grant CCR-9308003 and grant OSR-9452892, and the Office of Naval Research of the USA under grant N00014-98-1-0591.

References

[1] M. Abadi and L. Cardelli: *A Theory of Objects*, Springer-Verlag, 1996.

[2] R. Bastide: "Approaches in Unifying Petri Nets and the Object-Oriented Approach", *Proc. of the 1st Workshop on Object-Oriented Programming and Models of Concurrency*, Torino, Italy, 1995.

[3] O. Biberstein and D. Buchs: "Structured Algebraic Nets with Object-Orientation", *Proc. of the 1st Workshop on Object-Oriented Programming and Models of Concurrency*, Torino, Italy, 1995.

[4] E. Battiston, A. Chizzoni, and F.D. Cindio: "Inheritance and Concurrency in CLOWN", *Proc. of the 1st Workshop on Object-Oriented Programming and Models of Concurrency*, Torino, Italy, 1995.

[5] G. Booch, *Object-oriented analysis and design with applications*, Benjamin/Cummings, 1994 (2nd ed.).

[6] G. Booch, J. Rumbaugh, and I. Jacobson: *Unified Modeling Language User Guide*, Addison-Wesley, 1997.

[7] K. Chandy and C. Kesselman: "CC++: A Declarative Concurrent Object-Oriented Programming Notation", in *Research Directions in Concurrent Object-Oriented Programming* (eds. G. Agha, P.Wegner, and A. Yonezawa), MIT Press, 1993, 281-313.

[8] P. Coad and E. Yourdon: *Object-Oriented Analysis*, Yourdon Press, 1991 (2nd ed.)

[9] H. Ehrig and B. Mahr, *Fundamentals of Algebraic Speification 1 - Equations and Initial Semantics*, Pringer-Verlag, 1985.

[10] E. Gamma, R. Helm, R. Johnson, and J. Vlissides, *Design Patterns - Elements of Reusable Object-Oriented Software*, Addison-Wesley, 1995.

[11] D. Harel: "On visual formalisms", *Communications of the ACM*, vol.31, 1988, 514-530.

[12] X. He and Y. Ding: "Object-Oriented Specification Using Hierarchical Predicate Transition Nets", *Proc. of the 2nd Int'l Workshop on Object-Oriented programming and Models of Concurrency*, Osaka, Japan, 1996, 72-79.

[13] X. He: "A Method for Analyzing Properties of Hierarchical Predicate Transition Nets", *Proc. of the 19th Int'l Computer Softw. and Applications Conf.*, Dallas, August, 1995, 50-55.

[14] X. He: " A formal definition of hierarchical predicate transition nets", *Proc. of the 17th International Conference on the Application and Theory of Petri Nets (Lecture Notes in Computer Science, vol. 1091*, June, Osaka, Japan, 1996, 212-229.

[15] X. He and J.A.N. Lee: "A methodology for constructing predicate transition net specifications", *Software - Practice and Experience*, vol.21, no.8, 1991, 845-875.

[16] X. He and C.H. Yang: "Structured analysis using hierarchical predicate transition nets", *Proceedings of 16th Int'l Computer Software and Applications Conf.*, Chicago, 1992, 212-217.

[17] X. He and W. Yao: "Translating Hierarchical Predicate Transition Nets to CC++ Program Skeletons", *Proceedings of 21st Int'l Computer Software and Applications Conf.*, Washington, D.C., 1997, 64-69.

[18] K. Jensen: *Coloured Petri Nets: Basic Concepts, Analysis Methods and Practical Use - Volume 1: Basic Concepts*, EATCS Monographs in Computer Science, vol.26, Springer-Verlag, 1992.

[19] C. Kan and X. He, "A method for constructing algebraic Petri nets", *Journal of Systems and Software*, vol. 35, 1996, 15-27.

[20] G. Kappel and M. Schrefl: "Using an object-oriented diagram technique for the design of information systems", *Dynamic Modeling of Information Systems*, Elsevier Science Publishers, 1991, 121-164.

[21] C. Lakos: "From colored Petri nets to object Petri nets", *Proceedings of the 16th International Conference on the Application and Theory of Petri Nets*, Torino, Italy, 1995.

[22] C. Lakos: "The Object Orientation of Object Petri Nets", *Proc. of the 1st Workshop on Object-Oriented Programming and Models of Concurrency*, Torino, Italy, 1995.

[23] Y.K. Lee and S.J. Park: "OPNets: An object-oriented high-level Petri net model for real-time system modeling", *Journal of Systems and Software*, vol.20, 1993, 69-86.

[24] S. Matsuoka and A. Yonezawa: "Analysis of Inheritance Anomaly in Object-Oriented Concurrent Programming Languages", in *Research Directions in Concurrent Object-Oriented Programming* (eds. G. Agha, P.Wegner, and A. Yonezawa), MIT Press, 1993, 107-150.

[25] W. Reisig: "Petri nets in software engineering". *Lecture Notes in Computer Science*, vol.255, Springer-Verlag, 1987, 63-96.

[26] M. Shaw and D. Garlan: *Software Architecture*, Prentice-Hall, 1996.

[27] B. Stroustrup: *The C++ Programming Language*, 2nd edition, Addison-Wesley, 1991.

[28] E. Yourdon, *Modern Structured Analysis*, Prentice Hall, 1989.

CoOperative Objects:
Principles, Use and Implementation

C. Sibertin-Blanc

Université Toulouse 1/ IRIT
Place A. France, F-31042 Toulouse Cedex
sibertin@univ-tlse1.fr

Abstract. It is no longer relevant to praise the qualities of the Object-Oriented Approach and the Petri Net Theory. Each of them has proved to be a powerful framework in its field of application. However it is a challenge to associate them into a conceptual framework which combines the expressive power of both approaches while maintaining their respective merits. Moreover, it must be shown that such a formalism can be implemented in a sound and efficient manner.

This paper is a comprehensive presentation of the CoOperative Objects formalism. This formalism embraces the theoretical and pragmatic features of both the Petri net and the Object-Oriented approaches by thoroughly integrating their concepts. It is as well-adapted to the specification and the validation of open distributed systems as to their implementation. The basic idea is that a Petri net processes data objects as tokens, while the behaviour of an active object is defined by a Petri net. This paper also proposes a CoOperative Object solution to the dynamic dining philosophers problem, and tackles implementation issues through the presentation of SYROCO, a CoOperative Objects compiler.

1. Introduction

Petri nets (**PNs**) are one of the formal models of concurrency. They are successfully used to deal with concurrent discrete event systems and in particular with distributed systems such as operating or manufacturing systems, business or software development processes. They are applied to existing or planned systems for performing various tasks: requirements analysis, specification, design, test, simulation and formal analysis of the behaviour [ISO 97]. Projects concerning such systems most often lead to the development of software that either serves as a tool supporting the system's activities, or controls its behaviour, or constitutes its final implementation. In such casees, Petri nets are used only during the early steps of the project and not during the software implementation steps, because they are not a programming language. This restriction causes a change in the conceptual framework used to consider the system; this rupture is error prone, entailing additional works and making the project's traceabibity difficult.

G Agha et al. (Eds.): Concurrent OOP and PN, LNCS 2001, pp 216-246, 2001.

The original aim of CoOperative Objects (**COOs**) is to provide a PN-based formalism bridging the gap between the early steps of software development processes and the final steps (detailed design, programming, test). Such a formalism should provide all the people involved in a project with a single conceptual framework supporting the tasks contributing to the development of the software. The main requirements for such a formalism are as follows.

- PNs fail to account for the data processing dimension of systems. Indeed, most of the operations which cause a state change of a system also process some data, and thus it is necessary to consider tokens of a PN as data structures. Consequently, PNs have to be associated with a language permitting to define data structures and to describe how they are processed. Languages based on the Object-Oriented (**OO**) approach seem to be appropriate since passive objects and tokens have many properties in common.
- Modularity is an essential principle of System Engineering, and PNs fail to structure the model of a system as a collection of interacting components. As a consequence, it is necessary to introduce concepts which on the one hand provide each PN with an interface and on the other hand define how nets interact through their respective interfaces. Once again, the OO approach offers concepts which have proved to be efficient and a PN may be viewed as an active object.
- The main advantages of PNs are their cognitive simplicity (even though it is difficult to think about concurrent systems), their wide range of use (thanks to their abstract nature), and their suitability for behavioural formal analysis. An increase in the expressive power of PNs must no be offset by a decrease in their valuable features. Namely, even if PNs are supplemented with mechanisms that are beyond the scope of the behavioural analysis techniques, it should still be possible to keep these additions apart if we want to continue applying the analysis techniques.

According to these requirements, a PN is transformed into a CoOperative Object (**O**bject for short) by the following improvements.

1. The definition of a PN comes with the definition of object classes, using a sequential OO Programming Language, and tokens are instances of these classes. To process tokens, each transition is provided with a piece of code: when a transition occurs, this code is applied to the tokens involved. In addition, an Object may be provided with a data structure accessible by all the transitions of the PN.
2. Objects communicate through an asynchronous client/server, or request/reply protocol supported by token sending: when an Object C requests a service from an Object S, (1) C sends an argument-token into the appropriate place of S, (2) S processes this token as soon as the service is available and produces a result-token. The communication ends when (3) C retrieves this token. In addition, an Object may synchronously access public elements of the data structure of other Objects.

The introduction of these two basic tricks must result in a formalism facilitating the respect of the main principles of Software Engineering: rigor and formality, separation of concern, modularity, abstraction, anticipation of change, generality and incrementality [Ghezzi...91]. This requires thoroughly combining the fundamental concepts of the PN and OO approaches.

It turns out that the COO formalism solves another problem, which could also have been a guideline leading to this formalism.

Thanks to encapsulation, the OO approach seems to be well adapted for dealing with distributed systems. However, the state of the art proves that dealing with concurrency is difficult within the OO framework [Agha...93]. Indeed, many OO languages allow for inter-object concurrency, but very few allow for intra-object concurrency. In the lack of this latter feature, an object is a passive component since its activity consists of executing its methods upon request. Moreover, communications among objects are synchronous, since an object blocks when it waits for a reply, and this causes the behaviours of the objects of a system to be tightly coupled. On the other hand, intra-object concurrency turns objects into proactive and autonomous components able to concurrently process several tasks and to communicate asynchronously. This feature significantly improves the concurrency within the whole system, and objects gain a stronger cohesion because they are relieved of useless synchronisation constraints. Another problem raised by concurrency within the OO framework is a general and formal definition of inheritance, one of the key OO concepts [Wegner 88].

The essential requirement when enhancing OO languages in this way is to base concurrency within and among objects on a powerful and formal model of concurrency(notice that Occam, which is based on Hoare's CSP paradigm [Hoare 78], meets this requirement [May 87]). Now, an object is transformed into a CoOperative Object by the following additions.

1. The definition of an (active) object class comes with the definition of a PN which determines the behaviour of any instance of this class, so that the activity of an object consists of executing this control structure net. Namely, this net defines the availability of the services offered by the object.
2. The PNs of objects communicate through an asynchronous client/server protocol supported by token sending.

Thus, the COO formalism integrates the PN and OO paradigms into a single conceptual framework. From a PN point of view, it is a High-Level Petri Net formalism [Genrich...81, Jensen 85] which can handle the data processing dimension of systems and allow to structure models according to the OO principles. From an OO point of view, the COO formalism is a formal and fully concurrent language where Objects are proactive and able to concurrently process several tasks. We believe that combining the PN and OO approaches must produce a formalism which extends each of the two approaches, in order to retain the respective benefits of both. In order to achieve this, PNs have to be introduced into objects and conversely objects have to be introduced into PNs. The COO formalism works in this way: PNs are integrated into objects to make them active, and objects are integrated into PNs to provide them with data processing capabilities. The purpose of this paper is to give a comprehensive introduction to CoOperative Objects, thus it does not provide theoretical justifications of the design decisions. We simply stress the fact that any integration of the PN and OO approaches has to reconcile tightness with looseness. In order to retain the expressive power of both approaches, a tight integration accounting for the mutual dependence between the data structure and the control structure of Objects is necessary. On the other hand, reaping the pragmatic and theoretical benefits of both approaches requires a distinct integration which does not confuse their respective mechanisms.

The second chapter of this paper surveys the expressive power of the COO formalism. The third chapter details the definition of an Object and of a COO system, and addresses the COO's model of concurrency along with the dynamic creation and deletion of Objects. The static case of the dining philosophers problem is used as an example. The fourth chapter proposes a solution for the dynamic case of the dining philosophers problem which turns out to be a quite complex one. The fifth chapter presents the inheritance relationships among COO classes and the conditions for a COO class to be a subtype of another class; thanks to this subtype relationship, the COO formalism allows for polymorphism, which means that an instance of a subtype may be safely substituted when an instance of a type is expected. The sixth chapter indicates how various semantics of the COO formalism are defined and how the analysis techniques founded on the PN theory may be applied.

The last chapter presents an environment for the development of COO systems. SYROCO (an acronym for SYstÈme RÈparti d'Objets CoOpÈratifs) is mainly a COO compiler: it translates each COO class into a C++ class so that an instance of a COO class is implemented as an instance of the corresponding C++ class. Each Object is provided with an interpreter, a Token Game Player which executes the Petri net defining its behaviour; thus an Object is actually implemented as an autonomous process. Thanks to these features, SYROCO is efficient in space and time. The COO formalism may be viewed either as a tool for the specification, the design and the analysis of complex systems, or as a Concurrent Object-Oriented Programming Language. Accordingly, SYROCO is intended to be used either as a simulation tool or as a programming environment, and it provides users with a number of facilities for both purposes. Some of these facilities are pragmatic and intended to ease the development of complex COO systems. Other facilities allow a fine control of the behaviour of each Object.

2. COOs: An Overview

This chapter gives a general presentation of the structure and the behaviour of a COO system. The COO formalism views a system as a collection of active Objects, each Object being an instance of its COO class. While a COO system is running, the set of its member Objects may vary since Objects may be dynamically created and deleted. The behaviour of a COO system results from the concurrent behaviour of its Objects, each one processing its own activity. This activity involves communicating with other Objects according to a request/reply protocol.

The structure of an Object has two parts - a *Data Structure* and a *Control Structure* (Cf. Figure 1).

The Data Structure of an Object complies with the usual concept of an object. It includes a set of *attributes* and a set of functions, referred to as *operations*. The *public* elements of this Data Structure may be accessed in a *synchronous* way by other Objects, namely by the body of their operations.

The Control Structure of an Object makes it active. It includes the declaration of a set of *services* and a High-Level Petri Net referred to as its *OBCS* (for OBject Control Structure). This net defines the Object's behaviour. Its places serve as *state variables* of the Object; thus, the value of a place is a set of data objects. The transitions of the OBCS correspond to *actions* that the Object is able to perform; thus the current state of an Object determines the enabling of its actions, and the occurrence of an action produces a change of state. In order to process the data objects staying in places, each transition may include calls for operations or more generally a piece of code which has access to the Data Structure of the Object and to the public elements of the Data Structure of other Objects (Cf. transition t5). As for services, they allow *asynchronous request/reply* communications among Objects. Each service is supported by transitions, and is available only when (at least) one of these transitions is enabled (Cf. transition t1 or t2); if this is not the case, a request for a service is delayed until one of its associated transitions becomes available. In order to request a service from another Object, the net includes a pair of transitions: the first of these transitions issues the request, while the second one becomes enabled upon reception of the result of the request (Cf. transitions t3 and t4).

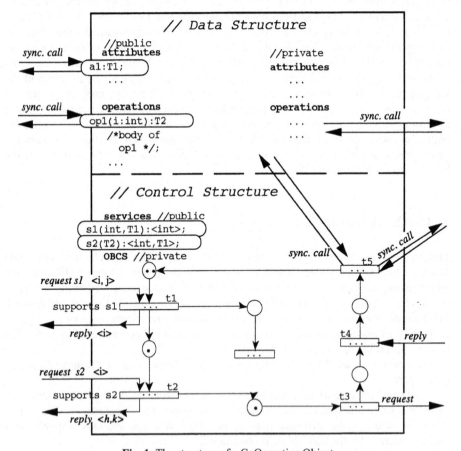

Fig. 1. The structure of a CoOperative Object

Thus, the activity of an Object consists in executing its OBCS as a background task, while processing the calls for its public operations upon request. The behaviour of a COO system results from the concurrent activity of its member Objects. The main distinctive features of the COO formalism are probably the *autonomy* of Objects (from an OO view) and the *dynamism* of COO systems (from a PN view). Autonomy requires (and is achieved by) two related mechanisms. The first one is a high-level communication protocol allowing each Object to distinguish its internal activity from its interactions with other Objects, so that its contribution to the activity of the whole system is clearly determined; the second one is the capability of each Object to have several tasks in hand at one time and thus to organise its activity according to its constraints and goals. Concerning dynamism, the possibility for COO systems to introduce new Objects or to remove member Objects without centralised control means they can cope with systems having a variable structure, and this feature significantly extends the class of systems with may be considered using PNs. Thanks to these features, the COO formalism is appropriate for the class of Open Distributed Systems [Gasser 91].

Due to lack of space, we cannot discuss to what extent the COO formalism fulfils the principles of Software Engineering mentioned above, although such a discussion is essential to validating this formalism. Therefore, we leave it to the reader, and just give a hint ! A formalism is mainly a cognitive tool allowing engineers to understand the system which they are faced by developing a model, a description, or a design of that system. For this purpose, a formalism has to provide high-level abstractions which are as close as possible to the concepts used to discuss the structure and the behaviour of systems, so that the model's architecture mirrors the organisation of the system as it is viewed by the designer.

In order to meet this requirement, the COO formalism relies upon the following meta-model: a system is embedded in an *environment* with which it interacts through an *interface*, and it may be viewed as being located in a four-dimensional space. These dimensions concern the *Entities* processed by the system, the *Operations* whose executions constitute the processing, the *Actors* which manage the entities and perform the operations, and the *Control Structure* which defines the system behaviour. Together, the current Entity instances, Operation occurrences and Actor instances of a system define the *state* of the system. These three dimensions constitute the static part of the system: as long as it is not provided with a Control Structure, a system carries out no work and its state never changes. Indeed, a change in the system's state only results from the occurrence of an *action*, that is the actual execution by an Actor of an Operation involving some Entities, and the role of the Control Structure is just to determine under which state(s) an action may occur. Once a system has been analysed according to this meta-model, it becomes easy to draw a COO model of this system through using the following correspondences: Entity/token, Operation/code associated to a transition, Actor/CoOperative Object, Control Structure/OBCS.

3. The CoOperative Objects Formalism

We shall now introduce the CoOperative Objects formalism using the static case of the dining philosophers as an example: the philosophers are steadily seated around the table. First, a centralised solution by means of a single Object of the class `PhiloTable` will be presented, allowing the structure and the semantics of an isolated Object to be introduced. Then, a decentralised solution modelling each philosopher as an Object of the class `SPhilo` will be presented, allowing the structure of a COO system and the interactions among Objects to be introduced.

3.1 A CoOperative Object in Isolation

The definition of COO classes is based on a sequential OO language, referred to as the *data language*, and SYROCO uses the language C++ for this purpose. This data language is used to define the types of tokens, attributes and parameters of COO classes, and also to define external functions and constants. These definitions are located in appropriate files and are shared by the classes of a system. In the case of the `PhiloTable` class, the C++ declarations given in Figure 2 are assumed. The data language is also used to code the operations of Objects as well as the pieces of code associated to transitions and places of the control structure net.

```
class Any;

class philo: Any {        //philo inherits Any
public:               //items accessed by the net
philo* rn;             //the right and left neighbours
philo* ln;
                 //counts how many times the philo eat
short nbeating;
void eat() {
  nbeating = nbeating + 1;
  };
void init(philo* l, philo* r){
  ln = l; rn = r;
  nbeating = 0;
  }
};

typedef int fork;

const nbphil = 4;
```

Fig. 2. External declarations used by the `PhiloTable` COO class

The `PhiloTable` class, shown in Figure 3, is a centralised solution to the dining philosophers problem. An Object of this class considers philosophers as data entities and it controls the behaviour of them all; when a philosopher is in a given state, the place corresponding to this state contains a token referring to this philosopher.

```
Class PhiloTable specification;
operations
 Init() ///            //set the neighbourgs of each philo
thephilos[1].init (thephilos[nbphil], thephilos[2]);
   for (int i=2; i<=nbphil-1; i++)
     thephilos[i].init (thephilos[i-1], thephilos[i+1]);
   thephilos[nbphil].init(thephilos[nbphil-1], thephilos[1]);
                    //set the initial marking of the OBCS
   for (i=1; i<=nbphil; i++) {
   RFork.ADDTOKEN (RFork.MakeToken(&thephilos[i], i));
   NoLFork.ADDTOKEN (NoLFork.MakeToken(&thephilos[i]));}
 ///
end.

Class PhiloTable implementation;
attributes
 thephilos: array[nbphil] of philo;
OBCS
```

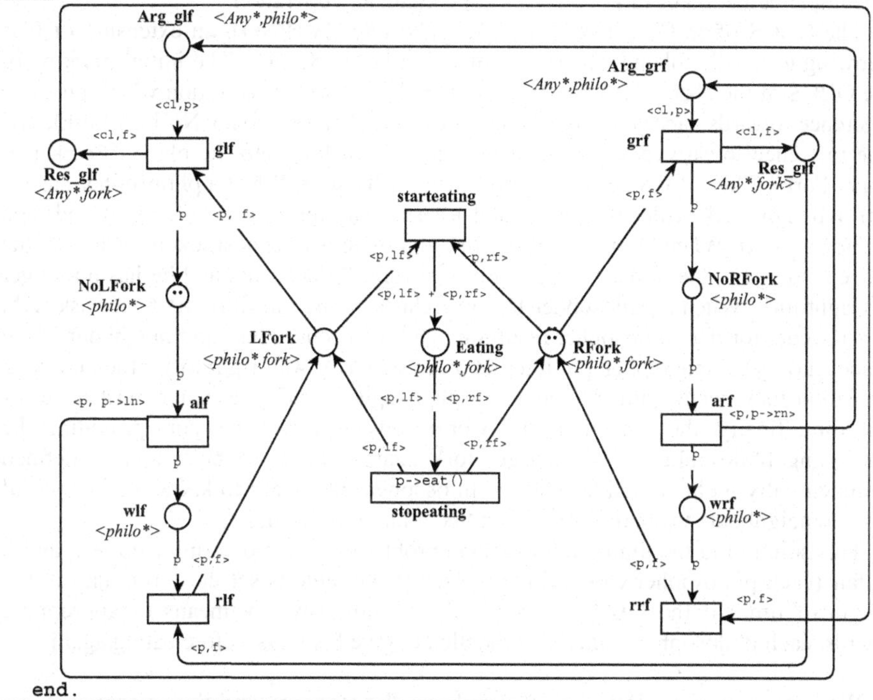

Fig. 3. Definition of the PhiloTable COO class

The definition of a COO class consists of a specification part and an implementation part. The *specification* contains what must be known about the class in order to use it properly. It includes the definition of the items of its interface - attributes, synchronous operations and asynchronous services- and it may be completed with an OBCS. This specification OBCS is only intended to document the observable behaviour of class instances. As for the *implementation*, it includes the definition of private items and of the actual OBCS of the class instances.

The PhiloTable class has a very simple specification, since it is intended to run in isolation and to have no communication with other Objects. It only features one special operation, named Init, which allows to initialise the attributes and the OBCS's marking. In the general case, operations are functions accepting arguments and returning a value computed from these arguments and the Object's attributes. Syntactically speaking, the signature of an operation follows a Pascal-like syntax, while its body (written in the data language) is enclosed between occurrences of the character string '/ / /'

The implementation of the PhiloTable class comprises one attribute, which is an array of philo's. In the general case, the type of an attribute is any type of the data language, namely a scalar type (e. g. Integer, fork or any enumerated domain), an object class (e. g. philo), or a *reference* towards an object class (e. g. philo*). The implementation of this class does not contain any operation.

The OBCS of an Object is a *Petri Net with Objects* (**PNO**), an extension of PNs allowing to handle tokens which are objects [Sibertin 85, 92]. The initial marking of the OBCS of the class PhiloTable is given by the Init operation which puts one reference towards each philosopher into the place NoLFork (for No Left Fork), and one reference toward each philosopher along with a fork into the place RFork (for Right Fork). This OBCS defines the following behaviour. When a philosopher has his left and right forks (places LFork and RFork), he may starteating and then stopeating. When he has his right fork and there is a request token for him in the place Arg_grf, the transition grf (for give right fork) occurs and he has no longer his right fork. When a philosopher has no right fork, the transition arf (for ask right fork) occurs for that philosopher resulting in a token sent to his right neighbour in the place Arg_glf and a token put into the place wrf (for wait right fork); then the glf transition may occur with the requested philosopher, since this latter must have his left fork; finally, the transition rrf (for receive right fork) occurs providing the requesting philosopher with his right fork again. The same behaviour is defined symmetrically on his left side. Thus a philosopher only needs to know the identity of his two neighbours to share the forks on his left and right sides.

This solution to the dining philosophers problem is a non-deterministic one, and it is fair (each philosopher eats infinitely often if the table is set up during an infinite length of time) if the OBCS is executed in a fair way. By means of appropriate guards, each philosopher could be compelled to give his forks before eating again.

We shall now provide technical details on the structure and the semantics of an OBCS.

The type of a *Place* (written in italic characters) is a list of types of the data language, and its value, or its *marking,* is the (multi-)set of tokens it contains. At any moment, the OBCS's state (or its marking) is defined by the distribution of tokens onto places.

According to the length of the type of the place, a *token* is either a raw-token with no specific value if the place's type is empty, a value belonging to the single data type of the place, or a list of values. Such a value is either a constant (e. g. 2 or 'Hello'), an instance of a class of the data language, or a reference to such an instance. As an example, the place RFork contains tokens which are made up of a reference towards a philo and an integer. One may wonder what the difference is between using an object class or a reference to this class in the type of a place. In the former case, the class instances are accessed 'by value', while in the latter case they are accessed 'by reference'. Thus, if an object appears among the values of a token, it makes no sense for another copy of this object to appear in another token of the same marking ([Sibertin 85] provides a structural condition to avoid this kind of *ubiquity* of objects); such a situation would go against an essential principle of the OO approach according to which each object is unique. On the other hand, a reference towards the same object may appear in several tokens of a marking; for instance, the transition grf is enabled if the variable p is bound to an object reference which appears both in a token lying in the place RFork and in a token lying in the place Arg_grf.

If the data language supports a subtyping relation, the type of a token lying in a place may be a subtype of the place's type.

A *Transition* is connected to places by directional arcs.

Each *arc* is labelled with a list of variables having the same length as the type of the place to which the arc is connected. These variables serve as formal parameters for the transition and define the flow of token values from input places to output places. A transition *may occur* (or is *enabled*) if there exists a *binding* of its input variables with values of tokens lying in its input places. The *occurrence* (or the *firing*) of an enabled transition changes the marking of its surrounding places: tokens bound to input variables are removed from input places, and tokens are created and put into output places according to variables labelling output arcs. As is usual in High-Level Petri nets, an arc may be labelled with a formal sum of lists of variables (cf. transition starteating), and an expression may be found instead of an output variable (cf. transition arf).

The type of variables labelling an arc is defined componentwise by the type of the corresponding place. If the same variable appears in the labelling of several arcs surrounding a transition, simple rules prevent the occurrence of a 'type mismatch error' [Sibertin 92, Syroco 95]. For instance, the variable p of the alf transition occurs both on the arc from the place NoLFork and on the arc to the place Arg_grf, providing respectively the types philo* and Any*. No type problem occurs since philo is a sub-type of Any and NoLFork is an input place while Arg_grf is an output place; thus the type of p is philo*.

A transition may be guarded by a *Precondition*, a side-effect free Boolean expression involving the transition's input variables and the Object's attributes or operations (Cf. transition leave in Fig. 7 below; syntactically, a transition of the OBCS refers to an attribute or operation of the Data Structure using the prefix _S->). In this case, the transition is enabled by a binding only if this binding evaluates the Precondition to true.

A transition may also include an *Action,* which consists of a piece of code in which the transition's variables and the Object's operations or attributes may take place (Cf. transition stopeating). This Action is executed at each occurrence of the transition and it allows the values of tokens to be processed. If an output variable of the transition does not appear on any input arc, the Action must assign a value to this variable in order to extend the binding which enables the transition; thus, if the type of the variable is an object class, each occurrence of the transition causes the creation of a new data object. Conversely, if an input variable does not appear on any output arc while its type is an object class, each occurrence of the transition entails the loss of the data object bound to the variable.

Finally, a transition may include a set of *Emission Rules*, which are side-effect free Boolean expressions involving its variables and the Object's attributes or operations. In this case, each output arc of the transition is connected to one of the Rules, and an occurrence of the transition causes the depositing of a token into the connected output place only if the Rule evaluates to true. For instance, an occurrence of the transition leave in Figure 7 puts tokens into places p1 and p3 if the expression ok is true and into places NoLFork and RFork if not. This trick makes the graphical representation of the OBCS simpler since, if the Rules are contradictory, each Rule is equivalent to one transition having this Rule as a Precondition. Conceptually, providing a transition with Emission Rules makes explicit the fact that putting tokens into some places is a choice resulting from the transition occurrence.

The activity of an isolated Object consists of executing its OBCS, i. e. repeatedly firing transitions which are enabled under the current marking. When the marking of an Object concurrently enables several transitions, the Object's activity includes several threads of control, or *tasks,* which can progress concurrently. Although PNOs support semantics for the concurrent enabling and occurrence of transitions [Sibertin 92], the default semantics of an OBCS is the interleaving one: only one transition occurs at once, so that the on going tasks progress in turn. Arguments for this choice are developed in another paper [Sibertin 97a]. To summarise, from a conceptual point of view these semantics relieve the designer of difficulties entailed by the sharing of data (attributes and tokens referring to objects), and from an implementation point of view there is no real improvement of performance when the system includes several Objects.

3.2 Interactions among CoOperative Objects

The behaviour of a COO system results from the concurrent activity of its actual Objects, and we shall now introduce cooperation among Objects through a decentralised solution of the static dining philosophers problem. Each philosopher will be viewed as an autonomous actor modelled by an instance of the COO class SPhilo shown in Figure 4, and a table is a set of instances of this class.

In order to be able to set communications, an Object must store references to other Objects. To this end, the types of attributes, places and parameters are allowed to be COO class references in addition to the types of the data language. But COO classes are not allowed in order to prevent the nesting of Objects. When an Object has a reference towards another Object, it has access to the public elements of that Object.

The public elements provided by an Object - attributes, operations and *services* - are defined in the specification of its class and they have different purposes. Attributes and operations are called from pieces of code of the data language and they are intended to support synchronous *data flows* between Objects. Services are called from OBCSs and are intended to support asynchronous *control flows* between Objects.

```
class SPhilo specification;
operations
 Init(n: COONAME, l: SPhilo*, r: SPhilo*, f: fork) ///
      setname(n);          //assign a unique name to the Object
      ln = l; rn = r; nbeating = 0;
      NoLFork.ADDTOKEN (NoLFork.MakeToken());
      RFork.ADDTOKEN (RFork.MakeToken(f)); ///;
services
 GiveLFork(): <fork>;
 GiveRFork(): <fork>;
OBCS                   //for documentation purpose only

end.

class SPhilo implementation;
attributes
 rn: SPhilo*;                //the right and left neighbours
 ln: SPhilo*;
 nbeating: short;     //counts how many times the SPhilo eats
operations
 eat() /// nbeating = nbeating + 1; ///;
OBCS
```

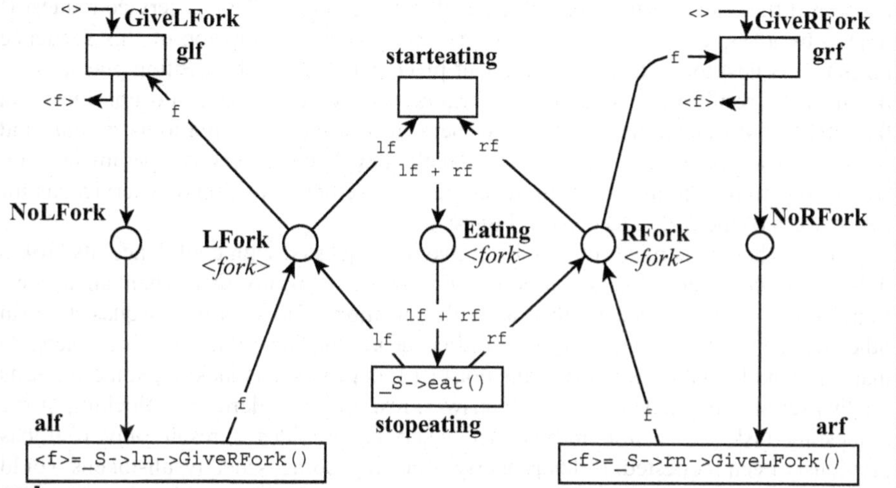

Fig. 4. Definition of the SPhilo COO class

Attributes and operations appearing in a specification are defined in the same way as in an implementation. They may be called by Preconditions, Actions and Emission Rules of other Objects. They may also be called by the code of operations, but this may cause synchronisation problems due to the synchronous semantics of operation calls. Indeed, an operation is assumed to always be available, and this property is not guaranteed if an operation calls an operation of another Object which in turn calls ... In fact, operation calls are synchronous because they are considered to be attributes whose value is computed upon request. In any case, public attributes and operations may always be removed from the interface and replaced by services. When the system under consideration is distributed and the Objects are intended to run on different computers, it is much more safe that their specifications include only services.

Services support *control flows* between Objects, that is to say interactions which have an effect upon the behaviour of the addressee and/or the applicant of the communication. A service request is asynchronous since the server may need some time to provide a result, either because its current state disables the service or because processing the request requires a lot of work. Requesting or rendering a service implies synchronisation constraints and relates to the behaviour of Objects; thus service requests and service executions are defined within OBCSs, and the specification of a class only contains the signatures of its services.

Each service provided by a COO class is implemented by a pair of places of its OBCS: an *argument-place* intended to receive tokens which are requests for the service, and a *result-place* in which the client retrieves the result of its request. (In fact, result-places may be implemented on the client side, so that the server sends the result-token to the client; but from a theoretical point of view, the contract of a server is only to make a result available for each accepted request; for instance, it is of no concern if the client disappears and leaves the result-token). Thus a service request is treated by a sequence of transition occurrences: the first transition of this sequence takes the request token from the argument-place, while the last transition puts a token down in the result-place. Graphically, argument- and result-places do not appear in the OBCS, but transitions connected to them (respectively referred to as *accept-* and *return-transitions)* bear a dangling arc labelled with the respective parameters. Of course, one transition may be both the accept- and return-transition of a service, as for the services of the SPhilo class in Figure 4.

In order that the services of an Object may be requested with confidence, its OBCS must be *honest* and *discreet* [Sibertin 93]. *Honesty* means that when an accept-transition of a service occurs, then a result-transition of this service is quasi-live (in other words: whatever transitions occur after the accept-transition, there is a reachable marking which enables a return-transition). If this property is lacking, some requests for this service may never receive an answer, resulting in a definitive blocking of the requesting task. *Discretion* means that a service provides a result only if it has previously been requested. If this property is lacking, some issued result-tokens would never be consumed. Honesty and Discretion are equivalent to the fact that any sequence of transitions rendering a service may be reduced to a single transition. In Figure 5, OBCS1 is not honest, OBCS2 is not discreet, while OBCS3 is neither of them. A third property that an OBCS must satisfy, reliability, will be introduced in section 5.

A service request takes place only in the Action of a transition (Cf. transitions alf and arf in Fig. 4). When such a *request-transition* occurs, a request-token grouping together the in-parameters of the request is put into the argument-place of the server's service; when the result-token of this request is available, the transition retrieves it from the corresponding result-place and completes its occurrence. Thus, it is possible that the occurrence of a transition requesting a service lasts for some time, until the server provides the result. However, other transitions may occur in the meantime so that a client is not blocked by a service request. Indeed, the formal semantics of a request-transition splits the transition into two: one transition for sending the request-token and another for retrieving the result-token, connected by a *waiting place* which holds the request identifiers (as an illustration, compare the transition `alf` of a `SPhilo` with the transitions `alf` and `rlg` of a `PhiloTable`).

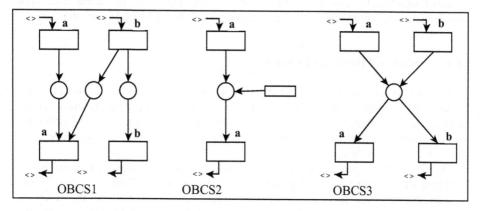

Fig. 5. OBCS1 is not honest, OBCS2 is not discreet, OBCS3 is neither honest nor discreet

The communication amongst Objects through services implements an asynchronous request/reply protocol, which allows designers to concentrate on conceptual issues without necessarily being concerned with the details of interactions. The expressive power of this protocol extends the asynchronous message sending protocol of Actors [Agha 86] since a transition may request a service using the "without reply" mode (Cf. transition `t3` in Fig. 7 below). In this particular case, only a request-token is sent to the server: this latter does not produce a result-token and the request-transition is not split. This protocol also extends the synchronous request/reply and rendezvous protocols, since a request-transition may disable all the transitions of the OBCS except the one retrieving the result-token, and in this way block the Object until communication is completed.

The OBCS of an Object determines both its internal behaviour and its asynchronous communications with other Objects, that is to say:

- its spontaneous activity (transitions `starteating` and `stopeating`), as an autonomous Object,
- the service requests it issues to other Objects (transitions `alf` and `arf`), as a client, and
- the availability of its services and the way requests are processed (transitions `glf` and `grf`), as a server.

Indeed, the internal behaviour of an Object can not be dissociated from its communications with others, since synchronisation constraints introduce a mutual dependence between these two dimensions: the state of an Object determines when requests are issued, accepted and answered, and the availability of requests and replies determines the behaviour of the Object. Nevertheless, the request and the processing of services are implemented by specific items of OBCSs, so that at the syntactic level, the 'synchronisation code' is clearly isolated [Matsuoka... 93]. When dealing with the conceptual design or the behavioural analysis of an Object, it is easy to focus either on its server side, on its client side, or on its internal behaviour.

As far as intra-Object concurrency is concerned, the model of the COO formalism is the interleaving semantics (Cf. 3.1). Concerning the inter-Object level of concurrency, the COO formalism adopts the true parallelism semantics, i. e. any number of Objects may fire a transition of their OBCS at the same time. The implementation issues caused by these semantics will be addressed in chapter 7.

3.3 Creation and Deletion of Objects

The COO formalism supports the dynamic creation and deletion of Objects. This feature is illustrated by the DPhilo class which models the dynamic dining philosophers problem (Cf. Fig. 7): each occurrence of transition t7 brings about the introduction of a new DPhilo around the table, and a DPhilo disappears when firing its transition dead.

An Object of a system introduces a new Object into this system by creating a new instance of a COO class and calling its Init operation. Then, the new Object becomes active; more precisely, it starts the execution of its OBCS after the execution of its Init operation. Since introducing a new Object into a system concerns the behaviour of the whole system, it must take place in the Action of a transition.

The deletion of an Object is more problematic. An Object is considered as being dead, and can thus be deleted, only when no other Object has a reference to it (so it can no longer receive an operation or service call) and in addition it has reached a dead marking (so it has nothing to do). A DPhilo satisfies this requirement, since firing the transition dead produces a dead marking. An implementation of COOs could include a Garbage Collector based on this principle. However, it is better if the deletion of an Object explicitly occurs in the Action of a transition, since it greatly concerns the behaviour of the whole system.

In any case, the deletion of an Object has to comply with the constraints of the asynchronous request/reply protocol:

1. when an Object, as a server, has received a request-token which enables an accept-transition, it cannot be deleted before it provides a result for this request. On the other hand, if the request-token does not enable any accept-transition, it is possible to consider that the Object has been misused by the issuer of the request.
2. when an Object, as a client, has issued a request for a service, it cannot be deleted before it retrieved the result of this request.

Anomalies resulting from the breaking of these rules may be considered both from the Petri Net and programming language points of view. From the PN point of view, in the first case a token will be blocked in the waiting place of the client, and in the second case a token will be blocked in the result-place of the server. Considering the COO formalism as a programming language, a task of the client will be blocked for ever in the first case, and in the second case an "invalid reference" run-time error will occur when the server sends the result. SYROCO does not include a Garbage Collector, but the function coodelete returns an error status and does not delete the Object if the two rules mentioned above are not satisfied.

4. The Dynamic Philosophers Case Study

A decentralised solution for the dynamic philosophers problem is proposed in this chapter, in order to illustrate the expressive power of the COO formalism with regard to the dynamicity of Objects. In this case, philosophers may join or leave the table. The corresponding COO system includes one instance of the class Heap in charge of keeping the forks used by the philosophers, and a dynamic set of instances of the class DPhilo; an instance of this set can create new instances of the class DPhilo (and so introduce new guests) and can also delete itself (and so leave the table).

To ensure that a table of dynamic philosophers remains bounded, the number of forks is limited, and they are kept by an Object of class Heap shown in Figure 6. The service Give is intended to supply a fork from the Heap, while the service Take stores a fork in the Heap. A new philosopher may be introduced at the table only if a request for the service Give has returned a fork, and a departing philosopher gives his fork back by requesting for the service Take. Transition t3 returns nothing if no fork is available. Thanks to this transition, the service Give is always available either through transition t2 (when there are forks in the place FreeForks) or through transition t3 (when there are none). Thus, requests for service Give are never delayed; this prevents the system from blocking when no fork is free and all the philosophers around the table simultaneously request a fork.

As far as eating and the exchange of forks are concerned, a dynamic philosopher behaves as a static one, so that the class DPhilo shown in Figure 7 inherits the attributes, the services and the OBCS of the class SPhilo. With regard to joining and leaving the table, the logic of the dynamic philosophers is much more complex. In fact, the instances of class DPhilo have to build a ring of Objects which is:
• consistent: each Object knows its current left and right neighbours and communicates only with them,
• dynamic: Objects may leave or join the ring,
• decentralised: there is no global supervisor who knows the setting of the table.

According to the statement of the dynamic dining philosophers problem, the DPhilo class must be designed in such a way that the two following requirements are ensured at any moment:

R1: one and only one fork is shared by two adjoining Philosophers;

R2: on both the left and right sides of a philosopher, every request sent to the neighbour is received by the real neighbour, and every reply is received by the real neighbour; for instance, the following must never happen: a request is sent to a philosopher who has left, a philosopher leaves with a pending request which will be never answered, a philosopher leaves without waiting for the result of a request, or a new philosopher is introduced between the sender and the addressee of a request.

Requirement R1 is easy to ensure, for instance by means of the following rule:
(1) a philosopher joins or leaves the table
 - with a right fork (while his right neighbour has no left fork), and
 - without a left fork (while his left neighbour has a right fork).

```
class Heap specification;
operations
  Init(nbfork: int) ///
    setname ('The Heap');
    philnum = 0;
    for (fork i=1; i<=nbfork; i++)
        FreeForks.ADDTOKEN (FreeForks.MakeToken(i));
    ///;
services
  Take(f: fork);
  Give(): <int, fork>;
end.

class Heap implementation;
attributes
  philnum: int;    //to give a number to each Philosopher
OBCS
```

Fig. 6. Definition of the COO class Heap

Requirement R2 is more difficult to ensure in the absence of centralised control. It implies that there is no concurrent moves at two adjacent places around the table. For instance, let us consider two adjoining philosophers *pl* and *pr*, and *p1* and *p2* be concurrently introduced between them; *p1* and *p2* do not know each other, so that they will both have a reference towards *pl* as their left neighbour and towards *pr* as their right neighbour, and this is clearly wrong. Thus, if *pl* is at the left side of *pr*, the following cases must be avoided:

c1: two philosophers are simultaneously introduced between *pl* and *pr*,

c2: *pl* and *pr* concurrently leave the table,

c3: *pl* leaves the table while a philosopher sits down on his right, and

c4: *pr* leaves the table while a philosopher sits down on his left.

Several policies ensure that such cases never occur; among them, we choose the following one:

(**2**) a Philosopher joins the table only if he is introduced by an already installed colleague, and he sits down on his lefthand side (so c1 is avoided);

(**3**) a Philosopher does not concurrently introduce a new guest and leave (c4 avoided);

(**4**) when a philosopher intends to leave, he asks his right neighbour for the permission to leave; this latter refuses if he himself is leaving or introducing, and when he accepts, he is prevented from leaving and introducing until he knows his new left neighbour (c2 and c3 avoided).

The OBCS of the class DPhilo implements this policy (among the elements inherited from the OBCS of the class SPhilo, only places NoLFork and RFork appear in Figure 7).

When joining, rule (1) is ensured by the initial marking produced by the operation Init and the fact that a philosopher tries to introduce a new guest only when he has no left fork; when leaving, (1) is ensured by the fact that places NoLFork and RFork are input places of the transition leave.

Rule (2) is ensured by the Action of the transition t7.

The mutual exclusion of leaving and introducing (rule (3)) is enforced by place NoLFork, since a DPhilo may introduce a new guest or leave the table only when his place NoLFork contains a token.

As for rule (4), it is ensured by the fact that the service LNMayLeave is also in mutual exclusion with introducing and leaving by means of place NoLFork. This service is supported by transitions t4 and t5; thus it is permanently available and this prevents the system from blocking when all philosophers simultaneously intend to leave.

```
class DPhilo specification;
inherits SPhilo;
operations
  Init(id:int, l:DPhilo*, r:DPhilo*, f:fork, hp:Heap*) ///
        setname (îPhilosopherî,id);
        ln = l; rn = r; h = hp;
        delayquit= rand()%4; delayint= rand()%6; nbeating= 0;
        NoLFork.ADDTOKEN (NoLFork.MakeToken());
        RFork.ADDTOKEN (RFork.MakeToken(f)); ///;
services
  LNMayLeave(): <Bool>;
  NewLN(phil: Dphilo*); NewRN(phil: DPhilo*);
end.

class DPhilo implementation;
attributes
  rn, ln: DPhilo*;              //redefined as a Dynamic Philosopher
  h: Heap*;
  delayquit, delayint: short;  //to make DPhilos a bit steady
OBCS
  inherits SPhilo;
//only places NoLFfork and Rfork of the inherited OBCS are shown
```

Fig. 7. Definition of the COO class Dphilo

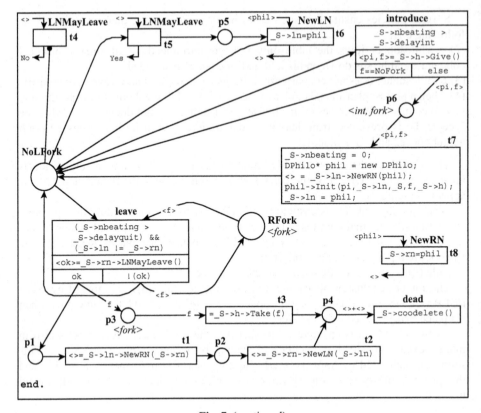

Fig. 7. (continued)

The Precondition of the transition `leave` guarantees that at least two philosophers remain around the table.

5. Inheritance and Sub-typing

Inheritance is one of the main concepts of the OO approach as it is both a cognitive tool which eases the design of complex systems and a technical support for software reuse and change [Meyer 88]. However, it has been pointed out that inheritance within concurrent OO languages entails many difficult problems or anomalies [Matsuoka...93]. These difficulties are due to the fact that, for a long time, inheritance has been confused with the concept of type [America 90]. *Inheritance* refers to the reuse of the components of a class by another one, so that the derived class includes elements inherited from its parent class together with its own elements. As for the concept of *type*, it is not specific to the OO approach and relies on the substitution principle: *s* is a subtype of *t* if an instance of type *s* may be substituted when an instance of type *t* is expected [Wegner 88]. Inheritance is a matter of structure sharing and it mainly relates to implementation issues, while typing is a matter of polymorphism and it mainly relates to the specification (the observable behaviour) of objects. These two viewpoints are quite close when sequential OO languages are considered, but there is no denying that they are far from each other when PNs are taken under consideration.

The COO formalism supports three kinds of inheritance.

Specification inheritance is aimed at supporting the subtype relation. In this case, the derived class includes the declaration of the public attributes of the inherited classes, and the signature of their public operations and services. In addition, it is possible to redefine (or *override)* the type of the arguments of operations and services according to the contravariant rule, and the type of their result according to the covariant rule. The *contravariant* rule specifies that the type of a parameter may be replaced by a supertype, and the *covariant* rule says that the type of a parameter may be replaced by a subtype. Thus, a class derived by specification inheritance offers an interface which is compatible with those of the base classes, but it does not share their code.

Implementation inheritance strengthens specification inheritance and enables the derived class to share the code of operations. In this case, the derived class includes the declaration of the attributes and services of the base classes, as well as the definition of their public and private operations; but the OBCS is not inherited and the derived class has its own one.

In the *OBCS inheritance* case, the derived class fully inherits another class; it inherits its specification and its implementation, including the OBCS. Multiple inheritance is not allowed, in order to prevent problems arising from the merging of OBCSs.

Subtyping includes two aspects: when a call for an operation or a service is addressed to an instance of the subtype instead of an instance of the supertype,

1. the type of the formal parameters of the subtype has to match the type of the actual parameters of the call, and
2. the requested service has to be available at the subtype instance if it would have been available at the supertype instance.

The first aspect warrants the safety of data types, while the second aspect warrants the safety of the behaviour. For instance, the class DPhilo is not a subtype of the class Sphilo, since a SPhilo provides its GiveLFork and GiveRFork services for ever, while a DPhilo stops as soon as it leaves. Conversely, class SPhilo is not a subtype of class DPhilo since this latter offers additional services.

Specification inheritance guarantees that the interface of the derived class offers the same possibilities of interaction as the interface of the base class, and a class B may be a subtype of a class A only if B inherits the specification of A. The authorised redefinitions of signatures preserve this compatibility, and it is commonly agreed that they maintain the type-safety of data exchanges [Liskov...93].

Figure 8 shows three classes which are in specification inheritance relationship but which feature different behaviours: a client of an instance of class A will successfully request the sequence of services "c b c b ...", while it will block if it attempts to do the same with an instance of class B, and it will succeed if it does the same with an instance of class C. Class C is a subtype of class A, while class B is not because it is unable to simulate the behaviour of class A.

In [Hameurlain...99], conditions upon the OBCS of two COO classes are given ensuring that one of them is a subtype of the other. The first condition is that any sequence of service requests which is accepted by the supertype is also accepted by the subtype. The second condition, referred to as *reliability*, concerns only the OBCS of the subtype; it requires that, whatever the marking reached by the Object, the set of services which are available under this marking only depends on the sequence of service requests which have been previously accepted. In other words, if two markings are reached while accepting the same sequence of services then these markings enable the same service requests. Figure 9 shows typical examples of OBCSs which are not reliable, while all the COO classes presented within this paper have a reliable OBCS. If the OBCS of a class does not satisfy this property, this class is a subtype of no COO class. In fact, any COO class should have a reliable OBCS; if not, the class has an unsafe behaviour, since a client has no means to know whether its next request will be accepted or not. It is worth noting that it can be decided whether a COO class is a subtype of another one [Hameurlain...99].

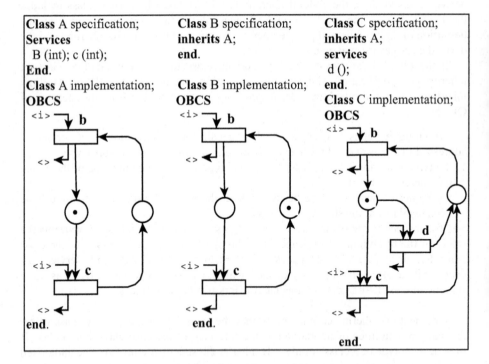

Fig. 8. Class C is a subtype of class A, while class B is not

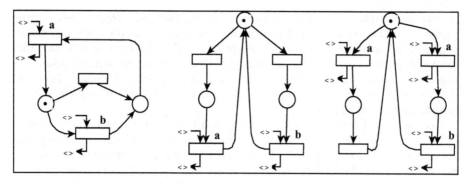

Fig. 9. Typical examples of unreliable OBCSs

6. Formal Semantics of the COO Formalism

As is to be expected, COOs enjoy a semantics which complies with the PN theory.

A basic feature of the COO formalism is that it makes a data language intervene in the definition of COO classes. Thus, any definition of a formal semantics of the COO formalism requires the data language to be provided with a formal semantics, and the COO's semantics results from the integration of the latter with the PN theory. However, we also need a pure PN semantics which would enable one to apply the PN analysis techniques to a COO system. Such a semantics would have to ignore the data processing part of Objects, and to account only for their behavioural part. The COO formalism enjoys such a semantics, which abstracts from the distinctive features of the data language and does not deal with the value of the data processed and sent by Objects. The possibility to define such a semantics stems from the fact that COOs integrate the PN and OO approaches while not confusing their respective mechanisms.

Modularity is another basic feature of the COO formalism, which is so provided with a two-levels semantics: the semantics of an isolated Object which mainly concerns the local processing of tokens, and the semantics of a COO system which deals with the communication among Objects.

The semantics of an isolated Object formalises how the values of tokens and attributes are taken into account by the Preconditions (the enabling rule) and processed by the Actions (the occurrence rule) of transitions. This may be fully achieved only if the data language has a formal semantics. For instance, [Sibertin 92] gives a formal semantics for isolated Objects, using a formalism based upon Abstract Data Types [Ehrig...85] as the data language.

Whatever the data language, it is possible to define a pure PN semantics which accounts for the flow of tokens across the places of the OBCS but ignores their structure, their value and how they are processed by the transition occurrences. For this purpose, a COO class is modified in such a way that all the types of places are references towards classes of data objects. Thus tokens become (lists of) references to objects and include neither scalar values nor objects. These new object classes remain abstract; thus Preconditions and Actions become useless and disappear, as well as attributes and operations, since tokens are pure identifiers referring to no value. After some additional transformations accounting for the dynamic creation and deletion of Objects, a COO class may be viewed as a Well Formed Coloured Petri Net [Chiola...90], the set of identities of the instances of each object class being considered as a Colour domain.

One may wonder whether applying these modifications to a COO system changes its behaviour; in other words, to what extent a semantics of COOs, which is partial because it is based only upon the PN theory, is an accurate semantics, and to what extent the results obtained by the PN analysis techniques of a COO system really describe the behavioural properties of this system? The answer depends on the use of Preconditions. If they are used jointly with the net structure of the OBCSs to define the Objects' behaviours -as for example a Precondition which evaluates to false whatever the value of tokens- then the PN semantics is not accurate. On the other hand, this semantics is accurate if Preconditions are used only as they are intended for solving conflicts among transitions or tokens; in this case, for any run of an Object stripped of its data part, there exists an initial value of this Object yielding this run [Sibertin 85] (The converse clearly holds in any case). Thus, a designer may trust the results obtained by the PN analysis techniques as far as he/she respects the 'separation of concern' principle.

As for the cooperation among the Objects of a system, [Sibertin 94] gives formal semantics which ignore the types of tokens. In fact, this paper provides two semantics for the cooperation. The first one is straightforward and defines the change in a COO system produced by a set of Objects which concurrently fire one transition. The second semantics may be viewed as a global, or centralised one; it is defined by means of an algorithm translating a COO system into a single Object which is isolated (it provides or requests no service) and equivalent (more precisely: bisimilar) to the whole system. As an example, this algorithm translates a system of four SPhilo Objects (cf. Figure 4) into an Object which is quite close to an instance of the PhiloTable class given in Figure 3 (the OBCS of the PhiloTable class accounts for the sending of tokens caused by service requests, but it does not include the elements for the dynamic creation and deletion of Objects, nor for the inheritance relation among COO classes, nor for the identification of service requests which is needed to distinguish requests concurrently issued or received for the same service).

Associating one of the semantics of communication among Objects with one semantics of isolated Objects provides the COO formalism with a complete formal semantics. The Coloured PN-based semantics are of a special interest, since they allow behavioural analysis techniques to be applied. The first way to analyse a COO system is to globally analyse the isolated Object to which the whole system is equivalent. The second way is to analyse each COO class in isolation while ignoring the specific treatment of accept-, return- and request-transitions. Then, some compositional results [Sibertin 93, Hameurlain...99] allow to incrementally deduce the behaviour of the whole system from the behaviour of its component COO classes.

7. SYROCO

SYROCO is an environment for designing COO classes and executing COO systems. It is intended to be used either as a simulation tool or as a programming environment [Syroco 96, Sibertin 97b].

A COO class is edited either using a tailored version of *MACAO*, the generic Petri net graphic editor developed at the MASI [Mounier 94], or as a text file compliant with a very simple syntax. In the former case, the MACAO file is translated into a text file before applying the following main utilities:

- *newproj*, to create the working environment (required directories and files) for a new system,
- *gencode*, to generate the C++ code associated with a COO class,
- *gentest*, to generate a main program for an instance of a COO class,
- *genmake*, to generate makefiles.

SYROCO is based on the language C++ in two ways. On the one hand, C++ is the data language used to write the code of the files containing the external declarations (as shown in Figure 2), and also to write the body of Object operations and the pieces of code embedded in the transitions and places of OBCSs. On the other hand, SYROCO generates C++ code for each COO class and implements any Object as an instance of a C++ class. The choice of C++ is contingent, and COOs can be implemented using another OO Programming Language. However, the crucial point is to use the same programming language both as the data language and as the implementation language. If two programming languages are used, the integration of the data processing and behavioural dimensions of Objects is difficult and results in poor performances, and in addition the designer-defined code embedded in an Object can access neither the kernel of SYROCO nor the Object's implementation.

Although SYROCO is mainly a COO compiler, the OBCS of each Object is not flattened into a static control structure. It is interpreted by a generic ìtoken game playerî which repeatedly looks for transitions which are enabled under the current marking and fires one of them. This feature allows the non deterministic nature of Petri nets to be retained and provides each Object with a powerful symbolic debugger. It also allows dynamic changes of the value of some parameters of the interpreter. In this way, each Object may control how its OBCS is executed. In addition, interpreting OBCSs avoids burdening each COO++ class with a large piece of code.

SYROCO provides users with a number of facilities both for the simulation and the final implementation of complex systems. Some of these facilities are pragmatic and intended to ease the development of COO systems, while others extend the expressive power of the formalism and are intended to allow a fine control of the behaviour of each Object. They will not be addressed here, but we shall discuss the main design decision of an implementation of the COO formalism. Details about the structure and the functionalities of SYROCO may be found in [SibertinÖ95a, 97b] and [Syroco 96].

7.1 Implementation of an Object

Each COO class gives rise to the generation of a C++ class, referred to as its COO++ class, in such a way that each instance of a COO class is implemented as an instance of the corresponding COO++ class. First, the compiler provides a COO class with two operations for each service, in charge of sending respectively request- and result-tokens. The compiler also transforms the OBCS of a COO class in order to implement the formal semantics of service requests (namely, argument- and result-places are added, request-transitions are split, and the sending of tokens is added to the Action of the appropriate transitions, cf. section III.2). After these modifications, the OBCS of each Object fully implements the PN semantics of the communications protocol through services.

Then, a COO class is compiled into a static data structure of the implementation language, and this is one essential reason for the efficiency of SYROCO. Only the tokens, which are implemented as instances of generated C++ classes, are stored in dynamic linked data structures. The structure of a COO++ class is very similar to the one of its COO class. A COO++ class has one member attribute for each attribute of the COO class and one member function for each operation, and also one member attribute for each place and transition of the OBCS. The class of a place attribute inherits functions for the management of tokens and it also includes specifically generated functions (the body of which is provided by the designer) to order the tokens of the marking or to be triggered upon the arrival or the departing of tokens. Similarly, the class of a transition attribute inherits general purpose functions, and it includes specifically generated functions to test its Precondition and to execute its Action. These classes of place and transition attributes are also equipped with attributes which determine the policy of the ordering of tokens into places, the priority of transitions, the delays associated with arcs, or even the hiding of tokens and transitions; an Object may change the value of these attributes while it is running and thus gains some reflexivity.

The implementation of the inheritance and subtyping relationships among COO classes heavily depends on the possibilities of the implementation language. Indeed, any reasonable means to implement the polymorphism among Objects has to be based upon mechanisms of the implementation language. Thus, SYROCO relies on the C++ inheritance mechanisms for supporting the three inheritance relationships among COO classes (Cf. chapter V). To achieve this, a COO class gives rise to the generation of three C++ classes: one for its specification, a derived class for the attributes and operations of its implementation, and a third derived class, the COO++ class, accounting for its OBCS. The specification, implementation and OBCS inheritance relationships among COO classes are translated into inheritance relationships among the corresponding C++ classes. When implementing the COO formalism, inheritance is the only aspect which depends on the implementation language, and SYROCO suffers from the limitations of C++ in this regard.

7.2 Execution of an Object

Each Object has its own OBCS interpreter; more precisely, it inherits a generic PNO interpreter which locally executes its OBCS. Thus, each Object is actually implemented as an autonomous process that encapsulates its behaviour. As a consequence, the behaviour of a COO system features the same modularity as its structure, and there is no difficulty in taking advantage of the resources of a multiprocessor computer or in distributing the Objects of a system over a network of computers [Sibertin 97a]. Of course, the concurrency among the Objects of a system requires some synchronisation between the interpreters of these Objects, but the resulting overhead is negligible. These Objects are weakly coupled (in the same way as in Actor languages [Agha 86]), since conflicts only occur on accesses to argument- and result-places for sending or receiving tokens. In any case, this solution is much more efficient than a single distributed Petri net interpreter (e.g. [B‚tlerÖ89] among others) which, being unaware of the dynamics of the Objects, has to check in all cases if a synchronisation is needed.

The default strategy of the OBCS interpreter is to fire only one transition at once, according to the *interleaving semantics*. Thus, an Object may have several on going tasks (according to the number of transitions enabled under the current marking), but each progresses in turn. The main reason for this choice is a conceptual one. Due to these semantics, the action of each transition is a critical section, and the occurrence of a transition is atomic with regard to other transitions. In this way, the designer is relieved of ensuring the mutual exclusion of transitions, even if their Actions refer to the same attribute(s). As a consequence, an Object needs to synchronise with other Objects, but it does not need to synchronise with itself since it never concurrently accesses its own data structure. From a performance point of view, the execution of one Object requests the computing environment to spawn only one process.

The principle retained for the algorithm of the interpreter is the *triggering place* mechanism [Colom...86]: to each transition is associated one of its input places, and the enabling of a transition is tested only if the marking of its triggering place is not empty. (The efficiency of this principle results from the fact that most transitions of a PN have one input place which controls their occurrence, while the other input places correspond to a synchronisation of resources. When a transition belongs to the path of the support of a place invariant, the input place of the transition which belongs to this support is a good candidate). This algorithm raises the problem of fairness: it selects the first enabled transition found and this transition occurs with the first tokens found, instead of randomly choosing a <transition, binding> couple among the enabled ones. Fairness is gained by the random nature of the search over the sets of transitions and tokens.

7.3 Concurrency among Objects

As far as concurrency among the Objects of a system is concerned, the default strategy is the *true parallelism semantics*; in other words, several Objects may be active at the same time and fire a transition of their OBCSs. This strategy causes no synchronisation overhead, thanks to the low coupling of OBCSs by token sending; if the interpreter of an Object finds an enabled transition, it may fire it without considering the choices made by the interpreters of other Objects. In conjunction with the interleaving semantics for the intra-Object concurrency, this choice makes use of the resources of the computing environment in a straightforward manner: the execution of a COO system needs as many processes as the current number of Objects [Sibertin 97a].

However, dealing with true parallelism entails implementation issues. One solution is to develop a specific runtime system which provides the required functionalities. The other solution is to rely upon the underlying operating system, using its capabilities as much as possible, as does SYROCO. As a consequence, SYROCO has to account for the actual services offered by operating systems, and thus it generates COO++ classes for sequential computing environments, for environments supporting *threads* (or lightweight processes), and also for COOL (the Chorus Object-Oriented Layer [Chorus 94]), a distributed computing environment compliant with *CORBA*.

The sequential version: The sequential version of SYROCO implements the interleaving semantics among Objects. Thus, the concurrency among Objects is only virtual. To this end, SYROCO includes a scheduler which randomly selects one Object of the system, and then activates this Object by calling its interpreter for a given number of transition occurrences. In this version, an interpreter returns as soon as no transition is enabled, since only the activation of another Object can enable a transition.

The threaded version: SYROCO implements the true parallelism semantics by delegating the actuation of Objects to the underlying operating system, according to the computing resources. In the threaded version, all Objects run in the same system process, but each one has its own thread of control. More precisely, making an Object active consists of spawning a new thread and calling the Object interpreter inside this thread, referred to as its *main thread*. As a consequence, each Object is accessed by several threads of control:

(1) its main thread,
(2) the threads of client Objects calling public operations or reading the value of public attributes,
(3) the threads of client Objects sending tokens into accept-places in order to request services,
(4) the threads of server Objects sending tokens into result-places as replies for previous service requests.

Each Object includes a mutual exclusion lock for ensuring that threads (1), (3) and (4) do not concurrently access the marking of places. As for the Object's attributes, they are accessed by threads (1) and (2) and SYROCO sets a specific lock during the execution of the Actions of transitions; thus, it is left to the designer to protect them against concurrent accesses by including appropriate get and release statements in the code of public operations, according to the logic of the system.

Contrary to the sequential version, an interpreter does not return when no transition is enabled, instead it blocks until it receives a token from a thread (3) or (4).

The CORBA version: This version is based on the same principles as the threaded version, and in addition it enables the Objects of a system to run on different nodes of a network. Due to this fact, it is advisable that Objects do not communicate by synchronous communications (i. e. public attributes and operations). This version relies on COOL for the remote function calls.

From a software development point of view, every one of these three versions of SYROCO is useful for finalising a COO system. The sequential version allows the designer to test and validate the behaviour of each Object as well as the cooperation among Objects; but issues related to concurrency among Objects are excluded. In the threaded version, the execution of an Action may be interrupted by operations and vice versa, and tokens may arrive at any moment. Thus, this version allows the designer to focus upon the concurrency among Objects. The issues related to the distribution of Objects are accounted for only by the CORBA version.

7.4 Performance Issues

COO++ classes include a lot of *compilation conditions* which may be selected by the user according to his/her aims. The choice among the three versions above as well as the actual use of the additional features of SYROCO (the delivery of traces and statistics, local clocks, keeping the names of places and transitions, ...) depends on these conditions. Thus, either a COO++ class is equipped with many facilities and features, or it is an optimised implementation of a COO class.

With regard to size issues, the size of the code shared by the COO++ classes of a system is around 60 K, whereas a class own code is mainly the embedded code provided by the designer. The memory required to allocate a new instance is a few Ks; for instance, the memory size required to allocate an instance of our DPhilo example (including 11 places and 17 transitions) varies from 2,5 to 3,6 K according to the number of compilation conditions. Thus, the PN-component of Objects is sparing of memory.

With regard to performance issues, the OBCS interpreter is quite efficient, and the overhead resulting from the concurrency is negligible with regard to the execution of Actions (in as much as they need some amount of computation). The occurrence of one million transitions of the DPhilo example takes about 4mn 30 on a Sun SPARCclassic station and 1mn on a Pentium 120 Windows NT PC with Visual C++. This speed does not depend on the size of the net, but it is likely to grow with the number of tokens of the marking. Indeed, the number of bindings of the input variables of a transition with tokens is equal to the product of the size of the markings of the transition's input places; so, the interpreter may have to build and test many bindings before finding out one which enables the transition.

Conclusion

The COO formalism belongs to the family of High-Level Petri Nets which attempt to overcome the inadequacy of PNs with regard to modularity and the data processing dimension of systems. This formalism draws its inspiration from the OO approach, and it introduces all the concepts of this approach into PNs while remaining within this theoretical framework. As a result, it widens the field of use of PNs in several directions:

- The modelling and analysis of systems where the data structure plays a crucial role, such as Information Systems, Workflow or Business Procedures;
- The modelling and analysis of *open* distributed systems which include a varying number of processes communicating in an asynchronous manner;
- The use of a single PN-based conceptual framework throughout the software development process, from the system analysis and specification steps to the implementation.

From an OO view, the COO formalism is a Concurrent OO Language which supports intra-object concurrency and provides objects with a large degree of autonomy. This formalism is based on a formal model of concurrency, PNs, and it makes the techniques of this theory applicable to the analysis of the behaviour of concurrent systems. As a result, it widens the field of use of the OO approach in several directions:

- Coping with critical concurrent systems, where the formal validation of the system's behaviour is essential;
- Coping with Multi-Agent Systems or Distributed Artificial Intelligence applications, which requires turning objects into autonomous and capable agents [Gasser 91];
- To serve as a reference model for Concurrent OO Languages.

As for SYROCO, it tends to prove that formalisms integrating Petri nets and the O-O approach may be implemented in a rigorous and efficient way, provided that the principle of "separation of concern" is obeyed.

Acknowledgement

SYROCO has been developed thanks to a grant awarded by CNET and the contribution of many people: R. Bastide, W. Chainbi, L. Dourte, N. Hameurlain, H. Kria, P. Palanque, A. Saint Upéry, P. Touzeau, J.-M. Volle. I am also grateful to C. Hanachi for useful comments on a preliminary version of this paper.

References

[Agha 86] G. AGHA
   Actors: A Model of Concurrent Computation in Distributed Systems. MIT Press, Cambridge, 1986.

[Agha...93] G. AGHA, P. WEGNER, A. YONEZAWA
   Research directions in Concurrent Object Oriented Programming, MIT Press, 1993.

America 90] P. AMERICA
   Designing an Object-Oriented Programming Language with Behavioural Subtyping. In Foundations of Object-Oriented Languages, J.W. de Bakker, W.P. de Roever, G. Rozenberg Eds., LNCS 489, Springer-Verlag, 1990.

[Bütler...89] B. B.TLER, R. ESSER, R. MATTMANN
   A Distributed Simulator of High Order Petri Nets. In Proceedings of the 10th Int. Conference on Application and Theory of Petri Nets, Bonn (G), June 1989.

[Chiola...90] G. CHIOLA, C. DUTHEILLET,G. FRANCESCHINIS, S. HADDAD
   On Well-Formed Coloured Nets and their Symbolic Reachability Graph. In Proceedings of the 11th International Conference on Application and Theory of Petri Nets, Paris (F), June 1990.

[Chorus 94] CHORUS SYSTEM.
   COOL V2 Programmer's Guide. Paris, Feb. 1994.

[Colom... 86] J.M. COLOM, M. SILVA, J.L. VILLARROEL
   On software implementation of Petri nets and colored Petri nets using high-level concurrent languages. In Proceedings of the 7th European Workshop on Application and Theory of Petri nets, Oxford (GB), July 1986.

[Ehrig...85] H. EHRIG, B. MAHR
Fundamentals of Algebraic Specifications. Springer-Verlag, 1985.
[Gasser 91] L. GASSER
Social Conception of knowledge and action: DAI foundations and open systems semantics. Artificial Intelligence 47, Elsevier, 1991.
[Ghezzi...91] C. GHEZZI, M. JAZAYERI, D. MANDRIOLI
Fundamentals of Software Engineering. Prentice-Hall International Editions, 1991.
[Genrich...81] H. GENRICH, K. LAUTENBACH
System modelling with High Level Petri Nets. Theoretical Computer Science 13, North Holland, 1981.
[Hameurlain...99] N. HAMEURLAIN, C. SIBERTIN-BLANC
Behavioural Types in CoOperative Objetcs. In Proceedings of the 2th Int. Workshop on Semantics of Objects as Processes, SOAP '98, within the 13th European Conf. on Object-Oriented Programming, ECOOP'99, June 1999, Lisboa, Portugal.
[Hoare 78] C.A.R. HOARE
Communicating Sequential Processes. Communication of the ACM, 21(8), 1978.
[ISO 97] Comittee Draft ISO/IEC 15909
High-level Petri Nets - Concepts, Definition and Graphical Notations. ISO SC7, October 1997.
[Jensen 85] K. JENSEN
Coloured Petri Nets. In Petri Nets: Applications and Relationships to Other Models of Concurrency Part I, W. Brauer, W. Reisig and G. Rozenberg Eds, Lecture Notes in Computer Science Vol. 254, Springer-Verlag,1985.
[Liskov 93] B. H. LISKOV
A New Definition of the Subtype Relation. In Proc. 7th European Conf. on Object-Oriented Programming, Kaiserlautern (G), Springer-Verlag, 1993.
[May 87] D. MAY
Occam 2 language definition. INMOS Limited, March 1987.
[Matsuoka...93] S. MATSUOKA, A. YONEZAWA
Inheritance anomaly in Object-Oriented Concurrent Programming Languages. In Research Directions in Concurrent Object-Orineted Programming, G. Agha, P. Wegner and A. Yonezawa Eds, MIT Press, 1993.
[Meyer 88] B. MEYER
Object-Oriented Software Construction. Prentice Hall, 1988
[Murata 89] T. MURATA
Petri Nets: Properties, Analysis and Applications; Proc. of the IEEE, vol 77, n∞ 4, April 1989.
[Mounier 94] J.-L. MOUNIER
The MACAO reference Manual. MASI Laboratory, Paris (F), june 1994.
[Sibertin 85] C. SIBERTIN-BLANC
High Level Petri Nets with Data Structure. In Proceedings of the 6th european Workshop on Application and Theory of Petri Nets; Espoo (Finlande), juin 1985.
[Sibertin 92] C. SIBERTIN-BLANC
A functional semantics for Petri Nets with Objects. Internal Report, University Toulouse 1 (F), October 1992.
[Sibertin 93] C. SIBERTIN-BLANC
A Client-Server Protocol for the Composition of Petri Nets. In Proceedings of the 14th International Conference on Application and Theory of Petri Nets, LNCS 691, Chicago (Il), June 1993.
[Sibertin 94] C. SIBERTIN-BLANC
Communicative and Cooperative Nets. Proceedings of the 15th International Conference on Application and Theory of Petri Nets, Zaragoza (Sp), LNCS 815, Springer-Verlag, 1994.

[Sibertin...95] C. SIBERTIN-BLANC, N. HAMEURLAIN, P. TOUZEAU
SYROCO: A C++ implementation of CoOperative Objects. In Proceedings of the Workshop on Object-Oriented Programming and Models of Concurrency; Turino (I), June 1995.

[Sibertin 97a] C. SIBERTIN-BLANC
Concurrency in CoOperative Objects. In Proceedings of the Second International Workshop on High-Level Parallel Programming Models and Supportive Environments, HIPS'97, Geneva (S), IEEE Society Press, April 1997.

[Sibertin 97b] C. SIBERTIN-BLANC
An overview of SYROCO. In Proceedings of the Tool Presentation Session, 18th International Conference on Application and Theory of Petri Nets, Toulouse (F), June 1997.

[Syroco...96] C. SIBERTIN-BLANC et Al
SYROCO : Reference Manual V7. University Toulouse 1 (F), Oct 1996. © 1995, 97, CNET and University Toulouse 1. SYROCO is available for non commercial use through the site http://www.daimi.aau.dk/PetriNets/tools.

[Wegner 88] P. WEGNER
Inheritance as an Incremental Modification Mechanism, or What Is and Isn't Like. In Proc. ECOOP 88, Oslo (Norway), Springer-Verlag.

OB(PN)²: An Object Based Petri Net Programming Notation

Johan Lilius

Åbo Akademi University
Department of Computer Science and
Turku Centre for Computer Science
FIN-20520 ÅBO
FINLAND
Johan.Lilius@abo.fi

Abstract. In this paper we present the object-based language OB(PN)², together with its semantics, defined as a translation into a class of high-level Petri nets. This translation defines the semantics of a OB(PN)² program as a net, which can be analyzed using existing reachability analysis tools. The OB(PN)² language is an extension of B(PN)² as defined by Best and Hopkins, and the semantics is inspired by the B(PN)² semantics defined in terms of M-nets. The semantics is interesting from two points of view: it lays the foundations for the development of automatic verification methods of concurrent programs written in object-oriented languages, and it can be seen as a set of rules for the translation of object-oriented specifications written in an object-oriented specification formalism into Petri-nets. The translation relies on the CCS-like composition operators defined for M-nets. Each program construct is translated to a box (a special kind of net) or an operation for combining boxes. Thus in essence each program is translated into an expression in the algebra of boxes.

1 Introduction

Concurrency is a natural feature of objects: given two objects, there is no reason why the execution order of their methods should be dependent on any other concept than synchronization. Indeed many object-oriented languages contain concurrency-primitives [21]. However adding object-oriented features to languages or formalisms specifically designed to model concurrent systems eg. Petri nets is a non-trivial task. It is not at all clear what the object-oriented part in such a formalism should be. Thus several proposals exist, eg. CLOWN [2,8] and CO-OPN/2 [7].

In this work we take a novel approach. Instead of extending Petri nets with new features, we view a kind of Petri nets, M-nets as an implementation formalism for an object-oriented language OB(PN)². The objects of OB(PN)² are inherently parallel, meaning that each object may service method requests in parallel, and the variable access semantics is a mutual exclusion semantics, which

G. Agha et al. (Eds.): Concurrent OOP and PN, LNCS 2001, pp. 247–275, 2001.
© Springer-Verlag Berlin Heidelberg 2001

is enforced by the resource-consciousness of Petri nets. The objects can communicate either synchronously through a hand-shake or asynchronously through a channel with an arbitrary capacity[1]. The statements in an $OB(PN)^2$ program are atomic actions, that is variable accesses, variable assignments and, procedure/method calls. Moreover since M-nets are provided with process-algebra like operations, the semantics of $OB(PN)^2$ is compositional. Currently only the scalar data-types integers and booleans are provided.

The semantics is interesting from two points of view:

1. it lays the foundations for the development of automated verification methods of concurrent programs written in object-oriented languages, and
2. it can be seen as a set of rules for the translation of object-oriented specifications written in an object-oriented specification formalism into Petri-nets.

The intended application area of the techniques presented in this paper is the analysis and verification of programs for safety-critical embedded systems. It is generally felt that direct analysis of programs is infeasible because of the state-space explosion problem. For example the work on static analysis of ADA [19] aims at abstracting away details of the program so that reachability graph generation becomes feasible. However we feel that recent advances in new analysis techniques, like the stubborn set method [20] as implemented in PROD [12] or the model-checking algorithms developed in [9] and implemented in the PEP-tool [11], make the full analysis of programs possible. Also we think that the area of safety-critical embedded systems, where the programs are usually relatively small, provides an application area where the full and automated analysis of programs will not only be practical but also cost-effective.

The presented formalism $OB(PN)^2$ is based on $B(PN)^2$, a Basic Programming Notation for Petri Nets [6], which was given a semantics in terms M-nets in [5]. In [17] $B(PN)^2$ was extended with procedures that allowed parameters to pass by value. In this paper we extend $B(PN)^2$ with *objects* and with *references*. In $B(PN)^2$ a variable is translated into a *data-box*. A data-box is a small net that consists of a place, where the value is stored, and some transitions, that are used to access the place. However for procedures this structure is no longer adequate, because with recursive and/or parallel procedure calls, several instances of the same variable need to be stored in the data-box. To accommodate this, the data-box was extended in [17] to store a *frame-index* together with the value, that marks the owner of the value. The owner of the value is one of possibly several concurrently and/or recursively active instances of the procedure in which the variable is declared. When one introduces objects this is no longer enough, because now we need to distinguish between instances of member variables that belong to different objects. For this we introduce an extra identifier, the *object-identifier*, which can be assigned at compile-time. The object-identifier, frame-index pair is called a *scope-index*. The scope-index is sufficient to always unequivocally distinguish between different instances of a variable. This means that we can store all the values of variables in one place,

---

[1] Channels are not described in this paper.

and use the scope-index together with the identifier to access the value. Thus we have in essence created *references*.

We make a distinction between objects and values like JAVA [1]. The built in types (integers and booleans) are passed by value, while objects are passed by reference. So like in JAVA we have an implicit notion of pointers that is transparent to the user.

The semantics of OB(PN)² differs substantially from the original B(PN)² M-net semantics [5]. We have had to extend M-nets to be able to correctly treat procedure call instances. Also the treatment of reference parameters requires changes in the original translation of variables.

The current version of the language does not provide synchronization constraints on the methods of the object, nor does it provide inheritance, because the combination of these two features may lead to problems, the "inheritance anomaly" [18]. Essentially this means that in the presence of inheritance, constraints put on the concurrent behavior of a subclass may break encapsulation and require redefinitions in the parent class. However a proposal for the translation of the language extended with inheritance and synchronisation constraints has been presented in [16]. Also note, that some of the restrictions on the language are done so that the translation can be done in a recursive descent fashion. We will point out these restrictions at the appropriate point.

The paper is structure as follows. Section 2 describes the syntax of OB(PN)². Section 3 gives a short introduction to M-nets, emphasizing the differences to more traditional net formalisms. In Section 4 we give an introduction to the basic ideas relevant for the translation. Section 5 gives the translation.

Acknowledgements

This research has been partially supported by the DAAD through the Konrad Zuse Programm, while the author was at Institut für Informatik, Universität Hildesheim, but most of it has been done while the author still was at Digital Systems Laboratory, Helsinki University of Technology. An extended abstract of this work has appeared as [15].

I would like to thank Eike Best, Elisabeth Pelz and Hanna Klaudel for helpful discussions during the early stages of this work.

2   OB(PN)²

The syntax of OB(PN)² is given in figure 1. It is the syntax of B(PN)² as given in [6], extended with procedure and class declarations. The language is a Pascal-like language extended with constructs for concurrency ||, non-deterministic choice (□), and classes.

Most of the syntax is self-explanatory, but let us point out that statements are encapsulated into *atomic actions* (< AtAct >), ie. an assignment $x := x + 1$ is written as $< x' = {}'x + 1 >$ meaning that the new value of x (x') becomes the old value ($'x$) incremented by one, in one atomic step. Atomic actions are

$$\begin{aligned}
\langle program \rangle &\longrightarrow \texttt{program } programname \langle body \rangle \\
\langle body \rangle &\longrightarrow \texttt{begin} \langle block \rangle \texttt{end} \\
\langle block \rangle &\longrightarrow \langle decl \rangle ; \langle block \rangle \mid \langle comlst \rangle \\
\langle decl \rangle &\longrightarrow \langle typedecl \rangle \mid \langle vardecl \rangle \mid \langle objdecl \rangle \\
&\quad \mid \langle classdecl \rangle \mid \langle procdecl \rangle \\
\langle typedecl \rangle &\longrightarrow \texttt{type } tname : \langle constructor \rangle \\
\langle vardecl \rangle &\longrightarrow \texttt{var } vname : tname \\
\langle objdecl \rangle &\longrightarrow \texttt{obj } oname : cname \\
\langle classdecl \rangle &\longrightarrow \texttt{class } cname \langle cdecls \rangle \texttt{endc} \\
\langle cdecls \rangle &\longrightarrow \langle cdecl \rangle ; \langle cdecls \rangle \mid \langle cdecl \rangle \\
\langle cdecl \rangle &\longrightarrow \langle typedecl \rangle \mid \langle vardecl \rangle \mid \langle objdecl \rangle \\
&\quad \mid \langle methoddecl \rangle \\
\langle procdecl \rangle &\longrightarrow \texttt{proc } pname (\langle arglst \rangle) \langle pbody \rangle \\
\langle methoddecl \rangle &\longrightarrow \texttt{method } mname (\langle arglst \rangle) \\
&\qquad \langle pbody \rangle \\
\langle pbody \rangle &\longrightarrow \texttt{begin} \langle pblock \rangle \texttt{end} \\
\langle pblock \rangle &\longrightarrow \langle pdecl \rangle ; \langle pblock \rangle \mid \langle comlst \rangle
\end{aligned}$$

$$\begin{aligned}
\langle pdecl \rangle &\longrightarrow \langle vardecl \rangle \mid \langle objdecl \rangle \\
\langle arglst \rangle &\longrightarrow \langle arg \rangle ; \langle arglst \rangle \mid \langle arg \rangle \\
\langle arg \rangle &\longrightarrow arg : type \mid \texttt{ref } arg : type \\
\langle comlst \rangle &\longrightarrow \langle com \rangle ; \langle comlst \rangle \\
&\quad \mid \langle comlst \rangle \| \langle comlst \rangle \\
&\quad \mid \texttt{do} \langle alt\text{-}set \rangle \texttt{od} \mid \langle pbody \rangle \\
&\quad \mid \langle com \rangle \\
\langle alt\text{-}set \rangle &\longrightarrow \langle alt\text{-}set \rangle \ \square \ \langle alt\text{-}set \rangle \\
&\quad \mid \langle comlst \rangle ; \texttt{exit} \\
&\quad \mid \langle comlst \rangle ; \texttt{repeat} \\
\langle com \rangle &\longrightarrow \langle AtAct \rangle \mid \langle ProcCall \rangle \\
&\quad \mid \langle MethodCall \rangle \\
\langle AtAct \rangle &\longrightarrow \langle expr \rangle \\
\langle expr \rangle &\longrightarrow \langle var \rangle \mid \langle classvar \rangle \mid \langle const \rangle \\
&\quad \mid \langle expr \rangle \langle op \rangle \langle expr \rangle \\
&\quad \mid \langle op \rangle \langle expr \rangle \\
\langle classvar \rangle &\longrightarrow this . \langle var \rangle \\
\langle var \rangle &\longrightarrow {'v} \mid v'
\end{aligned}$$

Fig. 1. The syntax of $OB(PN)^2$.

translated into transition guards and can thus be combined with the logical and operator. Variables that are members of a class are distinguished from global or procedure variables by the prefix *this*, however whenever the context is obvious we will leave this prefix out.

We make a restriction to the language: class and procedure declarations may not be nested, ie. they may only appear at the global scope.

3   M-Nets

This section gives the basic definitions of M-nets and tries to explain the basic intuitions behind the composition operators for M-nets. For a deeper explanation the reader is referred to [4].

High-level nets are nets where the places and transitions are annotated. The choice of formalism for these annotations determines the firing-rule of the model. Common to all these formalisms is, that they can also be unfolded into low-level nets by substituting values for all the variables in the annotations [13]. M-nets [4] extend the notion of high-level nets by introducing *parameterized labels* on transitions, and these labels are used to compose larger nets using composition operators. A CCS-like algebra for nets annotated with labels was already introduced by [3]. In [4] this idea is extended to allow parameters (variables) in the labels. This makes it possible to view the synchronization of complementary labels as communication between subnets [14].

3.1   The Definition of M-Nets

An M-net is a triple (S, T, ι), where S is the set of *places*, T is the set of *transitions*, and ι is a function called *inscription* that associates:

- to every place $s \in S$, $\iota(s)$ a pair $(\lambda_s \mid \alpha_s)$ where λ_s is the place label and α_s is the place annotation,
- to every arc $f \in (S \times T) \cup (T \times S)$, $\iota(f)$ a finite multi-set of typed variables or tuples of variables,
- to every transition $t \in T$, $\iota(t)$ a pair $(\lambda_t \mid \alpha_t)$, where λ_t is the transition label and α_t is the transition annotation.

The place label is an element of $\{e, i, x\}$ (e = entry place, i = internal place, and x = exit place). This label is used to determine how the places are to be treated when composing nets as explained below. The place annotation α_s is used to determine the *color* or *type* of the place. It is a nonempty set of values. Although the universe of values is unspecified, we shall assume that it contains at least the special value $\bullet$, the sets of integers and booleans, and the sets of finite tuples of integers and booleans. In figure 2 the place p_0 in N_1 is labelled $(i \mid Int)$. This means that the place is an internal place, and that it may contain any integer. On the other hand the places p_1 and p_2 of N_2 in figure 2 are labelled $(e, \{\bullet\})$, and $(x, \{\bullet\})$ respectively. Both places may only contain a black token ($\bullet$), and p_1 is an entry place, while p_2 is an exit place. We do not allow multiple instances of the same token in a place.

A transition label λ_t is a multiset of parametrised actions. A parametrised action is a term of the form $A(a_1, \ldots, a_n)$, where A is the action symbol and $a_1, \ldots, a_n$ are either variables or values. Each action symbol A has an arity $ar(A) = n$, and the set of action symbols is equipped with a bijection — for which $\forall A : (A \neq \overline{A}) \wedge (\overline{\overline{A}} = A) \wedge (ar(A) = ar(\overline{A}))$. For example in N_2 $x(k, l)$ is a parametrised action with two variables k and l, while in N_1 $\overline{var}(oval, nval)$ is its conjugation. Thus the variables in the conjugate may be different. Each transition has also a transition annotation, a *predicate* associated with it. The predicate is used to establish when a transition is *enabled* and thus also often called *guard*.

Let $Var(t)$ consists of the set of variables occurring in $\iota(s, t)$, and $\iota(t, s')$, for $s, s' \in S$ and of the free variables in $\iota(t)$. An assignment σ of values to the variables in $Var(t)$ is *enabling* iff the predicate α_t evaluates to **true**. A marking M is a function that associates to each place s a finite multiset over α_s. A transition t is *enabled* at a marking M iff there exists an enabling assignment of values σ of the variables in $Var(t)$ and the multiset of values $\iota(s, t)[\sigma]$, ie. $\iota(s, t)$ with the variables instantiated with values given by σ, is a submultiset of $M(s)$. The successor marking M' is then calculated as: $M'(s) = (M(s) - (\iota(s, t)[\sigma])) + (\iota(t, s)[\sigma])$. Eg. in the net N_2 an assignment $\{k \mapsto 1, l \mapsto 2\}$ would be an enabling assignment for t_2. At the marking $\{p_1 \mapsto \bullet\}$ the transition t_2 would be enabled. It could then fire and the resulting new marking would be $\{p_2 \mapsto \bullet\}$.

A predicate may contain operations on values. What the set of allowed operations is, is left open. The idea is that this set will be specified together with the universe of values. We shall assume that at least the normal operations on integers and booleans are available. Our definition above differs from the one given in [4] in that we allow arbitrary types for entry and exit places and tuples of variables as arc annotations.

3.2   The Composition Operations

In this section we describe five composition operations for M-nets, *parallel composition, transition synchronization, restriction, sequential composition,* and *choice.* The atomic M-nets used as arguments to the composition operators will often be called *boxes.*

The simplest of these operations is parallel composition of two nets $\Sigma_1 \| \Sigma_2$. It is formed as the disjoint union of the nets. Restriction is also straightforward, the result of Σ **rs** A is the net where all transitions whose labels contain the action symbol A or its conjugate $\overline{A}$ are deleted, together with the connecting arcs.

Transition synchronization is defined in two steps. The synchronization of a net Σ on an action symbol A, denoted Σ **sy** A is defined as the smallest net satisfying:

- The set of places of Σ and Σ **sy** A are the same.
- Every transition of Σ is also a transition of Σ **sy** A.
- If t, t' are transitions of Σ **sy** A and t'' arises from t and t' through a *basic synchronization* step on $(A, \overline{A})$ then t'' is also a transition of Σ **sy** A.

A transition t'' arises trough a basic synchronization of two transitions t, t' if:

1. The label of t contains a term $A(\ldots)$ and the label of t' contains a term $\overline{A}(\ldots)$.
2. Variables in $A(\ldots)$ and $\overline{A}(\ldots)$ are unified and consistently renamed, and the renaming is applied to all the labels and annotations of t and t'.
3. The label of t'' equals the multi-set sum of the labels of t and t', minus the two terms that were unified in step 2.
4. The arcs surrounding t'' are the multi-set of the arcs surrounding t and t'.
5. The annotation of t'' is the logical conjunction of the annotations of t and t'.

The intuition of the synchronization is best explained with an example. In figure 2, place p_0 of N_1 contains a value that may be read and written through the transition labeled $\overline{x}(oval, nval)$ (it can be thought of as a memory-cell for storing the value of a B(PN)^2 variable 'x'). N_2 is the result of translating the statement $x = x + 1$ into an M-Net. We first unify the parametrised actions $x(k, l)$ and $\overline{x}(oval, nval)$ and obtain $x(n, m)$ and $\overline{x}(n, m)$. This means that the variables on the arcs in N_1 must me relabelled $oval \mapsto n$ and $nval \mapsto m$, while the predicate of t_2 becomes $m = n + 1$. After synchronisation and restriction over $\{x\}$ we get as a result the net in figure 2(c). Transition t_3 will increment the value of the

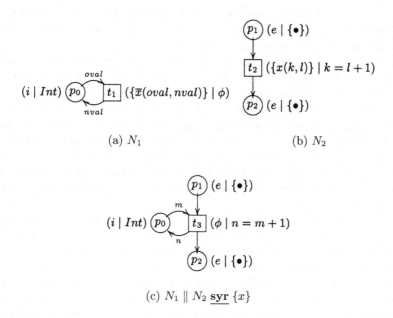

(a) N_1

(b) N_2

(c) $N_1 \parallel N_2 \; \mathbf{\underline{syr}} \; \{x\}$

Fig. 2. An example of a synchronization.

token x by 1 when fired. Synchronization and restriction are often done over the same set of labels, and we abbreviate this as $\mathbf{\underline{syr}}\{var\} = \mathbf{\underline{sy}}\{var\}\mathbf{\underline{rs}}\{var\}$.

All the nets that we will apply our operations on will have single entry and exit places. Thus the operations of sequential composition and choice are very simple. The sequential composition of two nets $\Sigma_1; \Sigma_2$ is obtained by merging the exit place of Σ_1 with the entry place of Σ_2. For choice $\Sigma_1 \square \Sigma_2$ we merge the entry places of Σ_1 with those of Σ_2 and the exit places of Σ_1 with those of Σ_2.

4   The Class Box Concept

The fundamental concept that will be used throughout the translation is that of a box. A box is a M-net with the following properties [3]:

- a distinguished set of *input*-places with no incoming arcs,
- a distinguished set of *output*-places with no outgoing arcs, and
- every transition should have at least one pre-place and one post-place (although we will relax this requirement).

The idea of the box-algebra is to give a set of operations, namely sequential and parallel composition, choice, iteration, recursion, and synchronization together with restriction, for composing boxes. As a result we obtain a net calculus that combines the advantages of Net theory (eg. causal semantics) with the advantages of process algebra (eg. compositionality). In the translation we will give boxes for all the basic objects of the language: classes, object, variable, types, procedures, methods, etc., while the control of the program will be translated as the operations of the box-calculus.

An object can be seen as an abstract data-type together with an internal state and an identity. The object has a set of internal variables, and a set of methods for manipulating its state. As the idea in the translation is to create a M-Net (a *box*) for each concept, we obtain one *variable-box* for each internal variable, one *procedure-box* for each procedure and method, and a *classid-box* to store the identity of the object. These boxes are then composed in parallel and synchronized and restricted over all internal labels. The resulting net is called a *class-box*. We will start the discussion by first introducing the ideas behind the concept of a procedure-box. Then we will look at the class box and finally introduce references.

The main problem in the translation of procedures is the need to distinguish between the values of the local variables of different instances of the procedure. A solution to this problem using *frame-indices*[2] was described in [17], and we shall briefly recall the idea here.

Let us start by looking at the execution of a recursive procedure in a parallel programming language. Assume that we have a program with some number of parallel threads where in two threads we will call a recursive procedure $P()$ with no parameters. In the case of a sequential language one would for each call allocate a stack frame that contains the information identifying the instance, but in a concurrent language each thread must have its own call stack. In terms of high-level nets this would suggest that our formalism should support a stack ADT. However if we *fold* the different instances of the procedure into one high-level *procedure box*, and use colored tokens, in this case integers, as indices to distinguish between the instances, it turns out that we can simulate the stacks by simply using pairs of integers. Each procedure instance has an index and this index stored by the caller. At the return of the callee the caller then knows which index to expect, and so the right thread is continued.

An example will make this idea clear. As our example we will use the short $OB(PN)^2$ program 3(a). The program consists of a single procedure declaration and a main program in which the procedure is called in parallel. The procedure may either call itself recursively, or return by an empty (*skip*) statement. The net N_1 of figure 3(b) is the procedure box of the procedure P, while the net N_2 of figure 3(c) is the parallel composition of the two calls to P. We now claim that the parallel composition of these nets N_1 and N_2 synchronized and restricted over the call and return actions $\{P_c, P_r\}$, will achieve the wanted behavior. To understand the idea it is easier to first consider the procedure box in figure 3(b) alone, and think of the synchronizations and restrictions as happening "on-the-fly" while the net is executed. Let us look at the structure of the procedure box N_1. The box is initialized at the beginning of the program by transition t_0 labeled by *init*, which stores a token 0 in place p_1. This place always

---

[2] In compilers the information needed by a single execution of a procedure is managed by an activation record or *frame*

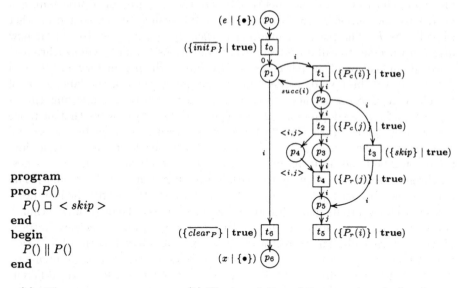

```
program
proc P()
  P() □ < skip >
end
begin
  P() ∥ P()
end
```

(a) The program.

(b) The translation of the procedure declaration.

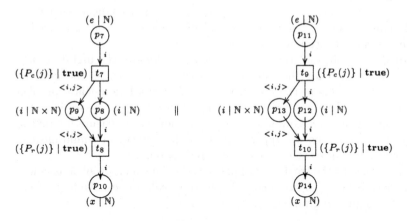

(c) The translation of the body of the program.

Fig. 3. A small OB(PN)² program and parts of its translation.

contains the next free frame-index. The box is now initialized and transition t_1, modeling procedure entry and labeled $\overline{P_c}(i)$, can synchronize with the either of the transitions labelled $P_c(x)$ in N_2. When this synchronization happens, t_1 fires resulting in a marking $\{p_1 \mapsto 1, p_2 \mapsto 0\}$. The token 0 in p_2 represents the fact that *instance* 0 of P is now ready to execute. In this case the body of the procedure consists of a choice between a *skip* transition t_3, or a procedure call t_2. Let us now suppose that before instance 0 continues its execution another thread calls P. The resulting marking is then $\{p_1 \mapsto 2, p_2 \mapsto \{0,1\}\}$. If now instance 1 decides to call itself recursively, transitions t_1 and t_2 will synchronize. Through unification of the parameters of the labels, the parameter x of the label $P_c(x)$ of t_2 will be bound to the value of the parameter i of the label $\overline{P_c}(i)$ of t_1. The pair $< x, i >$ is then stored in p_4, and we obtain the marking $\{p_1 \mapsto 3, p_2 \mapsto \{0,2\}, p_3 \mapsto 1, p_4 \mapsto < 1, 2 >\}$. The marking at p_3 means that instance 1 is waiting for a return. The instance of this return is given by the second component of the pair $< 1, 2 >$. If instance 2 now decides to do a *skip* action, the token 2 will end up in place p_5. Transition t_5, the exit transition, can now synchronize with t_4, resulting in the withdrawal of the pair $< 1, 2 >$ from p_4. The marking then becomes $\{p_1 \mapsto 2, p_2 \mapsto 0, p_5 \mapsto 1\}$ or in other words the instance 1 is ready to return to its main thread. As the execution of the program continues, new pairs $< i, x >$ are stored in p_4, and then retrieved through successive returns of the recursive calls. The correctness of this concept is based on the observation that the pairs $< i, x >$ in p_4 form disjoint ascending chains of natural numbers. In other words, the pairs $< i, x >$ "simulate" independent stacks for the different threads of execution. The transition t_0 (*init*) and t_6 (*clear*) are used to initialize the procedure box at the beginning of its scope and to clear the procedure box at the end of the scope. The reader may wonder about the fact that p_3 is redundant with p_4 (as is p_8 (p_{12}) with p_9 (p_{13})). The place p_4 is there to indicate that it can be refined. This is useful if one wants to implement an asynchronous procedure call, where the caller continues execution after the call.

We should also point out that we no longer can think of the high-level nets as short-hand notation for the unfolded low-level nets, because the unfolded nets will are infinite, due to the use of integers. This is a fundamental problem that cannot be solved without restricting the set of active procedure instances somehow. Fleischhack and Grahlmann [10] have devised a scheme for reusing frame-indices, where given an upper bound on the number of simultaneously active procedure instances the unfolding can be made finite.

Let us next turn to the treatment of parameters. The main problem is how to modify the data-boxes of [5] to handle the different call frames. We first extend our running example program with a parameter x:

proc $P(x : \{0,1\})$
    begin
        $P(x)\square < skip >$
    end

The procedure-box of figure 3(b) can be used mostly as such, only the labeling of the transitions has to be modified. The data-box of [5] has to be modified so that we can distinguish between the different instances of the variables. This is done by storing the frame-index along with the value of the variable. It is also convenient to change the definition of the data-box slightly. In the original version there is no distinction between the first use and the initialization of the

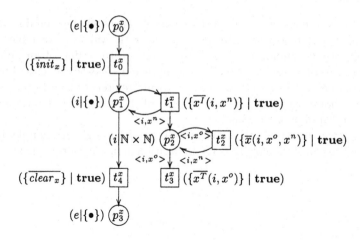

Fig. 4. The Databox of x.

variable. In the case of local variables in procedures and procedure parameters it is advantageous to have a separate initialization of each variable. In figure 4 the new version of the data box is shown. The transitions labeled *init* and *clear* will initialize and clear the box at scope begin and end respectively.

Next we need to transfer the value of the parameter into the data-box. This is accomplished through the labeling of transition t_1 in figure 3(b). Parameter transfer between caller and callee is done through unification of the corresponding parameters during synchronisation. In our case the label of t_1 will be as follows:

$$\{\overline{P_c}(i,x), x^I(i,x^n) \mid x = x^n\}$$

It is now easy to see, that when synchronizing and restricting the parallel composition of the procedure box with the data boxes, the parameters will be copied and transfered.

Let us now turn to the translation of classes and objects. In [17] we only needed frame-indices to distinguish between different instances of a variable. When one introduces objects this is no longer enough, because now we need to distinguish between instances of member variables that belong to different objects. To accommodate this we introduce an extra identifier, the *object-identifier*, which can be assigned at compile-time. The object-identifier together with a frame-index are sufficient to always unequivocally distinguish between different instances of a variable. The object-identifier, frame-index pair is called a *scope-index*. Since the frame-indices are generated dynamically it suffices to generate all the object-identifiers at compile time.

The crucial point to notice now is that the scope-index together with the identifier of the object or variable is a globally unique way to identify them. This means that we can store all the values of variables in one place, and use the scope-index together with the identifier to access the value. Thus we have in essence created *references*. In our translation references will be stored in variable-boxes, and special *type-boxes* will be use to store the values of variables.

For objects we cannot use type-boxes, because the value of an object is a much more complex entity. Instead we introduce a *class-box*. A class-box is the parallel composition of a *classid-box* for storing references to all objects of the class, of method-boxes, one for each method of the class, and of variable-boxes, one for each variable of the class.

The best way to explain the translation is again through an example. Figure 6 is a part of the translation of the class declaration in figure 5 after synchronisation

class A
    var $x : Int$;
    method $inc()$
    $\underline{d}$
    $< x' ='\, x + 1 >$
    $\underline{fin}$
endc

Fig. 5. A simple class A declaration in $OB(PN)^2$.

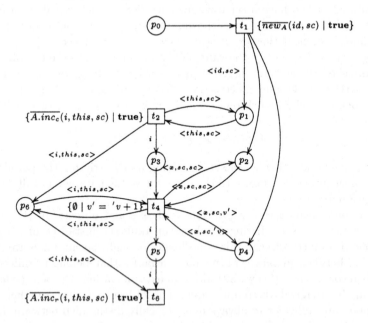

Fig. 6. N_3, Parts of the translation of the class A.

and restriction. The places p_0, p_1 and the transition t_1 are generated by the classid-box. The method inc generates $t_2, p_3, t_4, p_5, t_6, p_6$, and place p_2 represents the variable x, while p_4 will store the values of x. We have left out those parts of the translation that are only needed for the initialisation of the nets, togheter with some labels that would have cluttered the presentation

Transition t_2, labeled $\overline{A.inc}(i, this, sc)$ is the *method-call* transitions. The parameter i is the frame-index that is used to distinguish between different instances of the method, while *this* and *sc* form a reference to the object. Assume that somewhere in the program the object a is declared, and in the same scope later on a call $a.inc()$ is made. The initialisation of the object a is done by transition t_1 that store the object-identifier, scope-index pair for a in place p_1. At the same time the local variable x is created. The variable x is stored in p_2 while the initial value of x is stored in p_4. The call will be translated like a normal procedure call, but an implicit extra parameter *this* will be added. A reference to the object a will be passed as the parameters *this* and *sc*, and the call $a.inc()$ will synchronise with the transition t_2 in the net N_3. Firing t_2 results in a check that the object has really been created, ie. the correct identifiers must exist in p_1. At the same time the reference to a is copied into the local variable p_6 of the method. Transition t_4, which is the translation of the statement $v' =' v + 1$, then retrieves the right reference to a from p_6, and then using the scope information of a retrieves the reference to the variable x from the place p_2. This reference is finally used to retrieve the value of x from p_4 which is then incremented by one, and stored back into p_4.

5   The Translation into M-Nets

The translation will be given in terms of a function $\mathcal{T}[\![\,]\!]$ from the syntactic entities of OB(PN)^2 to M-nets. The general structure of the translation function is given in figure 7. Each syntactic construct in the grammar is given a translation into either a M-Net, or a M-Net expression over the translation of the components of the syntactic construct. Since the syntax is very simple, the translation function essentially defines a recursive descent translator.

5.1   Declarations

The $\langle$program$\rangle$ declaration defines the global scope. The global scope is used to restrict over the *init* and *clear* transitions of all declarations in the program. Thus all the boxes are initialized and cleared synchronously at the same time.

$\mathcal{T}[\![\texttt{program } programname \; \langle\text{body}\rangle]\!] = \mathcal{T}[\![\langle\text{body}\rangle]\!]$

$$\underline{\textbf{syr}} \; Init \; (\langle\text{body}\rangle) \; \bigcup Clear \; (\langle\text{body}\rangle)$$

The **begin-end** bracket is semantically transparent:

$$\mathcal{T}[\![\texttt{begin } \langle\text{block}\rangle \texttt{ end}]\!] = \mathcal{T}[\![\langle\text{block}\rangle]\!]$$

Creation and Termination of Scope. By the scope of a declaration we mean the part of the program to which a declaration applies. Hence the declaration *creates* a scope, ie. the declared entity becomes visible to the program. The scope

$$\mathcal{T}[\![\texttt{program } programname \ \langle body\rangle]\!] = \mathcal{T}[\![\langle body\rangle]\!] \ \textbf{syr } Init \ (\langle body\rangle) \ \bigcup Clear \ (\langle body\rangle)$$

$$\mathcal{T}[\![\langle decl\rangle\,; \ \langle block\rangle]\!] = \left(\mathcal{T}[\![\langle decl\rangle]\!] \parallel \left(\gamma^I \ (\langle decl\rangle) \ ; \ \mathcal{T}[\![\langle block\rangle]\!]; \ \gamma^T \ (\langle decl\rangle)\right)\right)$$
$$\textbf{syr } \delta \ (\langle decl\rangle)$$

$$\mathcal{T}[\![\texttt{type } tname:\ type]\!] = T \ (tname, type)$$

$$\mathcal{T}[\![\texttt{var } vname:\ tname]\!] = V \ (vname, tname)$$

$$\mathcal{T}[\![\texttt{class } cname \ \langle cdecls\rangle \ \texttt{endc}]\!] = C \ (name) \ \parallel \ \mathcal{T}[\![\langle cdecls\rangle]\!] \ \textbf{syr } \delta_C$$

$$\mathcal{T}[\![\texttt{proc } pname(\ \langle arglst\rangle\ ) \ \langle pbody\rangle]\!] = P \ (name, args) \ [Z \leftarrow \mathcal{T}[\![\langle pbody\rangle]\!]]$$
$$\parallel V \ (\langle arglst\rangle) \ \textbf{syr } \delta \ (\langle arglst\rangle)$$

$$\mathcal{T}[\![P_1 \| P_2]\!] = \mathcal{T}[\![P_1]\!] \| \mathcal{T}[\![P_2]\!]$$

$$\mathcal{T}[\![P_1;\ P_2]\!] = \mathcal{T}[\![P_1]\!] \ ; \ \mathcal{T}[\![P_2]\!]$$

$$\mathcal{T}[\![P_1 \ \Box \ P_2]\!] = \mathcal{T}[\![P_1]\!] \ \Box \ \mathcal{T}[\![P_2]\!]$$

$$\mathcal{T}[\![\texttt{do};\ alt-set \ \texttt{od}]\!] = \left[N^{\mathcal{I}}_{silent} * R \ (alt-set) * E \ (alt-set)\right]$$

$$\mathcal{T}[\![\langle expr\rangle]\!] = \left(\begin{array}{c} {\scriptstyle (e\ \mid\ \aleph)} \quad {\scriptstyle l} \quad {\scriptstyle (x\ \mid\ \aleph)} \\ \circ \xrightarrow{\ i\ } \Box \xrightarrow{\ i\ } \circ \end{array} \right)$$

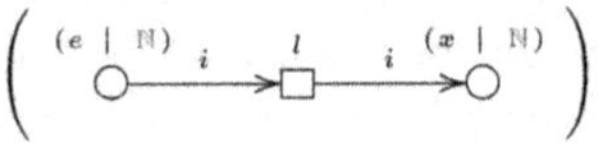

Fig. 7. The general structure of the translation function $\mathcal{T}[\![\ ]\!]$.

is *terminated* when the block after the declaration ends. In practice this means that a declaration is visible until the end of the program, procedure, class or method, in which it is declared.

The creation and termination of scope takes care of two things:

1. The creation and deletion of local variables that have been declared at the scope, and
2. the synchronization and the restriction over type, class, method and procedure names that have been declared at the scope.

The creation and deletion of local variables is done by synchronising with the transitions labelled new_{name}, $delete_{name}$, that created and delete the entity *name*. This is encapsulated in the functions $\gamma^I(< decl >)$, $\gamma^I(< decl >)$, of the nets N_I, and N_T below. The creation and termination of scopes in the translated M-Net is achieved by controlling the visibility of the transitions of the entity. This is done by synchronisation and restriction and is encapsulated into the function $\delta \ (\langle decl\rangle)$. The description of the translation of declarations now amounts to defining the functions $\mathcal{T} \ [[\langle decl\rangle]]$, and $\delta \ (\langle decl\rangle)$, for the different $\langle decl\rangle$'s in the grammar. Note that the initialisation of the type, class, method and procedure boxes is done by the program declaration.

$$\mathcal{T}[\![\langle decl\rangle\,; \ \langle block\rangle]\!] =$$

$$\left(\mathcal{T}[\![\langle decl\rangle]\!] \parallel (N_I \ (\langle decl\rangle) \ ; \ \mathcal{T}[\![\langle block\rangle]\!] \ ; \ N_T \ (\langle decl\rangle)) \ \textbf{syr } \delta \ (\langle decl\rangle)\right)$$

where,

$$N_I(< decl >) = \quad \overset{(e \mid N)}{\underset{i}{\bigcirc}} \xrightarrow{} \overset{\gamma^I(<decl>)}{\underset{i}{\square}} \xrightarrow{} \overset{(x \mid N)}{\bigcirc}$$

$$N_T(< decl >) = \quad \overset{(e \mid N)}{\underset{i}{\bigcirc}} \xrightarrow{} \overset{\gamma^T(<decl>)}{\underset{i}{\square}} \xrightarrow{} \overset{(x \mid N)}{\bigcirc}$$

Scope Identifiers. The correctness of the translation depends on the management of identifiers to distinguish different instances of variables. Essentially we need to maintain a "pointer" to the scope in which the variable was declared. There are 4 different types of scopes in OB(PN)^2 in which a variable can appear: the global scope, the scope of a procedures, the scope of class declaration, and the scope of a class member procedure (ie. method). We can view the global scope as a special anonymous or unnamed class declaration, so that the above 4 types of scope actually reduce to 2: the scope of class declaration, and the scope of a class member procedure. Since there may be several instances of a procedure or object active at the same time it is necessary to add an extra identifier, the *frame-index* to distinguish between the instances. Thus a *scope-index* is a pair $< sid, fid >$ where the left and right parts are interpreted as follows:

Global variable: Since there exists only one global scope, that is instantiated only once, we use the special pair $< \emptyset, \emptyset >$ to denote this.

Procedure variable: The scope-index is of the form $< \emptyset, fid >$, where $\emptyset$ denotes the fact that the procedure is declared in the global scope, while $fid \in \mathbb{N}$ is the frame-index of the procedure instance.

Class member variable: The scope-index is of the form $< oid, fid >$, where the $oid \in \mathbb{N}$ is the unique identifier of the object that the variable belongs to, and $fid \in \mathbb{N}$ is the frame-index of the object.

Class member procedure variables: The scope-index $< oid, fid >$ of variables declared in a class member procedure consists of the unique identifier of the object that the method belongs to, together with the frame-index of the instance of the method.

Thus the set of scope-indices (denoted *scid* in the figures below) is the set $(\mathbb{N} \cup \{\emptyset\} \times \mathbb{N} \cup \{\emptyset\})$. To simplify the translation we assume that all variables and objects that have been declared in the program have been assigned a unique id, and that this information is available at the beginning of the translation.

Type Declaration

$$\langle typedecl \rangle \longrightarrow \textbf{type } tname{:}\langle constructor \rangle$$

A type declaration is used to define subsets of the integers, although nothing prevents us to define other basic types together with some more complex type constructors. However since the basic types are the types provided by the underlying M-Net formalism, we have to remember to also extend the the M-Net formalism with the corresponding complex type.

Each type declaration is translated into a *type-box* (cf. figure 8). The type box is used to map variables to values. Thus all variables of the same type are stored in the same type box. More precisely, the type-box is used to store the id of the variable, the scope-index of the scope in which it was created (remember that we need the scope-index to correctly identify an instance of a variable), and its value.

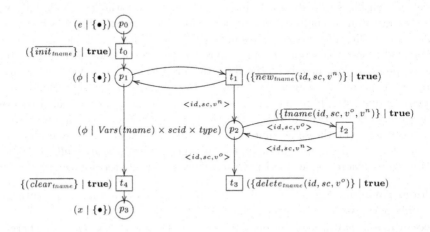

Fig. 8. The type box $T(tname, type)$ for storing values of variables of type $tname$.

The translation is given as follows:

$$\mathcal{T}[\![\textbf{type } tname: type]\!] = T(tname, type)$$
$$\gamma^I(\textbf{type } tname: type) = \emptyset$$
$$\gamma^T(\textbf{type } tname: type) = \emptyset$$
$$\delta(\textbf{type } tname: type) = \{new_{tname}, delete_{tname}, tname\}$$

Above, the set of variables of type $tname$ is denoted by $Vars(tname)$. Since we only allow subsets of integers, $type$ is of the form $\{l, \ldots, k\}$, where $l <= k$ and $l, k \in \mathbb{Z}$.

The basic types are the types defined in the M-Net formalism, ie. integers and booleans. We assume that a set of type boxes for the basic types (Int, Bool) are automatically declared directly at the beginning of the program.

Variable Declaration

$$\langle vardecl \rangle \longrightarrow \textbf{var } vname: tname$$
$$\langle cvardecl \rangle \longrightarrow \textbf{var } vname: tname$$
$$\langle mvardecl \rangle \longrightarrow \textbf{var } vname: tname$$

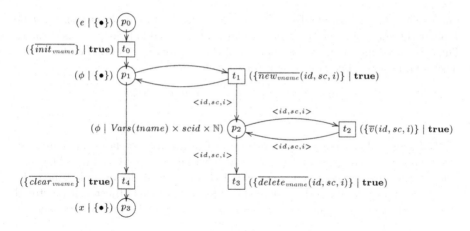

Fig. 9. A variable box $V(vname, tname)$.

For each variable declaration we generate a variable box (see figure 9). The variable box stores the unique id of the variable, the index to the scope that created it, and the index of the current owner. The pair $< varid, scid >$ conceptually is equivalent to a *pointer* in traditional programming languages like C. One still needs a frame-index to distinguish between the different instances of the variable.

The variable declaration must be treated differently depending on where in the program it appears (ie. global and procedure scope, class scope, or method scope), because the scopes of the variable are defined differently. We shall first give the definition for the global and procedure scope, and then discuss the creation and deletion for the other cases.

Global and procedure scope: When we declare a new variable we need to create a scope-index for it. If the scope is the global or procedure scope, the first element of the scope index is $\emptyset$. The second element is the frame-index i of the current procedure frame ($\emptyset$ again in the case of the global program). Remember that we assumed that all variables in the program have been assigned unique identifiers. The function $uid(vname)$ returns returns this id. The initial value (*initialvalue*) of the variable is **true** for booleans, 0 for integers, and l if the type is a range $\{l, \ldots, k\}$.

$$\mathcal{T}[\![\mathbf{var}\ vname:\ tname]\!] = V(vname, tname)$$
$$\gamma^I(\mathbf{var}\ vname:\ tname) = (\{new_{vname}(id, sc, i), new_{tname}(id, sc, v)\}\ |$$
$$id = uid(vname), sc =< \emptyset, i >,$$
$$v = initialvalue)$$
$$\gamma^T(\mathbf{var}\ vname:\ tname) = (\{delete_{vname}(id, sc, i), delete_{tname}(id, sc, i)\}$$
$$|\ \mathbf{true})$$
$$\delta(\mathbf{var}\ vname:\ tname) = \{new_{vname}, delete_{vname}, vname\}$$

Class scope: For a variable declared in a class, its creation and deletion must be done at the same time as the object to which the variable belongs to is created. Thus the handling of the scope index and the initial value is handled when the class is translated. The translation function is simply:

$$\mathcal{T} \, [\![\textbf{var } vname\colon tname]\!] = V \, (vname, tname)$$

Method scope: The creation and deletion of a variable declared in a method is similar to the creation and deletion of a variable in a procedure. The only difference is how the scope index is created. The owner of the variable is now the object to which the method belongs. The identifier of this object will be stored in a special place *this*, that we need to access. The translation then becomes:

$$\mathcal{T}[\![\textbf{var } vname\colon tname]\!] = V(vname, tname)$$
$$\gamma^I(\textbf{var } vname\colon tname) = (\{new_{vname}(id, sc, i), new_{tname}(id, sc, v),$$
$$this(oid, < sid, fid >, i)\} \mid$$
$$id = uid(vname), sc = < oid, i >,$$
$$v = initialvalue)$$
$$\gamma^T(\textbf{var } vname\colon tname) = (\{delete_{vname}(id, sc, i), delete_{tname}(id, sc, v),$$
$$this(oid, < sid, fid >, i)\} \mid sc = < oid, i >)$$
$$\delta(\textbf{var } vname\colon tname) = \{new_{vname}, delete_{vname}, vname\}$$

Object Declaration

$$\langle objdecl \rangle \longrightarrow \textbf{obj } oname\colon cname$$
$$\langle cobjdecl \rangle \longrightarrow \textbf{obj } oname\colon cname$$
$$\langle mobjdecl \rangle \longrightarrow \textbf{obj } oname\colon cname$$

An object declaration is basically like a variable declaration and thus the discussion about the differences between the different scopes applies here too. The difference is that the object should be created in a class box (defined below) instead of a type box, thus no initial value is needed. The translations are as follows.

Global and procedure scope:

$$\mathcal{T}[\![\textbf{obj } oname\colon cname]\!] = V(oname, cname)$$
$$\gamma^I(\textbf{obj } oname\colon cname) = (\{new_{oname}(id, sc, i), new_{cname}(id, sc)\} \mid$$
$$id = uid(oname), sc = < \emptyset, i >)$$
$$\gamma^T(\textbf{obj } oname\colon cname) = (\{delete_{oname}(id, sc, i), delete_{cname}(id, sc)\}$$
$$\mid \textbf{true})$$
$$\delta(\textbf{obj } oname\colon cname) = \{new_{oname}, delete_{oname}, oname\}$$

Class scope:

$$\mathcal{T}[\![\text{obj } oname:\ cname]\!] = V(oname, cname)$$

Method scope:

$$\mathcal{T}[\![\text{obj } oname:\ cname]\!] = V(oname, cname)$$
$$\gamma^I(\text{obj } oname:\ cname) = (\{new_{oname}(id, sc, i), new_{cname}(id, sc),$$
$$this(oid, <sid, fid>, i)\} \mid$$
$$id = uid(oname), sc = <oid, i>)$$
$$\gamma^T(\text{obj } oname:\ cname) = (\{delete_{oname}(id, sc, i), delete_{cname}(id, sc),$$
$$this(oid, <sid, fid>, i)\} \mid sc = <oid, i>)$$
$$\delta(\text{obj } oname:\ cname) = \{new_{oname}, delete_{oname}, oname\}$$

Procedure Declarations

$$\langle\text{procdecl}\rangle \longrightarrow \textbf{proc } pname(\ \langle\text{arglst}\rangle\ )\ \langle\text{pbody}\rangle$$
$$\langle\text{pbody}\rangle \longrightarrow \textbf{begin } \langle\text{pblock}\rangle\ \textbf{end}$$
$$\langle\text{pblock}\rangle \longrightarrow \langle\text{pdecl}\rangle\ ;\ \langle\text{pblock}\rangle \mid \langle\text{comlst}\rangle$$
$$\langle\text{pdecl}\rangle \longrightarrow \langle\text{vardecl}\rangle \mid \langle\text{objdecl}\rangle$$
$$\langle\text{arglst}\rangle \longrightarrow \langle\text{arg}\rangle\ ;\ \langle\text{arglst}\rangle \mid \langle\text{arg}\rangle$$
$$\langle\text{arg}\rangle \longrightarrow arg:\ tname \mid \textbf{ref } arg:\ tname$$

A procedure is translated into an instance of the net in figure 10, together with a set of variable boxes to keep the arguments of the procedures. The semantic function is the given by:

$$\mathcal{T}[\![\textbf{proc } pname(\ <arglst>\ )\ <pbody>]\!] =$$
$$P(pname)\,[Z \leftarrow \mathcal{T}[\![<pbody>]\!]]\ \|$$
$$V(<arglst>)\ \underline{\textbf{syr}}\ \delta(<arglst>)\ ,$$

where $P(pname)$ denotes the net in figure 10. In the procedure box the translation of the procedure body is inserted for the transition Z.

The translation of the arguments is relatively straightforward: we create one variable box for each argument and set up labels on the procedure creation transition so, that correct variables and values are copied into these variable boxes. The only thing we need to take care of is that the variable boxes are initialised correctly. For value parameters a new variable is created, while for reference parameters the reference is copied into the variable box. We only allow objects to be passed by reference. We shall assume that this restriction is enforced by some external means so that the translation becomes more manageable.

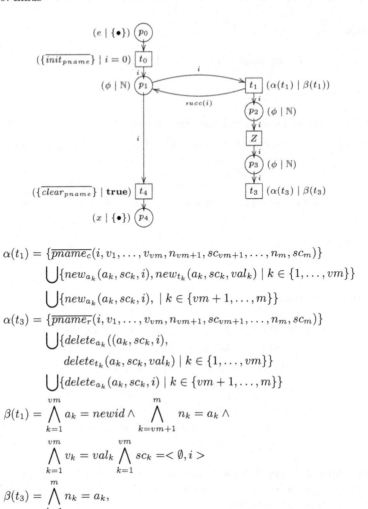

$$\alpha(t_1) = \{\overline{pname_c}(i, v_1, \ldots, v_{vm}, n_{vm+1}, sc_{vm+1}, \ldots, n_m, sc_m)\}$$
$$\bigcup\{new_{a_k}(a_k, sc_k, i), newt_{t_k}(a_k, sc_k, val_k) \mid k \in \{1, \ldots, vm\}\}$$
$$\bigcup\{new_{a_k}(a_k, sc_k, i), \mid k \in \{vm+1, \ldots, m\}\}$$
$$\alpha(t_3) = \{\overline{pname_r}(i, v_1, \ldots, v_{vm}, n_{vm+1}, sc_{vm+1}, \ldots, n_m, sc_m)\}$$
$$\bigcup\{delete_{a_k}((a_k, sc_k, i),$$
$$delete_{t_k}(a_k, sc_k, val_k) \mid k \in \{1, \ldots, vm\}\}$$
$$\bigcup\{delete_{a_k}(a_k, sc_k, i) \mid k \in \{vm+1, \ldots, m\}\}$$
$$\beta(t_1) = \bigwedge_{k=1}^{vm} a_k = newid \wedge \bigwedge_{k=vm+1}^{m} n_k = a_k \wedge$$
$$\bigwedge_{k=1}^{vm} v_k = val_k \bigwedge_{k=1}^{vm} sc_k =< \emptyset, i >$$
$$\beta(t_3) = \bigwedge_{k=1}^{m} n_k = a_k,$$

Fig. 10. The procedure box $P(pname)$ and the labelings.

The function V constructs a variable box for each reference variable that is declared in the parameter list:

$$V(\langle arglst \rangle) = \|_{k \in \{1 \ldots m\}} V(t_k, a_k),$$

while the function δ gives the initialization and termination labels of the variable boxes defined by the procedure declaration:

$$\delta(\langle arglst \rangle) = \{new_{a_k}, delete_{a_k} \mid k \in \{1, \ldots, m\}\}.$$

To ease the definition of the labelings in figure 10 we introduce an ordering on the parameters, so that the value parameters come before the reference parameters. Let vm be the number of value parameters, and m the number of all parameters in the declaration. Using this numbering, the parameters $1 \ldots vm$ are value parameters, while the parameter $vm + 1 \ldots m$ are reference parameters. We use a_i, t_i to refer to the i-th parameter.

The labelling of the procedure call transition t_1 consists of a parametrised action $\overline{pname_c}(i, v_1, \ldots, v_{vm}, n_{vm+1}, sc_{vm+1}, \ldots, n_m, sc_{vm_1})$ for synchronising with the procedure call, and a set of parametrised actions for setting up the arguments. The parameters $v_1 \ldots, v_{vm}$ are the values that are passed from the caller, while the parameters $n_{vm+1}, sc_{vm+1}, \ldots, n_m, sc_v$ are the references (n_k is the identifier while sc_k is the scope of the reference parameter k). Then for each of the arguments we need to initialise the corresponding variable box: For a value parameter a_k of type t_k, we need to initialise both the variable box with $new_{a_k}(a_k, sc, i)$ and the type box with $new_{t_k}(a_k, sc_k, val_k)$. A new identifier is created for a_k, and value val_k is assigned the value v_k from the parameter list, while sc_k is the scope for procedures, $< \emptyset, i >$ as discussed above. For a reference parameter a_r we just copy the identifier and the scope into the new box, ie. initialise the variable box with $new_{a_r}(a_r, s_r, i)$, and assign to a_r the identifier n_r, and s_r gets the value sc_r from the argument list. When the procedure finishes we we need to delete the corresponding variables from the local variable boxes. The details are encapsulated into the annotation functions α and β in figure 10.

Note that we only allow variable and object declarations in the procedure. This means that no nested procedures declarations and no local classes are allowed.

Class Declaration

$$\langle\text{classdecl}\rangle \longrightarrow \textbf{class } cname \,\langle\text{cdecls}\rangle\, \textbf{endc}$$

$$\langle\text{cdecls}\rangle \longrightarrow \langle\text{cdecl}\rangle \,;\, \langle\text{cdecls}\rangle \mid \langle\text{cdecl}\rangle$$

$$\langle\text{cdecl}\rangle \longrightarrow \langle\text{cvardecl}\rangle \mid \langle\text{cobjdecl}\rangle \mid \langle\text{methoddecl}\rangle$$

$$\langle\text{cvardecl}\rangle \longrightarrow \textbf{var } vname\text{: } tname$$

$$\langle\text{methoddecl}\rangle \longrightarrow \textbf{method } mname(\, \langle\text{arglst}\rangle\,) \,\langle\text{pbody}\rangle$$

A class declaration consists of local variable declarations and the declaration of some methods. The variable and method declarations are translated individually and composed in parallel with the class box (cf. figure 11). When an object is created, the class box has to initialise all the local variables. Thus the transitions t_1 and t_3 need to be labelled with the creation and destruction actions of the local variables. Methods are translated as procedures, except that the first parameter is a reference to the classid-box that stores the objects (this correspond to the implicit *this* or *self* parameter in object-oriented languages). In this way a method always references the right class-variables. What remains visible of the box are the object creation and deletion transitions together with the method call and return transitions.

$$\mathcal{T}\, [[\textbf{class } cname \,\langle\text{cdecls}\rangle\, \textbf{end}]] = C\,(cname) \underset{\parallel}{} \mathcal{T}\, [[\langle\text{cdecls}\rangle]] \, \underline{\textbf{syr}} \, \delta_C$$

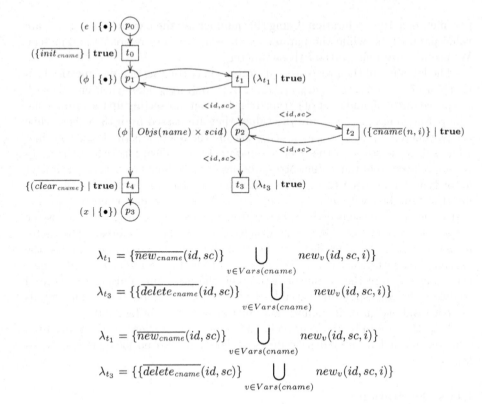

$$\lambda_{t_1} = \{\overline{new_{cname}}(id, sc)\} \bigcup_{v \in Vars(cname)} new_v(id, sc, i)\}$$

$$\lambda_{t_3} = \{\{\overline{delete_{cname}}(id, sc)\} \bigcup_{v \in Vars(cname)} new_v(id, sc, i)\}$$

$$\lambda_{t_1} = \{\overline{new_{cname}}(id, sc)\} \bigcup_{v \in Vars(cname)} new_v(id, sc, i)\}$$

$$\lambda_{t_3} = \{\{\overline{delete_{cname}}(id, sc)\} \bigcup_{v \in Vars(cname)} new_v(id, sc, i)\}$$

Fig. 11. The classid-box $C(cname)$

$$\gamma^I \text{ (class } cname \langle cdecls \rangle \text{ end)} = \emptyset$$
$$\gamma^T \text{ (class } cname \langle cdecls \rangle \text{ end)} = \emptyset$$
$$\delta \text{ (class } cname \langle cdecls \rangle \text{ end)} = \{new_{cname}, delete_{cname}, cname\}$$
$$\delta_C = \bigcup_{v \in Vars(\langle cdecls \rangle)} \{new_v, delete_v, v\}$$

Finally we also need to define the translation of the class declarations as a parallel composition of these (compare this to the translation at global and procedure scope in 5.1).

$$\mathcal{T} [[\langle cdecl \rangle ; \langle cdecls \rangle]] = \mathcal{T} [[\langle cdecl \rangle]] \| \mathcal{T} [[\langle cdecls \rangle]] .$$

Method Declaration

$$\langle methoddecl \rangle \longrightarrow \text{method } mname(\langle arglst \rangle) \langle pbody \rangle$$

A method declaration will basically be treated as a procedure call with one extra reference parameter, the object that it acts upon. This parameter will be called *this*. The translation is then given in terms of the procedure box of figure 10.

$$\mathcal{T} \, [[\text{method } \textit{mname} (\, \langle \text{arglst} \rangle \,) \, \langle \text{pbody} \rangle]] \, =$$
$$P \, (\textit{mname}, \textit{arglst}) \, [Z \leftarrow \mathcal{T} \, [[\langle \text{pbody} \rangle]] \,] \, \|$$
$$V \, (\langle \text{arglst} \rangle) \, \underline{\textbf{syr}} \, \delta \, (\langle \text{arglst} \rangle)$$

However the functions V and δ must now be modified to cope with the extra variable box for *this*.

$$V \, (\langle \text{arglst} \rangle) \, = V \, (\textit{cname}, \textit{this}) \, \|_{k \in \{1...m\}} \, V \, (t_k, a_k) \, ,$$
$$\delta \, (\langle \text{arglst} \rangle) \, = \{ \textit{new}_{\textit{this}}, \textit{delete}_{\textit{this}} \} \bigcup \{ \textit{new}_{a_k}, \textit{delete}_{a_k} \mid k \in \{1, \ldots, m\} \}.$$

We add a prefix *cname.* to the call and return transition labels, to allow for overloading of names. The labels of the net are given in figure 12.

5.2   Commands

$$\langle \text{comlst} \rangle \longrightarrow \langle \text{com} \rangle \; ; \langle \text{comlst} \rangle \mid \langle \text{comlst} \rangle \parallel \langle \text{comlst} \rangle$$
$$\mid \; \textbf{do} \, \langle \text{alt-set} \rangle \, \textbf{od} \mid \langle \text{pbody} \rangle \mid \langle \text{com} \rangle$$
$$\langle \text{alt-set} \rangle \longrightarrow \langle \text{alt-set} \rangle \, \square \, \langle \text{alt-set} \rangle$$
$$\mid \; \langle \text{comlst} \rangle \; ; \texttt{exit}$$
$$\mid \; \langle \text{comlst} \rangle \; ; \texttt{repeat}$$
$$\langle \text{com} \rangle \longrightarrow \langle \text{AtAct} \rangle \mid \langle \text{ProcCall} \rangle$$
$$\mid \; \langle \text{MethodCall} \rangle$$

A command is either a block, and atomic action or procedure/method call, a sequential or parallel composition of commands or an iteration construct. Each of these well be described below.

Atomic Actions. An atomic action is defined by the following grammar:

$$\langle \text{AtAct} \rangle \longrightarrow \langle \text{expr} \rangle$$
$$\langle \text{expr} \rangle \longrightarrow \langle \text{var} \rangle \mid \langle \text{classvar} \rangle \mid \langle \text{const} \rangle \mid \langle \text{expr} \rangle \, \langle \text{op} \rangle \, \langle \text{expr} \rangle \mid \langle \text{op} \rangle \, \langle \text{expr} \rangle$$
$$\langle \text{classvar} \rangle \longrightarrow \textit{this.} \, \langle \text{var} \rangle$$
$$\langle \text{var} \rangle \longrightarrow \, 'v \mid v' \mid v$$

$$\alpha(t_1) = \{\overline{cname.mname}(i, this, sc_{this}, v_1, \ldots, v_{vm},$$
$$n_{vm+1}, sc_{vm+1}, \ldots, n_m, sc_m)\}$$
$$\bigcup\{new_{a_k}(a_k, sc_k, i), new_{t_k}(a_k, sc_k, val_k) \mid k \in \{1, \ldots, vm\}\}$$
$$\bigcup\{new_{a_k}(a_k, sc_k, i), \mid k \in \{vm+1, \ldots, m\}\}$$
$$\alpha(t_3) = \{\overline{cname.mname}(i, this, sc_{this}, v_1, \ldots, v_{vm},$$
$$n_{vm+1}, sc_{vm+1}, \ldots, n_m, sc_m)\}$$
$$\bigcup\{delete_{a_k}((a_k, sc_k, i),$$
$$delete_{t_k}(a_k, sc_k, val_k) \mid k \in \{1, \ldots, vm\}\}$$
$$\bigcup\{delete_{a_k}(a_k, sc_k, i) \mid k \in \{vm+1, \ldots, m\}\}$$
$$\beta(t_1) = \bigwedge_{k=1}^{vm} a_k = newid \wedge \bigwedge_{k=vm+1}^{m} n_k = a_k \wedge$$
$$\bigwedge_{k=1}^{vm} v_k = val_k \bigwedge_{k=1}^{vm} sc_k = <oid, i>$$
$$\beta(t_3) = \bigwedge_{k=1}^{m} n_k = a_k,$$

Fig. 12. The labels of the method declaration

Atomic actions are the arithmetic expressions of $B(PN)^2$. An atomic action is translated into a single transition, and is thus executed atomically. The variables $'v, v'$ mean old and new value of v respectively. They correspond to the labels v^o, v^n respectively, v is a shorthand for the extra constraint $'v = v'$, while the operations are the usual arithmetic and boolean operations. The labels of the actions are constructed from the labels of the variables, together with equations that force the wanted equalities at synchronization. Let $expr$ be an expression. The semantics of the expression is defined recursively with the help of a function $Ins(expr)$. The function collects into a triple $(AS/E/SC)$ the parametrised actions in $\mathcal{T}[\![< expr >]\!]$ (AS), the expresion $expr$ with the variables renamed as above (E), and some possible extra conditions.

$$\mathcal{T}[\![< expr >]\!] = (\overset{(e|N)}{\underset{i}{\bigcirc}}\!\!\longrightarrow\!\!\overset{l}{\underset{i}{\square}}\!\!\longrightarrow\!\!\overset{(x|N)}{\bigcirc}),$$

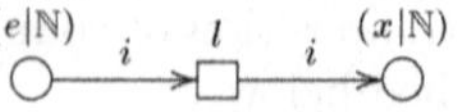

with $l = (AS|E \wedge SC)$.

$Ins(expr)$ is defined by the following rules, where t is the type of the variable v. For local or global variables:

$$Ins('v) = (\{t(id, sc, v^o, v^n), v(id, sc, i)\}/v^o/\emptyset)$$
$$Ins(v') = (\{t(id, sc, v^o, v^n), v(id, sc, i)\}/v^n/\emptyset)$$
$$Ins(v) = (\{t(id, sc, v^o, v^n), v(id, sc, i)\}/v^n/v^o = v^n)$$

For class variables:

$$Ins(this.'v) = (\{t(id, sc, v^o, v^n), v(id, sc, i), this(sc, i)\}/v^o/\emptyset)$$
$$Ins(this.v') = (\{t(id, sc, v^o, v^n), v(id, sc, i), this(sc, i)\}/v^n/\emptyset)$$
$$Ins(this.v) = (\{t(id, sc, v^o, v^n), v(id, sc, i), this(sc, i)\}/v^n/v^o = v^n)$$

For constants and expressions:

$$Ins(const) = (\phi/const/\phi)$$
$$Ins(e_1 \ op \ e_2) = (AS_1 \cup AS_2/E_1 \ op \ E_2/SC_1 \wedge SC_2) \ \text{for}$$
$$Ins(e_i) = (AS_i/E_i/SC_i)$$
$$Ins(op \ e) = (AS/op \ E/SC) \ \text{for} \ Ins(e) = (AS/E/SC)$$

Thus the translation of the expression $'x = x' + 1 \wedge x' = 3$ becomes (assuming that x is of integer type),

$$Ins('x = x' + 1 \wedge x' = 3) = \{Int(id, sc, x^o, x^n), x(id, sc, i) \mid x^o = x^n \wedge x^n = 3\}.$$

This means that the action can only be executed if the value of x is 3 in which case after execution of the atomic action the value of x will be 4.

There is one little refinement that has to be done to the above translation rules. If the variable v is declared at the global scope, then we cannot use i as the frame index, because the instance of v does not live in the frame i, but in the global frame. In this case the we must substitute $\emptyset$, the frame index of the global scope, for the index i in the translation above.

The Procedure Call. The standard form of a procedure call is the synchronous call. The net given in figure 13 is inserted for each procedure call. The procedure is called by transition t_1 after which the caller waits "in place p_1" until the procedure returns, at which point t_2 is fired. The pairs $< x, i >$ stored in p_2 are used to secure that the thread i continues after the right procedure return. We use the same numbering scheme for the parameters as defined for the procedure call.

$$\mathcal{T}[\![(pname(a_1, \ldots, a_m)]\!] = P_C(pname, a_1, \ldots, a_m)$$

We again assume that the arguments are ordered, and that a_k stands for the k:th argument, and t_k for its type.

The labelings are then given by:

$$\alpha(t_1) = \{pname_c(j, v_1, \ldots, v_{vm}, n_{vm+1}, sc_{vm+1}, \ldots, n_m, sc_m)\}$$
$$\bigcup \{a_k(a_k, s_k, i), t_k(a_k, s_k, val_k^o, val_k^n) \mid k \in \{1, \ldots, vm\}\}$$
$$\bigcup \{a_k(a_k, s_k, i), \mid k \in \{vm + 1, \ldots, m\}\}$$

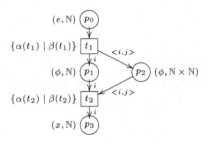

Fig. 13. Procedure call $P_C(pname, v_1, \ldots, v_m)$

$$\alpha(t_2) = \{pname_r(j, v_1, \ldots, v_{vm}, n_{vm+1}, sc_{vm+1}, \ldots, n_m, sc_m)\}$$
$$\bigcup\{a_k(a_k, sc_k, i), t_k(a_k, sc_k, val_k^o, val_k^n) \mid k \in \{1, \ldots, vm\}\}$$
$$\bigcup\{a_k(a_k, sc_k, i), \mid k \in \{vm+1, \ldots, m\}\}$$
$$\beta(t_1) = \bigwedge_{k=1}^{vm} v_k = val_k^o \bigwedge_{k=vm+1}^{m} (n_k = a_k \wedge sc_k = s_k)$$
$$\beta(t_2) = \bigwedge_{k=vm+1}^{m} v_k = val_k^n$$

An asynchronous procedure call is a call where the caller doesn't wait for the callee to return before proceeding. Instead the caller signals explicitly at some later point that it is ready to accept a return of the procedure. This semantics can also be implemented by extending the syntax of $OB(PN)^2$ with an $\langle accept \rangle P$ statement. The semantics of this is basically the same, except that the statements between the call and accept are inserted for p_1 between t_1 and t_2.

Method Call. A method call is handled analogously to a procedure call remembering that there is an implicit extra parameter (the object). The procedure call net is substituted in place of the call together with the labels as given below.

$$\mathcal{T}[\![obj.pname(v_1, \ldots, v_m)]\!] = pname_C(obj.pname, v_1, \ldots, v_m, this)$$

5.3   Control Connectives

The control connectives are directly translated into the corresponding M-net operations.

$$\mathcal{T}[\![P_1 \| P_2]\!] = \mathcal{T}[\![P_1]\!] \| \mathcal{T}[\![P_2]\!]$$
$$\mathcal{T}[\![P_1; \ P_2]\!] = \mathcal{T}[\![P_1]\!] \ ; \ \mathcal{T}[\![P_2]\!]$$
$$\mathcal{T}[\![P_1 \square P_2]\!] = \mathcal{T}[\![P_1]\!] \ \square \ \mathcal{T}[\![P_2]\!]$$
$$\mathcal{T}[\![\text{do } \langle \text{alt-set} \rangle \text{ od}]\!] = [N_{silent} * R(\langle \text{alt-set} \rangle) * E(\langle \text{alt-set} \rangle)] \ ,$$

where

$$R\left(\langle\text{alt-set}_1\rangle\,\square\,\langle\text{alt-set}_2\rangle\right) = R\left(\langle\text{alt-set}_1\rangle\right)\,\square\,R\left(\langle\text{alt-set}_2\rangle\right)$$
$$E\left(\langle\text{alt-set}_1\rangle\,\square\,\langle\text{alt-set}_2\rangle\right) = E\left(\langle\text{alt-set}_1\rangle\right)\,\square\,E\left(\langle\text{alt-set}_2\rangle\right)$$
$$R\left(\langle\text{comlst}\rangle\,;\,\texttt{repeat}\right) = E\left(\langle\text{comlst}\rangle\,;\,\texttt{exit}\right) = \mathcal{T}\left[\left[\langle\text{comlst}\rangle\right]\right]$$
$$R\left(\langle\text{comlst}\rangle\,;\,\texttt{exit}\right) = E\left(\langle\text{comlst}\rangle\,;\,\texttt{repeat}\right) = N_{stop},$$

and

$$N_{silent} = (\ \overset{(e|\text{N})}{\bigcirc}\xrightarrow{\ i\ }\overset{\{\}}{\square}\xrightarrow{\ i\ }\overset{(x|\text{N})}{\bigcirc}\ )$$

$$N_{stop} = (\ \overset{(e|\text{N})}{\bigcirc}\qquad\overset{(x|\text{N})}{\bigcirc}\ )$$

6 Conclusions

In the paper we presented OB(PN)² and its semantics in terms of M-nets. The semantics is interesting from two points of view:

1. it lays the foundations for the development of automatic verification methods of concurrent programs written in object-oriented languages, and
2. it can be seen as a set of rules for the translation of object-oriented specifications written in an object-oriented specification formalism into Petri-nets.

However, the semantics as presented here poses a big challenge for traditional verification tools. The problem stems from our use of the integers as the set of frame-indices. As this set is infinite, tools that rely on the existence of a finite unfolding of the high-level net model (eg. PEP) cannot be used. However in the case of where it is feasible to restrict the number of concurrently active procedure instances, it is possible to devise a refinement of the procedure box concept that reuses frame-indices [10], which will suffice to keep the unfolding finite. On the other hand a tool that does not necessarily rely on the unfolding (eg. PROD), but on the symbolic firing rule, may still be able to generate the reachability graph if for a given initial state the program does not have a non-terminating recursive call.

Clearly some things remain to be done. First the special data-structures of B(PN)² (channels and stacks) should be incorporated in the framework. Secondly and more importantly inheritance together with a mechanism for dynamic dispatch should be introduced. As mentioned in the introduction, a preliminary proposal for these features exists [16]. A challenge would be to incorporate a return statement. But this means that we can pass reference not only down the call hierarchy, but also up, which implies that we would need some form of garbage collection. On the other hand it might be possible to harness the approach of [10] for this.

A more intriguing avenue for research is the use of program analysis techniques, as developed for compiler optimization, in the translation to reduce the model, and in the model-checking phase to guide the model-checker.

References

1. Ken Arnold and James Gosling. *The Java Programming Language, Second Edition.* Addison-Wesley, 1998.
2. E. Battiston, A. Chizzoni, and F. De Cindio. Inheritance and concurrency in CLOWN. In *Proc. of the Workshop on Object-Oriented Programming and Models of Concurrency*, 1995.
3. E. Best, R. Devillers, and J. Hall. The box calculus: a new causal algebra with multi-label communication. In *Advances in Petri Nets 1992*, volume 609 of *Lecture Notes in Computer Science*, pages 21–69. Springer Verlag, Berlin, 1992.
4. E. Best, H. Fleischhack, W. Fraczak, R. P. Hopkins, H. Klaudel, and E. Pelz. A class of composable high level Petri nets with and application to the semantics of $B(PN)^2$. In G. De Michelis and M. Diaz, editors, *Application and Theory of Petri Nets 1995*, volume 935 of *Lecture Notes in Computer Science*, pages 103–120. Springer Verlag, 1995.
5. E. Best, H.Fleischhack, W.Fraczak, R.P. Hopkins, H. Klaudel, and E. Pelz. An M-net semantics of $B(PN)^2$. In Jörg Desel, editor, *Proceedings of STRICT'95*, Workshops in Computing, pages 85–100, Berlin, 1995. Springer Verlag.
6. E. Best and R. P. Hopkins. $B(PN)^2$ - a Basic Petri Net Programming Notation. In A. Bode, M. Reeve, and G. Wolf, editors, *Proceedings of PARLE-93*, volume 694 of *Lecture Notes in Computer Science*, pages 379–390, Berlin, 1993. Springer Verlag.
7. D. Buchs and N. Guelfi. COOPN2: An object oriented specification language for distributed system development. submitted article.
8. A. Chizzoni. CLOWN: Class orientation with nets. Master's thesis, Dept. of Computer Science, Univ. of Milano, 1994.
9. Javier Esparza. Model checking using net unfoldings. In *TAPSOFT'93*, volume 668 of *Lecture Notes in Computer Science*, pages 613–628, Berlin, 1993. Springer Verlag.
10. Hans Fleischhack and Bernd Grahlmann. A Petri net semantics for $B(PN)^2$ with procedures. In *Proceedings of PDSE'97 (Parallel and Distributed Software Engineering)*, pages 15 – 27. IEEE Computer Society, May 1997.
11. Bernd Grahlmann. An introduction to the principles, the functionality and the usage of the **PEP**-system. In *Proceedings of PEP: Programming Environment Based on Petri Nets Workshop*, number 14/95 in Hildesheimer Informatik Berichte, 1995.
12. Peter Gr nberg, Mikko Tiusanen, and Kimmo Varpaaniemi. PROD - A Pr/T-Net reachability analysis tool. Technical Report B11, Digital Systems Laboratory, Berlin, 1993.
13. K. Jensen and G. Rozenberg, editors. *High-Level Petri Nets.* Springer Verlag, Berlin, 1991.
14. Hanna Klaudel and Elisabeth Pelz. Communication as unification in the Petri Box Calculus. In *Proc.of FCT'95*, volume 965 of *Lecture Notes in Computer Science*, pages 303–312, 1995.
15. Johan Lilius. $OB(PN)^2$: An Object Based Petri Net Programming Notation. In Luc Bougé, Pierre Fraigniaud, Anne Mignotte, and Yves Robert, editors, *Proceedings of Euro-Par'96*, volume 1123 of *Lecture Notes in Computer Science*, pages 660–663, Berlin, 1996. Springer Verlag.
16. Johan Lilius. $OOB(PN)^2$: An Object-Oriented Petri Net Programming Notation. In *Proceedings of the Second Workshop on Object-Oriented Programming and Models of Concurrency*, 24 June 1996.

17. Johan Lilius and Elisabeth Pelz. An M-net semantics for B(PN)² with procedures. In V. Atalay, U. Halici, K. Inan, N. Yalabik, and A. Yazici, editors, *Proceedings of the Eleventh International Symposium on Computer and Information Sciences*, pages 365–374. Middle East Technical University, 1996.

18. S. Matsuoka, K. Wakita, and A. Yonezawa. Synchronisation constraints with inheritance: What is not possible. so what is? Internal report, Tokyo University, 1990.

19. R. N. Taylor. A general purpose algorithm for analyzing concurrent programs. *Communications of the ACM*, 26(5):362–376, 1983.

20. Antti Valmari. Stubborn sets for reduced state space generation. In *10th International Conference on Application and Theory of Petri Nets, Supplement to the Proceedings*, pages 1–22, 1989. On proceedingsit.

21. Akinori Yonezawa and Mario Tokoro. *Object-Oriented Concurrent Programming*. Computer Systems Series. MIT Press, 1987.

On Formalizing UML with High-Level Petri Nets

Luciano Baresi and Mauro Pezzè

Dipartimento di Elettronica e Informazione – Politecnico di Milano
Piazza Leonardo da Vinci, 32 – 20133 Milano, Italy
tel.: +39-2-2399-3400 – {baresi|pezze}@elet.polimi.it

Abstract. Object-oriented methodologies are increasingly used in software development. Despite the proposal of several formally based models, current object-oriented practice is still dominated by informal methodologies, like Booch, OMT, and UML. Unfortunately, the lack of dynamic semantics of such methodologies limits the possibility of early analysis of specifications.

This paper indicates the feasibility of ascribing formal semantics to UML by defining translation rules that automatically map UML specifications to high-level Petri nets. This paper illustrates the method through the hurried philosophers problem, that is first specified by using (a subset of) UML, and then mapped onto high-level Petri nets. The paper indicates how UML specifications can be verified by discussing properties of the hurried philosophers problem that can be verified on the derived high-level Petri net.

1 Introduction

Object-oriented methodologies are increasingly used in software development [19]. Despite the proposal of several formally based object-oriented methods ([10,6]), current industrial practice is still dominated by informal notations, such as OMT ([27]), Booch ([7]), Jacobson ([16]), which have merged into the *Unified Modeling Language* (UML, [11]). The success of these methodologies is due to user-friendly intuitive graphical representation, good tool support, and traditional skepticism of practitioners against formal methods.

Unfortunately, the lack of dynamic semantics of such notations limits the capability of analyzing defined specifications. CASE tools ([18,28,20]), which support this kind of object oriented methodologies, provide powerful analysis capabilities as to syntactic and static semantic properties, but they do not address dynamic semantic analysis, that is, execution, testing, reachability analysis, whose benefits have been widely recognized [31].

Researchers are trying to introduce dynamic analysis capabilities early in the requirements specification by either proposing new formally defined requirements specification notations or adding formality to existing informal notations. New formally defined specification notations ([13]) have been widely experimented, but they succeeded only in specific industrial sectors, such as the design of telecommunication protocols or safety critical applications, where the high costs

G. Agha et al. (Eds.): Concurrent OOP and PN, LNCS 2001, pp. 276–304, 2001.

of failures deeply modify the cost-benefit tradeoff. Attempts to add formal semantics to existing informal notations ([6,32]) barely modify end-user interfaces and interaction modalities, thus overcoming one of the major obstacle for breaking into much larger industrial sectors. The most popular approaches consist in defining translation algorithms that provide given specification notations with fixed semantics. By applying these algorithms users define a formal model that is equivalent to the original informal specification. Unfortunately, the results are particular formalizations of some notations, which, even if well suited for some application domains, cannot easily be generalized.

Recently, we investigated a new rule-based approach that allows users to associate different semantics with the same notation. Such an approach, called *CR approach* ([3], Customization Rules Approach), allows users to fit their interpretations of an informal notation by defining particular set of rules. Users do not only ascribe their semantics to the notation they are familiar with, but they can exploit all the benefits of a formal engine (simulation, analysis) without even knowing that it exists. This approach has been successfully validated by ascribing formal semantics to several data flow-based notations, such as different dialects of structured analysis ([2]) and to new special-purpose notations ([4,23]).

Object-oriented notations impose further requirements to the *CR approach*. Specific features such as inheritance, polymorphism, multiple overlapping views of the same features, and the emphasis on non-strictly operational aspects present new challenges. The goal of this paper is to demonstrate the suitability of the *CR approach* for the definition of the dynamic semantics of UML. The paper does not present the set of rules that formally define UML. It illustrates rather intuitively the automatic mapping of UML to high-level Petri nets through the hurried philosophers problem. The paper also indicates how we can prove important properties of the problem by analyzing the high-level Petri net derived from the UML specification.

This paper is organized as follows. Section 2 briefly sketches the *CR approach*. Section 3 discusses a possible use of the *CR approach* for formalizing UML. Section 4 shows the applicability of the approach to the hurried philosophers problems. Finally, Section 5 concludes by indicating the ongoing work.

2   CR Approach

This section describes a flexible framework for defining the syntax of graphical notations, for expressing their operational semantics through a mapping onto an operational formal model, and for presenting the results of dynamic analysis of the formal model in terms of the graphical notation.

A notation is defined by means of three sets of rules, that specify the abstract syntax, the semantics, and the visualization of dynamic analysis results, respectively. Abstract syntax rules define the elements of the graphical notation and their connectivity. Semantic rules define the dynamic behavior of the notation as a transformation in a formal operational model. Visualization rules describe the presentation of the results of dynamic analysis on the formal model in terms of the chosen graphical notation. Concrete syntax – not considered here – can easily be inherited from existing CASE technology.

Figure 1 illustrates the approach. The general framework comprises a graphical (specification) notation and a formal model. Application specialists define their models by using the graphical notation, for example UML. User models are scanned to automatically build the formal representation that corresponds to the graphical specification. The corresponding formal model gives semantics to user models and provides dynamic analysis capabilities. Semantic properties of the graphical specification can be checked on the formal model. Visualization rules translate the analysis results in terms of the graphical notation. In this way, domain experts exploit the benefits of formality without having to care about it. They can execute and analyze their models without being proficient in the formal model.

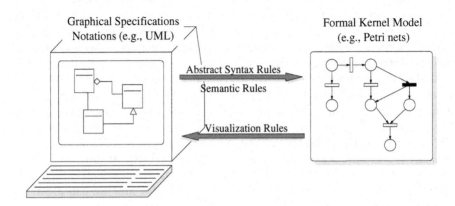

Fig. 1. The *CR approach.*

The definition of the abstract syntax and semantics of graphical notations requires the capability of designing rules to specify graphical languages. The framework proposed in this paper refers to the well known theory of graph grammars [22].

Two graph grammars, called *Abstract Syntax Graph Grammar (ASGG)* and *Semantic Graph Grammar (SGG)*, define the abstract syntax and the semantics of the notation. Each ASGG production corresponds to a SGG production. User modifications on models defined with the graphical notation are captured by means of ASGG productions; the associated SGG productions describe how to automatically update the corresponding formal model. By applying a pair of ASGG and SGG rules, we define also a correspondence between the (abstract) elements of the user model and the elements of the corresponding formal model. Such a correspondence is used by visualization rules.

A sample pair of ASGG and SGG rules is illustrated in Figure 2. The two graph grammar rules are given in a graphical style. A rule is a directed graph whose nodes are divided in three parts. The bottom left part indicates the ele-

ments on which the rule applies; the bottom right part indicates the elements
introduced by the application of the rule; the top (embedding) indicates how the
new elements have to be connected to the graph. Graph grammar nodes correspond to syntax elements of the specified notation, that is, either nodes or arcs.
For example, in the bottom right part of Figure 2(a) nodes 1 and 2 represent
nodes of type *object*, while node 3 represents an arc of type *method invocation*.
Graph grammar arcs indicate relations between the elements of the notation.
For example, the graph grammar arcs from node 1 to node 3 and from node 3
to node 2 represent relations *connect* between the nodes. Arcs belonging to the
bottom left part of a rule define the connections (relations) that must exist -
among selected nodes - to apply the rule. Arcs in the bottom right part of a rule
define the relations established by applying the rules. Arcs crossing the border
between the bottom left and top parts of a rule select the nodes in the existing
graph to which the newly created nodes (bottom right part) will be connected.
Arcs crossing the border between the top and the bottom right parts of a rule
set the connections between the added subgraph and the nodes identified so far.

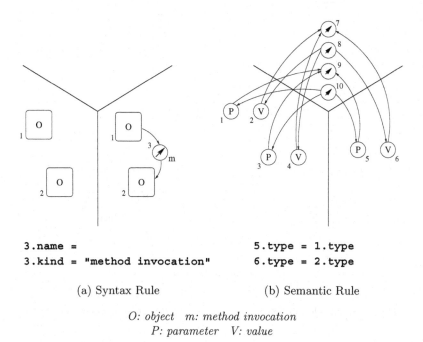

3.name =   5.type = 1.type
3.kind = "method invocation"   6.type = 2.type

(a) Syntax Rule   (b) Semantic Rule

O: object   m: method invocation
P: parameter   V: value

Fig. 2. *AddMethodInvocation*: abstract syntax and semantic rules.

The pair of rules of Figure 2 corresponds to adding a method invocation
between two objects (*UML Collaboration Diagrams*, addressed in Section 3.1).
The abstract syntax rule (2(a)) can be applied to two nodes of type *object* (nodes

1 and 2 of type O, in the bottom left part of the rule) and results in adding an arc of type *method invocation* between them (node 3 of type m in the bottom right part). The same numbers associated with the nodes in the bottom left and bottom right parts indicate that the two objects are kept in the graph together with their connections. The semantic rule (2(b)) applies to the two pairs of nodes of type *parameter* and *value* that correspond to the two *objects* selected by the abstract syntax rule (pairs $\langle 1, 2 \rangle$ and $\langle 3, 4 \rangle$, of nodes of type P and V). The four places are "fused" into two new places of type P and V, that represent the actual invocation of the method. The "fusion" is modeled by removing the four places and introducing two new ones. The connections with the embedding force the new P node to be connected to the same arcs to which the two old P nodes were formerly connected, and similarly for the new V node. Figure 3 shows a sample application of the rule of Figure 2.

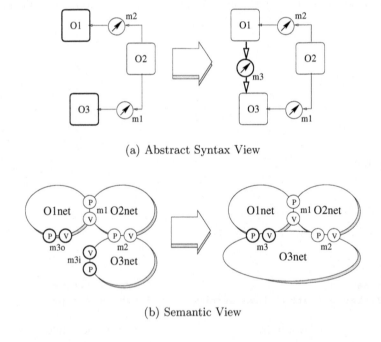

(a) Abstract Syntax View

(b) Semantic View

Fig. 3. A sample application of rule *AddMethodInvocation* of Figure 2.

Often graphical notations are textually annotated. Textual annotations must suitably be translated into the operational formal model. For example, classes of *UML Class Diagrams* are annotated with attributes and methods. Textual parts of graph grammar rules are used to define the correspondences between textual attributes of the graphical notation and the elements of the formal model. For example, the textual rules of the ASGG production of Figure 2(a) indicate that

the name of node 3 is supplied by users and its type is *method invocation*. In the
SGG production of Figure 2(b) the type of the new places 5 and 6 is equal to
the type of the removed places 1 and 2, respectively.

Visualization rules translate states and events of the semantic model in
proper representations of the graphical elements. A textual grammar describes
the mapping by referring to the types of the objects introduced by syntax and
semantic rules. Figure 4 shows the visualization rule that maps the firing of a
transition tId of type *stateTransition* to a set of (abstract) visualization events
on UML elements. Readers not familiar with C-like syntax can simply skim
through Figure 4: The rule, triggered by the firing of a transition tId of type
stateTransition, selects:

- All places of type P (parameter) in the preset of tId and associates them with
  the *readParameters* visualization event to visualize all the actual parameters
  of the invoked method.
- All places of type S (state) in the preset of tId and associates them with
  the *leaveState* visualization event to visualize the current state of the finite
  state machine and the fact that it is about to change.
- All places of type V (value) in the postset of tId and associates them with the
  writeValues visualization event to visualize all values produced by executing
  the method.
- All places of type S (state) in the postset of tId and associates them with
  the *enterState* visualization event to visualize the new current state of the
  finite state machine.

Customization rules can syntactically be checked to verify the existence of
a rule for each construct of the modeled notation and to verify the syntactic
matching of the rules. Formal semantic checks are not possible, since a set of
customization rules capture a particular interpretation, and thus, it should be
validated with respect to the "idea" of the notation implicitly assumed while
writing the rules. However, customization rules can be verified by inspection,
testing, and animation. Inspecting the high-level Petri nets produced on relevant
examples can reveal problems in the customization rules; execution and analysis
of well understood benchmarks could reveal unexpected behaviors of the high-
level Petri nets that may be caused by erroneous customization rules.

As in the case of compilers, customization rules can produce erroneous semantic interpretations that reflect in spurious behaviors. Only user experience
can distinguish between errors in the specification and errors in the customization rules. Such errors are the price paid to the flexibility of the *CR approach*:
the more unstable (flexible) the semantics of the informal notation is, the more
error-prone the customization rules can be. However, previous experiences in cus-
tomizing structured analysis allow us to believe the customization rules for an
industrial-strength informal specification notation, such as UML, can effectively
be validated through inspection, testing, and animation.

– *triggering semantic event: firing of a transition* **tId** *type* <u>*stateTransition*</u>

```
if (getType(tId) == ''stateTransition") then
   begin
```

– *mark UML elements corresponding to places of type* **P** *in the preset of* **tId** *with –* *visualization event* <u>*readParameters*</u> *(the event uses the values associated with – such elements)*

```
      foreach p1P (getType(p1P) == ''P") in preset
         begin
            entityId = getAbsId(p1P);
            eventType = ''readParameters";
            eventPars = [("value", compute(p1P))];
         end
```

– *mark UML elements corresponding to places of type* **S** *in the preset of* **tId** *with – visualization event* <u>*leaveState*</u> *(the event causes the system to leave the state – represented with such elements)*

```
      foreach p1S (getType(p1S) == ''S") in preset
         begin
            entityId = getAbsId(p1S);
            eventType = ''leaveState";
         end
```

– *mark UML elements corresponding to places of type* **V** *in the postset of* **tId** *with – visualization event* <u>*writeValues*</u> *(the event defines the values associated with – such elements)*

```
      foreach p1P (getType(p1P) == ''V") in postset
         begin
            entityId = getAbsId(p1P);
            eventType = ''writeValues";
            eventPars = [("value", compute(p1P))];
         end
```

– *mark UML elements corresponding to places of type* **S** *in the postset of* **tId** *with – visualization event* <u>*enterState*</u> *(the event causes the system to enter the state – represented with such elements)*

```
      foreach p1S (getType(p1S) == ''S") in postset
         begin
            entityId = getAbsId(p1S);
            eventType = ''enterState";
         end
   end
```

Fig. 4. An example of *Visualization Rule*.

3   Formalizing UML

Formalizing UML with the *CR approach* presents new challenges and problems. Object-oriented features like inheritance, polymorphism, multiple overlapping views, and emphasis on non-strictly operational aspects must carefully be studied. Possible incompletenesses and inconsistencies of UML specifications – due to the possibility of defining different views of the model – complicate the task of defining formal dynamic semantics. In UML, several non-homogeneous views can capture different details of the same components and often overlap. For example, *Class Diagrams* can describe only a subset of the classes defining the application; *Interaction Diagrams* provide a snapshot of a group of objects cooperating to achieve a common goal. Formalization must be able to capture different details from different sources, integrate them, and – if possible – identify and solve inconsistencies.

This paper illustrates a possible use of the *CR approach* to ascribe semantics to consistent UML specifications by means of high-level Petri nets ([14])[1]. We do not present all the details of the customization rules (graph-grammar productions) that define the abstract syntax and the semantics of UML, but we illustrate the semantic aspects by describing the underlying high-level Petri nets. Due to space consideration, we show only a sample of the details (predicates and actions) associated with the high-level Petri nets in the Appendix.

3.1   UML

This section briefly introduces UML. Readers proficient in the notation can either skip or skim the section. An UML specification ([11]) comprises 7 kinds of diagrams:

Use-Case Diagrams describe user scenarios. They define the possible interactions between the system under development and the external actors.

Class Diagrams define the classes and/or packages (groups of classes) that compose the system and the relations among them:
  - *associations* (undirected arcs) represent conceptual relationships between classes. The relationships can be clarified by defining the roles played by each class in the relationship.
  - *aggregations* (arcs with a white diamond head) represent **part-of** relationships. Objects of the class at the diamond head "contain" (references to) objects of the class at the tail. *Compositions* (arcs with a black diamond head) are heavier aggregation relations. Composition restricts aggregation by either requiring an object to be part of one other object only, or by binding the life of the "contained" object to the life of the "container" object.
  - *generalizations* (arcs with a white arrow head) represent **inheritance** relationships.

---

[1] An introduction to high-level Petri net is presented in the Appendix.

Figure 7 shows the class diagram for the case study. The diagram shows associations, generalizations and aggregations.

Interaction Diagrams give snapshots of the interactions among objects. They describe how groups of objects cooperate in some behavior. In UML, there are two kinds of *Interaction Diagrams*:

Collaboration Diagrams: Objects are shown as icons, while messages (method invocations) are represented with arrows. The sequence among messages is indicated by numbering the messages. A sample collaboration diagram is shown in Figure 14.

Sequence Diagrams: Objects are shown as boxes at the top of vertical dashed lines and messages are drawn with (horizontal) arrows between the dashed lines (objects). Messages are ordered from top to bottom.

State Diagrams describe the dynamics of each class as a Statechart [15]. Events and actions correspond to invocation of services (methods, from an implementation-oriented perspective) that are exchanged among objects (i.e., the arcs among entities in the *Interaction Diagrams*). Messages exchanged among objects are treated according to the defined communication type: simple, synchronous, balking, timeout, and asynchronous. Sample state diagrams can be found in Figures 8, 9, 10, and 11.

Activity Diagrams combine SDL [29] and Petri net features to provide a means for specifying the behavior of tasks that present internal parallelism.

Package Diagrams describe the partitioning of the system into modules (UML packages).

Deployment Diagrams describe how packages and objects are allocated on processors (or on more general hardware components).

3.2   Towards Customization Rules

In this paper, we deal with *Class Diagrams*, *State Diagrams*, and *Interaction Diagrams*. From the semantics viewpoint, *Collaboration Diagrams* and *Sequence Diagrams* can be considered different concrete views of the same elements, and thus can be treated similarly. *Activity Diagrams*, even if they could be useful in deriving the dynamic behavior of designed systems, are not addressed in this work. *Package Diagrams* and *Deployment Diagrams* are not considered since they specify the system at the design level, whereas here we concentrate on requirements definition.

Class Diagrams. *Class Diagrams* indicate the classes that compose the system, the (static) relations among them (associations, aggregations, and generalizations), and the external interfaces of the classes, that is, the services provided by the classes. From the semantic viewpoint, classes identify the subnets that will define the whole model and the services provided by the classes. Services are represented with pairs of high-level Petri net places, which indicate service requests and returned values. The type of the first place models the types of the formal parameters associated with the service invocation; the type of the receiving place models the types of returned values. Figure 5 presents an example: the high-level Petri net of class *Fork*, which is discussed in detail in Section 4.

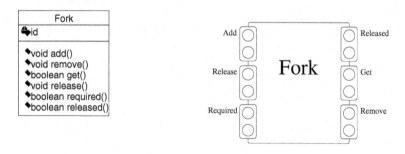

Fig. 5. The high-level Petri net corresponding to class Fork.

State Diagrams. The behavior of each class is described with a *State Diagram*, that is, a Statechart [15]. *State Diagrams*, modeling different classes, interact through exchanging invocations of services (messages). States are annotated with the identifier of the state and the action performed in the state. Transitions are annotated with labels that indicate the event that trigger the transition (above the line) and the actions produced by the triggered transition (below the line). An empty event indicates that the transition can fire spontaneously, while an empty action indicates that the transition produces no action. States are modeled with high-level Petri net places, and transitions with high-level Petri net transitions. Service invocations are modeled by connecting high-level Petri net transitions to the places modeling the service in the class's interface. Figure 6 illustrates the different translation schemas, corresponding to different kind of interactions. Examples of translations of *State Diagrams* are given in the case study in Section 4.

Interaction Diagrams. *Interaction Diagrams* identify the matching between the requests and the services. The same information could be obtained by matching the names of required and exported services in *Class Diagrams* and *State Diagrams*. *Interaction Diagrams* are used to make these connections explicit and to solve conflicts of method invocations due to late-binding and polymorphism. The services and their invocations are modeled with separate places, that are "merged" according to the information provided by *Interaction Diagrams*. The pair of graph grammar rules that define the merging of corresponding places is shown in Figure 2.

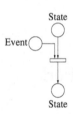

(a) Isolated event (not producing an action)

(b) Events producing an action

(c) "Spontaneous" actions (not triggered by an event)

(d) Method invocation requiring acknowledgment

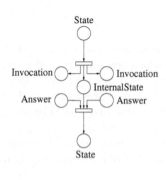

(e) Parallel method invocation requiring acknowledgment

(f) Sequential method invocations requiring acknowledgment

Fig. 6. Translation schemas for *State Diagrams*

4   Hurried Philosophers

In this section, we specify the hurried philosophers problem in UML, we illustrate the high-level Petri net derived from the UML specification with the *CR approach* and its use to prove fundamental properties.

The hurried philosophers problem extends the dining philosophers problem [9] by allowing new philosophers to be temporarily invited at the table. As in the original proposal, philosophers must get their left and right forks to eat. If forks are not available, philosophers can ask their neighbors for the forks. Some philosophers have the additional capability of introducing new philosophers, that sit around the table with a new plate and fork, taken from a (bounded) heap of shared resources. The newly arrived philosophers leave the table if asked to leave by the invitee, returning plate and fork to the heap of shared resources. At least two philosophers must be seated around the table.

4.1   UML Model

The *Class Diagram* of Figure 7 describes the main components of the hurried philosophers model and their relations. Class `Philosopher` describes the common features of all philosophers. They require interactions with forks and their neighbors to be able to eat. Class `ArrivingPhilosopher` inherits from `InvitingPhilosopher`, that inherits from `Philosopher`. Class `InvitingPhilosopher` adds service `introduceNewPhil` to introduce a philosopher and service `removePhil` to remove a philosopher; class `ArrivingPhilosopher` adds services for joining and leaving the table. Class `Butler` prevents deadlocks by restricting the number of philosophers that can concurrently compete for forks; it enforces fairness by granting authorizations to enter the forks competition. Philosophers ask the `Butler` before starting competing for forks. Class `Dispatcher` dispatches plates and forks to invited philosophers, and collects plates and forks from leaving philosophers. Since plates and forks are handled similarly, we explicitly model forks only. Class `Dispatcher` uses class `Fork` to add new forks.

Philosopher. The different types of philosophers (permanently or temporary invited) can naturally be modeled by means of inheritance. Figure 8 shows the *State Diagram* of class `Philosopher` and its subclasses `InvitingPhilosopher` and `ArrivingPhilosopher`. The properties of the subclasses are highlighted by gray background. Each philosopher starts from state `Thinking`, where he asks the `Butler` for permission to start competing for forks (action `askButler()`), and then move to state `Hungry`, where he waits for permission. When the `Butler` allows the philosopher to proceed (event `return(ACK)`), he tries to pick up his left fork (action `leftFork.get()`) and enters state `AFLF` (Asking For Left Fork). If the fork is available (event `return(true)`), the philosopher enters state `HLF` (Holding Left Fork). If the fork is not available (event `return(false)`), he asks the fork to notify its availability (action `leftFork.released()`), and he waits for the fork to be released in state `ANLF` (Asking Neighbor For Left Fork). As

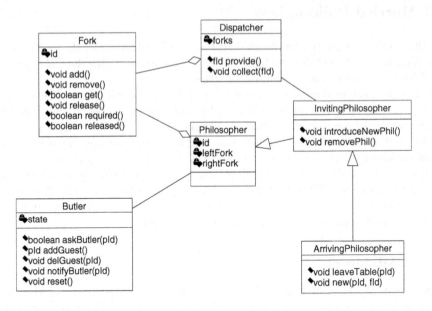

Fig. 7. Class diagram for the *Hurried Philosophers*.

soon as the fork is released (event `return(true)`), he acquires the fork (action `leftFork.get()`) and moves to state HLF. Notice that the request for service `leftFork.get()` from state Hungry to state AFLF could result in either a positive or negative answer. The obtained response is the event that determines the next step in the behavior of the philosopher. On the contrary, the invocation of the same service (`leftFork.get()`) from state ANLF to HLF can result only in a positive answer since the requested fork is in state Available (Figure 9). In fact, the request for `leftFork.get()` is modeled with a synchronous invocation (see Figure 14) and the fork answers only when it is actually available. Once acquired the left fork, the philosopher checks whether the fork has been required by his left neighbor (action `leftFork.required()`) and moves to state QLF (Querying Left Fork). If there is a pending request, the philosopher releases the fork and moves back to state Thinking. Otherwise he tries to acquire the right fork with a process analogous to the one followed to get the left fork, and moves to state Eating. When he finishes eating, after a given delay, he releases the forks and goes back to state Thinking.

An InvitingPhilosopher can introduce a new philosopher upon finishing eating and before releasing the forks (event `introduceNewPhil()`). The receipt of this event causes the philosopher to move to state WFD (Waiting For Dispatcher)). In this case, he interacts with the Dispatcher, that provides a new identifier for the fork to be put between the inviting and the arriving philosophers, if any is still available. An InvitingPhilosopher can also remove a

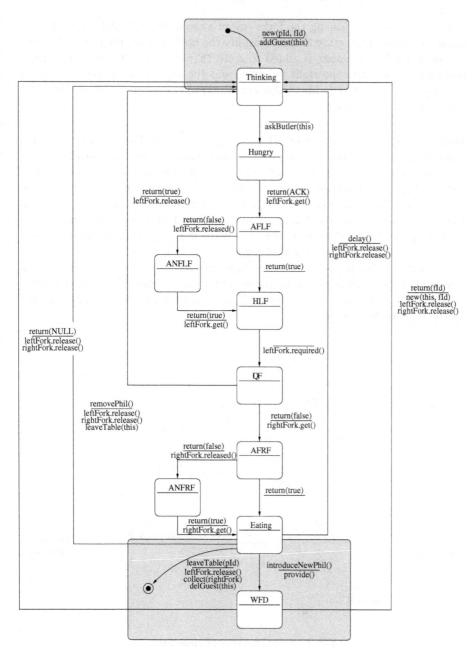

Fig. 8. State Diagram of class *Philosopher.*

philosopher before releasing the forks (event `removePhil()`). Adding or removing philosophers from state `Eating` simplify the updating of philosophers' states. An `ArrivingPhilosopher` is added to state `Thinking` (event `new(pId, fId)`). `pId` and `fId` identify the right neighbor and the right fork of the philosopher that has to be created (added). An `ArrivingPhilosopher` can leave the table only from state `Eating`.

Fork. Figure 9 shows the *State Diagram* of class `Fork`. Forks can be in either state `Available`, or state `Used`, or state `Required`. A new fork is added (created) on demand by the `Dispatcher`, that provides the identifier of the new fork. A first `get()` returns `true` to confirm the acquisition and it moves the fork to state `Used`. A second `get()`, not interleaved with a `release()`, returns `false` to notify the current unavailability of the fork and moves the fork to state `Required`. A `release()` in either state `Used` or state `Required` moves back to state `Available`. A `required` returns the current status of the fork. A `released()` can be served only in state `Available`. In this way, we define a synchronous communication with the asking object (philosopher), further illustrated by the *Collaboration Diagram* of Figure 14. Forks are removed (destroyed) by the `Dispatcher` on demand.

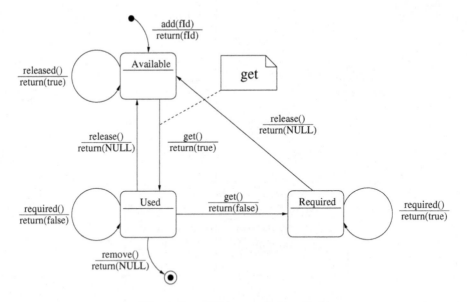

Fig. 9. State Diagram of class *Fork*.

Dispatcher. The *State Diagram* of Figure 10 describes the behavior of the `Dispatcher` that grants up to 2 forks. The `Dispatcher` serves `provide()` requests by invoking service `add()` of class `Fork` and `collect()` requests by invoking service `remove()` of class `Fork`. The `Dispatcher` of Figure 10 can easily be extended to n forks by either adding new states or introducing a state variable to record the number of currently available forks.

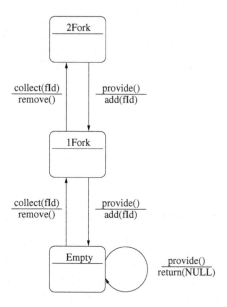

Fig. 10. State Diagram of class *Dispatcher*.

Butler. The *State Diagram* of class `Butler` is presented in Figure 11. The `Butler` keeps track of the identity of all the philosophers seated around the table. Events `addGuest()` and `delGuest(pId)` are used to keep the list of "active" philosopher up-to-date. `Philosophers` query the `Butler` by means of service `askButler()`. This event generates no action directly, but when predicate `readyToEat` evaluates to true, the `Butler` lets the `Philosopher` start eating. To ensure fairness, this version of the `Butler` adopts a very simple policy: each `Philosopher` can eat once, then he has to wait for the other philosophers to eat. Thus predicate `readyToEat` is satisfied if the philosopher has not already eaten. When a philosopher finishes eating, he notifies the `Butler` (event `notifyButler()`). When all the philosophers around the table have eaten once (event `readyToReset`), the `Butler` resets (action `reset()`) his state, and all the philosophers can start eating again. Other policies can be implemented without impacting on the approach described in the paper.

4.2   High-Level Petri Net Semantics

Figures 12 and 13 show the Petri nets produced by applying the *CR approach* to classes *Fork* and *Dispatcher*, respectively. They define the frameworks within which the instances of these classes (tokens) flow.

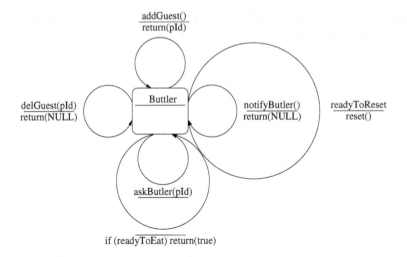

Fig. 11. State Diagram of class *Butler*.

External interfaces are derived from the *Class Diagram* of Figure 7 – together with *State Diagrams* – and are modeled with pairs of Petri net places. Each pair represents either an offered or a required service (method) and the corresponding answer. In Figures 12, and 13, interfaces are highlighted with rounded squares. States and transitions of the *State Diagrams* of Figures 9 and 10 are modeled with Petri net places and transitions, respectively. Each state is modeled with a Petri net place and each transition is modeled with a Petri net transition (see Section 3.1) connected to the places that represent the input and output states and to the places corresponding to the methods triggering the transition. For example, transition `get` from state `Available` to state `Used` of the *State Diagram* of Figure 9 is modeled with the Petri net transition `getT1` from place `Available` to place `Used` of the Petri net of Figure 12. Since the transition of the *State Diagram* is labeled with event `get()` and action `return(true)`, the corresponding Petri net transition "reads" from the P place of method `get()`, that represents the service invocation, and "writes" `true` in the V place of method `get()`, that represents the answer provided by the service.

New objects are instantiated by adding tokens, as done by the subnet corresponding to methods `add` of class `Fork` and method `new` of class `ArrivingPhilosopher`. Newly created tokens contain the identity of the newly created objects and the class the created object belongs to. The inheritance of properties is governed by the predicates of high-level Petri net transitions. Such predicates forbid tokens corresponding to objects of a superclass to enable transitions modeling services provided by a subclass. In the hurried philosophers problem, classes `Philosopher`, `InvitingPhilosopher` and `ArrivingPhilosopher` share the same subnet, as shown in Figure 15. However, tokens modeling `Inviting-Philosophers` are prevented to enable transitions that model methods of class `ArrivingPhilosopher`, for example, method `leaveTable`. The predicates of the

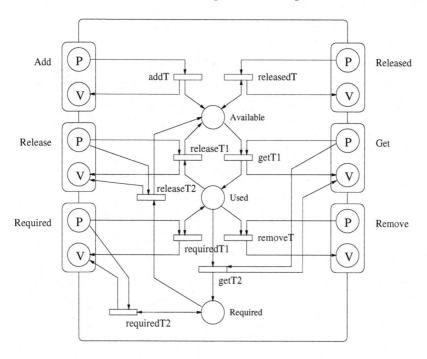

Fig. 12. Petri net semantics of class *Fork*.

corresponding high-level Petri net transitions require tokens to indicate class `ArrivingPhilosopher` as the class they belong to, to fire those transitions. Errors in method invocations (late binding) result in partial deadlocks, that is, tokens stacked in a place.

Objects can be deleted by removing the corresponding tokens, as done by methods `remove` of class `Fork`, or method `leaveTable` of class `ArrivingPhilosopher`.

Collaboration Diagrams define the connections among the subnets corresponding to the different classes. The customization rule that adds arcs in *Collaboration Diagrams* merge the places corresponding to the service in the interfaces of the two objects (classes) exchanging the message.

For example, the *Collaboration Diagram* of Figure 14 models the "starteating" task among a `Philosopher` P, the `Butler` B, and a `Fork` LF (the left fork of P). As indicated by the numbering of the arcs (service invocations), the `Philosopher` P ask the `Butler` B for permission (`askButler(P)`); the `Butler` B replies with a positive acknowledgment (`return(ACK)`); the philosopher tries to get his left fork (`LF.get()`); `Fork` LF replies false; the philosopher waits on service `LF.released()` (synchronous communication), and eventually receives a positive answer from the `Fork` LF. Special arrows indicate synchronous communications. Customization rules merge *P* and *V* places of services `askButler()`, `get()`, and `released()`.

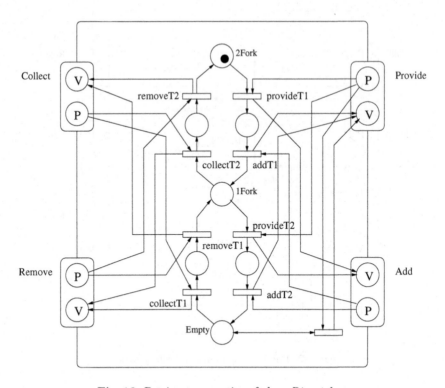

Fig. 13. Petri net semantics of class *Dispatcher*.

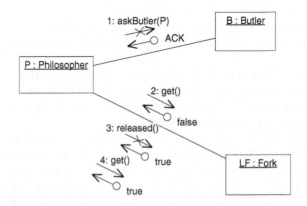

Fig. 14. A *Collaboration Diagram* for the *Hurried Philosophers* problem.

The result of merging the interface places of all the services used in all the *Collaboration Diagrams* of the example are shown in Figure 15. The figure illustrates the high-level Petri net produced by applying the customization rules, that formalize UML, to the specification of the hurried philosophers problem described in this paper. In Figure 15, subnets corresponding to the different classes are highlighted with boxes; only interface places lie on two boxes that correspond to the joined subnets. The Petri net is reported here to illustrate the semantics, but it is not shown to domain experts, who analyze the Petri net by interacting with the UML model as illustrated in the next section.

4.3   Analysis and Validation

The analysis of a UML specification formally defined by means of a high-level Petri net consists of three main steps:

- Define the properties on the Petri net that are equivalent to the properties of the UML specification that users would like to analyze.
- Analyze the Petri net with respect to the properties identified so far.
- Translate the analysis results on the Petri net in terms of UML.

Section 2 briefly illustrates the technique used in the *CR approach* to define the correspondences between states (markings) and events (firings) of Petri nets and states and events of UML specifications. The textual rules, sketched in Section 2, translate those analysis results that can be expressed with markings and events. This is not a heavy constraint since almost all results can be stated in this way.

This section describes how relevant properties of the hurried philosophers problem, defined in UML, can be mirrored on the corresponding high-level Petri net. We do not give the details of the algorithm that maps UML properties to high-level Petri net properties. The correspondence among the different sets of properties for the case study should illustrate the degree of difficulty of such a definition. We also indicate which analysis algorithms for high-level Petri nets can be used to verify such properties.

The original proposal of the hurried philosophers problem ([30]) explicitly draws our attention to three main properties: absence of deadlock, boundedness when resources (forks) are bounded, and mutual exclusion of the states of forks (*in any case a fork is either in the hand of a philosopher or in the fork heap*). Additional properties are required in the partially operational formulation of the problem: fairness (*philosophers eat in turn*), bounds on the states of the system (*there are at least two philosophers at the table*), and bounds on the behavior of the system (*a philosopher must eat at least once before leaving the table*). The UML specification (i.e., an object-oriented model) suggests yet other interesting properties: wrong method invocations, that is, methods invoked in the incorrect state, and presence of critical sequences.

In the following, we refer to few main analysis techniques of high-level Petri nets: checks on the structure of the Petri net or on the syntax of transition

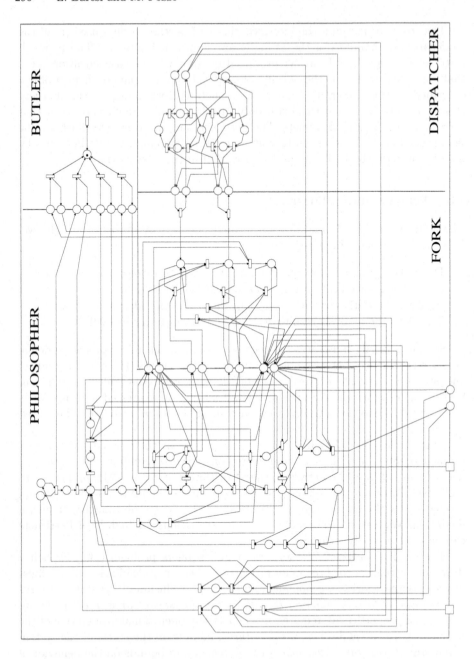

Fig. 15. The semantic high-level Petri net for the hurried philosophers problem produced with the *CR approach* illustrated in this paper.

predicates, net execution, and reachability analysis and model checking. Checks on the structure of the net or on the syntax of the predicates can straightforwardly be automated by means of widely available technology, for example, [8,1]. Execution of high-level Petri nets is currently provided by most CASE tools for high-level Petri nets [8,26]. Reachability analysis and model checking can be performed in many different ways. The reachability graph of the Place/Transition net obtained by ignoring the predicates and actions of the high-level net can be built fairly efficiently by using well-known algorithms. Model checking techniques can be used to prove properties on the reachability graph. Since the reachability graph of the Place/Transition net is only an approximation of the reachability set of the high-level net, only few properties can be proven in this way. Complex properties may require the construction of the reachability graph for the high-level net, such as the occurrence graph illustrated in [17][2]. We studied many of the properties discussed in this section by using execution and reachability analysis capabilities provided by Cabernet [26].

Absence of deadlocks. Absence of deadlock can be formulated by requiring that there exist no reachable UML states that prevent any method to be invoked eventually. Since method invocations are mapped on transition firings, this property can be formulated in Petri net terms as the absence of markings where no transition will ever be enabled again. Such a property can be tested by executing the Petri net and animating the corresponding UML specification. It can also be verified by building the occurrence graph of the high-level Petri net.

Boundedness. Boundedness can be formulated by requiring an upper bound to the number of objects that can be created according to the UML specification. Since objects correspond to tokens, this property can be mapped on the Petri net property that requires the number of tokens to be upper-bounded. Such a property can be tested by executing the Petri net and animating the corresponding UML specification. It can be verified by building the occurrence graph. The reachability graph of the Place/Transition net corresponding to the high-level net can be used instead each time the specification does not make use of predicates to bound resources. For example, boundedness of the number of forks in the system can be proven by means of the reachability graph of the Place/Transition net. This is why the number of tokens of type Fork is bounded by the number of states of class Dispatcher, or, equivalently, the number of Petri net places that model the internals of class Dispatcher. An extension to handle n forks, specified by using a UML state variable, mapped to a variable of the token modeling the Dispatcher, cannot be verified with the reachability graph of the Place/Transition net.

Mutual exclusion of the states of forks. Mutual exclusion of the states of Forks can be formulated by requiring each object of class Fork to be in one state of a

---

[2] Occurrence graphs have been originally defined for Coloured Petri nets, that have been shown equivalent to other classes of high-level nets [21].

given set, or, similarly, we can require each token of type Fork to be in one Petri net place of a given set. This can easily be checked statically on either the UML specification or on the corresponding high-level Petri net. The validity of such a property is a consequence of the operational style of both Statecharts, used to model the internals of UML classes, and Petri nets.

Fairness. Fairness can be expressed by requiring given sets of UML actions to happen infinitely often. This can easily be expressed by asking the Petri net transitions of the type corresponding to the set of actions to fire infinitely often, i.e., are live ([25]). Traces of Petri net executions can reveal possible problems; model checking on the occurrence graph can be used to prove the property. For example, checking that in none of the path of the occurrence graph the same philosopher eats twice before every other philosopher eats once can be used to prove that philosophers eat in turn.

Bounds on the states of the system. Bounds on the system states can be expressed by formulating properties on the states of UML objects (classes). Such properties can be mapped to properties on the markings of the places of the type corresponding to the bounded classes. They can be tested with execution and proven with reachability analysis. For example, asking at least two philosophers to be at the table, that is, asking at least two Philosopher object to be in one of the states characterizing class Philosopher, can be formulated by requiring at least two tokens of type Philosopher to belong to each reachable marking.

Bounds on the behavior of the system. Bounds on the behavior of the system can be expressed by stating properties on the sequences of UML events, that is, firing sequences of the corresponding Petri net transitions. Such properties can be tested by executing the net and can be proven on the occurrence graph. The reachability graph of the corresponding Place/Transition net can help in identifying possible wrong sequences, whose feasibility can be checked by executing the high-level Petri net. For example, the property that each philosopher must eat at least once before leaving the table can be formulated by asking that, between the invocation of methods addGuest and leaveTable, the philosopher must be in state Eating. Similarly, place Eating must be marked between the firing of transitions addGuest and leaveTable.

Wrong method invocations. Wrong method invocations can be stated by requiring that given methods will never be invoked in the wrong state. For example, the invocation of method get when the fork is in state required would cause a run-time error in the final implementation. Such a property corresponds to requiring that in no markings of the high-level Petri net both places get and required are marked. This is a typical property that can be proven by model checking on the occurrence graph. Animating the net execution cannot prove the validity of such a property, but can increase the confidence in the correct behavior.

Critical sequences. Critical sequences can be highlighted by requiring specific non-interruptible sequences of method invocations. For example, invocations of methods `released()` and `get()` on the same `Fork` cannot be interleaved with the invocation of other methods to obtain a correct behavior of the system. This translates on requirements on firing sequences that can be tested with execution and proven with model checking on the occurrence graph.

5 Conclusions

This paper presents an exercise on ascribing formal semantics to UML speci-fications. The method is illustrated by indicating a semantics for the hurried philosophers problem: the UML specification of the problem is mapped onto high-level Petri nets. The paper suggests a generalization of the mapping (a set of rules) that can be used to automatically produce formal semantics of UML specifications by means of high-level Petri nets. The paper indicates how users can analyze UML specifications by querying the corresponding high-level Petri net.

References

1. ARTIS s.r.l., Torino, Italy. *Artifex 3.1 – Tutorial*, 1994.
2. L. Baresi. *Formal Customization of Graphical Notations.* PhD thesis, Dipartimento di Elettronica e Informazione – Politecnico di Milano, 1997. in Italian.
3. L. Baresi, A. Orso, and M. Pezzè. Introducing Formal Methods in Industrial Prac-tice. In *Proceedings of the 20th International Conference on Software Engineering,* pages 56–66. ACM Press, 1997.
4. L. Baresi, M. Di Paola, A. Gargiulo, and M. Pezzè. LEMMA: A Language for an Easy Medical Models Analysis. In *Proceedings of IEEE Computer Based Medical Systems 97,* 1997. To appear.
5. L. Baresi and M. Pezzè. Towards Formalizing Structured Analysis. *ACM Trans-actions on Software Engineering and Methodology,* 7(1), jan 1998.
6. B.W. Bates, J.M. Bruel, R.B. France, and M.M. Larrondo-Petrie. Guidelines for Formalizing Fusion Object-Oriented Analysis Methods. In *Conference on Advanced Information Systems Engineering (CAiSE) 96,* pages 222–233, 1996.
7. G. Booch. *Object-Oriented Analysis and Design with Applications.* Benjamin Cum-mings, second edition edition, 1994.
8. S. Christensen, J. B. Joergensen, and L. M. Kristensen. Design/CPN — A Com-puter Tool for Coloured Petri Nets. *Lecture Notes in Computer Science,* 1217, 1997.
9. E.W. Dijkstra. *Co-operating Sequential Processes.* Academic Press, 1965.
10. E. H. Dürr and N. Plat. VDM++ Language Reference Manual. Technical report, IFAD - The Institute of Applied Computer Science, 1995.
11. M. Fowler and K. Scott. *UML Distilled: Applying the Standard Object Modeling Language.* Addison-Wesley, Reading, Mass., 1997.
12. H. Genrich. Predicate/transition nets. In W. Reisig and G. Rozemberg, editors, *Advances in Petri Nets,* LNCS 254-255. Springer-Verlag, Berlin-New York, 1987.

13. S. Gerhart, D. Craigen, and T. Ralston. Experience with Formal Methods in Critical Systems. *IEEE Software*, 11(1):21–28, January 1994.

14. C. Ghezzi, D. Mandrioli, S. Morasca, and M. Pezzè. A Unified High-Level Petri Net Model For Time-Critical Systems. *IEEE Transactions on Software Engineering*, 17(2):160–172, February 1991.

15. D. Harel, H. Lachover, A. Naamad, A. Pnueli, M. Politi, R. Sherman, A. Shtull-Trauring, and M. Trakhtenbrot. STATEMATE: A Working Environment for the Development of Complex Reactive Systems. *IEEE Transactions on Software Engineering*, 16(4):403–414, April 1990.

16. I. Jacobson. *Object-Oriented Software Engineering–A Use Case Driven Approach*. ACM Press/Addison Wesley, 1992.

17. K. Jensen. Coloured Petri Nets. In W. Reisig and G. Rozemberg, editors, *Advances in Petri Nets*, LNCS 254-255. Springer-Verlag, Berlin-New York, 1987.

18. Mark V Systems. *ObjectMaker User's Guide*, 1994. version 3.

19. B. Meyer. *Object-oriented Software Construction*. Prentice Hall, New York, N.Y., second edition, 1997.

20. MicroGold Software. *WithClass97 User's Guide*, 1997.

21. S. Morasca, M. Pezzè, and M. Trubian. Timed High Level Nets. *The Journal of Real-Time Systems*, pages 165–189, 1991.

22. M. Nagl. A Tutorial and Bibliographical Survey on Graph Grammars. In V. Claus, H. Ehrig, and G. Rozenberg, editors, *Graph Grammars and their Application to Computer Science and Biology*, volume 73 of *Lecture Notes in Computer Science*, pages 70–126. Springer-Verlag, 1979.

23. A. Orso. An Environment for Designing Real-Time Control Systems. Technical Report 97-56, Dipartimento di Elettronica e Informazione - Politecnico di Milano, 1997.

24. C. Petersohn, W.P. de Roever, C. Huizing, and J. Peleska. Formal Semantics for Ward & Mellor's Transformation Schemas. In D. Till, editor, *Proceedings of the Sixth Refinement Workshop of the BCS FACS*. Springer-Verlag, 1994.

25. J. Peterson. *Petri Net Theory and the Modeling of Systems*. Prentice-Hall, Englewood Cliffs, NJ, 1981.

26. M. Pezzè. Cabernet: A Customizable Environment for the Specification and Analysis of Real-Time Systems. Technical report, Dipartimento di Elettronica e Informazione, Politecnico di Milano, Italy, May 1994.

27. J. Rambaugh, M. Blaha, W. Premerlani, F. Eddy, and W. Lorensen. *Object-Oriented Modeling and Design*. Prentice Hall, New York, NY, 1991.

28. Rational. *Rational Rose: User Manual*.

29. O. Færgemand and A. Olsen. Introduction to SDL-92. *Computer Networks and ISDN Systems*, 26:1143–1167, 1994.

30. C. Sibertin-Blanc. Cooperative Nets. In R. Valette, editor, *Application and Theory of Petri Nets 1994, Proceedings of the 15th International Conference*, volume 815 of *Lecture Notes in Computer Science*, pages 206–218, 1994.

31. I. Sommerville. *Software Engineering*. Addison-Wesley, fifth edition, 1996.

32. E.Y. Wang, H.A. Richter, and B.H.C. Cheng. Formalizing and Integrating the Dynamic Model within OMT. In *Proceedings of the 19th International Conference on Software Engineering*, pages 45–55. ACM Press, May 1997.

A   High-Level Petri Nets

High-level Petri nets are Petri nets augmented with conditions and action on values associated to tokens. High-level Petri nets have been instantiated in several equivalent ways. The most well known classes of high-level Petri nets are Coloured Petri nets ([17]) and Predicate/Transition nets ([12]). In this paper we use HLTPNs, introduced in [14], since it is the model supported by Cabernet ([26]), which is the tool used in our experimental work with the *CR Approach*. [14] focuses mainly on the introduction of time in high-level Petri nets. In this paper, we refer to the untimed model; timing aspects will be considered in future extensions of our work. We refer to the Petri net extension use in this paper simply as *high-level Petri nets*.

High-level Petri nets are Petri nets, i.e., bipartite connected graphs, where places are associated with types; tokens are associated with variables and values, according to the type of the "container" place; transitions are associated with predicates and actions, according to the types of the places of their pre and post-sets. Variables that occur in predicates and actions of transitions are dynamically bounded to variables of tokens in the pre and post-sets of the transitions. A transition t is enabled by a tuple $tup - in$ of tokens in its preset if the predicate of t evaluates to *true* on the values of the tokens in $tup - in$. The firing of an enabled transition removes the enabling tuple from its preset and produces a new tuple $tup - out$ of tokens in its postset. The values of tokens in $tup - out$ are obtained by evaluating the action of t on the values of the variables of the enabling tuple.

Figure 16 shows a sample high-level Petri net, that corresponds to a subset of class `Butler` and the invocation of method `delGuest` by an `arrivingPhiloso-pher`. Class `Butler` is modeled with a place `butler` always marked, a transition `reset`, a pair of transitions for each method, and a set of interface places. Transition tt Reset "clears" the token in place `Reset` when all active philosophers have eaten once after the last reset. Figure 16 shows only the pair of transitions and the interface places corresponding to method `delGuest`.

Transitions `LeaveTable` and `LeftTable` model the invocation of and the answer from method `delGuest`, according to the schema illustrated in Figure 6. Places `leaveTableP`, `eating`, `leaving`, and `leaveTableV` are part of the model of class `philosopher`.

Figure 16 lists the types associated with the places as C++-like classes, the association of types to places as C++-like objects, predicates and action of transitions as C++-like predicates and functions, tokens as values for the attributes characterizing the type of the place the tokens belong to.

Place `butler`, of type `butlerState` contains a pair of boolean variables for each philosopher, indicating the state (`active` or not) and the condition (`hungry` or not). The token that initially marks place `butler` indicates two active, hungry philosophers, and two non-active philosophers. Transition `reset` fires when none of the active philosophers is hungry (i.e., all active philosophers have eaten once after the last reset), as stated by the associated predicate. The predicate compares field `active` with the negation of field `hungry`. This is true when all active philosophers (field `active` == `true`) ate after the last reset (field `hungry` = `false`) and all non-active philosopher (field `active` == `false`) are hungry (field `hungry` = `true`, the default value for non-active philosophers). The firing of transition `reset` sets all philosophers to `hungry`, as stated by the associated action.

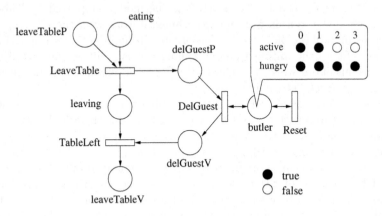

Fig. 16. A sample high-level Petri net.

Classes

```
class philosopher {                  class philPhil {
        string  type;                        philosopher state;
        integer id;                          philosopher param;
        integer leftFork;            };
        integer rightFork;
};                                   class butlerState {
                                             philInfo st[3];
class philInfo {                     }
        boolean active;
        boolean hungry;              class philId {
};                                           integer id;
                                     };
```

Places

```
philosopher leaveTableP, eating;
philId      delGuestP, delGuestV, leaveTableV;
butlerState butler;
philPhil    leaving;
```

Transitions

DelGuest
<u>predicate</u>
```
butler.st[delGuestP.id].active == true
```

<u>action</u>
```
delGuestV = delGuestP;
butler.st[delGuestP.id].active = false;
butler.st[delGuestP.id].hungry = true;
```

LeaveTable
<u>predicate</u>
```
(eating.rightFork == leaveTableP.leftFork) &&
(eating.type == ''arrivingPhilosopher")
```

<u>action</u>
```
delGuestP.id = eating.id;
leaving.state = eating;
leaving.param = leaveTableP;
```

Reset
<u>predicate</u>
```
(butler.st[0].active == !(butler.st[0].hungry)) &&
(butler.st[1].active == !(butler.st[1].hungry)) &&
(butler.st[2].active == !(butler.st[2].hungry)) &&
(butler.st[3].active == !(butler.st[3].hungry))
```

<u>action</u>
```
butler.st[0].hungry = true;
butler.st[1].hungry = true;
butler.st[2].hungry = true;
butler.st[3].hungry = true;
```

TableLeft
<u>predicate</u>
```
leaving.state.id == delGuestV.id
```

<u>action</u>
```
leaveTableV = leaving.state.id;
```

Transition LeaveTable models the request for removing a philosopher from the table. It is enabled by two neighbor philosophers: a philosopher in place leaveTableP asks his neighbor in place eating to leave. This is stated by the first term of the predicate that identifies neighbor philosophers by means of the shared fork. The philosopher that will leave must be of class ArrivingPhilosopher as asked by the second term of the predicate. This is the standard way of forbidding objects of a given class in a inheritance hierarchy to invoke methods of subclasses in the same inheritance hierarchy. Transition LeaveTable produces a token in place delGuestP with the id of the leaving philosopher, and a token in place leaving with both the philosophers involved in the execution of the method (fields state and param).

Transition `DelGuest` models the method execution. It removes the philosopher from the table only if `active`, as stated by the predicate. It "moves" the token from place `delGuestP` to place `delGuestV` and suitably sets the state of the token modeling the butler.

Transition `LeftTable` is enabled by two corresponding tokens in places `leaving` and `delGuestV`, as stated by the predicate. This predicate avoids wrong associations when several philosophers are concurrently executing the same method. Transition `LeftTable` produces a token in place `leaveTableV` that indicates the completion of the method execution.

Modeling a Groupware Editing Tool
with Cooperative Objects

Rémi Bastide, Philippe Palanque

LIHS-FROGIS, University Toulouse I,
Place Anatole France, 31042 Toulouse Cedex, FRANCE
{bastide | palanque}@univ-tlse1.fr

Abstract. This paper contains a solution to the case study proposed for the 2nd edition of the OO-MC workshop. In this paper, we merely recall the main features of the Cooperative Objects formalism, which is an object-oriented language, based on high-level Petri nets. We then include a Cooperative Object model describing the groupware editing tool described in the case study.

1 The Cooperative Objects Formalism

The Cooperative Objects (CO) formalism is a generic formalism dedicated to the modeling of concurrent and distributed systems. The main characteristic of CO is that the behavior of objects and their cooperation are modeled within the framework of the High Level Petri Nets (HLPN) theory.

The Cooperative Objects formalism has been presented in [2],[1], and several of its theoretical underpinnings can be found in [12]. We will therefore present here only the main features of the formalism, and the syntactical notations necessary to understand the treatment of the case study.

The goal of Cooperative Objects is to allow using efficiently the main features of object languages in the field of distributed systems. More precisely, we wish to provide an efficient notation allowing to describe the concurrent aspects of complex systems, in such a way that concurrency can be modeled inside the objects themselves as well as between objects. The specification of a CO class contains its *interface* (the list and signature of the methods it offers) and its *behavior*, modeled by HLPN and called the ObCS (Object Control Structure [9]). The ObCS defines the inner control structure shared by all instances of the class. The ObCS states the availability of methods according to the inner state of the object, and conversely the effect of methods execution on the object's state.

Obviously, even in a distributed system, purely sequential and algorithmic concerns remain very important. Thus CO do not aim at replacing conventional object languages, but can more adequately be considered as a host language for sequential OO languages such as Eiffel, C++ or Java. The CO formalism is supported by a tool

G. Agha et al. (Eds.): Concurrent OOP and PN, LNCS 2001, pp. 305–318, 2001.

called PetShop [6], which is designed to comply with CORBA [10], the Object Management Group's standard for distributed object systems. The interface offered by object classes is thus described in terms of the OMG's Interface Definition Language (IDL), and the details of the integration of CORBA IDL with the high-level Petri nets dialect used by CO is described in [5].

The integration of Cooperative Objects with a language such as Java is performed in the following way: ObCS are described by High-Level Petri nets, and the tokens that constitute the marking of the net can hold information, instead of being dimensionless entities like in conventional nets. In the PetShop tool (which is written in Java), the value of a token is a n-tuple of typed values, those values representing either:

- Any Java type (native type, or polymorphic class instance);
- A reference to another CO in the system.

Transitions in the net contain a precondition part and an action part, which are able to manipulate the tokens involved in the firing according to their type. Two quite different kinds of actions may be represented:

- A block of Java code, which is executed sequentially and in mutual exclusion with other transitions in the ObCS. This code can make use of Java objects involved in the firing, call their methods, dynamically create new Java objects, etc. This allows for an easy integration of CO with existing class libraries, and insures that COs are interoperable with more conventional approaches. For example we describe in [4] how to integrate CO within a User Interface Management System (UIMS).
- An *invocation* of another CO, i.e. the call of a method it offers; This call is executed in a concurrent and non-blocking way. The classical invocation strategies described in [8] are available (synchronous rendezvous, asynchronous message sending, time-out rendezvous), and each of these strategies is formally defined in terms of HLPN. The fact that both the inner behavior of objects and their communication primitives are modeled by HLPN allows providing a formal model of a system of communicating CO in terms of HLPN only.

1.1   Relationship between Interface and Behavior

The interface of a CO class is given in CORBA IDL, and the behavior of the instances is given by a HLPN. This behavior is related to the interface definition in the following way:

- For each service given in the interface, the ObCS net contains one *Service Input Port* (SIP) and one *Service Output Port* (SOP) that are special purpose places. The service input port is meant to receive the service invocations and their input parameters, while the service output port is the channel through which the service results will be provided by the object. A service input port can only have output arcs in the ObCS, and conversely a service output port can only have incoming arcs.

- The processing associated to a service is modeled by one or several macro-transitions related to the service's input and output ports. The transitions connected to the service input port are called *Accept Transitions*, while the transitions connected to the output port are called *Return transitions*.

Fig. 1 illustrates a simple CORBA-IDL interface, featuring only one service, *aService*, taking an integer as input parameter and returning a string. Fig. 2 illustrates an excerpt of a CO class definition. This CO class is defined to provide the behavioral specification for the *Example* interface (keyword *specifies*). The associated behavior is described in the ObCS: the subnet comprised between aService *accept* and *return transitions* (not detailed) is meant to compute a return string r according to the input parameter p.

```
interface Example {
     string aService(in integer a);
}
```

Fig. 1. A simple CORBA-IDL interface

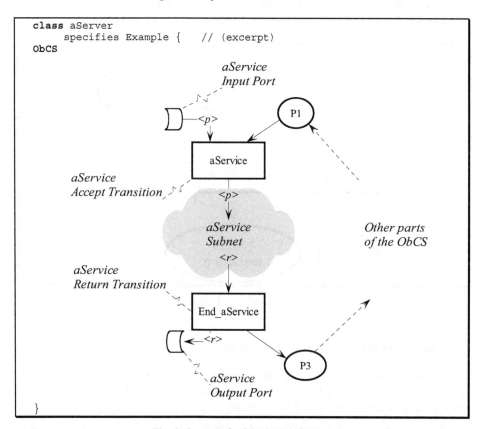

Fig. 2. Syntax of a CO class definition

The interface and ObCS together fully define the behavior of instances: A service request will begin executing when one of its associated *accept transitions* is enabled by the current marking of the ObCS. Conversely, the execution of a service is modeled by the occurrence of the associated macro-transitions, which states the side effect of the service execution on the object.

1.2 Timing in the ObCS Nets

The net dialect used to describe the ObCS uses timing on the transition's input arcs. An arc may feature an enabling time interval, which states the moments when a given token is available for the transition's firing. The origin for this time interval is the time when the token has initially entered the place.

Such a construct makes it easy to model „watchdogs", or time guarded actions that are frequent in distributed systems. Fig. 3 illustrates such a construct: Transition *T1* represents the normal processing to be applied to tokens that enter place *Waiting*. However, if such a processing does not occur before a given amount of time (*Delay*) has elapsed, A default action, modeled by transition *T2*, will be triggered.

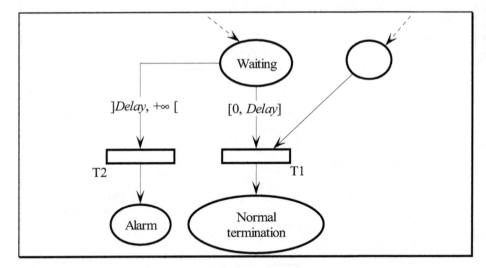

Fig. 3. Modeling a „watchdog"

1.3 Invocation Modes and Their Semantics

The ObCS are meant to describe the „server" behavior of objects (i.e. what their synchronization constraints are) but also their „client" behavior (i.e. how they request services from other Cooperative Objects in the system).

The communication between Cooperative Objects is syntactically expressed using the conventional dot notation, by actions in the ObCS transitions.

Such as service invocation is illustrated in Fig. 4: The class described here acts as a client of the *aServer* class described in Fig. 2. The ObCS descriptions includes the definition of the places type: for example, place PB is defined to hold references to instances of the *aServer* class, while the tokens contained in place PC will be 3-tuples holding an integer, a reference to an instance of *aServer* and a string.

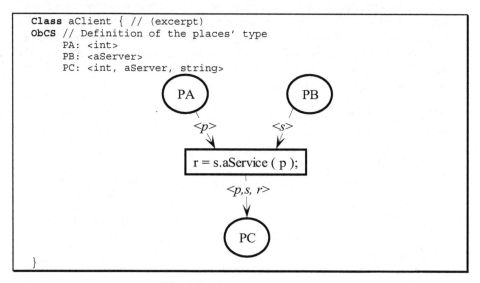

```
Class aClient { // (excerpt)
ObCS // Definition of the places' type
     PA: <int>
     PB: <aServer>
     PC: <int, aServer, string>
```

PA

PB

<p>

<s>

r = s.aService (p);

<p,s, r>

PC

```
}
```

Fig. 4. Synchronous Rendezvous

The variables on the arcs act as formal variables for the transition. The action of the transition is to call the service *aService* on the object bound to variable *s*, providing as a parameter the integer bound to variable *p*. A transition whose action is the invocation of another Cooperative Object is called an *invocation transition*. The default semantics for such a call is the *synchronous rendezvous*, whose operational semantics is described in Fig. 5.

The semantics of the synchronous rendezvous is given within the framework of high-level Petri nets, by enhancing the ObCS nets of both the client and the server of the rendezvous. Although the designer of the net only sees the ObCS descriptions as given in Fig. 2 and Fig. 4, the nets, before being actually executed in the running system, are expanded in the following way:

Client-side semantics
The following transformations have to be applied to each invocation transition:
- The invocation transition is considered as a macro-transition extending from the *request transition* to the *complete transition*. The *request transition* constructs a parameter token, including the original parameters of the service, the identity of the caller (variable *this*) and a globally unique call-identifier. This token is deposited in the *Invocation Parameter port*.

- A *waiting place* is introduced between the request transition and the complete transition. The presence of a token in this place indicates that a call is in progress.
- The results from the service call will be return to the client in its *Invocation Return Port*. The arrival of a return token will enable the *complete transition*, and terminate the service call on the client's side. It is important to note that the variable *id* is present on both input arcs of the *complete transition*: the transition is only enabled if a substitution is possible between the token values held in the *Waiting* and *Return Port* places, meaning that the same id is found in both tokens. This construct is necessary to allow a client to issue concurrently several invocations, and to enable the client to match the results it receives with the parameters it has initially provided.

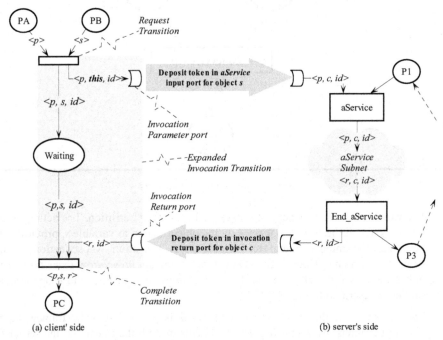

(a) client' side   (b) server's side

Fig. 5. Operational semantics of synchronous rendezvous

Server-side semantics

On the server's side, the structure of the net is not altered, but only the definition of the places' type and the inscriptions on the arcs. The only requirement for the server is to keep both the client's identity and the call-id within the service subnet, so that the results of the service can be properly routed back to the caller. The synchronous rendezvous is thus implemented as two unidirectional asynchronous message sending.

Only one primitive is required, and supposed to be provided by the implementation environment: the ability to deposit a structured token in a place of a remote Cooperative Object. This primitive is used to transport both the parameters and the results of an invocation. In the current implementation, this primitive is provided by a

CORBA compliant system, which takes care of the marshaling and unmarshaling of call arguments, and of the routing of tokens. However, other implementations could be easily substituted, such as Java RMI (Remote Method Invocation) or lower level solutions based on sockets.

Other invocation strategies can be described in the same way: The time-out rendezvous allows the client to specify the amount of time it accepts to wait for the result of the call, and to take a corrective action if the result is not provided within this delay.

The syntax of the time-out rendezvous and the associated semantics are described in Fig. 6: the semantics of the inhibitor arc adjacent to transition *discard* must be noted: the transition is enabled only if no substitution can be made on the variable *id*, i.e. if the token holding the call-id has already left the *Waiting* place.

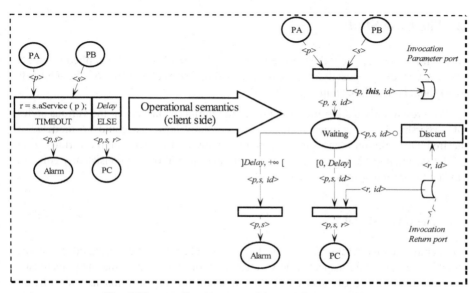

Fig. 6. Time-out Rendezvous and its associated semantics

Sometimes, the client may chose to simply ignore the possible results of a service, and to simply proceed as soon as the invocation has been issued, without any acknowledgment. This strategy is called asynchronous invocation, and its syntax and operational semantics are illustrated in Fig. 7.

It must be noted that whatever invocation strategy is chosen only impacts on the structure of the client's ObCS: the structure of the server's ObCS is not modified, and any service can thus be called with any invocation strategy.

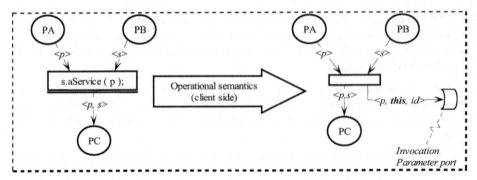

Fig. 7. Asynchronous invocation and its operational semantics

2   The Case Study

We will now present how the Cooperative Objects formalism can be applied to model the case study proposed as an exercise for the second edition of the OO-MC workshop, held in Osaka during ATPN'96.

We present only a simplified solution to the initial requirements: the various styles of ownership have not been retained, but we show how this could be modeled if desired. Our solution focuses instead on the communication between the various users of the system and on the resource-locking protocol that allows to consistently edit and update the shared shapes.

2.1   The UML Model

Cooperative Objects retain all of the features of classical object-oriented notations, notably their structural aspects (the various kinds of relationships that may relate two classes) and the use of inheritance as a fundamental organization and abstraction mechanism. To describe the classes involved in the solution of the case study and their relationship we will use the UML notation [11], which is essentially a derivative of the original OMT notation.

Fig. 8 illustrates the UML model of the case study, stating the interface of the classes and their relationships. The semantics and behavior of the classes will be discussed further in the following sections. Note that the UML notation uses a Pascal-like notation for the method's signature, while the CORBA-IDL that we use for the CO classes is more C++ - like.

In Fig. 8, the triangles denote generalization (the more general class being on top); The arrows between classes denote association, with the arrow end indicating the navigability of the association. The dotted arrows denote a dependency, stating that a class depends in some way on another, yet without holding references to instances of that class (in our case, the dependency is due to the method's signature).

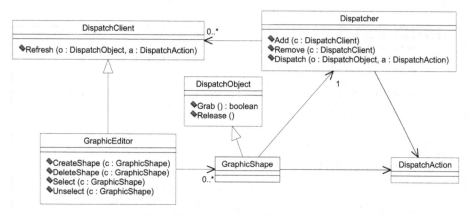

Fig. 8. The UML model of the case study

2.2 Multicasting

The distributed and cooperative nature of the case study requires special communication primitives between the software components. For example, each time a user adds or updates a graphic shape, this action must be forwarded to the other users. Likewise, each time a new user joins the editing session, he must be made known to all the others.

This kind of "one to many" communication is often known as **multicasting**: the same message is dispatched to a number of potential recipients, which will deal with it in their own specialized way. The Cooperative Object formalism does not include a special modeling construct to represent such communication. The formalism is strictly based on the client-server approach, where a client must know a reference to the server before calling a method, and where methods calls and answers are strictly one to one.

To emulate multicasting in the formalism, we define an interface called *Dispatcher*, and formally define its behavior through a CO class called *Dispatch_spec*. The purpose of this class is to maintain a list of potential clients, and to allow forwarding a same message to all the clients known by the dispatcher. The Cooperative Objects class for the dispatcher is given in Fig. 9.

Basically, when a Dispatcher object receives an invocation for its *Dispatch* method, it calls the method *Refresh* for each of the clients it holds, forwarding them an objet o, and an action a to execute on that object. The methods *Add* and *Remove* are meant to manage the list of clients.

The correct use of the *Dispatcher* class implies the use of polymorphic objects: Any objet that is to be called by the *Dispatcher* class must be an instance of a class that implements the *DispatchClient* interface (which features a *Refresh* method). This will be the case, for example, of the *GraphicEditor* class. Likewise, the objects to be dispatched must implement the *DispatchObject* interface, as this will be the case for the *Shape* class.

```
interface Dispatcher {
    void Add(in DispatchClient);
        // Add a client to the list of dispatch clients
    void Remove(in DispatchClient);
        // Remove a previously added client
    void Dispatch(in DispatchObject o, in DispatchAction a);
        // Dispatch the object o to all known clients,
        // to execute action a
}
class Dispatch_spec specifies Dispatch {
ObCS
    // Definition of the places' type
    Clients: <DispatchClient>
        // The Clients place holds the list of clients
    StillDispatching: <DispatchObject, DispatchAction>
    AlreadyDispatched: <DispatchClient>
```

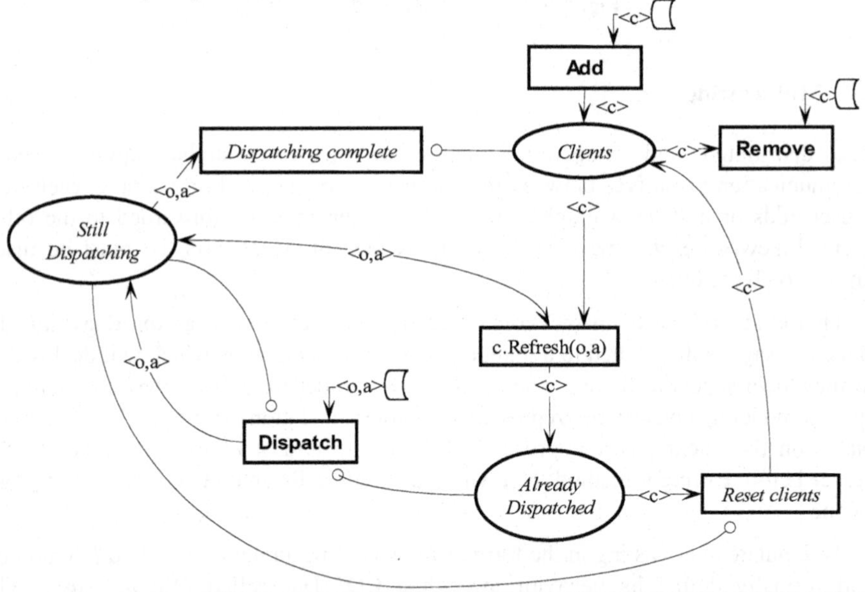

Fig. 9. The Dispatcher class

The behavior of the *Dispatch_spec* class can also be reused in other contexts: instead of holding a list of clients and dispatching an object to each of them, the same structure could be used to hold a list of shapes, and dispatch each of the shapes to one client. This would be useful, for example, when a new user joins the group to let him know of the current set of edited shapes.

2.3 The GraphicEditor Class

The *GraphicEditor* class models the behavior of the editing tool. Each user of the cooperative application will have a personal instance of *GraphicEditor*. The aim of this class is to allow users to create and manipulate shapes, while making sure that their actions are conform to the cooperation protocol defined in the requirements for the case study.

Several of the service input ports (for the services *CreateShape*, *DeleteShape*, *Select* and *Unselect*) have a special „button-like" representation: this is to distinguish those services as being *User Services*, i.e. services that are not meant to be triggered by other Cooperative Objects but directly by the user of the system, through some user-interface mechanism. The *Select* service, for example, could be triggered by clicking with the mouse on the on-screen representation of the shape. The implementation of that kind of user-services has been discussed in [3].

The GraphicEditor class, as shown above, deviates in some ways from the initial requirements, and makes some other requirements more precise, as shown in Fig. 10.

For example, at any moment an instance of the editor may own several shapes (their references are stored in the OwnedShape place). After an editor creates a shape, it initially owns it. Only owned shapes can be deleted. A free shape can become owned by selecting it, and conversely an owned shape becomes free by unselecting it. From the initial requirements, we have not retained the various styles of ownership, but they could be included by adding other services with the same pattern as Select / Unselect. For simplicity, we have not modeled the updating of the shapes by the editor: this could be easily done by adding one or more updating services, with the same activation precondition as the *DeleteShape* service.

2.4   The GraphicShape Class

The *GraphicShape* class essentially acts as a monitor, allowing to reliably lock and unlock a shape before deleting (or otherwise updating) it. The ObCS of the *GraphicShape* class is illustrated in Fig. 11 The shape is also responsible for forwarding the actions triggered on it to all the other editors, so that they are informed of the changing of the shape's status. To this end, the shape uses an instance of the *Dispatcher* class described above.

Two instances of the *GraphicEditor* class may concurrently try to select a shape, by calling it's *Grab* service; only one of them, however, will actually lock the shape (the *Grab* service returning TRUE) while the other one will be informed that it's locking attempt has been refused.

This kind of synchronization protocol is sometimes called „pessimistic locking", but an „optimistic locking" scheme such as the one described in [7] could be modeled as well.

```
interface GraphicEditor: DispatchClient {
    void CreateShape(in GraphicShape c);
            // Adds a new shape to the editor
    void DeleteShape();
            // Removes selected shape from the editor
    void Select(in GraphicShape);
            // Select a free shape, for later edition or deletion
    void UnSelect();
            // Unselect selected shape, freeing it for other editors
    void Refresh(in DispatchObject o, in DispatchAction a);
            // Refresh the shape o, for action a; inherited.
}
class Editor_spec specifies GraphicEditor {
ObCS        // Definition of the places' type
    OwnedShape: <GraphicShape>   // The shapes I own
    FreeShapes: <GraphicShape>   // The shapes owned by nobody
    UsedShapes: <GraphicShape>   // The shapes owned by someone else
```

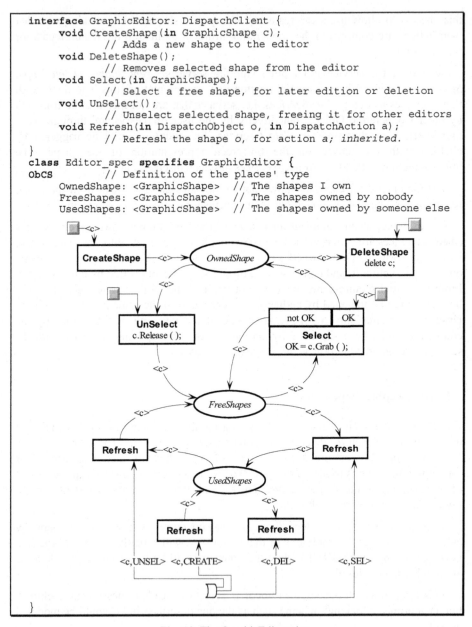

```
}
```

Fig. 10. The GraphicEditor class

Several of the requirements for the case study are not tackled in this presentation, for example the connection and deconnection of editors, or the hierarchical nature of the drawings. They could be handled by reusing the pattern of the Dispatcher class at various levels, so that a shape could for example hold references to all of its subshapes.

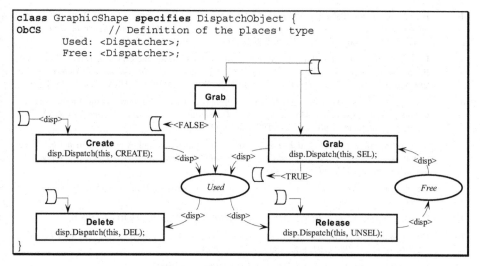

Fig. 11. The GraphicShape class

3 Conclusion

The language presented in this paper, aims at modeling concurrent and distributed systems following the OO approach. Objects are defined in a rather classical way, using most concepts of Object-Orientedness: class/object distinction, dynamic object creation, use relationship between objects by service call, inheritance and polymorphism. Each object may have a concurrent behavior, which is modeled by means of a High-Level Petri Net. Moreover, all objects may be simultaneously active and the service requests and the service results they send one to another synchronize them. Several important topics are not addressed in this paper for space reasons:

- As the language allows for inheritance with specialization semantics, it is necessary to ensure that a derived class is compatible with its ancestor, behavior-wise.
- Cooperative Objects are executable in essence, and several techniques for executing the models are devised in [1], at the system, the class or the object level.

References

1.   Bastide, Rémi. "Objets Coopératifs : Un Formalisme Pour La Modélisation Des Systèmes Concurrents." Ph.D. thesis, Université Toulouse III (1992).

2.   Bastide, Rémi, and Palanque, Philippe. "Cooperative Objects: a Concurrent, Petri-Net Based Object-Oriented Language." *Systems Engineering in the Service of Humans, IEEE-SMC'93*, Le Touquet, France, October 15-20, 1993. IEEE Press (1993)

3.   Bastide, Rémi, and Palanque, Philippe. "Petri Net Based Design of User-Driven Interfaces Using the Interactive Cooperative Objects Formalism." in *Interactive Systems: Design, Specification, and Verification, DSV-IS'94*. Fabio Paternò, Volume editor. Springer-Verlag (1994) 383-400.

4.   Bastide, Rémi, and Palanque, Philippe. "A Petri-Net Based Environment for the Design of Event-Driven Interfaces." *16ᵗʰ International Conference on Applications and Theory of Petri Nets, ICATPN'95*, Torino, Italy, June 1995. LNCS, no. 935. Springer (1995) 66-83.

5.   Bastide, Rémi, Palanque, Philippe, Sy, Ousmane, Le, Duc-Hoa, and Navarre, David. "Petri-Net Based Behavioural Specification of CORBA Systems." *20ᵗʰ International Conference on Applications and Theory of Petri Nets, ICATPN'99*, Williamsburg, VA, USA, June 21-25, 1999.

6.   Bastide, Rémi, Sy, Ousmane, and Palanque, Philippe. "Formal Support for the Engineering of CORBA-Based Distributed Object Systems." *Distributed Objects and Applications (DOA'99)*, Edinburgh, Scotland, September 5-6, 1999. IEEE (1999)

7.   Beaudouin-Lafon, Michel, and Karsenty, Alain. "Transparency and Awareness in a Real-Time Groupware System." *ACM Symposium on User Interface Technology (UIST'92)*, Monterey, CA, USA, November 1992. ACM Press (1992)

8.   Booch, Grady. *Object-Oriented Analysis and Design With Applications* Benjamin/Cummings (1994).

9.   European Space Agency. *HOOD Reference Manual.* 3.0, WME/89-173/JB ed. (1989).

10.   Object Management Group. *The Common Object Request Broker: Architecture and Specification. CORBA IIOP 2.2 /98-02-01*, Framingham, MA (1998).

11.   Rational Software Corporation. *UML Notation Guide.* 1.1 ed.1997.

12.   Sibertin-Blanc, Christophe. "Cooperative Nets." *15ᵗʰ International Conference on Application and Theory of Petri Nets, ICATPN'94*, June 1994. LNCS, no. 815. Springer (1994) 471-90.

Modelling Constrained Geometric Objects with OBJSA Nets

Maria Alberta Alberti, Paolo Evi, Daniele Marini

Dipartimento di Scienze dell'Informazione - Università degli Studi di Milano
Via Comelico 39/41 - 20135
e-mail: [alberti | marini]@dsi.unimi.it
Milano, Italy

Abstract. In this paper we introduce a formal specification of the problem of modelling geometric constrained objects adopting OBJSA nets, a high level Petri net. The geometric objects are defined imperatively while constructing them. The approach is innovative in that it solves constraints during manipulation, propagating messages among the objects involved in the geometric figure and it does not require numerical techniques. The formalization of the geometric constructions with OBJSA nets has been an important step to validate the system and in particular the constraint maintenance algorithm. Each class of the system is modelled by a OBJSA component and their compositions allows to describe a generic construction. The algorithm specified in OBJSA can be simulated in the ONE (OBJSA Net Environment) environment.

1 Introduction

In this paper we present a formal specification of an object-oriented approach to modelling geometric constrained objects. For this task constraints imposing is a delicate issue because it effects the way constraints are solved and can not be considered just a human-computer interface problem. For instance, constraints can be imposed in a numerical way, and this is very well suited for numerical solution method, or can be imposed by means of a logical language, and thereafter using logic programming approach to constraints solution. Our approach is based on a very natural way of creating a geometric figure: we adopt the paradigm of compass-and-ruler to instantiate lines or circles. New points are generated either directly or by intersections of lines and circles, and can thereafter be used to proceed in the figure construction. This approach mimics the compass-and-ruler drawing taught in schools in elementary geometry classes.

The innovation of our approach is to keep track of the construction process and to use it as the basis to solve the problem of constraint maintenance by a message propagation method. It is therefore important to find an effective way to represent the geometric construction process. A solution [1] that we adopted at first has been to define a bipartite oriented graph, the GCG, which resembles Petri nets in its graphical representation but differs from them in many aspects, that we will explore later on. In this paper we introduce a novel representation of the construction process by using

G. Agha et al. (Eds.): Concurrent OOP and PN, LNCS 2001, pp. 319-337, 2001.
© Springer-Verlag Berlin Heidelberg 2001

OBJSA nets (OBJ Superposed Automata [2]), a class of high level Petri nets. OBJSA nets formalise the geometric construction process, and moreover, allow to represent and simulate the message propagation algorithm which is used to maintain constraints when a point of the figure is interactively modified, in the simulation environment ONE (OBJSA Net Environment). This work allows us to give a sound background to the Eiffel based implementation, called GEObject [3], which has been used to explore the approach and create a large variety of examples of constrained constructions.

Constraint-based specification is not limited to geometric modelling, but can be extended to a variety of applicative areas; among them user interface design [4] [5], animation, industrial design [6], window management [7]. One of the earliest applications has been the constraint-based geometric modelling system developed by Sutherland [8], who implemented numerical methods for constraint solving. Also Nelson [9] adopted numerical methods (Newton-Raphson iteration technique) in the Juno system. In these systems objects are declared in terms of geometric relations between them and their definitions are subsequently translated into analytical descriptions of graphical primitives. Many fundamental relations, such as lines being congruent or parallel, lead to non-linear equations; moreover, it is often difficult to find initial conditions which allows the numerical method to converge. These facts imply that constraint solving by numerical techniques is not adequate for large systems and too slow for interactivity. Always in the area of geometric modelling, Fuller and Prusinkiewicz [10] implemented the L.E.G.O. system for 2D and 3D objects specification based on constructive methods: the constraints maintenance is assured by the same algorithms that express objects in geometric terms. Their approach is similar to ours, but while we explore the object-oriented paradigm, they exploit the Lisp system to express the imperative constraints. Constraints maintenance by constructive methods has also been adopted by Bruderlin [11] and Sohrt [12], who, using rewriting rules, discover and re-execute the construction steps. Van Emmerik [13] presented the CSG (Constructive Solid Geometry) system, where geometric relations are specified by constraints between local co-ordinate system and evaluated real-time. More recently, LinkEdit [14], overcomes the limits of the local propagation method SkyBlue [15] to solve cycles by calling a cycle avoidance subsystem, when a new constraint is introduced. LinkEdit presents an interactive solution to constraints declaration, as well. Our approach, by using the internal representation of a construction, does not requires algorithms for cycle avoidance and for numerical constraint solving. The construction becomes part of the system and keeps the relationships among objects. The algorithm dealing with transformations on objects is invoked when an object receives a transformation message. The transformation of a figure is essentially its reconstruction, obtained by propagating the transformation message to all descendants of an object, invoking the instance methods that have been activated during the construction.

In this paper, in §2 we outline the work background, in §3 we introduce a formalization of geometric constructions and transformations with a type of Petri nets, the OBJSA nets [2] and the motivation why a formalization is needed. In §4 we present how the construction and the transformation processes modelled by means of OBJSA nets have been simulated in the environment ONE [16] [17]. In §5 we give an outline of the object-oriented structure of GEObject system.

2 Geometric Constructions with Constraints

There is a limitation on the geometric objects which can be constructed by compass-and-ruler[1], but within these limits a lot of interesting figures and cinematic constructions are possible: for instance cinematisms such as connecting rod, crank shaft and piston, Hart inverter, or theorems such as the parallelogram inscribed in a quadrilateral. As an example of the compass-and-ruler constructions admissible by GEObject, let us now define the axis of a given segment. One could draw the segment, giving its end-points, and then two circles respectively centred in each end-point and radius equal to the segment itself. The axis of the segment will be the line passing through the intersections between the two circles (in Fig. 1 the segment is labelled s and the axis is line l). During the process one specifies some constraints among the entities involved in the figure: in the example, the line l turns out to be orthogonal to segment s, so this construction can also be used as a way to specify an orthogonality constraint between two lines. Other constraints in this figure are: point A and B are constrained to the segment s; circles c1 and c2 are constrained to be centred in A and B respectively and to have as radius the given segment; points C and D are constrained to the circles c1 and c2.

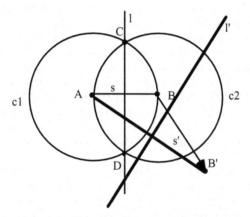

Fig. 1. Axis l of segment s and their transformation l' and s' obtained by translation of point B into B'

We are now interested in solving the problem of constraints maintenance during transformations: applying a translation to any point of the figure, we will have to

---

[1] Mathematicians since long time have been questioning whether a compass and a ruler were the minimum set of tools to solve geometric constructive problems and which problems could be solved by this method. Felix Klein, using Galois theory results (18) has clearly stated that the geometric figures that can be constructed using compass and ruler are those and only those that can be expressed in terms of algebraic equations that can be solved by applying a finite number of rational operations or second order radicals. These results establish the theoretical bound of the kind of problems that one can face in a constructive geometric system, like ours.

preserve the internal relationships among entities. For example, if we translate point B to position B', we obtain a new segment s' and a new axis l'.

In order to make explicit the process underlying a geometric construction and its successive manipulations, a bipartite graph, that we will call *GCG (Geometric Construction Graph)* [19] has been informally introduced and adopted. It describes the sequence of steps that enables us to generate a figure and the relationships among the constituent objects. GCG bare some resemblance to Petri nets, but it is definitely not: e.g. GCG do not consumes resources. GCG is a representation of a class of constructions, not of a single one: the uniqueness of a given figure derives from a specific instantiation and evolution of the GCG.

GCG's nodes are interpreted as geometric objects, represented by circles whose labels are the object names, and as constructive methods or actions, represented by rectangles whose labels are the action names, that can be the same for different classes of objects. In the graph input arrows to an action denote objects required as parameters of the action whose output arrows link to the generated objects. Arrows establish a partial temporal ordering among the operations that will lead to the figure. The ordering is defined starting from nodes without input arrows *(start nodes)* towards nodes without output arrows *(final nodes)*. It is supposed to have a *token* in any start nodes. An action can be executed only if all the input nodes connected to it contain a token. Action execution does not consume input tokens and produces an output token in all the corresponding output nodes. The construction process ends when all the final nodes contain a token. In Fig. 2 we show the GCG describing the construction of the segment axis: we start by instantiating two points, namely A and B, then we instantiate the segment s and the two circles c1 and c2, by sending appropriate action messages to the object points. We can select the derived object circles and send them a message to perform an intersection action, in order to generate new points C and D, that in turn will receive a message to build a line (see Fig. 1). The semantic of GCG is given in [19]. The order of the parameters is relevant: for example in Fig. 2 the do_circle method uses its first parameter as the centre and the second to compute the radius, making the two circles c1 and c2 different.

GCG's are helpful in describing informally and intuitively a problem which is sequentially solved by users but has intrinsic concurrent aspects. We observe, for instance, that in Fig. 2 do_circle and do_segment actions are independent from each other and the order of execution is not relevant, they can be executed simultaneously or in whatever order. This independence produces concurrency (actions can be executed simultaneously) and non-determinism (execution order is not completely pre-defined).

During the construction process, GEObject users actually perform actions sequentially, while concurrency can be exploited by the system, when it responds to the users' action to transform the figure. More precisely, users can translate one of the points involved in the figure and the system reacts by re-traversing concurrently the graph that describes the construction. For example, in the simple case of translation of a primitive point, as in Fig. 1 where point B is translated into B', the whole figure must be drawn again. In order to achieve the desired result the system has to repeat the operations specified by the GCG of Fig. 2, starting from the primitive objects, represented by the nodes labelled A and B, following the arcs and executing the specified actions do_segment, do_circle, intersect, do_line. The two actions do_circle, and do_segment can be executed concurrently. The action intersect can be executed only after the two do_circle actions, and the action do_line after the intersect one. Indeed, each action produces new objects that will in turn trigger other actions until the whole figure is re-drawn.

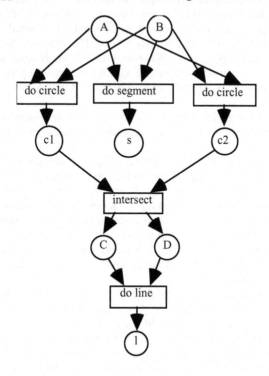

Fig. 2. GCG of the construction of a segment axis

When we apply a transformation to a point, the GCG graph at first is backward traversed to find unconstrained points in the hierarchy of the construction, then the transformation is applied to these free points and the GCG graph is forward traversed to re-apply the construction with transformed parameters. If no free points in the backward traversal are found, the whole figure is rigidly translated. If the effect of a transformation applied to intersecting objects is that they do not intersect any more, the intersection points become imaginary and are not generated. We can reach this state also by transforming a whole figure: for instance in a figure with intersecting circles by translating one of them so that they do not intersect anymore. In this last case, any object generated upon the intersection points can not be instantiated anymore, but it will be re-instantiated again as soon as the intersection points will become real. On the opposite, a point can be instantiated even if its location is outside the drawing area on the display window.

3 Representation of Geometric Constructions with OBJSA

The GCG graph is the data model adopted in GEObject, on which the transformation algorithm relies, but it is a rather informal tool to represent the relationships that are established during a geometric construction. The lack of a formal characterisation of this tool has moved us to a more formal model. Therefore we used OBJSA nets to describe the geometric processes. Whereas GCG, even if informal, is simple and straightforward to write and to understand, OBJSA nets have a precise semantics.

Among different types of Petri nets we chose OBJSA nets for the following reasons:

- the need of high level Petri nets to represent the information associated to geometrical entities (i.e. point co-ordinates, segment lengths, circle radius, angles...);
- OBJSA nets are suitable to represent object-oriented system: in particular for the data abstraction capabilities of the OBJ3 language;
- the availability of the simulation environment ONE (OBJSA Net Environment [16]) to support model definition and validation.

OBJSA nets were defined in order to describe generic distributed systems. To this extent they provide the notion of composition of the non deterministic components of the system. OBJSA are based on Superposed Automata Nets described in [20] which allow composition of subsystem models, and use OBJ3 language [21] to specify tokens holding information.

An OBJSA component is made of places and open/close transitions connected by oriented arcs: all elements are labelled. Open transition represents the interface of all components towards the external world, i.e., other components. To compose two or more open components (components with open transitions) one has to superpose the corresponding open transition of each open component. An open transition has a set of constraints (algebraically expressed by means of OBJ3 language) that must be satisfied during superposition. Token domain definition is also based on OBJ3 language and is divided in two parts: token name and token data. The token name represents the identity of the token and cannot be modified; the token data represents the information associated to the token and can be modified by transitions according to their label.

In order to model a system with OBJSA nets, one has to follow three steps:

1. identify logical independent entities in the system to model; each entity will be modelled by an *open OBJSA component*
2. define, for each open component, the token domain specification using OBJ3 language
3. compound open OBJSA components by means of their open transitions in order to obtain the complete system model.

We have defined an open OBJSA component for each class in GEObject. Therefore we defined component Geopoint, Geoline, Geosegment, Geocircle and so on (see § 5). Each of them is an OBJSA net with two places labelled respectively UNBORN and ALIVE and a set of transitions. In each component the place labelled UNBORN initially contains an infinite number of tokens whose token name is a unique positive integer. Token data is undefined. No tokens are initially present in the ALIVE place. Each transition corresponds to an action in the GCG, i.e., an operation that can be performed on a generic geometric object represented by a token. When a transition fires some required tokens are fetched from the ALIVE place, used, modified in their data part as necessary and re-placed in the ALIVE place of the same component, while other tokens are fetched from the UNBORN place, their data part is initialised, and placed in the ALIVE place of the same component.

If a geometric operation involves two or more object classes, the corresponding open components modelling those classes have an open transition. For example, to model the creation of a segment given two points we define an open transition called S-with-two-pts in any Geopoint and Geosegment component. Let us see how superposing these open transitions we obtain a new OBJSA component that models the creation of a segment by two points. In Fig. 3 it is shown a part of Geopoint component and in the following Fig. 4 a part of Geosegment component. Finally, in Fig. 5 we have composed Geopoint and Geosegment components by means of the superposition of open transition S-with-two-pts.

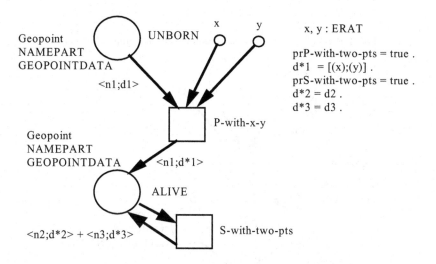

Fig. 3. Part of the Geopoint OBJSA component modelling a point

The definition of sort NAMEPART is given by the following OBJ3 code, where P, L, H, S, C, A, N stand for point, line, half-line, segment, circle, arc and angle respectively, and NAT is the set of natural number:

```
obj TYPE is sort Type.
  ops P L H S C A N: -> Type.
endo
obj NAMEPART is dfn Namepart is 2TUPLE [TYPE, NAT] * (
  op <<_;_>> to [_;_],
  op1*_to histype(_),
  op2*_to number(_) ).
endo
```

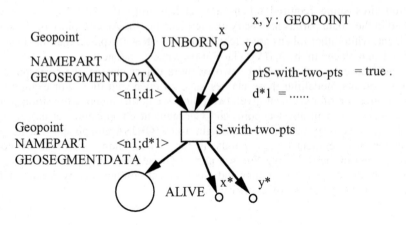

Fig. 4. Part of Geosegment OBJSA component modelling a segment

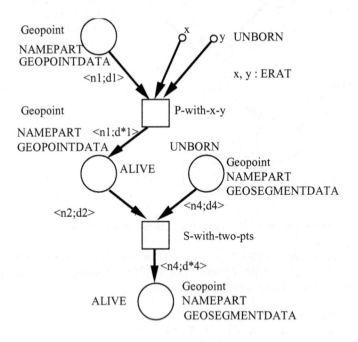

Fig. 5. Superposition of components modelling point and segment

The definition GEOSEGMENTDATA is built upon definitions of GEOPOINTDATA as shown in the following sketches of OBJ3 code:

```
obj POINT is dfn Point is 2TUPLE [ERAT, ERAT] * (
    op <<_;_>> to [_;_],
```

```
      op1*_to x(_),
      op2*_to y(_) ) .
    op imm: -> Point.
    op undef: -> Point.
  endo
  obj 2POINT is dfn 2Point is 2TUPLE [POINT, POINT]*(
    op <<_;_>> to [_;_],
    op1*_to first(_),
    op2*_to second(_) ) .
  endo
  obj GEOPOINTDATA is sort Geopointdata.
    pr POINT.
    subsorts Geopointdata < Point.
    ... operations omitted ...
  endo
  obj GEOLINEDATA is sort Geolinedata.
    pr 2POINT.
    subsorts Geolinedata < 2Point.
    ... operations omitted ...
  endo
  obj GEOHALFLINEDATA is sort Geohalflinedata.
    pr GEOLINEDATA .
    subsorts Geohalflinedata < Geolinedata.
    ... operations omitted ...
  endo
  obj GEOSEGMENTDATA is sort Geosegmentdata.
    pr GEOHALFLINEDATA .
    subsorts Geosegmentdata< Geohalflinedata.
    ... operations omitted ...
  endo
```

The sort ERAT represents an extension of the set of rational numbers necessary to define the square root operation rounded at 5 decimal digit (see [19]).

In Fig. 6 we show the OBJSA net modelling the construction of a segment axis. It is completed by the following Boolean predicate associated to each transition:

```
prP-with-x-y = (n1 == [P; 1]) or (n2 == [P; 2]).
prS-with-two-pts = (n14 == [P; 1]) and (n15 == [P; 2]).
prS-with-center-pt = ( (n4 == [C; 1]) and (n7 == [P; 1])
and(n8 == [P; 2])) or ((n4 == [C; 2]) and (n7 == [P; 2])
and (n8 == [P; 1]) ) .
prL-with-two-pts = (n5 == [P; 3]) and (n6 == [P; 4]).
```

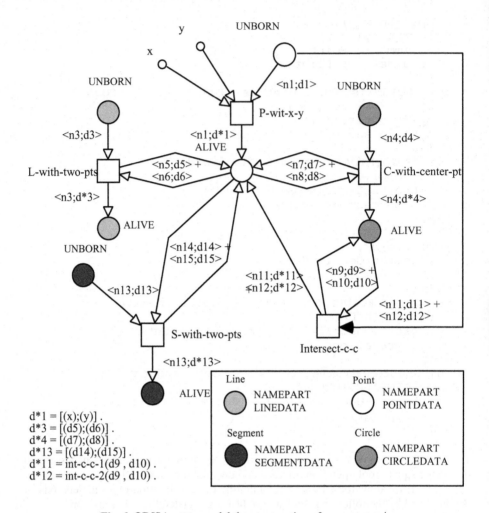

Fig. 6. OBJSA net to model the construction of a segment axis

Algebraic specification of operations `int-c-c-1` and `int-c-c-2` is given in [19]; they compute the co-ordinates of two points by intersecting two circles. The net in Fig. 6 is obtained by superposing four OBJSA open components: Geopoint, Geoline, Geosegment and Geocircle. We show only relevant transitions.

Initially, only UNBORN places contains some tokens, i.e., only `P-with-x-y` transition can fire: other transitions need at least 2 tokens from some ALIVE place. When two points are created (transition `P-with-x-y` fires two times) also transition `S-with-two-pts` and `C-with-two-pts` can fire. The transition `L-with-two-pts` can not fire because of the predicate associate to it: it requires to use points named [P; 3] and [P; 4] but existing points are called [P; 1] and [P; 2] because of predicate in transition `P-with-x-y`.

The two formalisms, GCG and OBJSA nets, to represent a geometric construction, briefly sketched above, are related; in fact if we unfold the OBJSA model, which represents the construction process of a figure, we get a GCG. The unfolding process that we apply to the OBJSA model is only a partial one. In fact a geometric figure can be seen as a couple: a *construction process* and a *set of values* to be associated to the geometric instances. The GCG model captures the construction process whereas the OBJSA model captures also the numeric value of geometric instances (i.e. points co-ordinates). This partial unfolding from OBJSA to GCG involves loss of information, whereas the reverse process requires addition of information. On the other hand GCG represents a construction in a more abstract way, i.e. it represents the class of all figures produced by that construction.

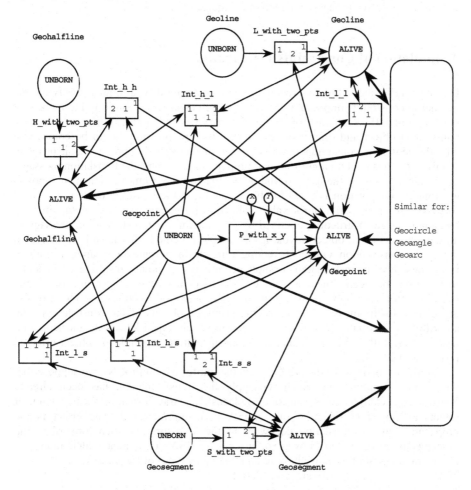

Fig. 7. General OBJSA net modelling any compass-and-ruler construction

We have modelled with an OBJSA net the whole structure of GEObject, as well, see Fig. 7. This net integrates all the OBJSA subnets which represent all objects and functionalities in GEObject, assuming that all predicates are true. The transitions model all basic creation methods and the places model elements of GEObject. All nodes labelled ALIVE initially have an infinite number of tokens, whereas those labelled UNBORN initially have no tokens. A geometric construction is represented by a specification of this general net, defining the necessary predicates of the transitions involved in the construction.

OBJSA has proved to be very appropriate to describe an object-oriented system as GEObject, where classes, that represent an abstraction of the geometric entities, are modelled by OBJSA components. Each Eiffel class is described by an OBJSA component and each creation method is modelled by a transition. The composition of different components, by means of their open transitions, enables to describe a generic constructive process of a geometric figure.

4 Simulation of Constraints Propagation

Our geometric modelling approach enables us to modify a figure satisfying the geometric constraints imposed at the time of the construction, limiting the numerical computation to the evaluation of intersections of lines and circles. When users select a constrained object and drag it on the screen, its new position has to be compatible with the geometric constraints imposed during the construction process. This problem is solved by means of an algorithm that propagates backward the translation message, given interactively by the user, to find points that are free to move. When such points are found their position is accordingly changed and the construction is re-executed forward with new parameter's value. The transformation of a figure therefore is essentially its reconstruction, obtained by propagating the transformation message to all descendants of an object, invoking the creation methods that have been activated during the construction. Objects receiving a transformation message can be primitive, i.e. points directly instantiated, or non primitive, e.g. points resulting form some intersection between lines or circles. A non primitive object receiving the transformation message broadcasts it to all its ancestors recursively. This process proceeds until the message arrives to the primitive objects of the construction. The algorithm is distributed among objects, because each object knows its ancestors and descendants, following the fundamental encapsulation principle of object-oriented programming.

In order to model with the OBJSA nets the tranformation process, we suppose that the geometric figure, on which we apply the transformation, has been already constructed (i.e. all tokens are in ALIVE places). The transformation starts when user selects an object (a token in the OBJSA net) and moves it. The effect is the propagation of the translation backward to the ancestors and then forward to the descendants as described above. So, in each OBJSA component which models a geometric class of objects, we have to add three new types of transitions:

move   to activate a translation
backward   to propagate the translation to all the ancestors
forward   to propagate the translation to all the descendants.

In order to handle the propagation of translations by means of these transitions, we need to extend the token domain, with data representing the state of objects during the translation propagation. The data to add are:

```
already-moved  specifies if the object represented by the token has undergone a
               backward propagation
move-state     standby   -  specifies an object not to be moved
               backward  -  specifies an object propagating the translation
                            to direct ancestors
               forward   -  specifies an object propagating the translation
                            to direct descendants
is-immaginary  specifies an immaginary object.
```

Tokens associated to points need the following token data:

```
is-primitive   specifies if the point has no ancestors (therefore generated by the
               method P-with-x-y).
is-anchored    the point is constrained to be fixed.
```

The move transition is always an open transition modelling interaction with users. It requires two external parameters: dx and dy representing the translation displacement. Its effect is to fetch the required token from the ALIVE place of the OBJSA component and to enable other transitions in order to propagate the transformation. The translation takes place if the object is not immaginary and it is not anchored (in case of a point). In Fig. 8 we have drawn the P-move transition for the Geopoint OBJSA component.

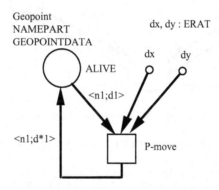

Fig. 8. *P-move* transition for the Geopoint OBJSA component

```
prP-move  =  not  isimmaginary  (pcomflag(d1))  and  not
isanchored (pointflag(d1)).
d*1 = if isprimitive (pointflag (d1))
      then psetforward (psetdelta (d1, [dx; dy]))
      else psetbackward (psetdelta (d1, [dx; dy]))
   fi.
```

We need a backward transition for each creation method involved in the geometric figure. In the example of the transformation of the segment axis (the construction of which is modelled by net in Fig. 6) we need a *backward* transition for the methods S-with-two-pts, L-with-two-pts, C-with-center-pt and Intersect-c-c. So, it is possible to link each object with all its ancestors, depending on the creation method used to instantiate the object. A *backward* transition is enabled when at least one of the input token has the variable *move-state* set to backward.

In the same way, we need a forward transition for each constructive method involved in the geometric figure. In the example of the transformation of the segment axis we need a forward transition for the methods S-with-two-pts, L-with-two-pts, C-with-center-pt and Intersect-c-c. So, it is possible to link each object with all its descendants, depending on the creation method used to create the descendants. A forward transition is enabled when at least one of the input token has the variable move-state set to forward. In Fig. 9 we have drawn the backward transition together with the forward transition for the S-with-two-pts method.

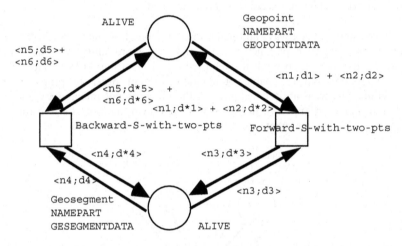

Fig. 9. Forward and backward transitions for the Geosegment component

```
prBackward-S-with-two-points = movestate(lcomflag(d4)) ==
backward.
  d*4 = lsetalreadymoved (lsetstandby(d4)).
  d*5 = if not alreadymoved (pcomflag(d5))
        then if isprimitive (pointflag(d5))
             then if not isanchored (pointflag(d5))
                  then psetalreadymoved
                             (psetforward (psetcharpt
                                    (d5,
  [(x(pcharpt(d5))+(ldelta (d4))));
                                      y (pcharpt(d5)               +
  y(ldelta(d4))))]
                                   )))
                  else    psetalreadymoved(psetforward(d5))
  fi
             else psetbackward(psetdelta(d5, ldelta(d4)))
  fi
        fi.
  d*6 = same as d*5
```

```
prForward-S-with-two-points  =   (movestate(pcomflag(d1))
== forward
                    or      movestate(pcomflag(d2))      ==
forward)
                    and (((isimmaginary(pcomflag(d1))
                         or
isimmaginary(pcomflag(d2)))
                      and                              not
isimmaginary(lcomflag(d3))))
                    or (lcharpt(d3) =/=
 [(pcharpt(d1);(pcharpt(d2))])) and
 lcharpt(d3)=/=[(pcharpt(d2))(pcharpt(d1))]))).
d*3      =    if       isimmaginary(pcomflag(d1))       or
isimmaginary(pcomflag(d2))
      then if not isimmaginary(lcomflag(d3))
          then lsetimmaginary(lsetforward(d3))
          else d3 fi
     else                 if                    lcharpt(d3)
=/=[(pcharpt(d1));(pcharpt(d2))] or
                        lcharpt(d3) =/=
 [(pcharpt(d2));(pcharpt(d1))]
         then               lsetforward(lsetcharpt(d3,
 [(pcharpt(d1));(pcharpt(d2))]))
          else d3 fi
     fi.
d*1 = d1.
d*2 = d2.
```

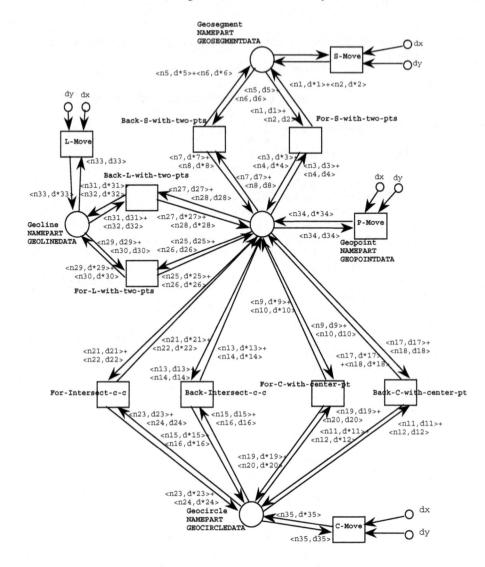

Fig. 10. Complete OBJSA net of segment axis construction and transformation

Eventually, the OBJSA net in Fig. 10 extends Fig. 6 modelling also components translation. Initially only move transitions are enabled (S-Move, L-Move, C-Move, P-Move).

In order to fire they need interaction: a user selects the object to move and activates the firing of the corresponding transition. For example, to move point [P; 1], a user has to select P-Move transition, chooses the point to move, and then specifies the translation amount dx and dy. The OBJSA net evolves automatically by means of successive firing of backward and forward transitions according to the algorithm described above. The algebraic specification of the OBJSA net in Fig. 10 is omitted. It can be derived from OBJSA nets in Fig. 8 and Fig. 9.

Both the construction and the transformation process modelled by means of OBJSA nets have been simulated in the environment ONE, allowing us to test the algorithms.

5 Object-Oriented Implementation of the System

GEObject was developed in the object-oriented Eiffel language (version 2.3), as an extension of its pre-defined classes. In particular we have defined classes to represent Euclidean Geometry, implementing some primitive geometric entities and the hierarchy among them (Fig. 11) [22].

At the first level of abstraction, each geometric object can be seen as an instance of the class `GeoNode`. This class characterises a generic geometric object defining its name and a set of deferred *interaction methods* (i.e. to apply geometric transformations, to modify the graphic aspect, etc.). Instances of the `GeoNode` class correspond to the nodes represented by circles in the GCG.

At the second level of abstraction, we have to distinguish between entities with a geometric nature *(GeometricObject)* and those with a numeric nature *(GeoViewer)*. The first ones have associated some geometric methods, such as *intersect*. The last ones are used to visualise numeric attributes of the geometric objects (i.e. segment length, angle width, etc.) and to compute further numeric values related to the figure. In fact, these objects have associated some algebraic methods like `+`, `-`, `*`, `/`, `sqr`, ...

At the third level of abstraction, we have to distinguish among different geometric entities (points, lines, circles, ...) and different numeric objects. At this level we supply the implementation of specific *creation methods*, which enable users to instantiate objects in different ways and of deferred *interaction methods*. For each class of object, we may find useful to have several *creation methods*: for instance a circle can be created given its centre and the radius, that is a second point, or given three different points. Class `GeoPoint` has only one creation method: a point is instantiated by specifying directly the (x y) co-ordinates; an object line of class `GeoLine` is specified by giving two points. Among interaction methods we distinguish among interface and geometric methods. The first ones allow users to interact with the object or with its graphic appearance; for example the method `hide`, that allow users to hide part of a construction, or `hide_name` to hide the object label. Among these methods the system provides also special methods to query the object numeric value: for instance to get a segment length or the actual point co-ordinates. The geometric methods are necessary to interact with the geometric object: for example, `translate` by a given distance or for segment objects a method could return the middle point. The most important polimorphic method in each class is `intersect` that computes the intersection points with other objects. The computation of intersecting points are the only actual numeric computation involved in the system.

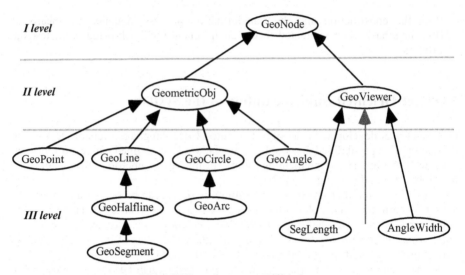

Fig. 11. Hierarchy among Eiffel classes

GEObject is an incremental geometric modelling system. Users can extend the set of primitive classes and of methods, by editing new classes and methods interactively while entering a construction. The construction is considered as a prototype, an example of the new procedure. Users do not have to write code to extend the system. Therefore this feature of GEObject is based on the *programming by example* paradigm.

For example, let us suppose we want to define the new class of equilateral triangles; we will have to construct an equilateral triangle starting from two vertexes so that the system can "learn" the construction generating automatically the Eiffel code corresponding to the new class `GeoEquiTriangle`. From now on users can draw a generic equilateral triangle without repeating the whole construction, but simply sending the creation method of the new class. To keep this new class in the system for future running sessions, the user must recompile the extended code.

The new class inherits the interaction methods from its ancestor class. For instance the class `GeoTriangle` can be defined as a specialisation of class `Polygon` and could be specialised into two different classes `GeoEquiTriangle` and `GeoRectTriangle`. If a user defines an interaction method called `median` for triangles, this is inherited by the two subclasses. We recall that interaction methods are constructive procedures producing geometric objects: in triangles medians and heights are generally obtained with different procedures, that will generate the same object in some special class, such as `GeoEquiTriangle`. Also it is important to remind that users do not specify properties but constructions; in the system one can verify properties, for instance by querying objects, but not prove theorems. There is no multiple inheritance. All classes derive from class `GenericObj`.

In the same way the system can be augmented by adding new construction methods: for instance, the segment-axis construction can be added to the system as a construction method of class `GeoSegment` or a new construction method of an existing class. Indeed, it is the user's responsibility to decide upon the hierarchy and to be consistent in terms of the geometry taxonomy.

6 Conclusion

In this paper we introduce a formalism to represent the problem of constrained geometric constructions, adopting OBJSA nets. The formalism allows us to define all possible geometric constructions by compass-and-ruler and their transformations. This work followed the implementation of an object-oriented system, GEObject, to build and transform figures by direct manipulation on the screen. The implementation of the formalism in the environment ONE allowed us to simulate both constructions and their transformations and contributed to the tuning and verification of the algorithms, particularly of the constraint maintenance by messages propagation. The OBJSA nets proved to be particularly suited to model a quite complex object-oriented system, such as GEObject.

This research has been partially funded by MURST 40% 1995-1996. The work of Paolo Evi has been supported by a fellowship of Università degli Studi di Milano and he would like to acknowledge the hospitality at Université Marne La Vallée.

References

1. Alberti, M.A., Bastioli E., Marini, D.: *Towards Object-Oriented Modelling of Euclidean Geometry*, The Visual Computer, vol. 11, (1995) 378-389
2. Battiston, E., De Cindio, F., Mauri, G.: OBJSA Nets: a class of high level nets having objects as domains, in *Advances in Petri Nets* 1988, G.Rosenberg (ed.), LNCS 340, Springer-Verlag, Berlin, (1988) 20-43
3. Alberti, M.A., Evi P., D. Marini http://www.dsi.unimi.it/~colos/GEOHT/rexecGEO.html
4. Borning, A.: The programming language aspects of ThingLab, A Constraint-Oriented Simulation Laboratory, *ACM Transactions on Programming Languages and Systems,* 3(4), (1981) 353-387
5. Borning, A., Duisberg, R.: Constrained-based tools for building user interfaces, *ACM Transactions on Graphics*, 5(4), (1986) 245-374
6. Roth, J., Hashimshony, R.: Algorithms in graph theory and their use for solving problems in architectural design, *Computer-Aided Design*, 20(7), (1988) 373-381
7. Cohen, E., Smith, E., Iverson L.: Constrained-based tiled windows, *IEEE Computer Graphics & Applications*, (1986) 35-45
8. Sutherland, I.E.: SKETCHPAD: a man-machine graphical communication system., *Proceedings Spring Joint Computer Conference*, (1963) 329-346
9. Nelson, G.: Juno, a constraint-based graphics system, SIGGRAPH Computer Graphics, 19(3), (1985) 235-243
10. Fuller, N., Prusinkiewicz, P.: Applications of Euclidean constructions to computer graphics, *The Visual Computer*, 5, (1989) 53-67
11. Bruderlin, B.: Constructing three dimensional geometric objects defined by constraints, *Workshop on Interactive 3D graphics*, Vol. 23-24, October, (1986) 111-129
12. Sohrt W.: *Interaction with constraints in three dimensional modeling*, Master's Thesis, Dept. Computer Science, The University of Utah, (1991)
13. van Emmerik, M.: A System for Interactive Graphical Modelling with Three-Dimensional Constraints, Proceedings of Computer Graphics Interactive '90, (1990) 361-376
14. Kwaiter, G., Gaildrat, V., Caubet, R.: LinkEdit: an interactive graphical system for modelling objects with constraints, *Proceedings of Compugraphics 96*, (1996) 211-219
15. Sannella, M.: The SkyBlue Constraint Solver, *Technical Report 92-07-02*, Dept. Comp. Science and Eng. University of Washington, (1993)

16. Battiston, E., De Cindio, F., Mauri, G.: *A class of Modular Algebraic Nets and its support Environment*, CNR, Progetto Finalizzato "Sistemi Informatici e Calcolo Parallelo", Rapporto n. 4/105, (1994)
17. Battiston, E., Tirloni, P.: *Guida all'ambiente ONE (OBJSA Net Environment)*, CNR, Progetto Finalizzato "Sistemi Informatici e Calcolo Parallelo", Rapporto n. 4, (1994)
18. Klein, F.: Famous problems of elementary geometry (1895), in: Klein et al. *Famous problems and other monographs*. New-York: Chelsea Pub. Co., (1980)
19. Evi, P.: *Euclidean Geometry Knowledge Representation with Object-Oriented Paradigm and Petri Nets*, Tesi di Laurea, A.A.. 1993-94, Università degli Studi di Milano, Dipartimento di Scienze dell'Informazione (in Italian) (1994)
20. De Cindio, F., De Michelis, G., Pomello L., Simone, C.: Superposed Automata Nets, Application and Theory of Petri Nets, C. Girault, W. Reisig (eds.), IFB 52, Springer-Verlag, Berlin, (1982) 189-212
21. Goguen, J.A., Winkler, T.: *Introducing OBJ3*, Report SRI-CSL-88-9, SRI International, Computer Science Lab., (1988)
22. Alberti M.A., Marini, D.: Knowledge Representation in a Learning Environment for Euclidean Geometry, in *The Design of Computational Media to Support Exploratory Learning*, C.Hoyles, A.DiSessa & L.Edwards (eds.), NATO ASI Series F, vol. 146, Springer-Verlag, Berlin, (1995) 109-126

An Object-Based Modular CPN Approach: Its Application to the Specification of a Cooperative Editing Environment*

Dalton Serey Guerrero[1], Jorge C.A. de Figueiredo[1], and Angelo Perkusich[2]

[1] Departamento de Sistemas e Computação
[2] Departamento de Engenharia Elétrica
Universidade Federal da Paraíba - Brasil
{dalton,abrantes}@dsc.ufpb.br, perkusic@dee.ufpb.br

Abstract. In this paper we discuss the modeling of a cooperative graphical editor by means of an object-based Petri net framework named G-CPN. Initially we introduce the G-CPN tool addressing both its informal and formal aspects. Then, the editor is informally specified, and based on such specification we introduce a multi-agent-based architecture to support its realization. After this, we define the different interactions in the system, in order to allow the computer supported cooperative work. Also, we show how the application of the introduced G-CPN can be applied to obtain an object-oriented model of some of the elements of the cooperative graphical G-CPN editor.

1 Introduction

One important aspect that should be considered in the basic tasks of software engineering, namely specification, design, project, coding, and testing is the possibility of providing environments supporting the concurrent cooperation among different people (users) in a design team, that might be geographically dispersed [9]. Another key aspect is the fact that the cooperative work contributes to reduce the overall costs to develop a software project [12]. Environments supporting a better interaction among members of a design team, and providing means, by which the software engineering tasks can be executed concurrently in a cooperative environment, may help in reducing development costs.

On the other hand, in the case of complex (parallel, concurrent, and distributed) system design, users need structuring tools allowing them to work with selected parts of the model, abstracting low level details of the other parts. The marriage between Petri nets [19] and the object-oriented software engineering approach had emerged as a solution with formal foundation to attack this problem. Among other formalisms in this direction we may cite [1,3,4,5,11,16], for a comprehensive discussion see [17]. G-Nets and G-Net systems were introduced as a Petri-net-based object-oriented framework to the specification, design, and

---

* The authors would like to thank CAPES/Brasil and CPNq/Brasil for the financial support.

G. Agha et al. (Eds.): Concurrent OOP and PN, LNCS 2001, pp. 338–354, 2001.

prototyping of distributed software systems [6,7]. The G-Net notation incorporates the notions of module and system structure into Petri nets and promotes abstraction, encapsulation and loose coupling among the modules. The definition adopted in this paper differs from previous work on G-Nets mainly due to the use of Coloured Petri Nets [15] as the formal basis. The resulting model is named G-CPN [13,14].

In this paper, the structural definition of the G-CPN tool as well as the modeling of a cooperative editing tool for hierarchical diagrams is presented. This editor is planned to be part of a modeling environment for simulation and analysis. First, we specify the editor, and based on it, we introduce a multi-agent-based architecture to support its realization. Then, we define the different interactions in the system, in order to allow the computer supported cooperative work (CSCW) [8,9,10,18]. Also, we show how to model this application by means of a G-CPN System.

This paper is organised as follows: in Sect. 2 we introduce G-CPN systems and the structural part of its formal definitions. Behavior/semantics of G-CPN modules and systems is defined informally as well as its graphical notation. In Sect. 3 we state the specification of the cooperative environment which is to be constructed and we introduce a multi-agent architecture used to the design of the environment. In Sect. 4 we present some of the G-CPN models corresponding to relevant parts of the system. Finally, in Sect. 5 we conclude the paper.

2   G-CPN

In this section we introduce G-CPN systems which are constructs of concurrent, cooperating and loose coupling objects, whose main purpose is the formal and executable modeling and specification of complex distributed software systems. G-CPN allows software systems specification to be developed in an incremental way. Thus, it encourages both software reuse and maintenance. G-CPN systems are defined in such a way that a CPN designer may easily model systems using G-CPN. This is possible because its formal definitions allow the use of CP-nets inside G-CPN modules with almost no modification.

A G-CPN System is a set of cooperating G-CPN modules. The services offered by a G-CPN system are the methods defined in each module. But a G-CPN system is more than the summation of all modules. A G-CPN system defines an *environment* and the model of interaction among the modules. The *environment* is a structure that provides means to keep track of every invocation and respective return. The environment must guarantee the correct (expected) sequence of events. A G-CPN module is the basic element from which G-CPN systems are constructed. Each module is an object with its own state and behavior. G-CPN modules provide a set of services or operations named methods that can be invoked by other modules of the system.

From a designer point of view, it is sufficient to create the specification of the modules integrating the system. However, the precise definition of G-CPN systems state more than the specification of all modules.

2.1   G-CPN Modules

The *interface*, also called *generic switching place* (GSP), is the element responsible for the communication between the IS and the rest of the system. To use a
G-CPN module is necessary to invoke one of its services. Only through the GSP,
other components of the system have access to methods and attributes encapsulated inside modules. Moreover, only through the GSP, the IS is able to access
the services of other G-CPN modules in the system. The GSP is an abstract
view of the module, in which it is not possible to see any detailed description.
Although we consider that the methods and attributes are part of the interface
itself, we can interpret the GSP as having a descriptor for each method and attribute. We can see it as a prototype of a function in a high-level programming
language.

To allow a G-CPN module to deal with multiple invocations concurrently
and because the CP-net definition does not provide any element for modeling
attributes and/or services invocation, it is necessary to give a new semantics to
some net elements. However, the IS is structurally identical to a CP-Net. So, a
designer familiar with CPN modeling will be able to work with G-CPN [13,14].

In a G-CPN module, each attribute declared in the GSP, is associated with a
place inside the IS. Thus, an instant value of any attribute is given by the marking
of an associated place. Since attributes model the state of an object, the places
associated with attributes keep their markings through different invocations.
Moreover, tokens inside an attribute place can only be used by one of the several
concurrent methods activation at a given time. The mutual exclusion needed to
access attributes may be modeled by the intrinsic dynamics of Petri Nets. The
type of any attribute is given by the color set associated with the attribute place.

As stated before, methods define the services provided by the module. There
are two places in the IS associated with each method declared in the GSP. The
first one, called the *starting place*, indicates the insertion point of values for the
method. The second, called the *goal place*, determines the termination for the
method activity, and tokens inside these places are seen as results of the methods.
Thus, each method has two associated color sets. The *starting place* color set
determines the type of the arguments, and the *goal place* color set specifies the
result type.

In order to allow multiple concurrent activations of a method in a G-CPN
module, we adopted the concept of *contexts*. Each activation of a method has
a different context and every token has a context associated with it. Thus, all
activations have their own sub-marking called the *context marking*, allowing to
distinguish tokens belonging to different activations. Therefore, it is possible to
keep the dynamic behavior to each one of them. It is important to notice that
to provide the concurrent access to the attributes, there is a special context
for attribute tokens. Thus, tokens in attribute places may be used in bindings
without taking into account the context of other tokens involved.

This way to define a G-CPN module allows to use CP-nets as the IS specification, considering simple modification. In fact, only some restrictions are imposed
to the structural definition, and intuitively, the behavior is kept the same as in

CP-nets. It is important to observe that any IS of G-CPN modules can be tested and analysed by means of the available software tools for CP-nets.

A G-CPN Module must provide some mechanisms for invoking other module services. The idea used here was derived from the G-Net structure [6,7]. An *instantiated switching place* (*isp*) is a special place used to invoke services provided by the methods of all modules in a system. An *isp* can be seen as an abstract view of the operation of a module of the system. In order to define an *isp*, the designer must specify a method declared in the GSP part of the module. The *isp* is associated with two normal places belonging to the IS, and with a service of the system. The first place, called the *upper place*, has only input arcs, and the second, called the *lower place*, has only output arcs. The associated service is called the *invoked method*. When a token is deposited in the *upper place* of an *isp*, we say that an *invocation* is enabled. If it occurs, the token will be taken from the *isp*, and the environment transports it to the *invoked method*. Whenever the method resumes, a result is sent back, and a *return* is enabled. If the *return* occurs, the result (with an appropriated token) will be deposited in the *lower place* of the *isp*. Clearly, we must associate each *isp* with two *color sets*: the type of the arguments to be passed (color set of the *upper place*), and the type of the results that will be returned (color set of the *lower place*). It is obvious to see that such *color sets* must agree with the colors of the *starting place* and the *goal place* of the invoked method. Therefore, the definition of *isps* does not change the nature of the IS. Thus, despite the restrictions imposed by the introduction of *isps*, the IS is still a strict CP-net.

Graphical Notation. The graphical notation for a G-CPN module is shown in Fig. 1. The GSP is the upper rounded corner rectangle. Inside the GSP, there are the name of the G-CPN module, the set of attributes and their color sets, the methods of the module and the color sets of their corresponding *starting* and *goal* places.

The IS is the lower rounded corner rectangle. The notation used is identical to CP-nets notation. However, every special place is distinguished by a graphical makeup. *Starting* places are named according to the related method. *Goal* places are distinguished by doubled border circles while *isps* are represented by ellipses. Inside the ellipse, an inscription indicates the *invoked method*.

The formal definition of the GSP includes more elements than those shown in the graphical representation. However, they are not expressed because all the elements may be deduced by a straightforward and intuitive procedure. We decided not to express them explicitly because they may be source of confusion and unnecessary complexity to the designer.

Formal Definition. We assume that the reader is familiar with the syntax and semantic of the language used to the net expressions, namely CPN-ML, although it is perfectly possible to use both CP-Nets and G-CPN in association with other language as well. Several auxiliary functions (such as *Type*()), and sets (as $\mathbb{B}$, the boolean set) needed to the formal definition, are also assumed to be known.

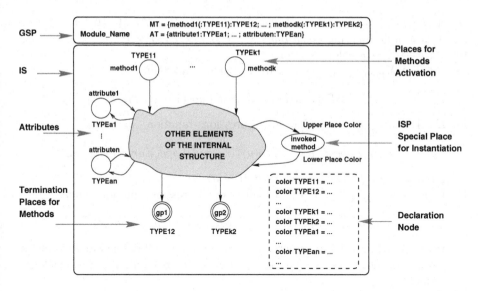

Fig. 1. Graphical notation for G-CPN modules

The use and the meaning of such objects is very intuitive, hence, they will not be described here as well. For a detailed information related to this topic see [15].

Definition 1. *A G-CPN module is a pair* $\langle GSP, IS \rangle$ *where GSP, also called the interface of the module, is a tuple* $\langle MT, AT, ISP, AP, GP, UP, LP, OBJ \rangle$ *and IS, the internal structure, is a tuple* $\langle \Sigma, P, T, A, N, C, G, E, I \rangle$, *satisfying the following:*

1. *MT is a finite non-empty set of* methods
2. $AT \subseteq P$ *is a finite set of* attributes
3. *ISP is a finite set of* invokers *or isps*
4. $AP : MT \rightarrow P$ *and* $GP : MT \rightarrow P$ *are the* activating *(or starting) and* terminating *functions where:*

$$\forall m \in MT \, [AP(m) = p_1 \Rightarrow I(p_1) = 0 \text{ and } GP(m) = p_2 \Rightarrow I(p_2) = 0]$$

5. $UP : ISP \rightarrow P$ *and* $LP : ISP \rightarrow P$ *are the* upper *and* lower *place functions for invokers, such that for any* $isp \in ISP$

$$UP(isp) = p_1 \text{ and } LP(isp) = p_2 \Rightarrow p_1^\bullet = {}^\bullet p_2 = \emptyset \text{ and } I(p_1) = I(p_2) = 0$$

6. $OBJ : ISP \rightarrow S$ *is the* objective *function for invokers, where S is a set of available methods*
7. Σ *is a finite set of non-empty* color *sets*
8. *P is a finite set of* places
9. *T is a finite set of* transitions

10. *A is a finite set of* arcs *such that* $P \cap T = P \cap A = T \cap A = \emptyset$
11. *N is a* node function, *such that:* $N : A \to P \times T \cup T \times P$
12. $C : P \to \Sigma$ *is a* color function
13. $G : T \to \Sigma_{Expressions}$ *is a* guard function, *such that:*

$$\forall t \in T \; [Type(G(t)) = \mathbb{B} \; and \; Type(Var(G(t))) \subseteq \Sigma]$$

14. $E : A \to \Sigma_{Expressions}$ *is an* arc expressions function, *such that*

$$\forall a \in A \; [Type(E(a)) = C(p(a))_{MS} \; and \; Type(Var(E(a))) \subseteq \Sigma]$$

where p(a) is the place component of $N(a)$
15. $I : P \to \Sigma_{Expressions}$ *is the* initialisation expression function, *where*

$$\forall p \in P \; [Type(I(p)) = C(p)_{MS} \; and \; I(p) \; is \; closed]$$

Some remarks are relevant here. To ensure that a well-defined module offers at least one service to the system, its set of methods is required to have at least one element.

The ISP set is a declaration of all invocation elements inside the net. Thus, it has as many elements as invocations places are needed inside the IS. Observe that the OBJ function maps each *isp* to a unique service in the system, implying that the same method is always invoked.

2.2   G-CPN Systems

In fact, we say that *invocations, activations, terminations* and *returns* are events that occur in the system, and causes similar events in the modules (see Fig. 2). Furthermore, the system (actually its *environment*) ensures the correct sequence of the events, provides transportation of information, and controls the creation of contexts. Clearly, the environment must have its own state.

The environment detects and activates enabled invocations in *isps*. When an invocation occurs, both the module and the environment change their states. The new state of the module is determined as presented before. The new state of the *environment* keeps information about the *pending invocation*, and about the context that is waiting for the invocation result. This is achieved by maintaining all associations between client and server contexts.

The activation of a method is enabled in a G-CPN system only when there is a *pending invocation* for that specific method in the state of the environment. Thus, if it occurs, the system changes its state. The new state of the activated module is calculated by adding a token to the activation place of the method. The new state of the environment is determined by removing the pending invocation.

In a similar way as *invocations*, an enabled termination is detected when there is a token in a *goal place*. If a termination occurs, the new state of the environment is determined by adding an information about a *pending termination*.

Returns are enabled events when there is a *pending termination*. If they occur, the new state of the environment is determined by removing the *pending termination* and the proper contexts association.

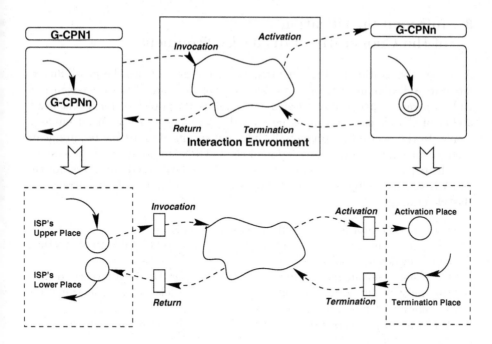

Fig. 2. Interaction between two objects

The graphical representation of a G-CPN system is depicted as the collection of all G-CPN modules in the system. The environment representation is not necessary, since it has a very intuitive behavior.

Formal Definition. Semantic definitions will not be presented in this paper due to lack of space, but they can be found in [14]. Nevertheless, from the previous informal introduction and the behavior of a CP-net, the precise behavior of a G-CPN module as well as of a G-CPN System can be very intuitively understood.

Definition 2. *A G-CPN system is a tuple* $\langle \Sigma, GCPN, S \rangle$ *where:*

1. *Σ is a finite set of non-empty types named* color sets
2. *GCPN is a finite and non-empty set of G-CPN modules with disjoint sets of places, transitions and methods. The set of types in each module is required to be a subset of the system's one:*

$$\forall G_i \in GCPN \, [G_i.\Sigma \subseteq \Sigma]$$

3. *S is the set of* services *of the system, defined as the union of the methods of all modules $S = G_1.MT \cup G_2.MT \ldots \cup G_n.MT$. If $G_i.OBJ(isp) = G_j.s$, for any $isp \in G_i$ and any $s \in G_j.MT$ then the color restriction must be satisfied:*

$$G_i.C(UP(isp)) = G_j.C(AP(s)) \text{ and } G_i.C(LP(isp)) = G_j.C(GP(s))$$

3 Informal Specification
of the Cooperative Editing Environment

A cooperative editing tool for hierarchical diagrams provides the possibility to integrate a development team to work over a network and allows the concurrent and cooperative specification, design and analysis of complex systems. The availability of CSCW is a primary issue related with modern development environments. In this section we present the informal specification or the requirements used as guidelines in the construction of the G-CPN specification of the system.

The requirements given below are derived from a case study proposed for the 2nd Workshop on Object-Oriented Programming and Models of Concurrency by Bastide et al [2]:

- Users may join or leave an editing session at will;
- The current members of an editing session ought to be visible to all, as their ownership over the objects;
- Graphical elements may be free or owned by a user. Different levels of ownership must be supported, including ownership for deletion, for encapsulation, for modification and for inspection.
- When an object is owned for deletion, encapsulation or modification, only one user must own it.
- Inspection ownership may be required by several users simultaneously;
- Only ownership for encapsulation is persistent;
- The creator of an element owns it for deletion until it is explicitly released.

Our approach is based on the definition of a multi-agent architecture. Different kinds of agents interact through the network in order to provide a cooperative service for a certain class of applications. Each agent is defined as a set of well defined software objects, and the communication among them is allowed by a well-defined high level protocol. Furthermore, each agent is internally designed as a multi-layer service, where each layer provides a refined service to its upper layer. There are two kinds of agents in the system: *user agent* and the *manager agent*.

A user agent is allowed to perform the following actions: to enter an editing session; to exit an editing session; to create objects; to delete objects owned by the user; to request ownership over objects; to free objects owned by the user.

The manager agent is allowed to perform the following actions: to create editing groups; to register new users; to remove registered users; to include a registered user in an editing group; to remove a registered user from an editing group; to modify any ownership as desired.

Architecture of the Environment. Each agent of the proposed architecture shown in Fig. 3 is defined as a layered entity. The upper layer, called *application* layer, is a Graphical Editor of hierarchical diagrams (G-CPN systems in this case) or a System Console to be used only by the *manager*. The intermediate layer is called *control* layer, and it is concerned with the correct manipulation of

all kind of objects in an editing session. The control layer performs the core of
the Computer Supported Cooperative Work. The intermediate layer used by the
manager is named *manager* layer. The lower layer is the *communication* layer,
and it is responsible for providing the correct message handling among the users,
by using the network and operating system services.

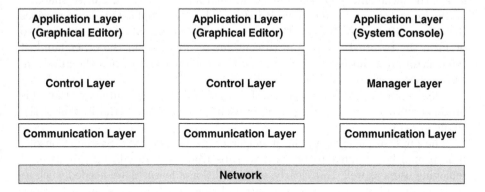

Fig. 3. Architecture of the environment

This approach allows the use of a simple graphical editor over the cooperative service stack. The graphical editor (the application) must guarantee that
the control layer will be asked/informed about any actions performed over the
graphical elements. Thus, if additional features are added to the editor, few
modifications, if any, will be needed in the control and communication layers.
Another advantage of this approach is that no locks over the objects are needed.
Basically, the system keeps track of "who is working on what".

Hierarchical diagrams can be managed by distributed or centralised data base
systems or even by a simple file system service. However, in order to separate
the object *management* problem and the *cooperative work* support, in this paper
we are not discussing a specific choice of either approach. This allows the use
of traditional graphical editors as well as the specification of new cooperative-
oriented graphical editors.

Application Layer. As stated before, in the case of a hierarchical diagram
cooperative editor, the application layer is a Graphical Editor allowing the access
of services offered by the control layer. The Graphical Editor offers the same
functionality as a stand-alone Editor, but when the user asks for editing a shared
object, its behavior will be oriented by the control layer. Thus, the user will be
prompted for identification and authentication as needed, and will be included
in a cooperative editing session being coordinated by a distributed algorithm.
The role of the manager is to provide a control over all editing groups. Thus, if
it is needed to include a new user in an editing group, it must be done by the

manager, through the System Console, an application that accesses the services offered by manager layer.

The Graphical Editor may be designed without any assumptions of locking algorithms, all it needs to do before performing any action over graphical elements is to ask the control layer if the user has the correct ownership for that element. If the user does not have the permission, the control layer itself tries to get the ownership by executing an internal algorithm to guarantee that nobody else in the system has done it before. Then the application layer receives the answer: if the user has got the ownership, the graphical editor is allowed to perform the required action and to continue on its job. On the other hand, if the control layer did not succeed in granting the ownership, then the graphical editor must stop the action it was asked to do.

Thus, no locks are really needed. The application layer guarantees that this protocol will be used: no object is used without a previous authorisation of the whole system.

The application layer in the manager agent is called the System Console. It is primarily intended to be used to subscribe users in editing groups, and removing users as well. But clearly, some additional features are needed, such as creating/dropping editing groups, control of rules in editing sessions, definitions of eventual hierarchical differences among users, etc. However, the system console operator is not required to be logged during an editing session.

Control Layer. When any operation is to be performed by the editor in a cooperative editing session, the control layer is asked about the possibility of such an action. Thus, the control layer must keep internal information about users, the graphical elements and ownership. The control layer verifies if such an information allows the user to perform the action without needing to contact the *cooperative partners*. If it does, the control layer answers immediately. On the other hand, if the control layer supposes the information is not up to date, or if it is needed to update the information of the cooperative agents, then it requests some service of the communication layer. When such a service is complete, the control layer is able to answer the request of the application layer correctly.

The control layer is sometimes requested to perform some actions by the communication layer. This happens when other agents in the system ask for information or when some event has happened in the rest of the system and this requires the modification of the information stored in the control layers of all cooperative partners. In fact, this models the arrival of messages from lower layers.

Manager Layer. The System Console Application functionality depends on the services offered by the manager layer. This layer allows the initialisation of groups and the registration of new users. Of course, primitives to destroy and update the information of such attributes are also needed. This layer is very similar to the user control layer.

The manager layer also uses the communication layer primitives in order to multicast information to all agents in the system. In the communication layer there are some reserved primitives that can be only used by the manager layer. Such primitives are representation of messages about the initialisation/deletion of groups or users.

Because the manager is a special user, the manager layer functionalities are simpler to work with. Thus, it is beyond the scope of this paper to detail these functionalities.

Communication Layer. The lower layer in the architecture proposed provide different communication services that in general are transformed in multicast messages. Furthermore, the communication layer must be able to receive messages from the network as well and to transform them into the correct service to be performed by the control layer. The communication layer may be implemented in many different ways depending on the services provided by the network and the operating system. Thus, if the system (and network) provides a reliable group communication service, then the communication layer is very thin. Otherwise, if the system has not any group communication service, and the only service provided is a simple message sending, then the communication layer needs to handle the reliable multicast algorithm itself.

In this paper, we are primarily concerned with the specification of the cooperative control of an editing session, thus no further discuss is related with communication layer issues.

4   Modeling the Objects of the Environment

In this section we show the specification of part of the environment presented before. As each layer uses services of the layer immediately below it, each service is modeled as a method in a G-CPN Module. Also the receiving of messages from the lower layers are modeled as methods invocations. The persistent information treated by each layer is modeled as an attribute.

4.1   Control Layer Services

Clearly what we need is a G-CPN module for each object and additional modules for the treatment of different services. We identified five objects, shown in Fig. 4: the *application-control interface*, responsible for providing access points to the application layer and that implements the services; a *communication-control* interface which provides insertion points for incoming messages from the network; an *users* module, responsible for all operations allowed over the users information; a *group* module, which encapsulates the information about groups and sessions (active groups) and respective allowed operations; and finally an *ownership* module, responsible for keeping information and operations about ownership granted over objects. The control layer must support all events related with the cooperative work. The methods Login and Logout are called when users decide to enter or leave an editing session. CreateObject and DeleteObject are called when a user creates or eliminates elements in/from the hierarchical diagram. GetOwnership and FreeOwnership allow users to ask or release some level of ownership of a given element.

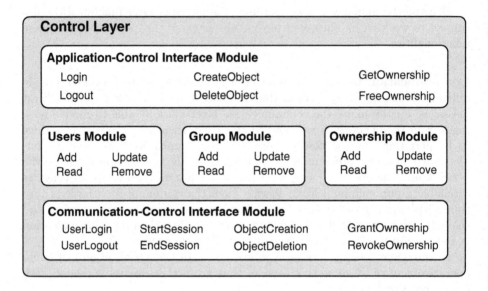

Fig. 4. Control layer objects

Other services are provided in the control layer to allow the communication layer to pass messages from the rest of the system. These services can be seen as the counterpart of those shown above, thus, if a given process receives a Login call, all cooperative processes must receive a UserLogin invocation. Other corresponding services offered to the communication layer are: UserLogout, ObjectCreation, ObjectDeletion, GrantOwnership, and RevokeOwnership. Two additional services, StartSession and EndSession, are provided to start and end sessions.

These services operate over some information modeled as attributes. The information kept by the control layer is about users, groups, sessions, objects (belonging to hierarchical diagram) and ownership.

4.2 Communication Layer Services

The communication layer provides multicasting services. Methods MC_UpdateGroup and MC_UpdateUser allow update messages. They are used by the Login and Logout services in the control layer. Methods MC_RequestOwnership and MC_ReleaseOwnership are used to deal with object ownerships. MC_CreateObject, MC_DeleteObject are used to indicate creation and destruction of objects.

Some methods cope with the arriving messages from the network. UpdateGroup, UpdateUser, RequestOwnership, ReleaseOwnership, ObjectCreation and ObjectDeletion are used as the counterpart of the multicast methods explained before.

Thus, if some user logs in the system, the sequence of methods invocations could be: the application layer invokes the method Login in the control layer; the Login service would need to invoke the MC_UpdateUser method in order to inform the cooperative processes that some user has passed from inactive to active; each process receives an UpdateUser message (it is seen as a method invocation); the UpdateUser message handler invokes the UserLogin method in the control layer. Thus, if a user logs in, every process in the system will be informed.

It is important to keep in mind that the Group Communication Services are required to be reliable. If a message is multicasted, then every process must receive it in the same order than any other process in the group. If d some message is lost, than no process at all receives the message.

4.3   G-CPN Models

Due to lack of space we concentrate on two methods of the *application-control* interface module, which is part of the control layer in the architecture. In order to simplify the presentation, we treat internal parts (detailed later) in an abstract way. In Fig. 5 we present the declaration of color sets and variables for the modules discussed here.

```
color USER_ID = int;
color GROUP_ID = int;
color OBJ_ID = int;
color Primitive = with ok | nok;
color STATE = with active | inactive;
color LEVEL = with inspection | modification| encapsulation | deletion;
color User = product USER_ID * GROUP_ID;
color USER_LIST = list user;
color USER_ID_LIST = list USER_ID;
color Obj_Request = product OBJ_ID * LEVEL;
color Request = product USER_ID * GROUP_ID * OBJ_ID * LEVEL;
color Object = product OBJ_ID * GROUP_ID * STATE * USER_ID_LIST;

var uid: USER_ID;
var ag, gid: GROUP_ID;
var obj: OBJ_ID;
var state: STATE;
var level: LEVEL;
var ul: USER_LIST;
var owners: USER_ID_LIST;
```

Fig. 5. Color sets declarations

In Fig. 6 we see the Login method. It works in a very straightforward way. If the user is registered in a given group, and the state of such a group is active (there is at least one user working already) then the new user is included by calling the MC_UpdateUser method of the communication layer. If a session is

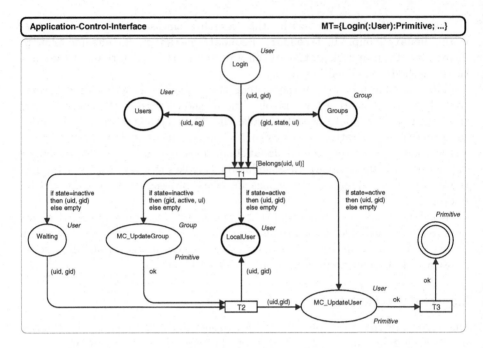

Fig. 6. Method Login

not already established, then the method MC_UpdateGroup is invoked in order to create it, and only after this the new user is included. There is also an attribute called LocalUser. It is also modified by the Login method. This attribute keeps the information about the user logged in the current point.

This G-CPN Module has only three transitions. The upper one, named T1, occurs only when the method is called and hence there is a token at the starting place Login. In fact, T1 only occurs if there are a user and a group satisfying the requirements stated in the inscriptions. That is, if the user identified by *uid* is included in the group identified by *gid*. If it happens, there are two possibilities: the state of the group gid is active, indicating that there is a working session of that group; or the state of the group is inactive, indicating that the group has not anybody working. If a session already exists, then a token is added in the upper place of the isp with inscription, that invokes the method MC_UpdateUser in order to multicast a message informing every agent in the system that a new user has logged in. If there is no working session of that group, a new session is created calling the method MC_UpdateGroup. After this, the transition T2 includes the user in the attribute place that represents the local user and adds a token in the isp MC_UpdateGroup. Transition T3 occurs when the user is correctly included in the session requested. Its occurrence puts a token in the goal place associated with the Login method.

Of course, this is a simplified version of the module itself. The actual module has not the attributes representing the Users, the LocalUser and the Groups inside it. These attributes are located in their own modules, encapsulated with their own methods. Furthermore, the actual module incorporates several methods inside the same Internal Structure.

In Fig. 7 we see the GetOwnership method. The method behaves as follows. When it is invoked, the application specifies the object and the level it requires (inspection, modification, encapsulation and deletion). If there is no ownership for that object the communication layer is required to multicast a request of ownership. This is done in order to guarantee that no other agent can get the ownership first. The actual modification in the *Ownership* place is done just by the activation of the GrantOwnership method meaning the arrival of a message from the communication layer. After this, the GetOwnership method proceeds, and it finishes by returning a confirmation or not to the application layer, depending on the state of the object ownership.

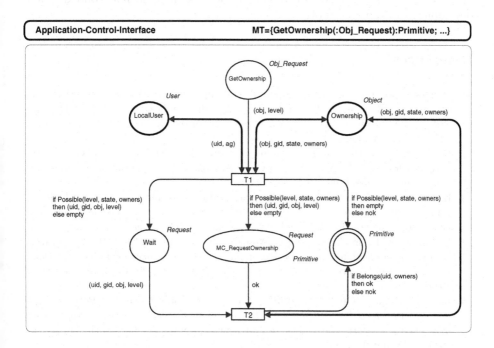

Fig. 7. Method GetOwnership

5   Conclusions

In this paper we have introduced a Coloured Petri Net object based class of Petri net named G-CPN, that was derived from a previous introduced model named G-Net. The new model introduces the use of CP-Nets as the internal notation to the modules and also formalises the interaction model among the modules.

The concept of a context is introduced in order to deal with multiple requests concurrently inside the module. Thus, G-CPN supports concurrency both among and inside objects in the system.

We have defined the architecture of a cooperative environment based on the concepts of multi-agent systems and computer supported cooperative work. Based on this architecture we have shown how to apply G-CPN to obtain the model for two of the objects of the proposed cooperative environment. Also, we are adapting the current implementation of a G-Net editor to support the new notation elements of G-CPNs, as well as to implement the primitives to access the cooperative layer of the environment.

Our main intention is to develop a complete tool and methodology to support the development of distributed, parallel and concurrent systems. Current work is being done on three different directions, namely: formal foundations of object-oriented Petri nets, application of the framework in more real contexts, and development of software tools.

References

1. M. Baldassari, G. Bruno, V. Russi, and R. Zompi. PROTOB a hierarchical object-oriented CASE tool for distributed systems. In C. Ghezzi and J.A. McDermid, editors, *ESEC'89*, volume 387 of *Lecture Notes in Computer Science*, pages 425–445. Springer-Verlag, 1989.
2. R. Bastide, C. Lakos, and P. Palanque. A cooperative Petri net editor. Case Study Proposal for the 2nd Workshop on Objetc-Oriented Programming and Models of Concurrency, June 1996.
3. E. Battiston and F. de Cindio. Class orientation and inheritance in modular algebraic nets. In *Proc. of IEEE International Conference on Systems, Man and Cybernetics*, pages 717–723, Le Touquet, France, 1993. IEEE.
4. E. Battiston, F. de Cindio, and G. Mauri. OBJSA nets: a class of high-level nets having objects as domains. In G. Rozenberg, editor, *Advances on Petri Nets 1988*, volume 340 of *Lecture Notes in Computer Science*, pages 20–43. Springer-Verlag, 1988.
5. D. Buchs and N. Guelfi. CO-OPN: A concurrent object oriented Petri net approach. In *Proceedings of the 12th International Conference on the Application and Theory of Petri Nets*, Lecture Notes in Computer Science. Springer Verlag, Aarhus, Denmark, 1991.
6. Y. Deng and S.K. Chang. Unifying multi-paradigms in software system design. In *Proc. of the 4th Int. Conf. on Software Engineering and Knowledge Engineering*, Capri, Italy, June 1992.
7. Y. Deng, S.K. Chang, J.C.A. de Figueiredo, and A. Perkusich. Integrating software engineering methods and Petri nets for the specification and prototyping of complex software systems. In M. Ajmone Marsan, editor, *Application and Theory of Petri Nets 1993*, volume 691 of *Lecture Notes in Computer Science*, pages 206 – 223. Springer-Verlag, Chicago, USA, June 1993.
8. P. Dewan and R. Choudhary. A high-level and flexible framework for implementing multiuser user interfaces. *ACM Transactions on Information Systems*, 10(4):345–380, October 1992.

9. P. Dewan and J. Riedl. Toward computer-supported concurrent software engineering. *IEEE Computer*, 26(1):17–27, January 1993.
10. C.A. Ellis, S.J. Gibbs, and G.L. Rein. Groupware: some issues and experiences. *Communications of the ACM*, 34(1):38–58, January 1991.
11. J. Engelfriet, G. Leih, and G Rozenberg. Net-based description of parallel object-based systems, or POTs and POPs. In J.W. de Bakker, W.P. de Roever, and G. Rozenberg, editors, *Foundations of Object-Oriented Languages*, Lecture Notes in Computer Science, pages 229–273. Springer-Verlag, 1990.
12. G. Forte and R.J. Norman. A self-assestment by the software engineering community. *Comm. ACM*, 35(4):28–32, April 1992.
13. D.D.S. Guerrero, J.C.A. de Figueiredo, and A. Perkusich. Object-based high-level Petri nets as a formal approach to distributed information systems. In *Proc. of IEEE Int. Conf. on Systems Man and Cybernetics*, pages 3383–3388, Orlando, USA, October 1997.
14. D.D.S. Guerrero, A. Perkusich, and J.C.A. de Figueiredo. Modeling a cooperative environment based on an object-based modular Petri net. In *Proc. of The 9th International Conference on Software Engineering and Knowledge Engineering*, pages 240–247, Madrid, Spain, June 1997.
15. K. Jensen. *Coloured Petri Nets: Basic Concepts, Analysis, Methods and Practical Use*. EACTS – Monographs on Theoretical Computer Science. Springer-Verlag, 1992.
16. Charles Lakos. Pragmatic inheritance issues for object Petri nets. Sem maiores indicações.
17. Charles Lakos. From coloured Petri nets to object Petri nets. In *Proceedings of the 16th International Conference on Applications and Theory of Petri Nets*, Turin, Italy, June 1995.
18. K.C. Lee, W.H. Mansfield Jr., and A.P. Sheth. A framework for controlling cooperative agents. *IEEE Computer*, 26(7):8–16, July 1993.
19. T. Murata. Petri nets: Properties, analysis and applications. *Proc. of the IEEE*, 77(4):541–580, April 1989.

KRON: Knowledge Engineering Approach Based on the Integration of CPNs with Objects[*]

J.A. Bañares, P.R. Muro-Medrano, J.L. Villarroel, and F.J. Zarazaga

Departamento de Informática e Ingeniería de Sistemas
UNIVERSIDAD DE ZARAGOZA
María de Luna 3, Zaragoza 50015, Spain
{Banares, PRMuro, JLVillarroel, Javy}@posta.unizar.es
http://diana.cps.unizar.es/iaaa

Abstract. This paper presents KRON (Knowledge Representation Oriented Nets), a knowledge representation schema for discrete event systems (DESs). KRON enables the representation and use of a variety of knowledge about a DES static structure, and its dynamic states and behavior. It is based on the integration of Colored Petri nets with frame based representation techniques and follows the object oriented paradigm. The main objective considered in its definition is to obtain a comprehensive and powerful representation model for data and control, and to incorporate a powerful modeling methodology. The communication model used in KRON is close to the generative communication model, which supposes an alternative to message passing. The inferences delivered from the DES behavioral knowledge are governed by a control mechanism based on a rule inference engine.

Keywords: Colored Petri nets, frames, knowledge engineering, DES.

1 Introduction

This paper is devoted to illustrate the main features involved in KRON (Knowledge Representation Oriented Nets). We starting creating KRON while we were working in the development of knowledge based models for DESs. It became clear in working with DESs the need to expand the power of our knowledge engineering representation schema with the integration of an adequate formalism to deal with discrete event system features.

A lot of integrations of Petri nets with different paradigms can be found in technical literature. These may be split into three main groups: 1) Extension of Petri nets with primitives to support methodological aspects (modularity, top-down and bottom-up design, ...); 2) Integration of Petri nets with algebraic specifications and 3) Integration of Petri nets with the frame/object paradigm. Several workshops about the integration of Petri Nets and objects are held regularly as part of prestigious conferences (Int. Conf. of Application and Theory

---

[*] This work was supported by the Spanish Interministerial Comission of Science and Technology (CICYT) under projects TAP95-0574 and TIC98-0587.

G. Agha et al. (Eds.): Concurrent OOP and PN, LNCS 2001, pp. 355–374, 2001.

of Petri Nets, IEEE Int. Conf. on Systems Man and Cybernetics, ...), this is a proof of the growing interest in this topic.

A HLPN extension belonging to the first group is HCPN (Hierarchical Colored Petri Net). HCPNs [21] provide a set of constructs to support modularity aspects. The idea behind HCPNs is to allow the construction of a large model by combining a number of small HLPNs into a larger net, and different structuring tools are proposed with this purpose. Posterior proposals extending HLPNs with structuring constructs can be found in [18] and [13]. Other works that propose different object oriented interpretations of HCPN constructs can also be found in [24] and [17].

The presentation of the most representative works on the PN integration with algebraic specifications (second group) can be briefly summarized as follows: Algebraic Nets [34], Many-sorted High-level Nets [7] and Petri Nets with structured tokens [29] are a result of the integration of HLPNs (used to describe the control structure of the system) and algebraic specifications (used to describe the data structure). These previous works have been the basis of many others, most of them also considering some object oriented focus. OBJSA Nets [4] and CO-OPN (Concurrent Object-Oriented Petri Nets) [10] are good examples. Its goal is to allow data abstraction and introduce net modularity.

Finally, the third group are the approaches based on a frame/object approach. From an engineering point of view, we consider them closer to human conceptual thinking than the ones based on algebraic specifications. What is required here, is a conceptual model which will enable engineers and computer scientists to describe domain concepts in a more intuitive way. Examples of this group are:

- In [15] high level Petri nets are integrated with the Entity-Relationship model to obtain the EER formalism. This model is revised in [16] incorporating object oriented concepts to increase expressiveness in data modeling. However, this approach is not extended to the process structure in order to provide an overall modeling framework. Finally, a second revision is done in [6]. In this last piece of work the internal behavior of each object is described by means of a Petri net (O-net). To obtain the global process structure partial nets are synchronized by another Petri net, the P-net. This P-net is not included in the object structure.
- Object Petri nets (PNO), which have been widely referred to in technical literature, were defined in [31] as High Level Petri Nets with Data Structures. Their objective is to incorporate the data modeling and updating into the net model by means of frame-like data structures. Starting from this seminal work, in [28] HOOD/PNO is proposed as a software engineering methodology that integrates PNO with the HOOD. In [32] two more extensions to PNO were introduced: Communicative and Cooperative nets. They enable the modeling of a system as a collection of nets that encapsulate their behavior while interacting by means of message sending and the client/server protocol.
- [1] presents PROTOB, an object oriented language and methodology based on PROT nets [9]. In this object oriented approach, objects communicate

by message passing and a hierarchical object decomposition like HOOD is allowed. However neither inheritance nor data representation aspects are considered.

- LOOPN++ [25] has mainly been used to describe network protocols. LOOPN is a textual language that supports object oriented structuring into HLPNs. The language has a formal semantics which makes it possible to transform OP-nets (Object Petri nets) into the simpler HLPN formalism. However, as it has been pointed out by the author, there is not a precise relation between OP-nets and the LOOPN++ language.

As the reader can see, there are a lot of integrated models. Most of the previous approaches concentrate on providing structuring tools in compliance with software engineering principles, by enforcing constraints that may result in a loss of freedom and flexibility. Most of them also extend the formalism of HLPNs. However, there is a great scope for further work in tailoring analysis techniques to extended HLPNs.

KRON is based on the integration of Colored Petri nets (CPNs) with the frame/object oriented paradigm. The integration model presented in this paper provides a close integration of HLPNs and the object model, and it does not extend the HLPN formalism. Frames and rules have been selected as a basis to support the representation aspects due to its power for knowledge representation. Additionally, we improved programming discipline by following an object oriented methodology obtaining important methodological advantages such as: 1) it supports conceptual models closer to human conceptualization and independent from implementation, thus the models are easier to understand; 2) it facilitates reusability and model extensibility based on encapsulation and inheritance characteristics.

The rest of the paper is organized as follows. Firstly, a brief presentation of KRON constructs that support the CPN formalism is presented. In the following sections, the case study of hurried philosophers is used to illustrate KRON. Section 4 shows the definition of dynamic and no dynamic entities and their relationships. Section 5 presents inheritance as a mechanism to share code. The communication model is presented in section 6. The paper finishes with a conclusions section.

2   KRON Constructs

Knowledge representation of DESs must involve the representation of information related to its dynamic behavior as well as more static information. From a conceptual point of view, the representation of a KRON model is based on semantic networks, whereas a frame implementation perspective has been adopted for its programming. In this programming context, the representation is structured around a set of conceptual entities with associated descriptions and interconnected by various kinds of associative links. However, in frame based representations, little attention has been paid to describing the coordination between objects in order to achieve collective behavior [19]. The application of frame/object based languages to the modeling of complex dynamic systems, has certain inconveniences due to the lack of a formalism to specify its dynamic behavior (concerning both, the states of the objects and the causal relationships between states and actions).

In addition to the programming features supported by frame/object oriented languages, our knowledge representation schema includes a set of primitives implementing the CPN formalism. CPNs provide the mechanism to describe the internal behavior of the dynamic entities and the interaction between them, with no necessity for a low level communication model. Figure 1 illustrates the frame hierarchy The KRON hierarchy can be decomposed into three important groups:

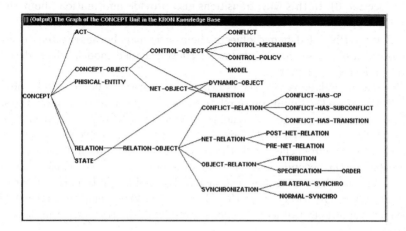

Fig. 1. KRON hierarchy.

1. **Net objects**. Dynamic entities in KRON are descendants of a specialized object called dynamic object. A dynamic object centralizes all the information related to a dynamic entity (abstract or real), and it is the repository of information about the entity states and activities. The behavioral description of a dynamic object class is represented by a CPN. The state will be mapped in a combination of CPN places and structured tokens, whereas the activities that produce state changes will be mapped in CPN transitions. The constitutive elements of the structure of a CPN are represented by individual concepts and dedicated object slots (Transitions, activity slots and state slots), which are aggregated or composed in dynamic objects to represent the behavior:

   − The state of a dynamic entity is represented by a set of state slots. To each state slot corresponds a single place of the CPN. State information in a CPN is represented by its marking, this means the places and the tokens located in the places. Tokens, which evolve by a CPN, are not mapped onto specialized objects in KRON. Any entity evolving through state slots plays the role of a *token*. The state of a dynamic entity is defined not only by the marking relations, but also by the token attributes (slots) that are relevant for that state. Structured tokens allow KRON to benefit from some CPN advantages like the aggregation of dynamic information to obtain more concise models.

- Activities producing state changes in a dynamic object are represented by transitions, and they are equivalent to the transitions of a CPN. Transitions that represent activities related to the same dynamic object are located in its `activity slots`. The interface of a KRON dynamic object is a subset of activity slots that hold transitions representing activities that must be carried out in cooperation with other dynamic objects (see section 6). In this way, transitions also provide information about the set of applicable services for the current state.

Finally, CPNs of dynamic objects themselves can be aggregated to create more complex nets in a high level structure called model, which describes the collective behavior.

2. **Relations**. Relations hold the information of interdependent KRON objects. KRON allows the definition of relations as an important concept at the same level as classes or objects. Generic relations are defined as a specialization of relation-object. When a relation is defined between two classes, a slot is created in the first class with the name of the relation, and another slot is created in the second class with the name of the inverse relation. Demons attached to these slots are responsible for making automatic updating of direct-inverse relations. From the CPN point of view, relations make possible the combination of objects in more complex data structures that represent tokens.

KRON also provides specific relations related to the description of dynamic behavior:

- Net relations support CPN arcs and expressions labeling them, and are used to specify connections between state slots and transitions. The information about net relations is stored in transitions.
- Synchronization relations provide a simple way to specify interconnection between dynamic entities, which is done by means of the synchronization of activities in the activity slots that constitute the interfaces of dynamic objects.

3. **Control objects**. These objects provide the mechanisms and policies used to implement the evolution rules of the underlying CPN (*token player* in Petri net terminology and *inference engine* in the knowledge representation terminology). The search for enabled transitions is carried out by an efficient matching algorithm [2].

A KRON model can be not fully deterministic, that is, there exist points in which decisions have to be taken in order to establish the model evolution. For the selection phase, transitions are grouped into conflicts by inspecting the net structure, and each one is provided with a particular control policy. Conflicts may also be related in order to provide them with a control policy.

Conflicts enable us to establish a simple interface between the model and a decision making system.

The interpretation of a model is carried out by the control-mechanism, which applies the corresponding control-policy to each conflict located in the model.

3   Relations to CPNs

The Petri net underlying a KRON model can be considered as a subset of a CPN with a special syntax. However, there still exist restrictions that are introduced to improve the modeling and simulation capabilities of CPNs to solve practical tasks. The formal analysis of properties was not a crucial issue in the KRON development. Our approach is closer to the work presented by Cherkasova et col. in [12], which combines CPNs with modeling by direct programming, than to works that extend the CPN formalism.

Following this pragmatic approach, the CPN formalism is not extended, but really, it is constrained to use simpler expressions. A KRON net differs from CPNs (as defined in [23]) in the following restriction: An Arc expression may only denote a unique token, but not set of tokens.

Another important difference with CPNs is introduced by the integration of CPNs with the object model. KRON tokens are entities, and their attribute values and relationships are considered in order to describe the system behavior. The nature of these tokens introduces a property called *ubiquity* [31]. Ubiquity concerns the token ability to have several occurrences in a marking. Formally, a KRON net with ubiquity is not a correct CPN because it produces the loss of the transition scope. Ubiquity produces the following undesirable effects: 1) it violates the partition and encapsulation of the state in dynamic objects. Moreover, it hides the way transitions modify the state because they have unlimited writing access to all token attributes. 2) Ubiquity is a property irreducible to algebraic analysis. This problem is not exclusive of the integration of the object model and Petri Nets. The problem arises in any representation language that allows different references (object pointers) to the same object. This property, which is known as *dynamic aliasing*, makes it difficult to prove the correctness of a system representation theoretically [27]. KRON allows the modeller to decide whether to avoid ubiquity in order to prove the correctness of the system representation, or to model in a more flexible way without to worrying about the ubiquity problem.

4   A KRON Model
for the Hurried Philosophers Case Study

In order to illustrate the representation schema, let us focus on "The hurried Philosophers" case study [32]. Since the proposal allowed free interpretation of philosopher behaviors and it was originally though for a message passing communication model, we state our interpretation first:

The case study is the very well known table of philosophers, with an extension: a philosopher may leave the table as he likes it, and new guests may be introduced. In the world of philosophy there are philosophers that may be thinking and eating. Moreover, philosophers must respect some rules of politeness. The philosophers interact in order to respect these rules. A philosopher who wants to be in the world of philosophy must be introduced in a common table, and must be sit in a chair, with a philosopher on his left side, and a philosopher on his right side. Philosophers share a fork with his right and one with his left neighbor. A philosopher only may start eating if he has a fork on his left and a fork on his right. Any philosopher may decide to leave the table if he is not eating. A philosopher leaves with the fork on his left side, which must be free, and he leaves his chair. A philosopher may be introduced between two philosophers if they are thinking (the fork between them is free). The guest philosopher takes a free chair and carries a fork on his left hand. Therefore, philosophers may interact to ask and give forks, and to enter and leave the table. In the following sections this case study will be completed.

The first step in the KRON modeling methodology is the identification of the dynamic and no dynamic entities which compose the system model at the chosen abstraction level:

```
{Chair                          {Fork                          {Philosopher
   is-a: Phisical-Entity           is-a: Phisical-Entity          is-a: Phisical-Entity
 ; Relations                    ; Relations                    ; Relations
   philo:                          left-chair:                    chair:
     attributeclass: seat            attributeclass: right-fork      attributeclass: seat
     valueclass: philosopher         valueclass: chair               valueclass: chair
   left-fork:                      right-chair:                 }
     attributeclass: left-fork       attributeclass: left-fork
     valueclass: fork                valueclass: chair
   right-fork:                   }
     attributeclass: right-fork
     valueclass: fork
   ....
}
  Seat                          { left-fork                     { right-fork
   is-a: attribution               is-a: attribution              is-a: attribution
   domain: Chair                   domain: Chair                  domain: Chair
     slot: philo                     slot: left-fork                slot: right-fork
     cardinality: 1                  cardinality: 1                 cardinality: 1
   range: Philosopher              range: Fork                    range: Fork
     slot: chair                     slot: right-chair              slot: left-chair
     cardinality: 1                  cardinality: 1                 cardinality: 1
   ....                            ....                           ....
}                               }                              }
```

Fig. 2. Entities and relationships.

Token Objects

In the problem description it can be identified the no dynamic entities Chair, Fork and Philosopher. It can also be identified the dynamic relation Seat representing the association between a chair and a philosopher, and the dynamic relations `left-fork` and `right-fork` representing the associations between a chair and his left and right forks. (In KRON dynamic relations are a specialization of the `attribution` relation). Figure 2 shows the frames that define these entities and relations. These entities and relations will not be considered dynamic entities from the model point of view. That means that their internal behaviors are not considered at this abstraction level, but they can complete the behavior of other entities playing the role of tokens.

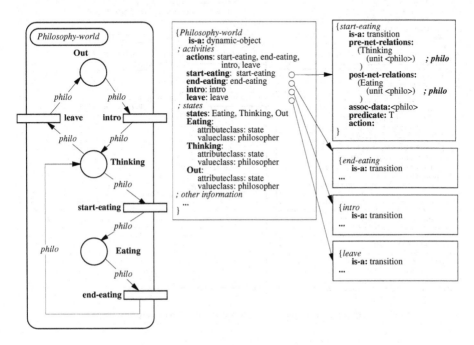

Fig. 3. The dynamic-object `Philosophy-world`.

Dynamic Objects

It is possible to identify the following dynamic entities from the problem description. The dynamic-object `philosophy-world` represents the activities of philosophers. They may enter and leave the world of philosophy, and may be thinking and eating. Figure 3 shows the dynamic-object representing the `philosophy-world`. It has a state slot for each CPN place. Each place in a CPN has an associated set of possible tokens. In the same way, each state slot has a constraint (`valueclass` metaknowledge) associated to the class of objects (tokens) that it can contain. In the `philosophy-world` all state slots hold philosopher instances.

We may consider the rules of politeness incrementally. Thus, first we define the dynamic-object `Chair-Politeness`. It represents that a chair may be introduced or removed from the table considering involved forks, but it does not

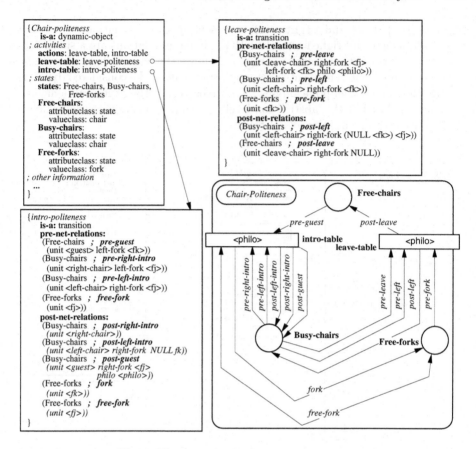

Fig. 4. The dynamic-object `Chair-Politeness`.

consider that philosophers may be eating or thinking. Figure 4 shows the `Chair-Politeness` dynamic object. The state slots `Free-chairs` and `Busy-chairs` hold `Chair` instances, and the state slot `Free-forks` hold `Fork` instances. Transitions `intro-table` and `leave-table` take into account the attributes `right-fork` and `left-fork` of `Chair` instances, and modify the relations between chairs and forks to introduce or remove a chair. The initial marking should, at least, hold two busy chairs and an arbitrary number of free chairs with their corresponding left fork. Transitions are parameterized with the `<philo>` variable, which represents that the `Chair-Politeness` must be synchronized with another dynamic-object to modify relations between chairs and philosophers.

Following, the `Chair-Politeness` may be specialized to consider that philosophers may be thinking or eating. Figure 5 shows the `Chair-Eating-Politeness`, which inherits state and activity slots from `Chair-Politeness`, and adds two new activities `start-eating` and `end-eating`, and a new state slot `Busy-forks`. It represents that a philosopher only may start eating if he has a free fork on his left and another one on his right side. When a philosopher is eating the corresponding forks are removed from the place `Free-fork`. In this way, his neighbors can not start eating, and guests may not be introduced next to him.

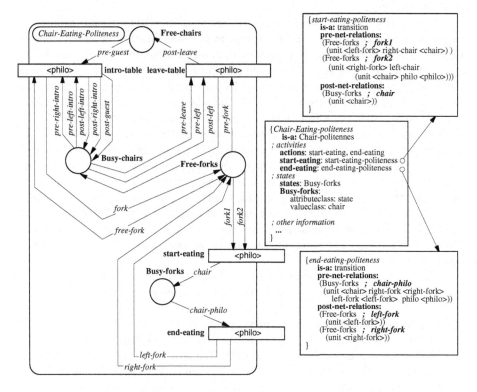

Fig. 5. The dynamic-object `Chair-Eating-Politeness`.

Transitions

Let us focus in the activity descriptions of previous **dynamic-objects**. From a discrete event system perspective, **transitions** carry out the specification and semantics of CPN transitions. Petri net arc information is supported in KRON by **net-relations** represented by two remarkable slots of transitions: Relations from **state slots** to **transition objects** working as enabling conditions are in the **pre-net-relations** slot, and relations from **transition objects** to **state slots** working as causal relations are in the **post-net-relations** slot.

From a knowledge representation perspective, information about activities can be considered as declarative knowledge in the "if/then" rule style (the similarities between CPN transitions and rules in rule based systems have been pointed out in several works [8], [2], [33]). The only difference is that in rule-based languages the enabling conditions on the left hand side of the rule (lhs), are clearly separated from the causal conditions on the right hand side (rhs). Nevertheless, the execution of a transition implies removing the enabling tokens from the input places and putting tokens in the output places according to the post-net-relations.

Expressions labeling the arcs are represented in KRON as arc expressions in pre and postconditions. An arc expression is a specification of restrictions on objects. These restrictions are represented, in a rule style, by a list of component pairs: the first component is the specification of a slot name or the string unit denoting an object instance; the second component, composed by one or two elements, is a partial pattern to match the slot value, it can be a variable, a specific constant value, a function or expression or another arc expression.

Following with the behavior representation of the Philosophy-world, the activity slots leave, intro, start-eating and end-eating, point to the corresponding transitions that represent activities producing state changes. To illustrate the internal structure of a typical transition, let us focus on a transition prototype from the philosophy-world, which is shown in figure 3 and called start-eating. The value in its pre-net-relations slot is: (Thinking (unit <philo>)). The first element identifies the state slot Thinking in the dynamic object. The second one represents the arc expression (unit <philo>) which is labeling the arc.

KRON variables are identified by angle brackets (e.g., <philo>). As it is general in rule based systems, variables play a double role:

Specify flow conditions. Arc expressions in the preconditions are interpreted as patterns that must be matched. They identify a token that must be in a place slot for a transition to be enabled. For example, the expression (unit <right-chair> left-fork <fj>) (label pre-right-intro in figure 4) defines a pattern that matches all chairs in place Busy-chairs having some value in the slot left-fork. There will be a binding between the variable <right-chair> and the matched instance, and there will be another binding between <fj> and the values in its slot left-fork. Additionally, these bindings can establish equality constraints on other arc expressions of the same transition with the same variable names.

Specify data flow. Values bound to variables in preconditions can be transferred to postconditions. Additionally, arc expressions in postconditions can specify modifications in the transferred data. Information of bound variables is also used to update slot values of the tokens involved in a firing.

Some particular features may be used in arc expressions to increase its expressiveness:

– An arc expression may appear as the second component of another arc expression. This is a pattern to match with the objects that are stored in the slot. For example, (see label fork2 in figure 5):
   (unit <right-fork> left-chair (unit <chair> philo <philo>))
   In this case <philo> is bound to the philosopher that is in the philo slot

of the `chair` stored in the `left-chair` slot of the `fork` instance bound to `<right-fork>`.

- A function call may appear as the second component of an `arc expression`. A function call is represented by a list whose first element is the symbol $, the second element is the function name, and the rest are the arguments. Functions may be used in postcondition for dynamic instantiation purposes. For example, (`unit` ($ `make-philosopher`)) may down a new instance of philosopher.

- To facilitate an incremental model design, KRON allows the use of incomplete transitions whose missing variables in preconditions must be provided by transition synchronization (see section 6). These variables play the role of parameters of the activities provided by the objects. For example, all activities of `Chair-Eating-Politeness` constitute its interface, and they have the parameter `<philo>`.

- The keyword NULL may appear in postconditions. NULL deletes all values from the slot. For example, (`unit` `<left-chair>` `right-fork` NULL `<fk>`), removes all values from slot `right-fork` before adding the new value bound to `<fk>` (label `post-left-intro` in figure 4). Additionally, a slot value can be replaced by another value using a list with NULL and the removed values. For example, (`unit` `<left-chair>` `right-fork` (NULL `<fk>`) `<fj>`) removes the value bound to `<fk>` from slot `right-fork` of object `<left-chair>`, then it adds the value bound to `<fj>` to this slot (label `post-left` in figure 4).

Additionally, each `transition` has a `predicate` associated. The `predicate` imposes a logical constraint on the transition enabling. It is a Boolean function, which can only contain those variables that are already in the expressions of the arcs connected to the transitions. The predicate is supposed to be `true` by default.

Sometimes it is useful to execute some action (execution of some particular subprogram). This is the purpose of a transition method called `action`. This method is called each time the transition is fired. The method receives the bindings of the transition variables as a parameter.

5   Instances, Classes and Inheritance in Dynamic Objects

The purpose of previous sections was to explain how the dynamic behavior of different kinds of entities is described in KRON. In this section we will focus on the use of inheritance as a mechanism to share code and representation. Therefore, we have considered the inheritance as a subclassing relation. Subclassing highlights redundancy within a system and the object-oriented decomposition yields smaller models through the reuse of common mechanisms, thus providing an important economy of expression. The subtyping relationship has not been considered. (Different approaches to formalize the behavior preservation between parent and descendant classes can be found in [10] and [3]).

Object oriented modeling starts by creating a hierarchy of classes, from more generic to more specialized, whose elements will be further instantiated to build a particular system model. Frame based languages make emphasis on inheritance issues and they provide not only support for traditional slots and method inheritance, but also allow the programmers the specification of additional types of inheritance (overriding, adding, unioning, wrappering, ...).

In our working context of discrete event system domain, entities with similar state space and behavior are grouped defining a hierarchy of dynamic object classes. A dynamic object class is a template to construct a composed object, whose instantiation implies the instantiation of the CPN structure that describes its behavior. All instances of a dynamic object class inherit the same Petri net with the same initial marking. Following the same process, transitions with similar structure and behavior are classified in a hierarchy tree of transition classes. Therefore, the behavior of a child class is obtained from the inherited Petri net by adding new transitions and state slots, or providing more specific details about them. For example, inherited state slots may be specialized with additional restrictions on the tokens they can hold.

The creation of the transition hierarchy requires more attention. Thus, a child transition class may be specialized in the following different ways:

Adding enabling conditions: The inheritance type of the `pre-net-relations` and `predicate` slots is *union*. This means that their values are derived by the *and* composition of the values that are in the subclass slot and the inherited values from its superclasses. Therefore, the net enabling conditions of a transition class is restricted by defining new values in the pre-net-relation slot of a transition subclass. Additional enabling conditions may be imposed on a transition class by adding new values to the predicate slot.

Adding new actions: The inheritance type of the `post-net-relations` and `action` slots is also *union*. When the transition is fired `pre-net-relations` and `post-net-relations` imply the modification of the respective state slot values. Therefore, new actions may be defined by adding new `pre` and `post-net-relations` values to a transition subclass. A transition firing also implies the execution of the action method. The `action` method may be specialized in a child transition class by wrapping code before, after or around the inherited code, or overriding it. Moreover, the code of action methods implies the execution of dynamic object methods. Therefore, the action method can be indirectly specialized by the specialization of dynamic object methods.

A transition instance is never created directly, but only through the instantiation of its dynamic object. Transition classes in activity slots are instantiated and replaced by their instances. An important feature of KRON is that the representation of an activity that is carried out in cooperation among different entities, is collected into only one `transition` instance. In this case, the state slots of pre- and post-conditions may belong to different dynamic objects. For this reason a transition instance inherits all slot values from the transition class, but `pre/post-net-relations` add to each inherited net relation a reference to the dynamic object instance.

A new dynamic object class can also be created by *multiple inheritance*. In this case, the subclass inherits several separated nets from their superclasses, which can be joined to build a more complex one. The connection can be made by adding transitions and places that model the control flow interaction between inherited nets. Multiple inheritance facilitates composition of incomplete representation behavior (virtual classes) during the model development. This means that the Petri net underlying a dynamic object class may be incomplete, and therefore this class should be refined to complete the behavior representation.

Finally, it is important to note that some problems have been detected with the integration of concurrency and inheritance. In concurrent object oriented languages, it is called *synchronization code* the code that selects the set of services that a concurrent object can execute and that depends on its state; that is, the enabling conditions of transition objects in KRON terms. The reuse of the synchronization code in concurrent object oriented languages has been considered difficult due to *inheritance anomaly*: synchronization code cannot be effectively inherited without non trivial class redefinitions [26]. S. Matsuoka and A. Yonezawa identify three kinds of inheritance anomaly: 1) *State partitioning anomaly* occurs when the subclass needs to make a partition of the set of states the superclass can have. 2) *History only sensitiveness of states*, appears when the methods in a parent class must be modified because the application of a method in a subclass depends on the history information, which does not manifest itself in the values of the inherited instance variables. 3) *State modification anomaly*, appears when the definition of a subclass requires the modification of inherited enabling conditions to account for a new action.

KRON mitigates some of the effects of the inheritance anomaly. A KRON model allows the appropriate separation of the synchronization code (enabling conditions) from the action (a piece of code) attached to transition objects. It makes the refinement of actions easier, allowing the inheritance mechanism to override the two parts separately. On the other hand, Petri nets have a guard based synchronization schema. Thus, the state partitioning anomaly pointed in [26] does not occur because the addition of new net conditions allows the differentiation of substates.

Following with the case study, the figure 5 shows how to specialize the dynamic-object Chair-Politeness. To complete the model, it must be created an instance of Philosophy-world and an instance of Chair-Eating-Politeness. Instances of Chair, Philosopher and Fork, and the initial relations between them, will be created by the constructors of dynamic-objects that execute its initial-marking method.

6   Dynamic Entity Connections

The construction of system models in KRON is done incrementally by first, designing isolated entities and then imposing the interactions between them to compose a bigger subpart of the system model. From a dynamic perspective, the behavior of an entity is represented by a CPN underlying a dynamic object.

The overall dynamic model will be homogeneous if the dynamic interactions between dynamic objects are described in the same terms as the internal behavior of objects. This means that if the interactions are described in terms of places, transitions and arcs, the behavioral model of the system will be a CPN and the advantages of using this formal view can be fully assumed. In KRON, the representation of interactions between dynamic objects becomes the same problem as the high level Petri net connections.

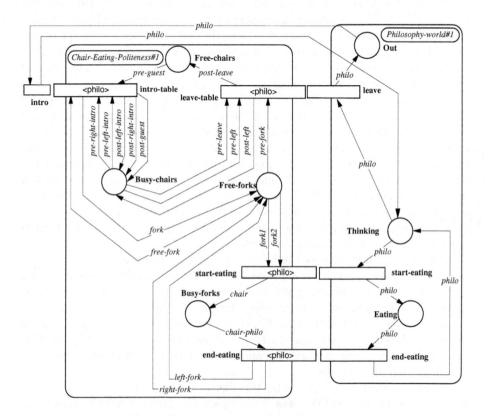

Fig. 6. Synchronization of dynamic-object instances.

CPNs may be connected by merging transitions or places, and by means of new arcs [5]. In KRON, transition merging has been selected as the main mechanism to represent the interactions between dynamic objects. The advantages of this approach will be pointed out at the end of this section. This approach provides a synchronous communication style that has been adopted by other works as modular-CPNs [14], CO-OPN [10] or OBJSA [3]. In this mechanism, an interaction between two or more objects can be interpreted as the execution of a joint activity, where each object has only a partial view of the real activity and its constraints. This interaction implies the synchronization of the internal behavior of those objects.

To illustrate the possibilities for dynamic entity connections we will synchronize the instances `philosophy-world#1` and `Chair-Eating-Politeness#1`. (See figure 6). When a philosopher enters or leaves from the philosopher world, new relations between chairs, forks, and philosophers are defined. The action `intro-table` of `Chair-Eating-Politeness#1` implies that a philosopher must occupy a chair, and the action `intro` of `Philosophy-world` must respect the politeness rules. Therefore these actions must be synchronized. In fact, they could be considered as partial views of the same activity. The result is a merging transition that synchronizes the behavior of both dynamic entities. In the same way, the activity `leave` of the `Philosphy-world` must be synchronized with the activity `leave-table` of `Chair-Eating-Politeness`. On the other hand, it is necessary to have two forks to start eating. Therefore, `Philosophy-world` and `Chair-Eating-Politenes` must synchronize its respective activities `start-eating` and `end-eating`.

To support this approach, the dynamic objects interface in KRON is a set of activity slots. Thus, the transitions of a dynamic object can be internal or interface transitions. Only the interface transitions can be externally synchronized. Connections between dynamic objects are established by naming the activity slots that must be related in some manner. The synchronization mechanism generates a new merged transition by multiple inheritance of the originals (for a complete transition merging, additional mechanisms are supported in KRON to specify the relations between variables from different transitions whose names have local scope). The transition generated by the merging replaces the originals in all activity slots involved in the synchronization. It allows the different dynamic objects related by a synchronization to maintain the same view over a transition after the merging.

Two types of synchronization (by transition merging) have been designed for KRON. *Normal synchronization* substitutes synchronized transitions by a unique transition with all pre and postconditions of the original ones. However, a slightly different case arises when a single activity may be synchronized with several alternative activities, and these other activities can not be synchronized with one another. This is the place for a *bilateral synchronization*, which supposes a replication process.

Some characteristics of the proposed approach for dynamic entity synchronization are:

1. Synchronized transitions are handled as a single object that belongs to cooperative objects. In this way, communication between different entities, from a dynamic perspective, is supported by the same formalism that defines the internal dynamic of each object.
2. Synchronized transitions provide a symmetric form of cooperation by an arbitrary number of entities, and no direction of communication is intended. Transition synchronization provides a higher level mechanism to communicate objects than the classical message passing. Collective behavior of objects

can be described without an implementation model of communication, and does not restrict the model to the client-server framework. Intuitively, transition objects are similar to the space tuples of generative communication [11]. Tokens may interact through transition objects by inserting tuples with the bindings produced by arc expressions. Communication may produce if there is pattern matching between tuples. (In the examples philosopher, chair, and fork entities interact through transition objects in this way). A comparison of Petri nets and the generative communication can be also found in [20], where Petri nets are used to specify the behavior of the hurried philosophers based on objects and the generative communication. In this approach the space tuple is represented by places that hold tokens, whereas in our approach the space tuple is represented by transitions that holds the bindings produced by tokens.

3. It may be argued that synchronized transitions violate the encapsulation, because they have access to the local state of cooperative dynamic objects. However, a synchronized transition denotes a relation between dynamic objects that defines the rules of this violation. As Rumbaugh points out in [30], a relation is not something to be hidden, but rather, to be specified abstractly, without imposing an implementation.

KRON also allows the definition of composite dynamic-objects. When a composite is instantiated, the different parts of the composite are instantiated and the synchronization relations between the parts are established. In this way, KRON may reuse through inheritance and aggregation.

7   Conclusions

In this paper we have presented KRON (Knowledge Representation Oriented Nets), a knowledge representation schema for discrete event systems (DESs). KRON is based on the integration of CPNs with frame based representation techniques and follows the object oriented paradigm. In addition to the features generally supported by object oriented languages, a set of primitives implementing the CPN formalism is included. CPNs provide the mechanism to describe the internal behavior of dynamic entities and the interactions between them. The Hurried Philosophers example has been adapted to highlight some relevant KRON capabilities.

Most of the approaches integrating objects and HLPNs, extend the HLPN formalism. The approach adopted here does not extend the CPN formalism. The frame-based representation of KRON supports the data and methodological aspects with no need to extend the CPN formalism. So, all the advantages of the use of this formalism can be profited from working with KRON.

KRON may reuse models through inheritance and aggregation. On the one hand, KRON uses inheritance as a mechanism to share code and representation. On the other hand, aggregation is supported by CPN composition. CPNs may be connected by merging transitions or places, and by new arcs. In KRON, transition merging has been selected as the main mechanism to represent the interactions between dynamic objects. This approach provides a synchronous communication style with all its advantages.

The semantics of the behavioral rules is supported in KRON by a so called control mechanism or interpreter. The control mechanism interprets the model to make the net evolve. The implementation of an efficient interpreter of KRON models may be found in [2]. In order to interpret the model, transitions are grouped into conflicts. The interpretation of the model is orthogonal to the model itself. An only interpreter may execute the model, or it may be attached an interpreter to each dynamic entity.

A prototype of a simulation tool with graphical display and animation facilities has been implemented on top of a known knowledge engineering environment called KEE [22] from Intellicorp.

References

1. M. Baldassari and G. Bruno. PROTOB: An object oriented methodology for developing discrete event dynamic systems. *Computer Languages*, 16(1):39–63, 1991.
2. J.A. Bañares, P.R. Muro-Medrano, and J.L. Villarroel. *Application and Theory of Petri Nets 1993*, chapter Taking Advantages of Temporal Redundancy in High Level Petri Nets Implementations, pages 32–48. Number 691 in Lecture Notes in Computer Science. Springer Verlag, 1993.
3. E. Battiston and F. de Cindio. Class orientation and inheritance in modular algebraic nets. In *Proc. of IEEE International Conference on Systems, man and Cybernetics, Le Touquet-France*, pages 717–723, 1993.
4. E. Battiston, F. de Cindio, and G. Mauri. *Advances in Petri Nets 1988*, chapter OBJSA Nets: a class of high-level Petri nets having objects as domains, pages 20–43. Number 340 in Lecture Notes in Computer Science. Springer Verlag, 1988.
5. B. Baumgarten. *Advances in Petri Nets 1988*, chapter On internal and external characterization of PT-net building block behavior. Number 340 in Lecture Notes in Computer Science. Springer Verlag, 1988.
6. G. Berio, A. Di Leva, P. Giolitto, and F. Vernadat. The m*-object methodology for information system design in cim environments. *IEEE Tran. on Systems, Man, and Cybernetics*, 25(1):68–85, January 1995.
7. J. Billington. Many-sorted high-level nets. In *Proc. of Third International Workshop on Petri Nets and Performance Models, Kyoto*, pages 166–179, 1989.
8. G. Bruno and A. Elia. Operational specification of process control systems: Execution of prot nets using OPS5. In *Proc. of IFIC'86, Dublin*, 1986.
9. G. Bruno and G. Marchetto. Process-translatable Petri nets for the rapid prototyping of process control systems. *IEEE transaction on Sosftware Engineering*, 12(2):346–357, 1986.
10. D. Buchs and N. Guelfi. CO-OPN: a concurrent object oriented petri net approach. In *Proc. of the 12th International Conference on Application and Theory of Petri Nets*, pages 432–454, Gjern (Denmark), June 1991.
11. N. Carriero and D. Gerlenter. Linda in context. *Communications of the ACM*, 32(4), April 1989.
12. L. Cherkasova, V. Kotov, and T. Rokicki. *Applications and Theory of Petri Nets 1993*, chapter Modeling of Industrial Size Concurrent Systems, pages 552–561. Number 691 in Lecture Notes in Computer Science. Springer Verlag, 1993.

13. S. Christense and N.D. Hansen. *Application and Theory of Petri Nets 1994*, chapter Coloured Petri Nets Extended with Channels for Synchronous Communication, pages 159–178. Number 815 in Lecture Notes in Computer Science. Springer Verlag, 1994.

14. S. Christensen and L. Petrucci. *Applications and Theory of Petri Nets 1992*, chapter Towards a Modular Analysis of Coloured Petri Nets, pages 113–133. Number 616 in Lecture Notes in Computer Science. Springer Verlag, 1992.

15. A. Dileva and P. Giolito. High-level petri nets for production system modelling. In *Proc. of the 8th European Workshop on Application and Theory of Petri Nets*, pages 381–396, Zaragoza (Spain), June 1987.

16. A. Dileva, P. Giolito, and F. Vernadat. Executable models for the representation of production systems. In *Proc. of the IMACS-IFAC Symposium on Modelling and Control of Technological Systems, IMACS MCTS 91*, pages 561–566, Lille (France), June 1991.

17. S. English. *Coloured Petri Nets for object-oriented modelling*. PhD thesis, University of Brighton, 1993.

18. R. Fehling. A concept for hierarchical petri nets with buiding blocks. In *Proc. of the 12th International Conference on Application and Theory of Petri Nets*, pages 370–389, Aarhus, 1991.

19. R. Fikes and T. Kehler. The role of frame-based representation in reasoning. *Communications of the ACM*, 28(9):904–920, September 1985.

20. T. Holvoet and P. Verbaeten. Using petri nets for specifying active objects and generative communication. In *Object-Oriented Programming and Models of Concurrency. A workshop within the 17th International Conference on Application and Theory of Petri Nets*, 1996.

21. P. Huber, K. Jensen, and M. Shapiro. Hierarchies in coloured petri nets. In *Proc. of the 10th European Workshop on Application and Theory of Petri Nets*, pages 192–209, Bonn, June 1989.

22. Intellicorp. *KEE User Guide*. Intellicorp, 1989.

23. K. Jensen. *Coloured Petri Nets: Basic Concepts, Analysis Methods and Practical Use*. EATCS Monographs on theoretical Computer Science, Springer-Verlag. Edited by W. Brauer, G. Rozenberg and A. Salomaa, Berlin Heidelberg, 1992.

24. C.A. Lakos. The role of substitution places in hierarchical coloured petri nets, thecnical report tr93-7. Technical report, Computer Science Department, University of Tasmania, August 1993.

25. C.A. Lakos and C.D. Keen. LOOPN++: A new language for object-oriented petri nets, thecnical report tr94-4. Technical report, Computer Science Department, University of Tasmania, 1994.

26. S. Matsuoka, K. Wakita, and A. Yonezawa. *Research Directions in Object-Based Concurrency*, chapter Inheritance anomaly in object-oriented concurrent programming languages. MIT Press, 1993.

27. B. Meyer. *Object-Oriented Software Construction*. Computer Science. Prentice Hall, Englewood Cliffs, N.J., 1988.

28. M. Paludetto and S. Raymond. A methodology based on objects and petri nets for development of real-time software. In *Proc. of IEEE International Conference on Systems, man and Cybernetics, Le Touquet-France*, pages 717–723, 1993.

29. W. Reisig. *Theoretical Computer Science 80*, chapter Petri Nets and Algebraic Specifications, pages 1–34. Elsevier Science Publishers B.V., 1991.

30. J. Rumbaugh. Relations as semantic contructs in an object-orientated language. In *Proc. of the ACM Object-Oriented Programming Systems, Languages and Applications, OOPSLA'87*, pages 466–481, October 1987.

31. C. Sibertin-Blanc. High-level petri nets with data structures. In *Proc. of Workshop on Applications and Theory of Petri Nets. Finland*, June 1985.
32. C. Sibertin-Blanc. *Advances in Petri Nets 1994*, chapter Cooperative Nets, pages 377–396. Number 815 in Lecture Notes in Computer Science. Springer Verlag, 1994.
33. R. Valette and B. Bako. Software implementation of petri nets and compilation of rule-based systems. In *11th International Conference on Application and Theory of Petri Nets*, Paris, 1990.
34. J. Vautherin. *Advances in Petri Nets 1987*, chapter Parallel Systems Specifications with Coloured Petri Nets and Algebraic Specifications., pages 293–308. Number 266 in Lecture Notes in Computer Science. Springer Verlag, 1987.

Modeling of a Library with THORNs

Frank Köster[1], Stefan Schöf[2], Michael Sonnenschein[1], and Ralf Wieting[3]

[1] Oldenburger Forschungs- und Entwicklungsinstitut für
Informatik-Werkzeuge und -Systeme (OFFIS)
Escherweg 2 · D–26121 Oldenburg (Germany).

[2] Software Design & Management (SD&M)
Region West · Am Schimmersfeld 7a · D–40880 Ratingen (Germany).

[3] Media Service AG (MSG)
Cloppenburger Straße 300 · D–26133 Oldenburg (Germany).

`{koester,sonnenschein}@offis.uni-oldenburg.de`

Abstract. THORNs combine the widely used object-oriented programming language C++ with various features of Petri nets for modeling concurrency and time. In this way complex distributed systems can be modeled in a detailed manner. THORNs can be transformed to C++ code and executed sequentially or concurrently by simulators for validation and experiments. This paper shows both, features of THORNs and their modeling approach by an example.

1 Introduction

Petri nets allow to model concurrency and conflicts in a simple manner. Modeling with objects allows the integrated view of values together with operations on it following the idea of abstract data types resp. data encapsulation; concepts of inheritance and polymorphism have to be added to this approach to get a fully object-oriented modeling technique. So it is an obvious idea to model complex distributed systems by combining the object-oriented approach with Petri nets. There are three different levels of combining these approaches:

- Objects are modeled as tokens of a net. A transition refers to objects consumed from its preset places, modifies attributes of these objects and produces these objects on its postset places. Consumed objects may also be deleted and produced objects may also be created by the transition. Objects can be active in parallel, but parallelism inside of an object is not allowed. This attempt underlies e.g. PACE [PAC89], DesignBeta [CT93], MOBY [FL93], and even THORNs [SSW95].
- Objects are modeled by subnets. The net for an object models enabling and disabling of methods and also possibly parallel execution of methods within an object. Examples of this class are defined by Battiston, Di Giovanni, Lakos, Sibertin-Blanc, Paludetto or Valk [BDM96,Di 91,Lak95,PR93,SB94,Val96]

G. Agha et al. (Eds.): Concurrent OOP and PN, LNCS 2001, pp. 375–390, 2001.
© Springer-Verlag Berlin Heidelberg 2001

- Objects are modeled by nets, methods and attributes are modeled by sub-nets. The net for a method models the internal state of an object together with the control flow of the method. So concurrent data streams within the execution of a method may be modeled, too. A member of this class of modeling techniques is GINA [Son93].

In a different attempt to combine Petri nets with the concurrent object-oriented approach, Petri nets can be used to model the behaviour of actor programs (see [ELR90,SVN91]).

After an informal introduction to the concepts of THORNs, this paper gives an example of both, features of THORNs and modeling with THORNs by specifying the behaviour of a library. In contrast to some other attempts THORNs also allow to model aspects of time. THORN models can be created using a graphical editor and executed by a sequential or distributed simulator; i.e. a THORN model of a distributed system may be simulated on a workstation cluster or a parallel computer [Sch95]. In this way, THORNs can also be used as a distributed, object-oriented programming language.

2   Timed Hierarchical Object-Related Nets

In this section we give an overview of the basic features of THORNs. THORNs integrate several known concepts of high-level Petri nets [JR91] with respect to individual tokens, additional net elements, timed transitions, and structuring mechanisms into one formalism.

Because of the complexity of the language, the description of the THORN semantics given below can not be complete with respect to all possible combinations of different features, but it will be sufficient to understand the concepts and the example of modeling with THORNs given in this paper. Another informal description of THORNs can be found in [SSW95], and a complete formal definition of a slightly different (initially) version of THORNs is given in [FLSW93].

2.1   Individual Objects

An obvious enhancement of high-level Petri nets is the introduction of individual tokens [Jen90,Gen87,LK91]. In THORNs we speak of objects instead of tokens because they are instances of classes of an object-oriented programming language (here C++). Thus, an object type is specified by a C++ class (cf. Section 3).

For the definition of object types the modeler can make use of all object-oriented concepts C++ provides, e.g. encapsulation, inheritance, polymorphism and dynamic binding. At the moment, one drawback of C++ is the absence of any run-time type information. Since this feature is essential for the distinction of objects of different classes in a set of objects marking a place, we already use a future solution proposed for C++ to this problem, i.e. the class of an object x can be determined by the function `typeid(x).name()`.

2.2   Places and Arcs

In THORNs we distinguish several kinds of places and arcs. Figure 1 shows the different place structures and arc types together with their graphical representation.

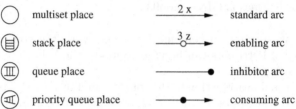

Fig. 1. Place Structures and Arc Types

A place has a *name*, a *type*, a *capacity*, and a *structure*. The type specifies the type of objects allowed on the place: all objects have to be of that type or a type derived (by inheritance) thereof. The capacity specifies an upper bound for the number of objects the place may hold in a state of the net (an infinite capacity is possible as well). Finally, the structure (cf. Fig. 1) determines how the objects on a place are organized, e.g. a queue place stores the objects according to FIFO principle.

An arc has a *name* and a *type* (cf. Fig. 1). Besides standard arcs three other arc types are provided to connect places and transitions. Enabling arcs influence the activation of the incident transition equally as standard arcs, but no object will be consumed at the firing of the transition. Inhibitor arcs disable the transition if objects reside on the incident place, and consuming arcs do not affect the activation of a transition, but all objects on the incident place are removed if the transition fires. Enabling and standard arcs may be labeled with *variable names* and *arc weights* in addition. A variable name is used to reference objects from the transition's code (see below). An arc weight models the number of objects consumed or produced on the incident place by firing of the transition. Arc weights are allowed only for arcs incident to multiset places.

A further kind of arcs called *hyper arcs* will be explained in this paper together with the concept of hierarchy in THORNs.

2.3   Transitions

The transitions of a THOR net play the most important role because they represent active parts of a model, like processes in a data flow model. Therefore, each transition is labeled with a *name*, an *activation condition*, an *action block*, a *delay time function*, a *firing time function*, and a *firing capacity*.

The activation condition is specified as a boolean expression over variables of incoming arcs. It has to be fulfilled by preset objects bound to the variables in order to enable the transition. Activation conditions must not have side effects. To avoid name conflicts, all variable names of arcs incident to the transition have to be different.

The action code of a transition is performed using consumed and produced objects like in/out value-parameters when the transition fires. It consists of several C++ statements where variables of all incident arcs can be used. If an incident arc has the weight $w > 1$, its variable name is interpreted as a name of an array of length w. The type of an ingoing variable is dynamically determined by the type of the bound object; it may be the type of the corresponding preset place or an inherited type thereof. The type of an outgoing variable follows from the inscription of the action block; it has to be conform to the type of the corresponding postset place. The action code may contain - in principle - any form of side effects as user interaction or storing and retrieving data in a data base. Such side effects are difficult to handle for the distributed simulation of THORNs and will not considered here further on.

The delay time function and firing time function can be composed out of constants, variables of incoming arcs, C++ operations, or even random number generators. Their domain has to be non-negative real numbers. The delay time of a transition indicates how long a transition must be enabled without interruption before it starts to fire, and the firing time specifies the time period necessary for executing the activity represented by that transition.

Finally, the firing capacity defines how often at most a transition may fire in parallel to itself (infinite firing capacities are possible as well).

2.4 Firing Rule

A transition combined with a binding of objects on preset places to incoming arc variables is called a firing event or *occurrence element*. Since a transition may fire with different occurrence elements, we speak of occurrence elements when we consider the firing rule.

An occurrence element will be delayed for the time given by the delay time function of the transition if the precondition is fulfilled by the according binding. Therefore all places connected by inhibitor arcs to the transition have to be empty, all preset places connected by standard or enabling arcs have to hold enough objects with respect to corresponding arc weights, and the activation condition has to be fulfilled by the binding. If an occurrence element has been delayed at least for the whole delay time without interruption it is called completely delayed. The delay time may be interrupted by firing of a transition consuming an object bound in the occurence element or by creating an object on a preset place connected to the transition of the occurence element by an inhibitor arc. An occurrence element is *enabled* (with respect to the weak firing rule [Bra84]) if it is completely delayed, the transition has free firing capacity, and places in the postset of the transition can store enough objects with respect to the weights of outgoing arcs.

An enabled occurrence element starts firing by consuming all objects bound to variables of standard arcs from preset places. Places connected by consuming arcs are emptied. After waiting a period determined by applying the firing time function, the transition finishes firing: its action block is executed and outgoing objects are produced on postset places. As long as an occurrence element fires, the transitions firing capacity is decremented by one and the required place capacity on postset places is reserved. Firing of a transition can not be interrupted or cancelled by other events.

At each moment of a net's execution one enabled occurrence element starts firing after another as long as enabled occurrence elements exist, and then in the same way all occurence elements finish firing one after another. After that time skips to the next event, which may be the end of a firing or delay time period. This strategy is called maximum direct firing rule.

The state of a THORN is defined at each moment of its execution by the set of stored and reserved tokens on all places, and the current remaining delay time and firing time of all occurence elements.

2.5   Hierarchy

THORNs combine the two well-known hierarchy concepts of *transition refinement* and *subnet invocation* [Jen92,CK81] and fusion of places that are discussed for coloured Petri nets in [HJS90]. So THORNs offer a general-purpose mechanism for dynamic hierarchical structuring of nets. We illustrate this concept on a small example. Figure 2 shows a calling transition with some border places and the invoked subnet. The border places are divided into input, output, and share places connected to the transition by ingoing standard arcs, outgoing arcs, and undirected hyper arcs (represented by dotted lines), respectively. They are corresponding to places (of equal types) in the called subnet in a one-by-one manner. Border places of the subnet have a special graphical representation. Enabling, inhibitor, and consuming arcs have the same effect to calling transitions as to standard transitions but do not affect the subnet, and their incident places do not correspond to places in the called subnet.

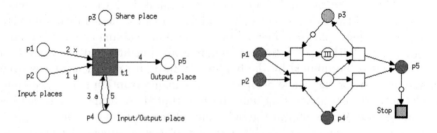

Fig. 2. Calling transition and called subnet

Calling a subnet is very similar to calling a procedure executing an imperative program. Input and output places correspond to formal and actual parameters of the procedure. Share places can be compared to global variables in the environment of the procedure.

Calling transitions are labeled similar to standard transitions except for their action block which consists only of a name of a subnet. The firing rule for occurrence elements of a calling transition is similar, as well. But at the end of a firing a new instance of the specified subnet is created, objects previously consumed via standard arcs are passed onto the corresponding input places of the subnet, and all share places of the calling transition are identified with the corresponding share places of the subnet (place fusion). So hyper arcs are only a graphical representation for the fusion of places in the border of a calling transition and the called subnet.

After initializing its border places the new subnet is treated like all other (sub)nets. Transitions may start firing according to the maximum direct firing rule discussed above until a stop transition of the subnet starts firing — stop transitions are also labeled similar to standard transitions but don't have an action block because their only purpose is the termination of a subnet. After a stop transition has fired and all currently firing transitions of the subnet have finished their firing, the subnet terminates: objects on output places are passed to output places of the corresponding calling transition (notice: the number of passed objects may vary between zero and the arc weight of the corresponding outgoing arc of the calling transition; if necessary, objects are selected in a nondeterministically way), reserved place capacities on output places are released, the firing capacity of the calling transition is increased, and the instance of the subnet and all further invoked subnets are deleted.

This combined hierarchy concept contains ordinary transition refinement as a special case for a calling transition without input and output places that calls a subnet without a stop transition.

Subnets created by refined transitions are the basic elements for load balancing in the distributed simulator for THORNs [Ree96].

3   Modeling with THORNs

In this section we discuss techniques for modeling complex systems with THORNs. First in general, the level of detail/abstraction of a model should be chosen with respect to its use. Then a more object-oriented or a more functional style of modeling can be applied.

The more object-oriented modeling approach can follow four steps:

1. Identify the basic kind of objects of the considered system and define according C++ classes with appropriate attributes and methods for them. Make use of object-oriented concepts like encapsulation, inheritance, or polymorphism in order to achieve reusable, maintainable, and understandable object descriptions.
2. Use places for the storage of objects. Places should represent passive parts of a model.
3. Describe processes or activities with transitions. Transitions should represent the dynamic part of model. Hence, they should be used for object manipulation, generation, or deletion by calling their methods in the transition code. Furthermore, the timing behaviour of a system should be expressed with the time labels of a transition.

4. Utilize transition refinement and subnet invocation for the structuring of models and the modeling of complex dynamic system behaviour. It is a good idea to represent bigger parts of a model on a higher level of abstraction by a refined transition and describe this part with more detail within the corresponding subnet.

This attempt starts with identifying classes and objects of the system as usual for object-oriented analysis (see e.g. [CY91]). If it is a major design goal to analyse the communication structures within a system, these structures have to be modeled explicitly. This can also be done by identifying first functional units that communicate by object exchange. Such a model is a combination of a data flow model [De 79] with an object oriented method for modeling data.

Within the THORN model of a library (see next section) we model communication structures of the system explicitly by the net structure. To do this, it is necessary to examine the system with respect to the following items:

1. detect the basic (communication) structure of the system and model it by a (hierarchical) net,
2. describe objects which are exchanged between structural components of the system (abstract messages as well as real objects, e.g. people), and
3. analyse the temporal behaviour of structural components.

With this information we are able to reproduce the structure of the system by using places (passive object holders), transitions (active object manipulators resp. processes), and edges (for exchanging objects). The temporal behaviour of the system can be modeled by the delay time and firing time of transitions. Realization of objects is supported by using C++ to label transitions and to describe objects. Inheritance and polymorphism are useful to describe objects and types in a maintainable and understandable manner.

4 THORN Model of a Library

In this section, a fictitious library is used to show how the behaviour of complex distributed systems can be modeled with THORNS. We discuss only a part of the complete model extracted from [Kös96]. The part of the model is chosen in a way that all modeling features of THORNS can be demonstrated.

The library is modeled as a building with floors, a staircase, an information desk, and an archive. Books in the main library or in one of four external libraries can be ordered, borrowed and returned by visitors asking clerks for service. One possible purpose of the model is to identify whether the number of clerks suffices or if visitors must wait too long for their requests to be granted. The model description begins with a short overview of the structure and services of the library. Thereafter, well-chosen subnets and C++ objects are described in more detail.

The basic structure of the library is determined by one main and four external libraries (cf. Fig. 3). Within the main library corridors and a staircase connect

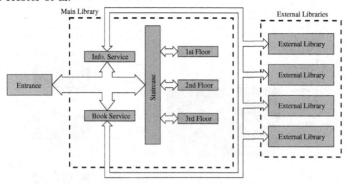

Fig. 3. Basic structure of the library

three floors and the entrance. The library users may only contact the main library. Books available at the main library may be borrowed by users. Books belonging to an external library must be called for by the main library before they are available there. Library clerks are engaged with lending books, giving information about books, and rearranging returned books into the shelves of the library.

After having given a coarse overview of the library, we now describe its modeling as a THORN. Figure 4 shows the so-called *design graph*, which is useful to represent the hierarchical dependencies within a THORN and which assists in browsing the net hierarchy within the THORN editor. The nodes of this graph represent the different subnets and the edges show dependencies (by transition refinement) between them.

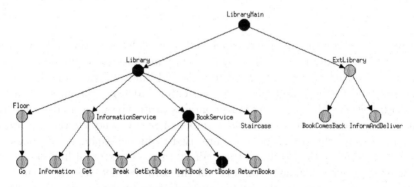

Fig. 4. Hierarchical structure of the library model

The subnet **LibraryMain** represents the whole library. It uses the subnets **Library** and **ExtLibrary** for the main and external libraries. The latter calls two other subnets to exchange books and information with the main library. The main library uses subnets for the staircase and the floors, which on their part use a subnet **Go** for moving clerks between different floors. In addition the net **Library** uses two subnets for providing its services. These two subnets call

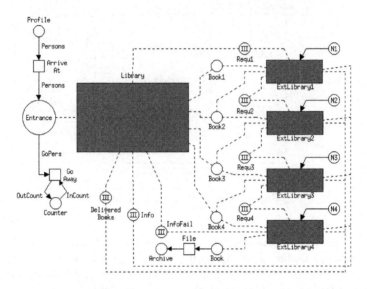

Fig. 5. Main net of the library

various other subnets to assist them. Especially note that subnet **Break** is used from two different subnets.

Only the subnets represented by black nodes will be discussed in in this paper.

4.1   LibraryMain

Figure 5 represents the main net, which is directly related to the system's structure (Fig. 3). The subnet called by **Library** models the main library while the subnets called by **ExtLibray***i* model the external libraries. The transitions **ArriveAt** and **GoAway** model how visitors enter and leave the entrance hall of the main library. The place **Profile** contains prototype data used to generate visitors. **Counter** is used to save the number of people who left the library. All other places and the transition **File** are used to model exchange of books and requests for information between the main library and external libraries.

All transitions **ExternalLibrary***i* call the same subnet, but the value on N*i* is used as a parameter to the subnet instantiation to distinguish the external libraries. The places **Requ***i* are used for exchange of book and information requests. These places are structured as queues because the requests should be handled according to a FCFS order. If a request can't be handled correctly by any external library it will be put on **InfoFail**. The places **Book***i* are used to handle exchange of books. If a book doesn't belong to any external library it will be put to place **Book** and finally, by firing of transition **File**, to the **Archive**. If a book/information request is treated successfully the book/information is put on **DeliveredBooks/Info**.

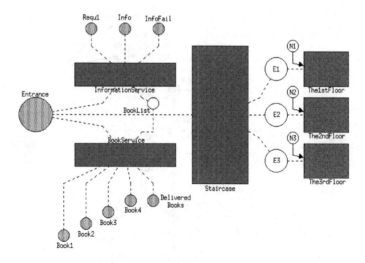

Fig. 6. Subnet Library

In this subnet there are places capable of storing different kinds of objects; for instance Book*i* and Book have the type LIB_BOOK_INF. It is used to store information on books and is derived from a base class STD_BOOK_INF, which contains some standard attributes of books (e.g. title, author, and ISBN). This information is completed by LIB_BOOK_INF with information specific to books inside a library, e.g. the shelf a book belongs to. The places Requ*i* have the type REQUEST, which is used for book queries and is derived from LIB_BOOK_INF. Request specific information is added here.

In the same way classes for visitors (VISITOR) and clerks (CLERK) are implemented. They are derived from a common base class PERSON, which contains several attributes for a person (e.g. his name). VISITOR and CLERK define additional attributes and redefine some virtual methods of PERSON in a visitor/clerk specific way. These classes will be discussed in more detail below.

4.2   Library

In Fig. 6 the subnet called by transition Library in Fig. 5 is shown. It models the main library. There is a direct mapping from the system's structure to net elements (cf. Fig. 3).

Within this subnet there are six refined transitions. The subnets called by transitions InformationService and BookService model how visitors are supported by staff in getting books or information, respectively — below we will discuss BookService in detail. The refined transition Staircase calls a subnet which models a staircase. This subnet enables clerks to change their location between entrance (Entrance) and the three floors (E*i*, The*i*stFloor). This part of the model can be used for describing path length of clerks walking through the library, but this will not be discussed in this paper.

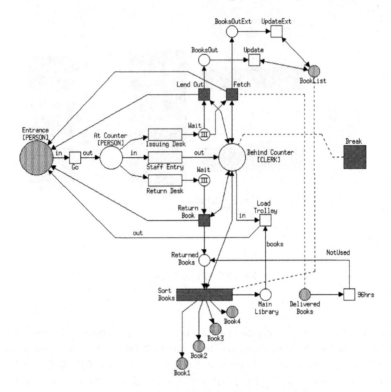

Fig. 7. Subnet BookService

The places Requ1, Info, InfoFail, Booki, DeliveredBooks, and Entrance are share places. By these places the subnet Library is permanently connected to the calling net (cf. Fig. 5). A short description of these places has already been given above. Place BookList contains all books which belong to the main library. These are modeled as instances of class CAT_BOOK_INF, which is derived from LIB_BOOK_INF and adds e.g. information whether a book is available or lent out. The concrete books stored in the library are part of the initial marking of this subnet.

Transitions The*i*stFloor use objects on places N*i* in the same way as transitions ExtLibrary*i* in Fig. 5 use similar places.

4.3 BookService

In Fig. 7 the subnet called by BookService in Fig. 6 is presented, which models interactions between visitors of the library and clerks.

The share places Entrance, BookList, DeliveredBooks, and Booki have already been described above (cf. Fig. 6). The transition 96hrs is a transition with a delay time of 96 hours, which fires if a book isn't fetched by a visitor within 96 hours after its delivery from an external library. In this case the book is sent

```
class PERSON {
public:
  virtual double Velocity() { return Normal(6 - 0.2 * NumberOfBooks, 1) };
  // in m/s; person is the slower the more books he carries

  // further public member functions, constructors, and destructors

private:
  int NumberOfBooks;

  // further private attributes and member functions.
};
```

Fig. 8. Declaration of class PERSON

back to its external location. SortBooks maps returned books (ReturnedBooks) to the different buildings they belong to (MainLibrary to the main library, Book$i$ to external libraries). Break enables clerks to take a rest.

IssuingDesk leads visitors to a queue, where they wait to borrow books. Transitions LendOut and Fetch have similar tasks. Always the first visitor in the queue place Wait is served by a clerk from place Behind Counter. Visitors are able to borrow books belonging to the main library (LendOut) or one of the external libraries (Fetch). This is reported to place BooksOut or BooksOutExt whereupon the transition Update or UpdateExt is able to fire, i.e. to update BookList.

Firing of ReturnDesk leads visitors to a queue, where they wait to bring back borrowed books by firing ReturnBook. Comparable to LendOut and Fetch the first person within the queue Wait will be served.

By the remaining transitions it will be shown how the different attributes and member functions of objects on preset places may be used in transition inscriptions. On that occasion, it is demonstrated how polymorphism may be used in THORN models. Prior to this, a more detailed (though only partial) view of class CLERK is given. It is derived from the abstract class PERSON, which defines common attributes of all kind of people in the library. In addition, clerks may use a trolley to carry more books than an ordinary person (e.g. a visitor), in which case they move slower. The class VISITOR, which isn't described further, is derived from PERSON as well. The (partial) definitions of PERSON and CLERK are given in Fig. 8 and 9.

All people in the library (visitors as well as clerks) move from Entrance to At Counter by firing of transition Go (cf. Fig. 7) executing the action code {out = in;}. The type of both places is PERSON and thus they may actually contain instances of class CLERK and VISITOR (there are no instances of PERSON, since it is an abstract class). By firing of transition Go, an object is generated on place At Counter with the same type and attributes as the object consumed from place Entrance, i.e. Go acts polymorphically on the objects. In addition, the duration of transition Go is determined by the function {10 / in.Velocity()} (entrance to desk distance is 10 meters). Since method Velocity is defined as virtual, the function actually called for a VISITOR object is different from the function called for a CLERK object, which must respect whether the clerk uses a trolley or not.

```
class CLERK : public PERSON {
public:
    bool HasTrolley() { return TrolleyAttr };

    void PutOnTrolley(LIB_BOOK_INF*);
    // put books (class LIB_BOOK_INF) on trolley

    virtual double Velocity() {
        if (HasTrolley())
            return Normal(4 - 0.1 * NumberOfBooks, 0.5);
        else
            return Normal(6 - 0.2 * NumberOfBooks, 1);
    }
    // with trolley clerk is slower than without

    // further public member functions, constructors, and destructors

private:
    bool TrolleyAttr;
    // true if clerk has trolley

    // further private attributes and member functions.
};
```

Fig. 9. Declaration of class CLERK

Transition `StaffEntry` models a clerk's way behind the counter. It uses the condition code `{typeid(in).name() == ''CLERK"}` to determine that only people may go to place `Behind Counter`, which have type CLERK. Here the run-time type information of an object is used explicitly to maintain compatibility of object and place types. Transition `LoadTrolley` will omitted here for reasons of simplicity.

4.4 SortBooks

The transition `SortBooks` in Fig. 7 is an example of a calling transition with only a partial production of objects on postset places. The corresponding subnet is shown in Fig. 10. Border places `Behind Counter`, `ReturnedBooks`, `Book`i, and `MainLibrary` already were described above. Transition `IsBack` modifies some attributes of a book to mark this book as returned. After this, the book is put on place `Sort` though transitions `To`i and `ToMain` get enabled. One of these transitions fires dependently on the attributes of the books. Firing of one of these transitions yields the production of a token on place `Okay`, which causes firing of the stop transition and hence termination of the net (cf. Sect. 2).

5   Conclusion

THORNs have been shown to be an appropriate object-oriented modeling approach for distributed systems. They combine the object-oriented features of C++ to model data with various features of Petri nets for modeling hierarchy, concurrency and time for processes. THORNs can be compiled to C++ code that can be executed by a sequential or distributed simulator. All features (except

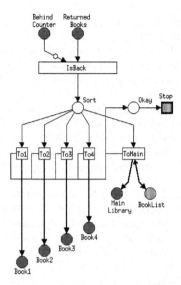

Fig. 10. Subnet SortBooks

inheritance, polymorphism and run-time type information of objects) shown in the example above are implemented by the simulators for THORNs. So THORNs have actually been used for modeling and simulation in such different areas as work flow, industrial plants, communication protocols or population dynamics.

References

BDM96.   E. Battiston, F. De Cindio, and G. Mauri. Modular algebraic nets to specify concurrent systems. *IEEE Transactions on Software Engineering*, 22(10):689–705, Oct. 1996.

Bra84.   W. Brauer. How to Play the Token Game? or Difficulties in Interpreting Place/Transition Nets. *Petri Net Newsletter*, 16:3–13, Feb. 1984.

CK81.   L. A. Cherkasova and V. E. Kotov. Structured nets. In W. Brauer, editor, *Proceedings of Mathematical Foundations of Computer Science*, volume 118 of *Lecture Notes in Computer Science*, pages 242–251. Springer-Verlag, Berlin, Germany, 1981.

CT93.   S. Christensen and J. Toksvig. DesignBeta V2.0.1 — BETA code-segments in CP-nets. Lecture Notes OO&CPN No 5, Computer Science Department, Aarhus University, Denmark, 1993.

CY91.   P. Coad and E. Yourdon. *Object Oriented Analysis*. Prentice Hall, Englewood Cliffs, second edition, 1991.

DD95.   G. De Michelis and M. Diaz, editors. *Proceedings of the 16th International Conference on Application and Theory of Petri Nets*, volume 935 of *Lecture Notes in Computer Science*, Torino, Italy, June 1995. Springer-Verlag, Berlin, Germany.

De 79.   T. De Marco. *Structured Analysis and System Specification*. Prentice Hall, Englewood Cliffs, 1979.

Di 91.   R. Di Giovanni. Hood nets. In G. Rozenberg, editor, *Proceedings of the 11th* *International Conference on Application and Theory of Petri Nets*, volume 524 of *Lecture Notes in Computer Science*, pages 140–160, Berlin, Germany, 1991. Springer-Verlag.

ELR90.   J. Engelfriet, G. Leih, and G. Rozenberg. Parallel object-based systems and Petri nets. Technical report, Leiden University, The Netherlands, 1990.

FL93.   H. Fleischhack and U. Lichtblau. MOBY – a tool for high level Petri nets with objects. In *Proceedings of the IEEE International Conference on Systems, Man and Cybernetics*, volume IV, pages 644–649, Le Touquet, France, 1993. IEEE.

FLSW93.   H. Fleischhack, U. Lichtblau, M. Sonnenschein, and R. Wieting. Generische Definition {hierarchischer} {zeitbeschrifteter} {höherer} Petrinetze. Bericht der Arbeitsgruppe Informatik-Systeme AIS–13, Fachbereich Informatik, Universität Oldenburg, Germany, Dec. 1993. In German.

Gen87.   H. J. Genrich. Predicate/Transition nets. In W. Brauer, W. Reisig, and G. Rozenberg, editors, *Petri Nets: Central Models and Their Properties*, volume 254 of *Lecture Notes in Computer Science*, pages 207–247. Springer-Verlag, Berlin, Germany, 1987. Auch in [JR91, pp. 3–43].

HJS90.   P. Huber, K. Jensen, and R. M. Shapiro. Hierarchies in coloured Petri nets. In Rozenberg [Roz90], pages 313–341. Auch in [JR91, pp. 215–243].

IEE93.   IEEE. *Proceedings of the IEEE International Conference on Systems, Man and Cybernetics*, volume II, Le Touquet, France, 1993.

Jen90.   K. Jensen. Coloured Petri nets: A high level language for system design and analysis. In Rozenberg [Roz90], pages 342–416. Auch in [JR91, pp. 44–119].

Jen92.   K. Jensen. *Coloured Petri Nets – Basic Concepts, Analysis Methods and Practical Use Volume 1*. EATCS Monographs on Theoretical Computer Science. Springer-Verlag, Berlin, Germany, 1992.

JR91.   K. Jensen and G. Rozenberg, editors. *High-Level Petri Nets – Theory and Application*. Springer-Verlag, Berlin, Germany, 1991.

Kös96.   F. Köster. Bewertung hierarchischer Petrinetze als Grundlage für Mapping-Verfahren bei der verteilten Simulation. Diplomarbeit, Universität Oldenburg, Germany, May 1996. In German.

Lak95.   C. A. Lakos. From Coloured Petri Nets to Object Petri Nets. In De Michelis and Diaz [DD95], pages 278–297.

LK91.   C. A. Lakos and C. D. Keen. LOOPN – language for object-oriented Petri nets. In *SCS Multiconference on Object-Oriented Simulation*, pages 22–30, Anaheim, CA, Jan. 1991.

PAC89.   Gesellschaft für Prozeßrechnerprogrammierung, Oberhaching (Germany). *PACE – graphisch-interaktives Simulations- und Prototypingwerkzeug*, 1989. In German.

PR93.   M. Paludetto and S. Raymond. A methology based on objects and petri nets for development of real-time software. In *Proceedings of the IEEE International Conference on Systems, Man and Cybernetics* [IEE93], pages 705–710.

Ree96.   G. Reents. Effizienzsteigerung bei der optimistischen verteilten THOR-Netzsimulation. Diplomarbeit, Universität Oldenburg, Germany, Nov. 1996. In German.

Roz90.   G. Rozenberg, editor. *Advances in Petri Nets*, volume 483 of *Lecture Notes in Computer Science*. Springer-Verlag, Berlin, Germany, 1990.

SB94.   C. Sibertin-Blanc. Cooperative nets. In R. Valette, editor, *Proceedings of the* 15th *International Conference on Application and Theory of Petri Nets*, volume 815 of *Lecture Notes in Computer Science*, pages 471–490, Zaragoza, Spain, June 1994. Springer-Verlag, Berlin, Germany.

Sch95.   S. Schöf. A distributed simulation engine for hierarchical Petri nets. In P. Schwarz, editor, *10. Workshop "Simulation verteilter Systeme und paralleler Prozesse"*, number 50 in ASIM-Mitteilungen, pages 153–159, Dresden, Oct. 1995.

Son93.   M. Sonnenschein. An introduction to GINA. In *Proceedings of the IEEE International Conference on Systems, Man and Cybernetics* [IEE93], pages 711–716.

SSW95.   S. Schöf, M. Sonnenschein, and R. Wieting. Efficient simulation of THOR nets. In De Michelis and Diaz [DD95], pages 412–431.

SVN91.   Y. Sami and G. Vidal-Naquet. Formalization of the behavior of actors by colored Petri nets and some applications. In *PARLE '91, LNCS 506*, pages 110–127, 1991.

Val96.   R. Valk. On the process of Object Petri Nets. Bericht Nr. 185 (FBI-HH-B-185), Universität Hamburg, Fachbereich Informatik, Vogt-Kölln-Straße 30, D-22527 Hamburg, Germany, June 1996.

Inheritance of Dynamic Behavior
Development of a Groupware Editor

Twan Basten[1,*] and Wil M.P. van der Aalst[2]

[1] Dept. of Electrical Engineering, Eindhoven University of Technology,
The Netherlands
tbasten@ics.ele.tue.nl
[2] Dept. of Computing Science, Eindhoven University of Technology, The Netherlands
wsinwa@win.tue.nl

Abstract. One of the key issues of object-oriented modeling is *inheritance*. It allows for the definition of subclasses that inherit features of some superclass. Inheritance is well defined for static properties of classes such as attributes and methods. However, there is no general agreement on the meaning of inheritance when considering dynamic behavior of objects. This paper studies inheritance of dynamic behavior in a framework based on Petri nets. The notions of an object life cycle and inheritance between life cycles are defined. The inheritance relation is based on two fundamental concepts, namely *blocking* and *hiding* method calls. Several transformation rules are given to construct subclasses from a given superclass, thus allowing reuse of life-cycle specifications during a design. To show the validity of the approach, the results are applied to the development of a groupware editor.

Key words: object orientation – inheritance – object life cycle – dynamic behavior – Petri nets – computer supported cooperative work (CSCW)

1   Introduction

In software-engineering practice, the popularity of object-oriented modeling and design is increasing rapidly. Three methods are in common use: OMT [12], OOD [4], and UML [6]. One of the key issues in any object-oriented method is *inheritance*. The inheritance mechanism allows the user to specify a subclass that inherits features of some other class, its superclass. A subclass has the same features as the superclass, but in addition it may have some other features.

The concept of inheritance is well defined for *static* features of an object class, i.e., its methods and its attributes. However, a class also contains a definition of the dynamic behavior of objects. That is, it specifies the order in which the methods of an object may be executed. In this paper, such a specification is called the *life cycle* of an object. OMT, OOD, and UML use state-transition diagrams for specifying life cycles. Such a diagram shows the state space of an

---

* This work was done while the author was employed at the Department of Computing Science, Eindhoven University of Technology, The Netherlands.

G. Agha et al. (Eds.): Concurrent OOP and PN, LNCS 2001, pp. 391–405, 2001.

object and the method calls that cause a transition from one state to another. Looking at the definition of inheritance in, for example, OOD and its informal explanation, a subclass that inherits features of some other class extends the *static structure* as well as the *dynamic behavior* of its parent class. However, in the further treatment of inheritance, OOD only defines inheritance of static features. It does not specify the meaning of inheritance of dynamic behavior. It is implicitly assumed that the behavior of a subclass is an *extension* of the behavior of its superclass. OOD does not further elaborate on the precise meaning of "extension."

In this paper, we use Petri nets (See for example [11]) for specifying the dynamics of an object class. There are several reasons for using Petri nets. First of all, Petri nets provide a graphical description technique which is easy to understand and close to state-transition diagrams. Second, concurrency and synchronization are easy to model in terms of a Petri net. Third, many techniques and software tools are available for the analysis of Petri nets. Finally, Petri nets have been extended with data (color), time and hierarchy [8,9]. The extension with data allows for the modeling of attributes and methods. The extension with time allows for the quantification of the dynamic behavior of an object. The hierarchy concept can be used to structure the dynamics of an object class.

In the Petri-net framework, we formalize what it means for an object life cycle to extend another life cycle. The results presented here are based on two earlier papers. In [3], inheritance of dynamic behavior is studied in terms of a simple process algebra. We believe that the essence of inheritance of dynamic behavior consists of *blocking* and *hiding* method calls, two notions which are well investigated in process algebra. In [1], the results from [3] are translated to Petri nets. In the current paper, we do not repeat all the obtained results. We restrict ourselves to the basic definitions concerning life-cycle inheritance and an informal explanation of some important results, being four transformation rules that allow designers to construct subclasses from some given superclass in a straightforward way. The main contribution of this paper is that it shows the validity of our approach by means of a small case study. The concepts as defined in this paper are applied to the development of a groupware editor. The results show that life-cycle inheritance stimulates the reuse of life-cycle specifications. A detailed study of the notion of inheritance of dynamic behavior, further elaborating on the results of [1,3] and the current paper, has appeared in [2].

The remainder of this paper is organized as follows. Section 2 introduces the notions of an object life cycle and inheritance of life cycles. Section 3 presents several inheritance-preserving transformation rules on object life cycles. In Section 4, the results are applied to the development of a groupware editor. Finally, Section 5 ends with some concluding remarks.

2   Object Life Cycles and Inheritance

Place/Transition nets. In this paper, we use a specific class of Petri nets known as Place/Transition nets or simply P/T nets. Let A be some universe of action labels, which in the remainder can be thought of as method identifiers. Action labels in A are the so-called *observable* actions. To denote *silent* or *unobservable* behavior, usually corresponding to internal method invocations, a special label τ is introduced. Let A_τ denote the union of A and $\{\tau\}$.

Definition 1 (Labeled P/T net). *An A_τ-labeled Place/Transition net N is a tuple (P, T, F, ℓ), where*

i) *P is a finite set of* places;
ii) *T is a finite set of* transitions *such that P and T are disjoint;*
iii) *$F \subseteq (P \times T) \cup (T \times P)$ is a set of directed arcs, called the* flow relation;
iv) *$\ell : T \to A_\tau$ is a labeling function.*

Some place p is called an *input place* of a transition t if and only if there exists a directed arc from p to t. Place p is called an *output place* of t if and only if there exists a directed arc from t to p. We use ${}^\bullet t$ to denote the set of input places for a transition t. The notations $t^\bullet$, ${}^\bullet p$, and $p^\bullet$ have similar meanings.

The following definition introduces a property of P/T nets that is useful in the definition of object life cycles. The reflexive and transitive closure of a relation R is denoted R^*; the inverse of R is denoted R^{-1}.

Definition 2 (Connectedness). *A P/T net (P, T, F, ℓ) is connected, if and only if, for every two places or transitions $x, y \in P \cup T$, $(x, y) \in (F \cup F^{-1})^*$.*

Places of a P/T net may contain zero or more *tokens*. The *state* or *marking* of a net is the distribution of tokens over the places. A marking is represented by a finite multi-set, or bag, of places. The following notations are used for bags. For the explicit enumeration of a bag, a notation similar to the notation for sets is used, but using square brackets instead of curly brackets and using superscripts to denote the cardinality of the elements. To denote individual elements of a bag, the same symbol "$\in$" is used as for sets. The sum of two bags X and Y is denoted $X + Y$; the difference of X and Y is denoted $X - Y$.

Definition 3 (Marked P/T net). *A marked P/T net is a pair (N, s), where N is a labeled P/T net (P, T, F, ℓ) and where s is a bag over P denoting the state of the net.*

Marked P/T nets have a dynamic behavior which is defined as follows. A transition is *enabled* if and only if each of its input places contains at least one token. An enabled transition can *fire*. If a transition fires, then it *consumes* one token from each of its input places; it *produces* one token for each of its output places. The visible effect of a firing is the *label* of the transition.

Definition 4 (Firing rule). *Let (N, s) be a marked P/T net with N equal to (P, T, F, ℓ). For any enabled transition $t \in T$, the firing of t is denoted $(N, s) [\ell(t)\rangle (N, s - {}^\bullet t + t^\bullet)$.*

The firing rule allows us to define the notion of reachable states.

Definition 5 (Reachability). *Let (N, s) be a marked A_τ-labeled P/T net. State s' is reachable from s, denoted $(N, s) [*\rangle (N, s')$, if and only if s' equals s or if for some $n \in \mathbb{N}$ there exist $a_0, \ldots, a_n \in A_\tau$ and markings $s_1, \ldots, s_n$ such that $(N, s) [a_0\rangle (N, s_1) [a_1\rangle \ldots [a_n\rangle (N, s')$.*

Since we are interested in comparing object life cycles, which are specified by P/T nets, it is necessary to have an equivalence for P/T nets. Nets with the same external behavior, but with possibly different silent behavior must be considered equal. For this purpose, *branching bisimilarity* is a suitable equivalence [7]. In this paper, the details of the definition of branching bisimilarity are not important and, hence, omitted. The reader is referred to [1].

Object life cycles. Using P/T nets for specifying object life cycles allows us to specify a partial ordering of method calls. However, not every labeled P/T net specifies a life cycle. A life cycle is a net having exactly one *initial* or *input* place i. It also has a unique initial transition t_{cr} which corresponds to the creation of an object. A life cycle refers to a *single object*. It suffices to consider just one object because multiple objects of the same class interact via the execution of methods and not directly via the life cycle. Since we focus on one object at a time, initially, place i contains a single token.

An object life cycle must also have a unique *final* or *output* place o. An object terminates when, and only when, it reaches the marking consisting of a single token in o. In addition, if a marking has a token in o, it must be the only token in the marking. This means that upon termination of an object, all information about the object is removed. Furthermore, we assume that it is always possible to terminate. However, this does not mean that an object is *forced* to terminate.

A final requirement of a life cycle is that it is connected. Given the other requirements, it simply makes no sense to add unconnected parts to a life cycle.

The following definition formalizes the notion of an object life cycle.

Definition 6 (Object life cycle). *Let (N, s) be a marked A_τ-labeled P/T net, where N equals (P, T, F, ℓ). Let s' be any marking reachable from s. (N, s) is an object life cycle if and only if the following conditions are satisfied.*

i) Object creation: P contains a place i and T a transition t_{cr} such that $^\bullet i = \emptyset$, $i^\bullet = \{t_{cr}\}$, and $^\bullet t_{cr} = \{i\}$;

ii) Single-object requirement: $s = [i]$;

iii) Object termination: P contains a place o such that $o^\bullet = \emptyset$; furthermore, if $o \in s'$, then $s' = [o]$;

iv) Termination option: $(N, s') [*\rangle (N, [o])$;

v) Connectedness: N is connected.

Figure 1 shows the, very simple, life cycle of a class **person**. A person comes into existence when he or she is born. While alive, a person may celebrate his or her birthday. Eventually, a person dies.

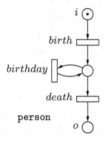

Fig. 1. A simple example of an object life cycle.

Life-cycle inheritance. At this point, it is possible to define inheritance of object life cycles. As we show in the remainder, two basic concepts play an important role in life-cycle inheritance. The first notion is the concept of *blocking* method calls. Let $(N_0, [i])$ and $(N_1, [i])$ be two object life cycles such that the set of methods of N_1 contains all methods of N_0. One possible, informal definition of inheritance is as follows.

> If it is not possible to distinguish the observable behavior of $(N_0, [i])$ and $(N_1, [i])$ when only methods of N_1 that are also present in N_0 are executed, then $(N_1, [i])$ is a subclass of $(N_0, [i])$.

In other words, $(N_1, [i])$ is a subclass of $(N_0, [i])$ if the observable behaviors of $(N_1, [i])$ and $(N_0, [i])$ are equivalent when methods of N_1 which are not present in N_0 are blocked.

The second basic concept is the concept of *hiding* method calls. Hiding a method call means that it is no longer observable. The notion of hiding inspires another definition of inheritance.

> If it is not possible to distinguish the external behavior of $(N_0, [i])$ and $(N_1, [i])$ when arbitrary methods of N_1 are executed, but when only the effects of methods that are also present in N_0 are considered, then $(N_1, [i])$ is a subclass of $(N_0, [i])$.

This means that $(N_1, [i])$ is a subclass of $(N_0, [i])$ if the observable behaviors of $(N_1, [i])$ and $(N_0, [i])$ are equivalent when hiding methods of N_1 which are not present in N_0.

The subtle difference between the two forms of inheritance is that in the second definition methods new in N_1 are executed without taking into account their effect, whereas in the first definition they are not executed at all. Below, we give an example to illustrate this difference.

To formalize inheritance of life cycles, the notions of blocking and hiding method calls must be translated to P/T nets. The technical terms for blocking and hiding actions are borrowed from process algebra, where they are called *encapsulation* and *abstraction*, respectively (see also [3]). Note that, in this paper, these two terms have a more specific meaning than usual in object-oriented design.

Definition 7 (Encapsulation and abstraction). *Assume that (N, s) is a marked A_τ-labeled P/T net with $N = (P, T, F, \ell)$.*

i) *For any $H \subseteq A$, the encapsulation operator ∂_H removes all transitions with a label in H from N. Formally, $\partial_H(N, s) = (N', s)$ where $N' = (P, T', F', \ell')$ such that $T' = \{t \in T \mid \ell(t) \notin H\}$, $F' = F \cap ((P \times T') \cup (T' \times P))$, and $\ell' = \ell \cap (T' \times A_\tau)$.*

ii) *For any $I \subseteq A$, the abstraction operator τ_I renames all transition labels in I to τ. That is, $\tau_I(N, s) = (N', s)$ where $N' = (P, T, F, \ell')$ such that for any $t \in T$, $\ell(t) \in I$ implies $\ell'(t) = \tau$ and $\ell(t) \notin I$ implies $\ell'(t) = \ell(t)$.*

Note that blocking a method is achieved by removing the corresponding transitions from the net structure.

Encapsulation and abstraction can be used to define four different inheritance relations between object life cycles. Of course, it is possible to formalize the above two informal definitions of inheritance. It is also possible to combine encapsulation and abstraction in the definition of inheritance. In this way, two more meaningful inheritance relations can be defined. For a detailed study of the four inheritance relations, the reader is referred to [1,2,3]. In this paper, we focus on the most general inheritance relation, called *life-cycle inheritance*. An object life cycle is a subclass of another object life cycle, its superclass, if and only if hiding some methods and blocking some other methods of the subclass results in an object life cycle equivalent to its superclass.

Definition 8 (Life-cycle inheritance). *Let $(N_0, [i])$ and $(N_1, [i])$ be two object life cycles; let A be the set of observable method identifiers. Life cycle $(N_1, [i])$ is a subclass of life cycle $(N_0, [i])$ under* life-cycle inheritance *if and only if there exist disjoint $H \subseteq A$ and $I \subseteq A$ such that $\tau_I \circ \partial_H(N_1, [i])$ is branching bisimilar to $(N_0, [i])$.*

As mentioned, in [1,2,3], life-cycle inheritance is investigated in detail. Among other things, it is shown that it has several desirable properties such as transitivity and reflexivity. Figure 2 gives four object life cycles of persons illustrating the definition of life-cycle inheritance.

Class person1 describes a person that, at some point during his or her life, decides to marry. A person that can decide whether to marry or not, as described by class person2, should be a subclass of person1. It is easy to see that *blocking* method *stay_single* means that the behavior of the two persons is identical. Hence, class person2 is a subclass of person1. Note that *hiding* the method *stay_single* does not yield the desired result. When hiding *stay_single* instead of blocking it, a possible behavior of a person of class person2 is *birth* followed by *death*. A person of class person1 can never exhibit such a behavior.

Class person3 shows a person that divorces after marriage. Hiding the new method *divorce* easily shows that person3 is a subclass of person1. In this case, *blocking* the new method in the subclass does not yield the desired result. Blocking *divorce* means that a person of class person3 will never die, which is clearly not true for persons of class person1.

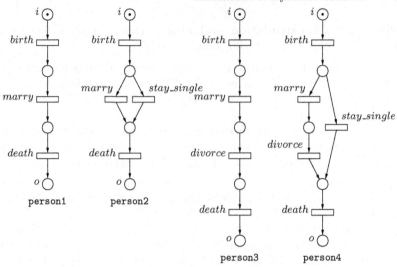

Fig. 2. An illustration of life-cycle inheritance.

Class `person4` extends class `person1` with both methods *stay_single* and *divorce*. In this example, the *combination* of encapsulation and abstraction is needed to show that `person4` is a subclass of `person1`.

3   Inheritance-Preserving Transformation Rules

It is possible to show that object life cycles can only have finitely many states. In Petri-net terminology, they are bounded [1]. This implies that checking branching bisimilarity between life cycles is decidable. However, in practical applications, it can be a complex and tedious task. Therefore, this section presents several inheritance-preserving transformation rules. These rules can be used to design subclasses of some given class, thus, supporting reuse of life-cycle specifications. Moreover, they show the essence of life-cycle inheritance. In this paper, we only give an informal explanation of the transformation rules. Formal definitions can be found in [1]. In [2], inheritance-preserving transformation rules are studied in more detail. In particular, attention is paid to developing a set of transformation rules that is effective in practical design situations.

Let N_q be a P/T net such that $(N_q, [i])$ is an object life cycle. Figure 3 shows four inheritance-preserving transformation rules transforming N_q into a subclass. For the sake of simplicity, we require that the result of each transformation is again an object life cycle.

The first transformation rule, for reasons explained in [3] called *PT*, can be formulated as follows. The *alphabet* of a P/T net denotes the set of all transition labels occurring in the net which are not equal to τ.

If N_p is a P/T net such that *i*) no transitions are shared between N_p and N_q and *ii*) all transitions of N_p with input places in N_q have a label which does

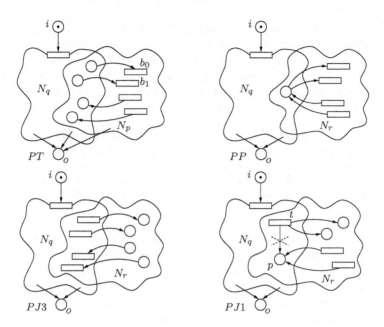

Fig. 3. Inheritance-preserving transformation rules.

not appear in the alphabet of N_q and is not equal to τ, then the union of N_p and N_q is a subclass of N_q.

The above transformation rule means that it is allowed to add a *choice*, or a new branch of behavior, to an existing object life cycle. In the example of Figure 3, it is crucial that methods b_0 and b_1 do not occur in the alphabet of N_q. It is not difficult to see that blocking b_0 and b_1 leads to a net whose behavior is equivalent to the behavior of N_q. Methods b_0 and b_1 act as so-called *guards*, separating the subclass extension from the original life cycle. In the example of Figure 2, the extension of class **person1** with method *stay_single* resulting in class **person2** is captured by transformation rule *PT*.

The second rule shown in Figure 3, called *PP*, is a special case of *PT* showing that *PT* captures a simple form of *recursion*. Rule *PP* states that it is possible to extend an existing class with an iteration. Under certain assumptions about the net N_r, it preserves a more restricted form of inheritance than *PT* does (see [1,2] for details).

The third transformation rule, called *PJ3* shows that it is possible to add *parallel behavior* to an object life cycle.

If N_r is a P/T net such that *i*) no places are shared between N_q and N_r, *ii*) all transitions of N_r have a label which does not appear in the alphabet of N_q and *iii*) transitions in N_q with input places in N_r obey the free-choice property, then the union of N_r and N_q is a subclass of N_q.

The second requirement means that only new methods are used in the extension of the original life cycle. This means that hiding the new methods yields a behavior equivalent to that of the original life cycle. The third requirement is needed for the correctness of the transformation rule. In most practical examples, such as the examples in the next section, it is satisfied. An introduction to the free-choice property and the class of free-choice Petri nets can be found in [5].

The fourth and final transformation rule, $PJ1$, shows that it is possible to *insert* behavior in between two parts of the original life cycle.

> If N_r is a net such that *i*) N_q and N_r share exactly one place p and one transition t with (t, p) in the flow relation of N_q, *ii*) all transitions of N_r other than t have a label which does not occur in the alphabet of N_q, and *iii*) N_r satisfies certain requirements with respect to liveness, boundedness, and the free-choice property, then the union of N_q and N_r *without* the arc (t, p) is a subclass of N_q.

This transformation rule says that it is allowed to replace an arc in an object life-cycle by an entire P/T net. Again, the requirement that all new methods must have a fresh identifier means that hiding these methods yields a behavior equivalent to the behavior of the original life cycle. The third requirement above is necessary to guarantee that the extension behaves properly. That is, if transition t fires and thus activates the extension N_r, it must be possible that eventually a token arrives in p without leaving any tokens in the other places of N_r. This means that the behavior of the original net N_q is not affected by the addition of N_r. By applying rule $PJ1$, it is possible to show that class **person3** in Figure 2 is a subclass of class **person1** in the same figure.

The four transformation rules presented in this section are not the only possible ones. However, they are characteristic for life-cycle inheritance. It is interesting to note that they capture important design constructs such as choices (PT), parallelism $(PJ3)$, sequential composition $(PJ1)$, and iteration (PP). This observation in particular has convinced us that the concepts presented in this paper touch upon the fundamentals of inheritance of dynamic behavior.

4   The Groupware Editor

Introduction. In this section, the concepts developed in the previous sections are applied to the development of a groupware editor. We do not give full specifications of the classes involved, but focus on the object life cycles. This means that we do not specify data types, class attributes, or method implementations. If necessary, an informal explanation is given.

The requirements for the groupware editor are as follows. The editor is meant to edit some kind of diagrams. Multiple users, possibly situated at different workstations, may be editing a single diagram in a joint editing session. Users may choose to either view or edit a diagram. They may join or leave a session at will. It must be clear to all users who is currently editing some given diagram. The diagrams under consideration may be complex, possibly consisting of multiple components. Components may introduce hierarchy in a diagram. It is possible to open or close a component revealing or hiding its details. Not just any user may view or edit any component in a diagram. Users must have permission to do so. Permissions are not fixed; to get permission, a user can simply select a component and then try the desired command. If there are no conflicting permissions, the command succeeds; otherwise, it fails and the user is notified. Users may explicitly surrender permissions. Most permissions are automatically reset if the user leaves the editing session; a few permissions may persist between sessions.

In the next paragraphs, life-cycle inheritance is used to design a groupware editor satisfying the abovementioned requirements. The development is split into three steps. First, a groupware *viewer* is designed. The design is fairly simple and the resulting system allows multiple users to view existing diagrams. Second, the viewer is transformed into a multi-user editor by adding editing functions to classes in the viewer design. Third, the multi-user editor is specialized to a groupware editor by adding permissions. The example shows how life-cycle inheritance can be used to structure an object-oriented design process. It also shows how object life cycles can be reused in a design.

The multi-user viewer. In the design of the viewer, four classes can be distinguished: `user`, `diagram`, `user_session`, and `edit_session`. The first two classes have straightforward interpretations. The third and fourth class are intended to structure viewing sessions. An `edit_session` object keeps track of all users viewing a particular diagram; an object of class `user_session` maintains all data involved in a particular session of a particular user. Figure 4 shows the life cycles for objects of the above four classes.

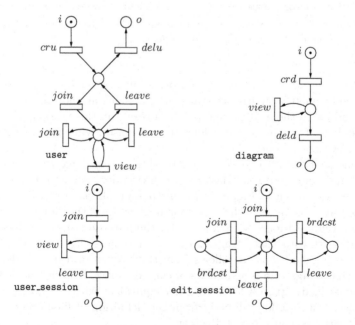

Fig. 4. A multi-user viewer.

Users can be created (*cru*) or deleted (*delu*). This means that they are added to or removed from the list of users of the viewer. Once a user exists, (s)he may join and leave editing sessions. If the user participates in at least one edit session, (s)he may issue *view* commands in order to view diagrams. It is not difficult to see that user satisfies the requirements of an object life cycle (Definition 6).

Class diagram has a straightforward life cycle. Diagrams can be created (*crd*), viewed (*view*), and deleted (*deld*). A user interacts with a diagram through a user_session object. Upon joining an editing session, a new object of class user_session is created. The only possible command a user can execute is the *view* command. If the user leaves the editing session, the user_session object is terminated.

Objects of class edit_session keep track of all users involved in a single session. The first user that "joins" an editing session for some diagram, actually creates a new edit_session object. If other users join a running session, this does not lead to the creation of a new object. The information is simply stored in the existing object. To fulfill the requirement that all users must know who is participating in the session, we assume that the implementation of *join* is such that the new user gets a list of users already present. Furthermore, the information about a new user is broadcast (*brdcst*) to all other users participating in the session. When a user leaves, this information is broadcast to all remaining users. The last user leaving the session terminates the edit_session object.

In a complete implementation of a system, it is clear from the code which methods interact with each other. Since we do not give method implementations, we make a few assumptions. First, user methods invoke methods with the same label in user_session, which, in turn, invoke methods of the same name in edit_session and diagram. Second, methods which do not have counterparts in one of the other classes are assumed to interact with (objects in) the environment. Examples are *cru*, *delu*, *crd*, *deld*, and *brdcst*.

The multi-user editor. This paragraph describes a specialization of the multi-user viewer, namely a multi-user editor. Permissions are not yet incorporated. They are added in the next paragraph. As a result, in this version of the editor, it is possible that a component is deleted by one user, while it is being changed or viewed by another one.

Editing facilities have been added to three classes, namely classes user, diagram, and user_session. Class edit_session does not need to change. The new classes are shown in Figure 5. In this paragraph, we argue that the three new classes are subclasses of the corresponding classes of the viewer. Some places in the life cycles of Figure 5 are labeled for the purpose of future reference.

Let us start with class diagram_e. It is now possible to modify diagrams by means of method *mod*. It follows from transformation rule PP depicted in Figure 3 that diagram_e is a subclass of class diagram. It is clear that blocking method *mod* in diagram_e leads to a net equivalent with diagram. So in this simple case, it also follows directly from the definition of life-cycle inheritance that diagram_e is a subclass of diagram.

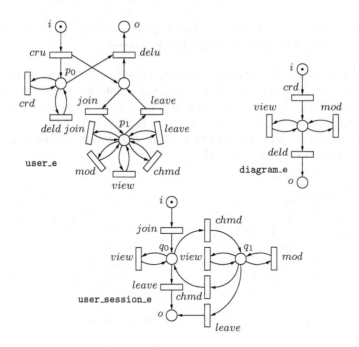

Fig. 5. A multi-user editor.

Class **user** has been extended to class **user_e**. Since we are developing an editor, users of the system are now responsible for creating and deleting diagrams. Creation and deletion of diagrams can be done independently of editing (other) diagrams. This means that we added parallel behavior to the life-cycle of **user** objects. It follows from rule $PJ3$ that the addition of place p_0 and methods crd and $deld$ yields a subclass of **user**. A second addition is that users can now modify diagrams. For this purpose, method mod has been added. Users have to choose whether they want to modify a diagram or whether they are satisfied with just the option to view it. Method $chmd$ (change mode) can be used to toggle between viewing and editing mode. The implementation of mod and $chmd$ can be such that only a subset of the users is allowed to enter editing mode. It simply follows from applying transformation rule PP that the addition of methods mod and $chmd$ to class **user** leads to a subclass. Therefore, the subsequent application of rule $PJ3$ and rule PP yields that **user_e** is a subclass of **user**.

For each editing session a user joins, a **user_session_e** object is created. This object keeps track of the mode in which the user is for this particular diagram. Initially, the user is in viewing mode. By invoking method $chmd$ the user can change to editing mode. This means that we have extended the life cycle of **user_session** objects with a choice. Transformation rule PT can be applied to show that **user_session_e** is a subclass of **user_session**. Method $chmd$ acts as the guard.

Summarizing, in this paragraph, we have applied three of the four transformation rules given in Section 3 to extend the viewer of the previous paragraph

to an editor. What is important is that we have *reused* the specifications for the
object life cycles of the viewer in the design process.

The groupware editor. In this paragraph, we extend the multi-user editor of
the previous paragraph with permissions. Permissions are needed to prevent all
kind of anomalies. For example, they guarantee that it is impossible that one
user deletes a component when another user is viewing it. Four new classes are
introduced, namely user_p, diagram_p, user_session_p, and edit_session_p.
Figure 6 shows two of these four classes.

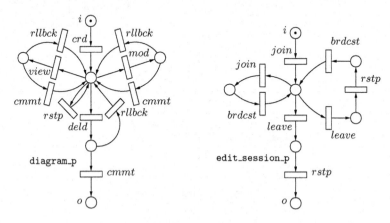

Fig. 6. A groupware editor.

The current set of permissions for editing a diagram is maintained in the
corresponding diagram_p object. This is the only feasible option since some permissions may persist between editing sessions. As stated in the requirements, a
user can get permission for some editing command by simply selecting a component and executing the command. Therefore, after every editing action *view*,
mod, or *deld*, diagram_p executes either a rollback (*rllbck*) or a commit (*cmmt*),
depending on whether or not the user has permission for the editing action. Another change when compared to class diagram_e is that the method *rstp* has
been added which can be used to reset permissions. The three other classes
must invoke *rstp* whenever appropriate.

Classes user_p and user_session_p are simple extensions of user_e and
user_session_e and, therefore, not shown in Figure 6. Since users may surrender
permissions at any time, class user_p is obtained from user_e of Figure 5 by
adding a loop consisting of a single transition labeled *rstp* to place p_1. Since users
interact with diagrams through user sessions, the life cycle of user_session_p
is constructed from the life cycle of class user_session_e by adding the same
loop as above to places q_0 and q_1.

Class edit_session_p is constructed from class edit_session of Figure 4.
As mentioned, permissions are reset when a user leaves a session. This means
that edit_session is extended with calls of method *rstp* after each invocation
of the *leave* method. The resulting class edit_session_p is shown in Figure 6.

It remains to be shown that the new classes are subclasses of the corresponding classes in the earlier designs. The addition of method *rstp* to classes user_e and user_session_e is captured by transformation rule *PP*. Hence, user_p and user_session_p are subclasses of classes user_e and user_session_e, respectively.

The addition of *rstp* to diagram_e is also captured by rule *PP*. Applying rule *PJ*1 three times shows that the addition of *cmmt* to diagram_e preserves life-cycle inheritance. To show that the addition of *rllbck* preserves inheritance, consider the intermediate result obtained after the previous additions. The desired result now follows easily from transformation rule *PT*. Hence, diagram_p is a subclass of diagram_e and, therefore, also of diagram.

To show that edit_session_p is a subclass of edit_session, we have to apply rule *PJ*1 twice simultaneously.

5   Concluding Remarks

In this paper, we have presented a definition for inheritance of dynamic behavior in the framework of Petri nets. In our opinion, the notions of *blocking* and *hiding* method calls are fundamental to inheritance of dynamic behavior. We have informally presented four inheritance-preserving transformation rules that allow for the straightforward construction of subclasses from a given superclass. To validate our approach, we have applied the transformation rules to the development of a groupware editor. However, we only considered life cycles. We did not give full class definitions nor did we give implementations of all the methods. A future challenge is to incorporate the results either in a full fledged object-oriented formalism based on Petri nets, such as OPN [10], or to incorporate them into a framework as OMT, OOD, or UML. In the latter case, one could choose to translate the notion of life-cycle inheritance to state-transition diagrams or one could choose to replace state-transition diagrams by Petri nets. An advantage of incorporating life-cycle inheritance into a Petri-net-based formalism as OPN is that one obtains an integrated framework with a sound theoretical basis. A disadvantage is that object-oriented languages based on Petri nets are not yet in common use. An advantage of incorporating the results in a framework as OMT, OOD, or UML is that it will be easier to get acceptance of the notion of life-cycle inheritance in practice, particularly when it is translated to state-transition diagrams.

References

1. W.M.P. van der Aalst and T. Basten. Life-Cycle Inheritance: A Petri-Net-Based Approach. In P. Azéma and G. Balbo, editors, *Application and Theory of Petri Nets 1997, 18th. International Conference, ICATPN'97, Proceedings*, volume 1248 of *Lecture Notes in Computer Science*, pages 62–81, Toulouse, France, June 1997. Springer, Berlin, Germany, 1997.

2. T. Basten. *In Terms of Nets: System Design with Petri Nets and Process Algebra.* PhD thesis, Eindhoven University of Technology, Department of Mathematics and Computing Science, Eindhoven, The Netherlands, December 1998.

3. T. Basten and W.M.P. van der Aalst. A Process-Algebraic Approach to Life-Cycle Inheritance: Inheritance = Encapsulation + Abstraction. Computing Science Report 96/05, Eindhoven University of Technology, Department of Mathematics and Computing Science, Eindhoven, The Netherlands, March 1996.

4. G. Booch. *Object-Oriented Analysis and Design: With Applications.* Benjamin/Cummings, Redwood City, CA, USA, 1994.

5. J. Desel and J. Esparza. *Free Choice Petri Nets*, volume 40 of *Cambridge Tracts in Theoretical Computer Science.* Cambridge University Press, Cambridge, UK, 1995.

6. M. Fowler and K. Scott. *UML Distilled: Applying the Standard Object Modeling Language.* Addison-Wesley, Reading, Massachusetts, USA, 1997.

7. R.J. van Glabbeek and W.P. Weijland. Branching Time and Abstraction in Bisimulation Semantics (extended abstract). In G.X. Ritter, editor, *Information Processing 89: Proceedings of the IFIP 11th. World Computer Congress*, pages 613–618, San Fransisco, California, USA, August/September 1989. Elsevier Science Publishers B.V., North-Holland, 1989.

8. K.M. van Hee. *Information Systems Engineering: A Formal Approach.* Cambridge University Press, Cambridge, UK, 1994.

9. K. Jensen. *Coloured Petri Nets. Basic Concepts, Analysis Methods and Practical Use,* volume 1, *Basic Concepts.* EATCS monographs on Theoretical Computer Science. Springer, Berlin, Germany, 1992.

10. C. Lakos. From Coloured Petri Nets to Object Petri Nets. In G. De Michelis and M. Diaz, editors, *Application and Theory of Petri Nets 1995, 16th. International Conference, Proceedings*, volume 935 of *Lecture Notes in Computer Science*, pages 278–297, Torino, Italy, June 1995. Springer, Berlin, Germany, 1995.

11. W. Reisig. *Petri Nets: An Introduction*, volume 4 of *EATCS monographs on Theoretical Computer Science.* Springer, Berlin, Germany, 1985.

12. J. Rumbaugh, M. Blaha, W. Premerlani, F. Eddy, and W. Lorensen. *Object-Oriented Modeling and Design.* Prentice-Hall, Englewood Cliffs, NJ, USA, 1991.

Object Coloured Petri Nets –
A Formal Technique
for Object Oriented Modelling

Christoph Maier[1] and Daniel Moldt[2]

[1] FAST e.V., Arabellastr. 17, D-81925 München, `maier@fast.de`
[2] University of Hamburg, Computer Science Department, Vogt-Kölln-Str. 30,
D-22527 Hamburg, `moldt@informatik.uni-hamburg.de`

Abstract. Object Coloured Petri Nets (OCP-Nets) are an extension of Coloured Petri Nets (CPN). OCP-Nets are well suited to model the dynamic aspects of a system. They supersede most techniques currently used in object oriented modelling such as Interaction Diagrams. This will be shown in an example formalising an informal use case description. With their formal semantics, graphical representation, means to model concurrency, and executability OCP-Nets lead to an improved Object Oriented Modelling approach.

1 Introduction

Object Oriented Modelling can be divided into two parts: modelling the static structure and modelling the dynamic aspects of a system. Class diagrams are the main technique to describe the structure of a system, i.e. the classes and their relationships. The dynamic aspects are mode-led with techniques like State Charts, Message Trace Diagrams or Object Interaction Diagrams (see for example [13,4,15,18]).

But while class diagrams are well understood, the techniques for dynamic aspects have two disadvantages. First, they lack a formal semantics like that of Petri nets. This often leads to ambiguous models and thus problems with the transition from analysis to design and implementation. The second major drawback of these techniques is their lack of appropriate means to model concurrency. Consider for instance the Concurrent Object Interaction Diagram technique, proposed by Rumbaugh in [14], which falls far short from the expressive power of Petri nets.

We propose Object Coloured Petri Nets (OCP-Nets) to supersede the aforementioned techniques. OCP-Nets are an object oriented Petri net formalism, based on Coloured Petri Nets (CPN) [9] enhanced by the fusion place concept of Hierarchical Coloured Petri Nets (HCPN).[1] OCP-Nets have a formal semantics

---

[1] In the rest of the paper, the signature of CPN as defined by Jensen in [9] is used. If not otherwise stated, the meaning is equal or similar to that in [9].

G. Agha et al. (Eds.): Concurrent OOP and PN, LNCS 2001, pp. 406–427, 2001.

and are capable of adequately modelling concurrency. Building on their graphical representation and operational interpretation, OCP-Nets lend themselves to a prototype based approach of software development.

This paper is organised in two parts with section 2 to 6 describing OCP-Nets and section 7 describing the object oriented modelling process with OCP-Nets. The first part starts with an informal introduction into OCP-Nets. After this, the formal definitions of the static structure are given. Then a few formal aspects of the dynamic structure and behaviour of OCP-Nets are described. The full definitions of these aspects can be found in [11]. The next section addresses inheritance and encapsulation in OCP-Nets. After this, the conversion of an OCP-Net into a static CP-Net is outlined. The second part describes the way, OCP-Nets can be used to give a formal specification of a use case and its objects and how a combined simulation of them can be obtained. Finally, some conclusions are given.

2   Informal Introduction into OCP-Nets

OCP-Nets adapt the concepts of object oriented programming to Petri Nets. Thus, an OCP-Net is made up of a set of *class nets*. Each class net can be used to offer one or more services (methods). Like classes in an object oriented programming language, class nets describe the static structure of an OCP-Net. To execute an OCP-Net, instances of the class nets are created. These instances are called *object nets*.

Object nets communicate through the exchange of tokens. This can be done asynchronously via *communication fusion places* and synchronously through a modified version of synchronous channels, proposed in [5]. Thus this concept is quite similar to the client/server concept of Cooperative Nets [16] when viewing the calling object as a client and the called object as a server. The main difference to Cooperative Nets is the use of CPN as the basic formalism and the explicit support for synchronous communication between transitions.

Now, an informal introduction to OCP-Nets is given. Fig. 1 shows the function *push* of a stack, modelled with a Coloured Petri Net (CP-Net). This stack can store elements of type *INT*. An element on place *elem* is pushed via the transition *push* to the stack which is represented as a list. The end of the operation is signaled with a token of type *UNIT* on the place *pushOK*. *UNIT* denotes the type of the anonymous token, *tok* denotes the anonymous token itself.[2]

This example is now transformed to an OCPN class net which offers the service push. Class nets communicate using so called *communication transitions*. There are four kinds of communication transitions. The IN- and OUT-transitions are used to offer a service and the INV- and REC-transitions are used to invoke a service. The tokens are sent from an INV-transition to an IN-transition and from an OUT-transition to a REC-transition. Fig. 2 shows a class net offering the service *push*. The element x is received via the IN-transition *push(x)*. After

---

[2] The correct multiset notation 1'tok is omitted for the sake of simplicity.

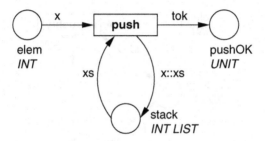

Fig. 1. Function stack modelled as a CP-Net

completing the service, the token *tok* is sent to the calling class net via the OUT-transition *push()*. The parameters are listed inside the brackets. The shortcut *push()* is used instead of *push(tok)*.

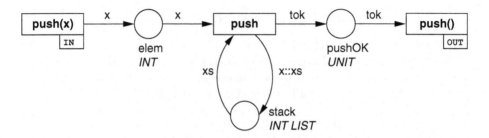

Fig. 2. Function push modelled as a service of a class net

The next example illustrates the two other types of communication transitions: INV- and REC-transitions. Fig. 3 shows a piece of a class net that calls the service *push*. An INV-transition (invoke) calls a service in a class net by sending it one or more tokens as parameters. In contrast, the result of a call is received via a REC-transition (receive). Since there may be more than one instance of the class net stack, an unique identifier is assigned to each object net. The type of the identifiers is *OID*. The class net shown stores the identifier of its stack in the place *myStack* and uses this identifier to call the service with the INV-transition *id.push(y)*. After the call, the net may perform some actions, depicted by a dotted line, until the service is completed.

3   Static Structure

We will now present the formal definition of the static structure of an OCP-Net. An OCP-Net is a set of class nets together with global fusion places and channel definitions for communication. First, the definition of a class net is given and after this the definition of an OCP-Net.

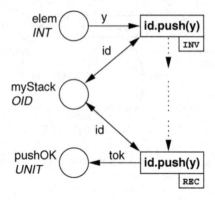

Fig. 3. Class net, calling the service push

Definition 1. *Class net*

A class net is a tuple $CN = (CPN, TN, TT, IF)$.

(i) CPN is a non hierarchical CP-Net $(\Sigma, P, T, A, N, C, G, E, I)$
  (a) Σ is the set of all types with $OID, MID, CID \in \Sigma$. OID denotes the set of all unique object net identifiers, MID the set of service identifiers and CID the set of all unique class net identifiers.
  (b) $T = T_{norm} \cup T_{in} \cup T_{out} \cup T_{inv} \cup T_{rec}$, with $T_{in} \cap T_{out} = \emptyset$
  (c) $P = P_{norm} \cup P_{in} \cup P_{out} \cup P_{inv} \cup P_{rec}$,
    with the sets $P_{norm}, P_{in}, P_{out}, P_{inv}$ and P_{rec} being pairwise disjoint.
    1. $\forall p \in P_{in} : in(p) = \emptyset \wedge out(p) = \{t\} \wedge t \in T_{in}$
    2. $\forall p \in P_{out} : in(p) = \{t\} \wedge out(p) = \emptyset \wedge t \in T_{out}$
    3. $\forall p \in P_{inv} : in(p) = \{t\} \wedge out(p) = \emptyset \wedge t \in T_{inv})$
    4. $\forall p \in P_{rec} : in(p) = \emptyset \ \wedge out(p) = \{t\} \wedge t \in T_{rec})$
  (d) N, C, G and E are defined as in [9].
  (e) I is defined as in [9], with:
    $\forall p \in P \backslash P_{norm} : I(p) = \emptyset$.
(ii) TN is a service naming function. $TN : T_{in} \cup T_{out} \rightarrow MID$, $MID \subseteq \Sigma$.
    $\forall t_i \in T_{in} \cup T_{out} : TN(t_i) \neq \epsilon$.
(iii) TT is a type function. $TT : T_{in} \cup T_{out} \rightarrow \Sigma^*$, with:
    $\forall t_i \in T_{in} \cup T_{out} : TT(t_i) \neq \emptyset$.
(iv) IF is a finite set of instance fusion sets with $IF \subseteq 2^{P_{norm}}$.
    $\forall fi \in IF : \forall p_1, p_2 \in fi :$
    $[C(p_1) = C(p_2) \wedge I(p_1)\langle\rangle = I(p_2)\langle\rangle]$

(i) A class net is a CP-net with the following restrictions:
  (a) There are three special sets of identifiers included in the type set Σ.
  (b) Normal transitions are differentiated from communication transitions. The sets of communication transitions of type IN and OUT must be disjoint.

(c) Normal places are differentiated from communication places and must all be pairwise disjoint. In addition, a communication place of type IN must not have an incoming arc and must be connected with exactly one transition of type IN. Similar restrictions are given for the other communication places.

(d) The node, colour, guard and expression functions are defined as for CP-Nets.

(e) The initialisation function is defined as for CP-Nets. Communication places must not have an initial marking.

(ii) TN assigns a service name to every communication transition of type IN and OUT. The set of service names is denoted by MID.

(iii) TT assigns a type to every communication transition of type IN and OUT. This is the type of the parameters of an IN transition and of the result of an OUT transition.

(iv) The instance fusion sets correspond to attributes of a class.

An OCP-Net $OCPN$ is the union of a set of class nets $CNS = \{CN_1, CN_2, .., CN_n\}$. The following conventions are used: P_{OCPN} denotes the set of all places, T_{OCPN} the set of all transitions and A_{OCPN} the set of all arcs of an OCP-Net:

$$P_{OCPN} = \bigcup_{CN \in CNS} P_{CN}$$

$$T_{OCPN} = \bigcup_{CN \in CNS} T_{CN}$$

$$A_{OCPN} = \bigcup_{CN \in CNS} A_{CN}$$

$$(1)$$

The set of global fusion places used for asynchronous communication is called P_{com}. It is defined as follows for an OCP-Net consisting of n class nets:

$$P_{com} = (\bigcup_{i=1}^{n} P_{in_i}) \cup (\bigcup_{i=1}^{n} P_{out_i}) \cup (\bigcup_{i=1}^{n} P_{inv_i}) \cup (\bigcup_{i=1}^{n} P_{rec_i})$$

The set of all communication transitions is named T_{com} and defined as follows:

$$T_{com} = (\bigcup_{i=1}^{n} T_{in_i}) \cup (\bigcup_{i=1}^{n} T_{out_i}) \cup (\bigcup_{i=1}^{n} T_{inv_i}) \cup (\bigcup_{i=1}^{n} T_{rec_i})$$

Definition 2. *OCP-Net*

An OCP-Net is a tuple $OCPN = (CNS, CS, CF, CLN, root)$ with

(i) CNS being a set of class nets CN_i, $1 \leq i \leq n$. $\forall CN_i, CN_j \in CNS$:
$CN_i \neq CN_j \Rightarrow (P_{CN_i} \cup T_{CN_i} \cup A_{CN_i}) \cap (P_{CN_j} \cup T_{CN_j} \cup A_{CN_j}) = \emptyset$.

continued

(ii) CS is a channel specification $CS = (CH, CT, \Delta, CE)$

    (a) CH is a finite set of channel identifiers with:
$(P_{OCPN} \cup T_{OCPN} \cup A_{OCPN}) \cap CH = \emptyset$

    (b) CT is a channel type function. $CT : CH \rightarrow \Sigma$

    (c) Δ denotes the kind of channel. $\Delta = \{\texttt{IN}, \texttt{OUT}, \texttt{INV}, \texttt{REC}\}$.

    (d) CE is a channel expression function. It is defined from T_{com} into finite sets of communication expression. A communication expression has the form: $(expr, (send, rec), \#, ch)$, with $expr$ being an expression, $send \in OID, rec \in OID, \# \in \Delta$ and $ch \in CH$. The channel expression function CE must obey the following restrictions:

        1. $\forall CE(t)$ with$\# = \texttt{IN} : [t \in T_{in} \wedge rec = \texttt{self}]$

        2. $\forall CE(t)$ with $\# = \texttt{OUT} : [t \in T_{out} \wedge send = \texttt{self}]$

        3. $\forall CE(t)$ with $\# = \texttt{INV} : [t \in T_{inv} \wedge send = \texttt{self}]$

        4. $\forall CE(t)$ with $\# = \texttt{REC} : [t \in T_{rec} \wedge rec = \texttt{self}]$

    (e) $\forall t \in T_{com}$ with $CE(t) \neq \emptyset$, $\forall (expr, (send, rec), \#, ch) \in CE(t) :$
$[Type(expr) \subseteq CT(ch) \wedge Type(Var(expr)) \subseteq \Sigma]$.
If $t \in T_{in} \cup T_{out} : TT(t) \subseteq CT(ch) \wedge ch = TN(t)$.

(iii) CF is a finite set of global fusion places for asynchronous communication with $CF \subseteq 2^{P_{com}}$. $\forall fc \in CF : \forall p_1, p_2 \in fc : [C(p_1) = C(p_2)]$, $\forall p \in P_{com} : \exists fc \in CF, p \in fc$.

(iv) CLN is a class net naming function, assigning to every class net CN_i an unique identifier from set $CID \in \Sigma$.
$CLN : CNS \rightarrow CID$.

(v) $root$ is a start function denoting the class nets that have to be instantiated at the beginning of a simulation.
$root : CNS \rightarrow BOOL$

(i) The elements of the class nets must be pairwise disjoint.

(ii) The channel specification is defined as follows:

    (a) The set of channel identifiers must be disjoint with the place, transition, and arc sets.

    (b) CT assigns a type to every channel.

    (c) The kind of channel denotes its purpose. This set is predefined and could therefore be removed from the OCP-Net definition. However, to give a definition, consistent to the one given by [5], the set has been included in the channel specification.

    (d) The channel expression of [5] has been extended with the (send, rec) tuple. A communication transition may have more than one channel expression.

    (e) The type of $expr$ must be a subtype of the channel type. The type of the variables used must be included in Σ. For communication transitions of type IN and OUT, the type assigned to the transitions must be a subtype of the channel type and the channel name must be the same as the one assigned to the communication transitions.

(iii) The global fusion places are used for asynchronous communication. Only communication places may be included in CF. All communication places are a member of a global fusion set.

(iv) CLN assigns every class net a unique name that is used by the function *new* to instantiate an object net.

(v) *root* denotes the class nets from which an object net is instantiated at the start of an execution.

The channel expressions will be described in detail together with an example in section 4.2. Finally, the functions used in [5] are changed according to the slightly different channel definition. For $t \in T$, $\# \in \Delta$ and $ch \in CH$, $Expr(t, \#, ch)$ is used to denote the set of all *(expr, (s, r))* tuple connecting t and ch in direction $\#$.

$$Expr(t, \#, ch) = \{(expr, (s, r)) \mid (expr, (s, r), \#, ch) \in CE(t)\} \tag{2}$$

The expression function E is extended to cover channel expressions. $E(t, \#, ch)$ denotes the multiset sum of all *(expr, (s, r))* tuples in $Expr(t, \#, ch)$:

$$E(t, \#, ch) = \sum_{(expr, (s,r)) \in Expr(t, \#, ch)} 1`(expr, (s, r)) \tag{3}$$

The set of variables of a transition is extended with the variables of the communication expressions:

$$\forall t \in T_{OCPN} : Var(t) = \{v \mid v \in Var(G_{CN}(t)) \vee$$
$$\exists a \in A(t) : v \in Var(E_{CN}(a)) \vee$$
$$\exists (expr, \#, (s, r), ch) \in CE(t) :$$
$$[v \in Var(expr) \vee v = s \vee v = r]\}. \tag{4}$$

4   Dynamic Structure and Behaviour

In this section the differences between the static and dynamic structure of OCP-Nets are explained. In addition, the asynchronous and synchronous way of communication are described including the extensions to the enabling rule of CP-Nets necessary for channel expressions.

4.1   Creation and Deletion of Object-Nets

OCP-Nets differentiate between a static model structure and a dynamic simulation structure. The static structure consists of class nets while the dynamic structure consists of *object nets* which are instances of class nets. The activation and firing rules refer to this dynamic structure. *Object nets* are created with the reserved function *new*. This function assigns a unique identifier to the object net and adds the new object net to the simulation structure. This way it is possible

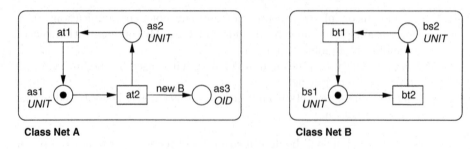

Class Net A   **Class Net B**

Fig. 4. Creation of Object Nets: Static Structure

to add new instances while executing an OCP-Net. The execution starts with creating an instance of a specific class net, called the *root net*. The creation of an object net is now illustrated using the two class nets in figure 4. It shows two class nets, A and B. Net A always creates a new instance of B when transition at2 occurs. The function *new* evaluates to the identifier of the created object net. The initial marking of both nets is given inside the places. Net A is the root net. The static net structure, omitting the complete arc definitions, colours etc., is given by $P = \{as1, as2, as3, bs1, bs2\}$, $T = \{at1, at2, bt1, bt2\}$ and $A = \{as1TOat2, \ldots\}$. Now to execute the system, an instance of the root net A is created with identifier 1^3 leading to the following simulation structure with PI denoting the place instances, TI the transition instances, AI the arc instances and M the current marking.

    (*i*) $PI = \{(as1, 1), (as2, 1), (as3, 1)\}$
    (*ii*) $TI = \{(at1, 1), (at2, 1)\}$
    (*iii*) $AI = \{(as1TOat2, 1), \ldots\}$
    (*iv*) $M = 1`((as1, 1), tok)$

The elements of the simulation structure are tuples which consist of the name of the element and the object net identifiers. After instantiating root net A the execution can begin with the occurrence of transition at2 creating an instance of B with identifier 2. This leads to the following simulation structure.

    (*i*) $PI = \{(as1, 1), (as2, 1), (as3, 1), (bs1, 2), (bs2, 2)\}$
    (*ii*) $TI = \{(at1, 1), (at2, 1), (bt1, 2), (bt2, 2)\}$
    (*iii*) $AI = \{(as1TOat2, 1), \ldots, (bs1TObt2, 2), \ldots\}$
    (*iv*) $M = 1`((as2, 1), tok) + 1`((as3, 1), 2) + 1`((bs1, 2), tok)$

The identifier *2* can now be used by the instance of *A* to access the new instance of *B*. An existing object net may be deleted from the simulation structure by the function *del* which removes all elements of the corresponding identifier. Note, *new* and *del* are not methods of a class net but defined globally.

---

³ Instead of the global identifier 1 a tuple consisting of the class name and a local counter could be used (see [12]).

4.2   Synchronous and Asynchronous Communication

There are two ways of communication in OCP-Nets, asynchronous and synchronous. It is therefore possible to model the way of communication common to distributed systems (asynchronous) and the one of most programming languages (synchronous). Both ways of communication are now explained in detail using the following convention: the reserved word *self* is always bound to the identifier of the corresponding object net.

Asynchronous Communication.   Asynchronous communication is realised with global fusion places as used in [9]. Figure 5 gives an example for the stack already shown in figure 2. One communication fusion place is connected to the IN communication transition (place *pushIN*) and one to the OUT communication transition (place *pushOUT*). A communication fusion place is denoted with *CF*. The place *sender* is used to store the identifier of the calling net. The message format is *((sender, receiver), parameter)*. Therefore, the colour of the fusion place is *(OID*OID)*INT*. The colour *OID*OID* is denoted by *OID*. By using *self* which is always bound to its own identifier the called object net ensures that it only removes token that belong to it. This is necessary because all instances of *pushIN* belong to the same global fusion set. The global fusion set represents a blackboard for the messages or a kind of broadcast where only the right net can pick up the messages directed to it.

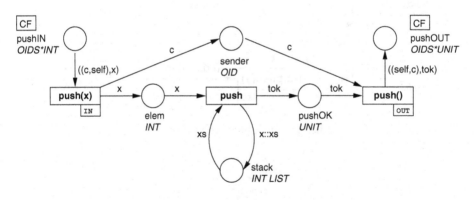

Fig. 5. Asynchronous Stack

Note that the service *push* as shown in figure 5 may be used by more than one calling net at the same time. This point will be addressed in section 7.1.

The calling net from figure 3 is extended in the same way as shown in figure 6. The rest of the calling net is indicated by dotted lines. The calling net may invoke the service *push* by placing a token on place *pushIN* consisting of its own identifier *self*, the identifier of the called net *id* and the parameter *y*. Only the called net with identifier *id* may withdraw this token from *pushIN* and perform the service, placing the result on its place *pushOUT*. The net that has invoked the service may then withdraw the resulting token from *pushOUT*. Again, the arc inscription *((c,self),tok)* of the calling net ensures that only this net can withdraw the result token. Note that *self* in figure 5 indicates the identifier of the called net while in figure 6 it indicates the identifier of the calling net.

Synchronous Communication.   Synchronous communication between nets is realised with a modified version of synchronous channels as described in [5]. With synchronous channels, transitions may exchange tokens in one corresponding step. The synchronisation and exchange of tokens is realised with communication expressions which are attached to transitions. The modified communication expression used in OCP-Nets consists of four elements: the multiset of tokens to be transferred, a tuple of identifiers denoting the sending and receiving net, a channel type (IN, OUT, INV or REC) and a channel name. Figure 7 shows the two channel expressions of the push service.

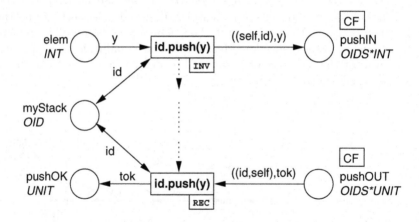

Fig. 6. Calling Stack Asynchronous

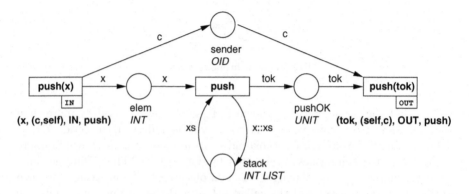

Fig. 7. Synchronous Stack

The according extensions to the calling net for synchronous communication are shown in figure 8.

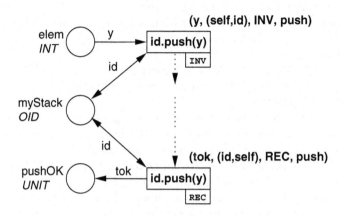

Fig. 8. Calling Stack Synchronous

The sending of a token via a synchronous channel is realised with an extension to the enabling rule of Coloured Petri Nets (see [9], [5]) which is defined in Def. 3. With t an instance of a transition is denoted, ch is a channel name. $E(t, INV, ch)$ is the multiset of all tuples, consisting of the parameter and sender/receiver tuple, of channel type INV and channel ch that are attached to transition t (see section 3). A marking M on places $p \in P_{norm}$ is defined as for CP-Nets (see [9]). For fusion places (IF and CF) the definitions of Hierarchical CP-Nets (see [9]) have been adopted accordingly to define their markings.

Definition 3. *Enabling rule*

A step Y is enabled in a marking M if:

(i) $\forall p' \in PIG$: $\displaystyle\sum_{(t,b)\in Y, p\in p'} E(p,t) < b > \le M(p')$.

(ii) $\forall ch \in CH$:

$$\sum_{(t,b)\in Y} E(t, INV, ch) < b > = \sum_{(t,b)\in Y} E(t, IN, ch) < b > \wedge$$
$$\sum_{(t,b)\in Y} E(t, REC, ch) < b > = \sum_{(t,b)\in Y} E(t, OUT, ch) < b >$$

The first part of Def. 3 is the same as for Hierarchical CP-Nets to incorporate the fusion places. The second part is the extension used for synchronous communication. This will be illustrated by an example using the nets shown in figure 7 and figure 8. Let us assume that the identifier of the net offering the service push is *idRec* and the identifier of the calling net is *idSend*. The calling net wants to send a token on place *elem* with value *100* to its stack. The stack itself is empty so far. Thus, the current marking is given by

$M = 1'((elem, idSend), 100) + 1'((myStack, idSend), idRec) +$
      $1'((stack, idRec), []).$

Under the binding

    $b_1 = \ < y = 100, self = idSend, id = idRec >$

the channel expression of the INV transition of the calling net evaluates to

    $(100, (idSend, idRec), INV, push).$

Under the binding

    $b_2 = \ < x = 100, self = idRec, c = idSend >$

the channel expression of the IN transition of the called net evaluates to

    $(100, (idSend, idRec), IN, push).$

Thus, the first two elements of both channel expressions evaluate under the binding $b = b_1 + b_2$ to the same expression: $(100, (idSend, idRec))$. The step

    $Y = 1'((push_{INV}, idSend), b1) + 1'((push_{IN}, idRec), b2)$

is therefore enabled and may occur, withdrawing the token *100* from place *elem* in the calling net and placing it on place *elem* in the called net.

This section has presented the two main ways for communication. Thus the modeller is enabled to use the most appropriate way of communication for the problem at hand.

5 Further Object Oriented Concepts

This section describes the way inheritance and encapsulation are realised in OCP-Nets.

5.1 Inheritance

Inheritance is an important mechanism of object oriented programming as it allows for reuse and polymorphism. Wegner distinguishes two basic ways of inheritance: subtyping and subclassing (see [17]). Using subtyping the subtype is compatible with its supertype which means it can be used in every context the supertype is used. This leads to the *contravariant* style of function declaration. A subtype may override a function of its supertype if its arguments are supertypes of the arguments of the supertypes function. Subclassing leads to the *covariant* style of function declaration. A subclass may override a function of its superclass if its arguments are subclasses of the arguments of the superclass. Both options have their advantages. Subclassing allows for a more extensive reuse of existing code while subtyping is somehow safer because a subtype may be used in *every* context the supertype is used.

OCP-Nets use subtyping because it allows for a more secure and thus robust computing. The subtype relation can be described using the notion of an interface. In [6] an interface is described in the following way.

> Every operation declared by an object specifies the operations name, the objects it takes as parameters, and the operations return value. This is known as the operations **signature**. The set of all signatures defined by an object's operations is called the **interface** of the object.   [6, S.13]

The interface of an object therefore specifies all operations an object can perform. The relation between the interface of an object and its type is characterised as follows:

> A **type** is a name used to denote a particular interface. We speak of an object as having the type "Window" if it accepts all requests for the operations defined in the interface named "Window". An object may have many types, and widely different objects can share a type. ... We say that a type is a **subtype** of another if its interface contains the interface of its **supertype**. Often we speak of a subtype *inheriting* the interface of its supertype.
>
> [6, S.13]

The signature of a service of a class net is defined by the service name, the parameter type of the IN transition and the resulting type of the OUT transition (see functions TN an TT in Def. 1). For example, the push service in figure 2 has the signature *(push, INT, UNIT)*. If a service has only an IN and no OUT communication transition the result type is $EMPTY$. For example, a push service without confirmation would have the signature *(push, INT, EMPTY)*. The interface of a class net is then defined as the set of all signatures of its services. This leads to the following subtype definition with SIG denoting the signature of a service and ITC denoting the interface of a class net.

Definition 4. *Subtype-Relation*

A class net CN_A is a subtype of a class net CN_B, denoted by $CN_A<:CN_B$, if:
$\forall SIG_{super} \in ITC_B : \exists SIG_{sub} \in ITC_A,$
$\qquad SIG_{super} = (name_{super}, arg_{super}, result_{super}),$
$\qquad SIG_{sub} = (name_{sub}, arg_{sub}, result_{sub})$ thus that:

(i) $name_{super} = name_{sub}.$
(ii) $arg_{super} \subseteq arg_{sub}.$
(iii) $result_{super} = result_{sub}.$

This definition states that for every signature in the supertype there must be one in the subtype with the following conditions: the name of both services must be the same, the arguments of the supertype service must be a subtype of the arguments of the subtype service and the result type must be the same. Then a class net is called a subtype of another class net. Note that this definition allows a class net to be the subtype of more than one supertype.

5.2   Encapsulation

OCP-Nets already provide for encapsulation because places of a net can only be accessed via services provided by the net. In addition, a private service, used by the net for internal computation, can also be protected from improper access in an easy way (see Figure 9). This net offers a public service *mthPublic* which in turn uses a private service *mthPrivate*. This private service is protected from outside access by using *self* instead of a variable for the sender in the channel expression. Therefore, only the net itself may access this service. An asynchronous service is protected in the same way by using the tuple *(self,self)* in the according arc expression.

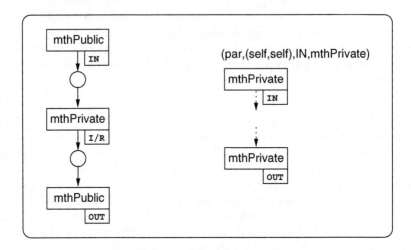

Fig. 9. Private Services in OCP-Nets

6   From OCP-Nets to Coloured Petri Nets

OCP-Nets have three aspects that distinguishes them from Coloured Petri Nets (CP-Nets) as described in [9]: fusion places, synchronous channels and a dynamic structure. The corresponding CP-Nets for nets with fusion places have been described in [9] and the ones for synchronous channels in [5]. Therefore, this section only describes the corresponding static structure for the dynamic structure of OCP-Nets. The basic idea is to enhance the class nets in a way that they behave like a set of object nets. This enhancement corresponds to folding the object nets into a class net as proposed in [3] and [12]. The colours of the places are enhanced with the type OID denoting the object identifier. This way a place in the class net can contain all tokens of the different object nets. The resulting net is shown in figure 10 for the asynchronous case. Corresponding to this the arc inscriptions are all extended with a variable for the identifier (id). This ensures that a transition does not mix tokens of different objects while firing. In addition the class net is extended with a place holding all currently valid object identifiers ($allIDs$). It is denoted in the figure as a page fusion place. This place is used as a side condition for communication transitions instead of the reserved identifier *self* to ensure that only valid instances are called. The last extension of the class net is adding the services *new* and *del* to enable the

creation and deletion of new objects (not shown in the figure). *New* creates all
tokens needed for an object according to the given initial marking. The service
del removes all tokens belonging to a given object identifier. This construction
corresponds to the one described in [3] and [10] and is described there in more
detail.

It is therefore possible to convert an OCP-Net in a Hierarchical CP-Net
which itself can be converted into a CP-Nets. The existing analysis techniques
can therefore be used for OCP-Nets.

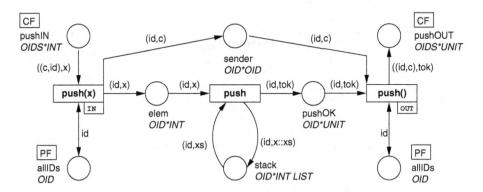

Fig. 10. Static Structure of an OCP-Net

7   Use Case Modelling with OCP-Nets

This section shows how to use OCP-Nets in object oriented modelling by formalising a use case. Before this, a few modelling conventions aimed to improved
the readability of the models are described.

7.1   Modelling Conventions

The intention of OCP-Nets is to use them as a basic formalism together with a
tool that hides tedious details from the user. A user only models the services of a
class net and has the choice to allow concurrency inside a service or not. Similar
he should be able to define a whole net as mutual exclusive or a set of its services.
The resulting synchronisation places can easily be added by an appropriate tool.
A tool can also hide the details of synchronous and asynchronous communication,
leaving to the user only the task to decide whether a service may be called
synchronous or asynchronous. This way the user will also be freed from specifying
the arc inscriptions or channel expressions for communication, removing a source
of error.

A few graphical conventions allowing for more readable models are already
used in this paper which are now described. Often, the calling net only waits until

the service is performed. Thus, a shortcut is used to simplify the net representation. This shortcut is shown in figure 11. The INV- and REC-transitions are combined to one transition. This transition is called I/R-transition (invoke/receive). Note, this is not a new type of communication transition. It is just an abbreviation for the INV- and REC-transition and the place *wait*.

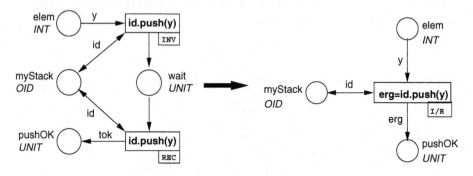

Fig. 11. Shortcut: I/R-transition to call a service

In addition to this shortcut, two more conventions will be used in the rest of this paper. First, the arc expression *tok* may be omitted. Second, an arc expression denoting a variable may be omitted. Instead the name of the place may be used in guards and method calls. Figure 12 gives an example.

Fig. 12. Conventions

7.2   Example

Use case modelling is a good way to get a feeling for an application area and to determine the requirements. They were introduced by Jacobson ([7]) and have been integrated in the Unified Modeling Language UML ([13]). A use case describes a way in which the user, called actor in [7], can use a system. This is similar to a *scenario*. But while a scenario describes only one possible flow use cases can have several flows of control. One can therefore think of an use case as a set of scenarios. A system is then described by the different use cases it offers. A use case consists of three parts; a natural language description with possible alternative flows of control, a diagram, showing its objects and optional an *Interaction Diagram* to model its dynamic behaviour. An *Interaction Diagram*, called *Sequence Diagram* in [13], has two drawbacks in that it is only a semi-formal technique and lacks adequate means to model concurrency. In this section we will show how to use *OCPN* to replace *Interaction Diagrams*.

A simplified version of the use case description for an automatic teller machine (ATM) that Jacobson uses in [8, pp. 254] will be used below to illustrate use case modelling with OCP-Nets. It describes the basic flow of control when a card is inserted into an ATM.

> *The use case starts when a bank card has been inserted. The Transaction Handler is invoked which reads the code of the card and checks whether the card is valid.*
> *If the card is not valid, the session is ended according to the Final Flow. If the card is valid, the Graphical User Interface is asked to request the actor to type in his/her PIN. When this has been done, the Card Transaction Handler reads the code and compares it with the code it received when the card was inserted. If the PIN is not correct the session is ended according to the Final Flow. If the PIN is valid the Card Transaction Handler asks the Graphical User Interface to present the options withdrawal, balance, transfer and deposit to the actor. Card Transaction Handler gets the customer's choice in return and starts the according use case.*
> Final Flow: *The bank card is ejected by the Card Input object; the use case is then terminated."*

The example is quite simple but sufficient to demonstrate the basic concepts of modelling with OCP-Nets in an object oriented way. The control flow is modelled as an OCP-Net in figure 13. The Card Input object is denoted by *ci* and the Graphical User Interface by *dis* in the transition inscriptions. The transaction starts with IN transition *trans()*. The card code is then read by *cd=readCode()*

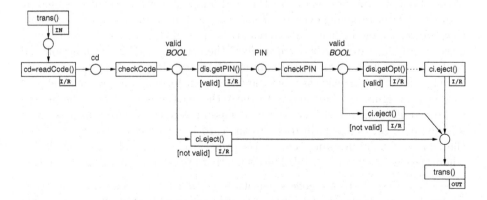

Fig. 13. Use Case Transaction Flow as OCP-Net

and after it the code is checked by an internal service.[4] If the code is valid, the PIN is requested from the display by transition *dis.getPIN()*. It is checked in turn and if the PIN is valid the user is asked for an option by *dis.getOpt()*. The chosen option is then performed (indicated by a dotted line) and after it the card is ejected by *ci.eject()*. If the code or PIN is not valid the card is ejected, too.

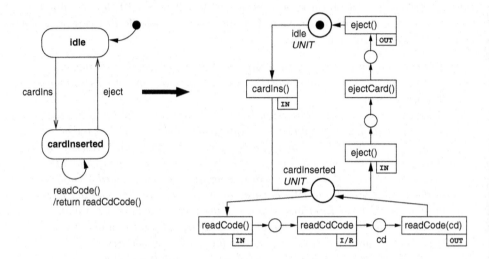

Fig. 14. State Chart CardInput and corresponding OCP-Net

The figure gives a clear and precise description of the flow of control and can also show possible concurrency, even if it is not the case in the example given. In addition, one can go one step further with OCP-Nets. The use case uses the two objects card input and display. The behaviour of these objects can be modelled with OCP-Nets in a way similar to *Statechart Diagrams* used in UML. Figure 14 shows the Statechart for the card input object and the corresponding OCP-Net. After receiving event *cardIns()* the card input object changes its state to *cardInserted* and is then ready to read the card code (*readCode()*). If it receives event *eject()* in state *cardInserted* it ejects the card by some internal action and confirms this action by the corresponding OUT transition.

The model for the use case and the one for card input can now be combined as shown in figure 15. This allows to check the consistency between the models and to simulate them together. The model assumes that a device that detects an inserted card sends event *cardIns()* to the card input object and activates then the transaction by invoking the service *trans()*. The figure shows a simulation where the card code is not valid. The events that are exchanged during simulation

---

[4] It is also possible to use code segments as in Design/CPN and attach the code directly to the transition. Of course, this only makes sense if the code is used only once.

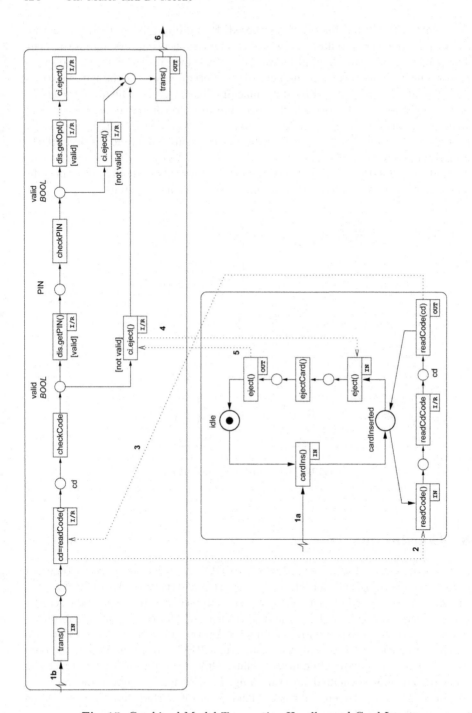

Fig. 15. Combined Model Transaction Handler and Card Input

are depicted by dotted lines and numbered. By adding a behavioural model for the display one can simulate the whole configuration using only one formalism. The principal structure behind figure 13 - 15 is shown in figure 16. In the upper part it shows a process layer, depicted by a Petri net and in the lower part an object model layer, depicted by state machine like symbols. The system consists of a set of objects and processes that are supposed to run on these objects. This structure is also found in workflow systems. A workflow is a part of a business process that is executed in a workflow system using different services of the underlying computer system (see also [1] and [2]). As this is the same structure as used here, OCP-Nets are also well suited for workflow modelling, especially when these workflows are executed on an object oriented system.

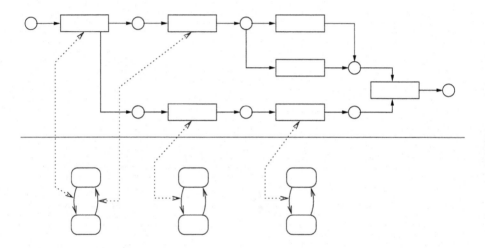

Fig. 16. Structure of a Specification

8   Conclusion

In this paper an informal introduction to OCP-Nets has been given together with a complete formal definition of their static structure. An OCP-Net is a set of class nets. A class net offers one or more services to other class nets by encapsulating it within an IN- and OUT-transition. This corresponds to methods in object oriented programming languages. Another class net calls a service via an INV-transition and receives the result via a REC-transition. This is the only way for class nets to communicate. Thus, OCP-Nets resemble the structures known from object oriented programming. In addition, an inheritance relation has been defined between class nets. Therefore, OCP-Nets can be classified as truly object oriented nets.

The dynamic structure and behaviour of OCP-Nets have been described to-gether with a few formal definitions. More about the definitions and their ap-

plication can be found in [11]. The dynamic structure of OCP-Nets is a new
and important aspect, because it gives the analysis and design models a semantics similar the programming languages used for implementation. However,
the transformation into Coloured Petri Nets provides the means to apply most
formal results from CP-Nets to OCP-Nets.

The formal specification of a use case model has shown how to use OCP-Nets
as a technique for object oriented specification. They provide appropriate means
to cover concurrency and synchronisation and allow for simulation of the models.
OCP-Nets also have a formal semantics and are based on a well known Petri net
standard.

In combination with the results from [3], [12] and [11] it is of special interest
how OCP-Nets can be integrated into object oriented analysis and design. A
rough idea can be derived from section 7 and figure 16 where the relation of
different views is sketched. Therefore, the combination of object orientation and
Petri nets seems to be very promising for object oriented analysis and design
and will be an important issue of our further research.

References

1. W.M.P. van der Aalst. Verification of Workflow Nets. In P. Azema and G. Balbo,
   editors, *Application and Theory of Petri Nets 1997*, volume 1248 of *Lecture Notes
   in Computer Science*, pages 407–426. Springer-Verlag, Berlin, 1997.
2. W.M.P. van der Aalst, H.M.W. Verbeek, and D. Hauschildt. A Petri-net-based
   Tool to Analyze Workflows. In *Petri Nets in System Engineering (PNSE'97) –
   Modelling, Verification and Validation*, pages 78–89. University of Hamburg (FBI-
   HH-B-205/97), 1997.
3. Ulrich Becker and Daniel Moldt.   Objektorientierte Konzepte für gefärbte
   Petrinetze. In Gert Scheschonk and Wolfgang Reisig, editors, *Petri-Netze im Ein-
   satz für Entwurf und Entwicklung von Informationssystemen*, Informatik Aktuell,
   pages 140–151, Berlin Heidelberg New York, 1993. Springer-Verlag.
4. G. Booch. *Object-Oriented Design*. Benjamin/Cummings Redwood City, CA, 2.
   edition, 1993.
5. Soren Christensen and Niels Damgaard Hansen. Coloured Petri Nets Extended
   with Channels for Synchronous Communication. In *International Conference on
   the Application and Theory of Petri Nets*, volume 815 of *Lecture Notes in Computer
   Science*, pages 159–178. Springer-Verlag, 1994.
6. Erich Gamma, Richard Helm, Ralph Johnson, and John Vlissides. *Design Patterns – Elements of Reusable Object-Oriented Software*. Addison-Wesley Publishing Company, Reading, Massachusetts, 1995.
7. Ivar Jacobson, Magnus Christerson, Patrik Jonsson, and Gunnar Övergaard.
   Object–Oriented Software Engineering – A Use Case Driven Approach. Addison–
   Wesley Publishing Company, 1992.
8. Ivar Jacobson, Maria Ericsson, and Agneta Jacobson. *The Object Advantage, Business Process Reengineering with Object Technology*. Addison-Wesley, 1995.
9. Kurt Jensen. *Coloured Petri Nets: Volume 1; Basic Concepts, Analysis Methods
   and Practical Use*. EATCS Monographs on Theoretical Computer Science. Springer-Verlag, Berlin Heidelberg New York, 1992.

10. Christoph Maier. Darstellung von Objektorientierten Konzepten mit Petrinetzen. Studienarbeit, Universität Hamburg, Fachbereich Informatik, Vogt-Kölln Str. 30, 22527 Hamburg, Germany, May 1996.

11. Christoph Maier. Objektorientierte Analyse mit gefärbten Petrinetzen. Diplomarbeit, Universität Hamburg, Fachbereich Informatik, 1997.

12. Daniel Moldt. *Höhere Petrinetze als Grundlage für Systemspezifikationen*. Dissertation, Universität Hamburg, Fachbereich Informatik, Vogt-Kölln Str. 30, 22527 Hamburg, Germany, 1996.

13. RATIONAL Software Corporation. *UML Notation Guide*, September 1997. Version 1.1. Available at http://www.rational.com/.

14. J. Rumbaugh. Modeling and design – OMT: The functional model. *Journal of Object Oriented Programming*, 8(1):10–14, 1995.

15. James Rumbaugh, Michael Blaha, William Premeralani, Frederick Eddy, and William Lorensen. *Object-Oriented Modeling and Design*. Prentice Hall, Englewood Cliffs, New Jersey 07632, 1991.

16. C. Sibertin-Blanc. Cooperative Nets. In *15th International Conference on the Application and Theory of Petri Nets*, volume 815 of *Lecture Notes in Computer Science*, pages 471–490, Zaragoza, Spain, 1994. Springer-Verlag.

17. Peter Wegner. The object-oriented classification paradigm. In B. Shriver and P. Wegner, editors, *Research Directions in Object-Oriented Programming*, Cambridge, 1987. Cambridge University Press.

18. Edward Yourdon, Whitehead Katharine, Jim Thomann, Karin Oppel, and Peter Nevermann. *Mainstream Objects: An Analysis and Design Approach for Business*. Yourdon Press computing series. Prentice Hall, New York, 1995.

An Actor Algebra
for Specifying Distributed Systems:
The Hurried Philosophers Case Study⋆

Mauro Gaspari and Gianluigi Zavattaro

Department of Computer Science, University of Bologna
Mura Anteo Zamboni 7, 40127 Bologna, Italy
{gaspari,zavattar}@cs.unibo.it

Abstract. In this paper, we introduce an actor language following a
"process algebra" notation. The idea is to define a formalism based on a
standard process algebraic approach which provides basic object-oriented
features, such as object identity, asynchronous message passing, implicit
message acceptance and dynamic object creation. This approach allows
us to reuse standard results of the theory of concurrency in a context
where an high level object oriented specification style is preserved. To illustrate the expressive power of our formalism, we provide a specification
of the Hurried Philosophers case study.

1 Introduction

The object-oriented research community developed techniques, tools, and environments that have been applied to several software development projects in
the context of a wide range of application domains. In particular, distributed
object-oriented programming is one of the most promising candidate paradigms
to build large scale distributed systems. OMG the Object Management Group
consortium, CORBA [24] the object-oriented standard for integrating applications running in heterogeneous distributed environments developed by OMG,
and Java [7], the new internet language developed by Sun Microsystems, are all
examples of such efforts.

On the other hand, most of the theoretical computer science efforts in the
theory of concurrency are oriented to study process algebras such as CCS [18]
or the π-calculus [19] which do not provide a direct representation of objects
as first class entities. In these formalisms processes are stateless entities which
communicate exploiting synchronous message passing and the representation of
an object involves a large number of processes [25].

As a consequence of this situation there is a big gap between theory and
practice: results developed from the theory of concurrent systems, such as the
theories of equivalence for process algebras, can hardly be applied to real object
oriented distributed systems.

⋆ This paper has been partially supported by the Italian Ministry of Universities
(MURST).

G. Agha et al. (Eds.): Concurrent OOP and PN, LNCS 2001, pp. 428–444, 2001.

The formal framework presented in this paper addresses this problem. In particular we introduce a process algebra based on an object-oriented model which has a clean formal definition, like the *pi*-calculus, while providing basic object-oriented features, such as object identity, asynchronous message passing, implicit message acceptance and dynamic object creation. This approach allows us to reuse standard results of the theory of concurrency in a context where an high level object oriented specification style is preserved. To illustrate the expressive power of our formalism, we provide a specification of the Hurried Philosophers case study.

The paper is organized as follows. In Section 2 we highlight the basic communication mechanisms of distributed object-oriented languages. In Section 3 we introduce a process algebra based on such mechanisms. In Section 4 we present a specification of the Hurried Philosophers case study. Finally, in the last section we briefly summarize our approach and put it in perspective.

2   Distributed Object Oriented Programming

Objects are the basic run-time entities in an object-oriented system. Objects have a local memory accessed through a set of attributes and a behaviour implemented as a set of methods, *i.e.,* a set of procedures and/or functions that defines the meaningful operations on the local memory. Moreover, there is a notion of object-identity [16]. Identity is that property of an object which distinguishes each object from all others.

In distributed object-oriented systems objects are autonomous units executing concurrently and interacting by message-passing. Several models for distributed object-oriented programming have been proposed over the past years. Two main approaches have been followed:

- The first is based on π-calculus [19,25,15,23], which can be seen as an extension of process algebras with the notion of naming and dynamic process creation.
- The second approach is based on the actor model [14,2], which for a long time has been the unique model of concurrent computation able to deal with dynamic aspects such as process creation and transmission (π-calculus was designed about 10 years later).

In the following, we try to individuate the communication mechanisms which are common to most of these models and, consequently, which can be considered basic interaction mechanisms in this context.

2.1   Object Identity

Support for object-identity can be achieved by associating to each object a unique name or address. This notion of object-identity based on names has a natural mapping in the actor model [2]. Actors are named objects with a behaviour which is a function of incoming communications. Each actor has a unique name (mail address) determined at the time of its creation. This name is used to specify the recipient of a message. Conversely, object-identity is not easily embeddable in formalisms such as CCS [18] or π-calculus [19], where message dispatching is performed by means of channels. In these formalisms the association address-process is not unique: a process may have several ports (channels) from which it receives messages and the same channel can be accessed by different processes.

2.2   Synchronous vs Asynchronous Message Passing

Objects can communicate by means of synchronous or asynchronous message-passing mechanisms: the synchronous mechanisms do not allow the buffering of messages, thus a message to be delivered needs a "hand-shake" between the sending object and the receiving object; this is the communication mechanism adopted by agents in CCS [18]. The asynchronous mechanism assumes that the buffering of messages is possible, thus it is not required an agreement with the receiving object to send a message; this is the approach adopted by actors [3] and by asynchronous process algebras [10].

Since our goal is to select a set of primitive interaction mechanisms, we need to specify which kind of message passing should be considered as primitive in this context. We claim that asynchronous message passing is the most suitable basic interaction mechanism for distributed object-oriented programming. This because synchronous communication can be modelled providing an adequate "rendez-vous" protocol, while, vice versa, if we want to model asynchronous message passing with a synchronous language, we need to introduce an extra entity to deal with the buffering of messages. However, the buffering of messages requires unbounded buffers; this assumption seems reasonable, expecially in all those sytems in which the storage media have a capacity which is largely greater than the average amount of pending messages.

2.3   Explicit vs Implicit Message Acceptance

Given that the basic message passing mechanism is asynchronous, and that the name of an object is used to specify the recipient of a message, we have still to establish which is the form of receive primitive which is more suitable to distributed object-oriented programming.

The receive primitive may be explicit or implicit. A receive operation is explicit when it appears in the program, while it is implicit when it does not correspond to an operation in the programming language and it is performed implicitly at certain points of the computation: typically in actor systems the receive is implicit and is performed only when the actor is idle.

We think that the receive should be implicit in a basic calculus for distributed object-oriented programming: in this way, the communication has the form of an asynchronous object method invocation. Hence, each object has several methods, each one managing possible received messages. By the way, asynchronous message passing with an implicit receive mechanism is a basic feature of the actor model.

2.4   Staticity vs Dynamicity

An aspect which is central in distributed object-oriented programming is dynamicity, i.e., the possibility of creating new objects at run time. Most of the proposals for new languages [8,6] handle dynamic issues by providing rigorous laws which regulate the creations of new objects and the transmission of channel and process identifiers.

2.5   Summary

As a result of the analysis presented above the basic interaction mechanisms, which are suitable to model distributed object-oriented, are close to those of the actor model. In fact, the actor model provides support for object identity, asynchronous message passing, implicit receive mechanism and support for dynamicity. Thus, the second approach seems to be more adequate for our purposes and we have chosen the actor model as a model for the process algebra presented in this paper.

2.6   The Actor Model

The actor model was introduced by Carl Hewitt about 20 years ago [14] . Actors are self-contained agents with a state and a behaviour which is a function of incoming communications. Each actor has a unique name (mail address) determined at the time of its creation. This name is used to specify the recipient of a message supporting object identity. Actors communicate by asynchronous and reliable message passing exploiting an implicit receive mechanism.

Actors make use of three basic primitives which are asynchronous and non-blocking: *create*, to create new actors; *send*, to send messages to other actors; and *become*, to change the behaviour of an actor [2].

The current formalizations of actors, such as [3,4,22], do not exploit a "process algebra" style, for this reason it is difficult to import in the actor model techniques and results developed for other asynchronous calculi [9,15,5]. Unfortunately most of current efforts on the semantics of concurrent languages do not consider the actor model at all. Although recently Robin Milner [20] suggested that it may worth to investigate in this direction.

3   An Algebra of Actors

Let $\mathcal{A}$ be a countable set of *actor names*: a, b, c, a_i, b_i,... will range over $\mathcal{A}$ and L, L', L'',... will range over its power set $\mathcal{P}(\mathcal{A})$ (*i.e.*, L, L', $L'' \subseteq \mathcal{A}$). Let $\mathcal{V}$ be a set of values $\mathcal{A} \subset \mathcal{V}$ containing, e.g., *NIL*, *true*, *false*. We assume value expressions e built from actor names, value constants, value variables, the expressions *self*, *state*, and *message*, and any operator symbol we wish. We will denote values with v, v', v'',... when they appear as contents of a message and with s, s', s'',... when they represent the state of an actor. $[\![e]\!]_s^a$ gives the value of e in $\mathcal{V}$

assuming that a and s are subsituted for *self* and *state* inside e; e.g. $[\![self]\!]_s^a = a$ and $[\![state]\!]_s^a = s$. The special expression *message* represents the contents of the last received message. Whenever a message is received, its contents is substituted for each occurrence of the expression *message* in the receiving actor.

Let C be a countable set of *actor behaviours*: C, D, ... will range over C. We suppose that every behaviour D is equipped with a corresponding definition $D \overset{def}{=} P$ where P is a program, that is a term defined by the following abstract syntax:

$$P ::= become(C, e).P \mid send(e_1, e_2).P \mid create(b, C, e).P \mid$$
$$e_1{:}P_1 + \ldots + e_n{:}P_n \mid \sqrt{}$$

Observe that we allow also the recursive definition of behaviours, e.g,

$$C \overset{def}{=} become(C, state).\sqrt{}.$$

Actor terms are defined by the following abstract syntax:

$$A ::= {}^{a}\mathbf{C_s} \mid {}^{a}[\mathbf{P}]_s \mid \langle \mathbf{a}, \mathbf{v} \rangle \mid \mathbf{A}|\mathbf{A} \mid \mathbf{A} \backslash \mathbf{a} \mid \mathbf{0}$$

An actor can be idle or active. An idle actor ${}^{a}C_s$ (composed by a behaviour C, a name a, and a state s) is ready to receive a message. When a message is received the actor becomes active. Active actors are denoted by ${}^{a}[P]_s$ where P is the program that is executed. The actor a will not receive new messages until it becomes idle (by performing a *become* primitive). The state s is sometimes omitted when empty (i.e. $s = \emptyset$). A program P is a sequence of actor primitives (*become*, *send* and *create*) and guarded choices $e_1{:}P_1 + \ldots + e_n{:}P_n$ terminating in the null program $\sqrt{}$ (which is usually omitted). An actor term is the parallel composition of (active and idle) actors and messages, each one denoted by a term $\langle a, v \rangle$ where v is the contents and a the name of the actor the message is sent to. Also a restriction operator $A \backslash a$ is used in order to allow the definition of local actor names ($A \backslash L$ is used as a shorthand for $A \backslash a_1 \backslash \ldots \backslash a_n$ if $L = \{a_1, \ldots, a_n\}$) while 0 stands for an empty actor.

The actor primitives and the guarded choice are described as follows.

- *send*:
  The program $send(e_1, e_2).P$ sends a message with contents described by e_2 to the actor indicated by e_1:
  $${}^{a}[send(e_1, e_2).P]_s \overset{\tau}{\longrightarrow} {}^{a}[P]_s \mid \langle [\![e_1]\!]_s^a, [\![e_2]\!]_s^a \rangle$$
  where τ represents an internal invisible step of computation.
- *become*:
  The program $become(C, e).P'$ changes the state of the actual actor from active to idle:
  $${}^{a}[become(C, e).P']_s \overset{\tau}{\longrightarrow} ({}^{d}[P'\{a/self\}]_s) \backslash d \mid {}^{a}C_{[\![e]\!]_s^a} \qquad \text{with } d \text{ fresh}$$
  The primitive *become* is the unique one that permits to change the state according to the expression e; we sometimes omit e if the state is left unchanged (i.e. $e = state$). The continuation P' is executed by the new actor ${}^{d}[P'\{a/self\}]_s$. This actor will never receive other messages (i.e. it is unreachable) as its name d cannot be known to any other actor. Indeed, the

Table 1. Operational semantics.

Send	$^a[send(e_1, e_2).P]_s \xrightarrow{\tau} {}^a[P]_s \mid \langle [\![e_1]\!]_s^a, [\![e_2]\!]_s^a \rangle$
Deliver	$\langle a, v \rangle \xrightarrow{\overline{av}\emptyset} 0$
Become	$^a[become(C, e).P']_s \xrightarrow{\tau} ({}^d[P'\{a/self\}]_s) \backslash d \mid {}^a C_{[\![e]\!]_s^a} \quad d \text{ fresh}$
Create	$^a[create(b, C, e).P']_s \xrightarrow{\tau} ({}^a[P'\{d/b\}]_s \mid {}^d C_{[\![e]\!]_s^a}) \backslash d \quad d \text{ fresh}$
Receive	$^a C_s \xrightarrow{av} {}^a[P\{v/message\}]_s \qquad\qquad \text{if } C \overset{def}{=} P$
Guard	$^a[e_1{:}P_1 + \ldots + e_n{:}P_n]_s \xrightarrow{\tau} {}^a[P_i]_s \qquad \text{if } [\![e_i]\!]_s^a = true$

Res	$\dfrac{A \xrightarrow{\alpha} A'}{A\backslash a \xrightarrow{\alpha} A'\backslash a}$	$a \notin n(\alpha)$
Open	$\dfrac{A \xrightarrow{\overline{av}L} A'}{A\backslash b \xrightarrow{\overline{av}L \cup \{b\}} A'}$	$a \neq b \;\wedge\; b \in n(v)$
Par	$\dfrac{A \xrightarrow{\alpha} A'}{A\mid B \xrightarrow{\alpha} A'\mid B}$	if $\alpha = \overline{av}L$ then $a \notin act(B) \wedge$ $L \cap fn(B) = \emptyset$
Sinc	$\dfrac{A \xrightarrow{av} A' \quad B \xrightarrow{\overline{av}L} B'}{A\mid B \xrightarrow{\tau} (A'\mid B') \backslash L}$	
Cong	$\dfrac{B \equiv A \quad A \xrightarrow{\alpha} A' \quad A' \equiv B'}{B \xrightarrow{\alpha} B'}$	

expression *self*, which is the unique one that returns the value d, is changed in order to refere to the name a of the initial actor.

- *create*:

  The program $create(b, C, e).P'$ creates a new idle actor having state s and behaviour C:

  $$^a[create(b, C, e).P']_s \xrightarrow{\tau} ({}^a[P'\{d/b\}]_s \mid {}^d C_{[\![e]\!]_s^a}) \backslash d \qquad \text{with } d \text{ fresh}$$

  The new actor receives a fresh name d. This new name is initially known only to the creating actor, in fact a restriction on the new name d is introduced.

- $e_1{:}P_1 + \ldots + e_n{:}P_n$:

  In the agent $e_1{:}P_1 + \ldots + e_n{:}P_n$, the expressions e_i are supposed to be boolean expressions with value *true* or *false*. The branch P_i can be chosen only if the value of the corresponding expression e_i is *true*:

  $$^a[e_1{:}P_1 + \ldots + e_n{:}P_n]_s \xrightarrow{\tau} {}^a[P_i]_s \qquad \text{if } [\![e_i]\!]^a = true$$

The function n returns the set of the actor names appearing in an expression, a program, or an actor term. Given the actor term A, the set $n(A)$ is partitioned in $fn(A)$ (the free names in A) and $bn(A)$ (the bound names in A) where the bound names are defined as those names a appearing in A only under the scope of some restriction on a. We use $act(A)$ to denote the set of the names of the actors in A. An actor term is well formed if and only if it does not contain two distinct actors with the same name. In the following we will consider only well formed agents, and we will use Γ to denote the set of well formed terms (A, B, D, E, F,... will range only over Γ).

We model the operational semantics of our language following the approach of Milner [19] which consists in separating the laws which govern the static relation among actors (for instance $A|B$ is equivalent to $B|A$) from the laws which rules their interaction. This is achieved defining a static structural equivalence relation over syntactic terms and a dynamic relation by means of a labelled transition system [21].

Definition 1. - Structural congruence, *is the smallest congruence relation over actor terms ($\equiv$) satisfying:*

(i) $^a[\sqrt{}]_s \equiv 0$ *(v)* $0 \backslash a \equiv 0$

(ii) $A|0 \equiv A$ *(vi)* $(A \backslash a) \backslash b \equiv (A \backslash b) \backslash a$

(iii) $A|B \equiv B|A$ *(vii)* $(A|B) \backslash a \equiv A|(B \backslash a)$ *where* $a \notin fn(A)$

(iv) $(A|B)|D \equiv A|(B|D)$*(viii)* $A \backslash a \equiv A\{b/a\} \backslash b$ *where* b *is fresh*

Definition 2. - Computations. *A transition system modelling computations in the actor algebra is represented by the triple $(\Gamma, T, \{\overset{\alpha}{\to} \mid \alpha \in T\})$. $T = \{\tau\} \cup \{av, \overline{av}L \mid a \in \mathcal{A}, v \in \mathcal{V}, L \subseteq \mathcal{A}\}$ is a set of labels, where τ is the invisible action standing for internal autonomous steps of computation; av and $\overline{av}L$ respectively represent the receiving and the emission of the message with receiver a and contents v. The set L in the label $\overline{av}L$ represents the set of actor names in the expression v which were initially under the scope of some restriction. $\overset{\alpha}{\to}$ is the minimal transition relation satisfying the axioms and rules presented in Table 1.*

The rules *Send*, *Become*, *Create* and *Guard* have been already discussed. Rule *Deliver* states that the term $\langle a, v \rangle$ (representing a message v sent to the actor a) is able to deliver its contents to the receiver by performing the action $\overline{av}\emptyset$. The corresponding receiving action labeled with av can be performed by the actor a when it is idle (rule *Receive*). The other rules are simply adaptation to our calculus of the standard laws for the π–calculus. The most interesting difference is due to the fact that in our calculus, more than one restriction can be extended by one single delivering operation. In fact, in our case the contents of a message is an expression instead of a unique name. This is the reason why we have added the set L to the label $\overline{av}L$.

Another difference is in the rule *Par*: the actor term $A|B$ can deliver a message inferred by A (*i.e.*, execute an emission action $\overline{av}L$), only if B does not contain the target actor (*i.e.*, $a \notin act(B)$).

3.1   Discussion

There are several differences with respect to the formal semantics of actors in [3,4] and in [22] which is worth to point out.

- We do not assume a fair message delivery mechanism as in [3,4] and in [22].
- The algebra of actors describes only communication and synchronization primitives, while in the semantics of Agha et al. actor primitives are embedded in a functional language. This enables us to focus on concurrency and interagent communication related aspects and not deal with issues concerning the sequential execution of programs inside actors.
- The operational semantics of the algebra of actors is defined by means of a labelled transition system instead of a simple reduction system as in [3] or the rewriting rules in [22]. This allows to use standard observational equivalences of process algebras *e.g.*, bisimulation, testing, failure or trace, without defining explicit observers.
- We have introduced the guarded choice as an alternative to the conditional which is present in previous formalization of actors [3].
- We provide an explicit representation of the state of an object while in Agha et al. the state of an actor is represented as part of its behaviour.
- We have introduced a mechanism to model termination of actors. Actors are not perpetual processes with a default behaviour as usual, but they can terminate: an actor terminates whenever it finishes its internal computation. This is not a limitation because a perpetual actor can always be obtained performing an explicit become operation for each internal computation.
- In the algebra of actors, actors are created exploiting a single basic primitive, while in the semantics of Agha et al. the creation process is composed of two basic operations, the creation of an empty actor and the initialization of its behaviour. The main advantage of our approach is that we do not need to restrict the possible computations to guarantee an atomic create operation.
- We introduce a restriction operator similar to the one of the π-calculus. This operator is more tractable with respect to the approach of [3] based on the specification of the sets of receptionists and external actors in actor configurations. On the other hand, the calculus presented in [22] uses the inverse operator indicating the actors which are reachable from the outside world explicitly.
- Finally, in the operational semantics of Agha et al. a receiving rule that is reminiscent of the rule *IN* of [15] is used. This rule (as discussed in [5]) has the disadtvantage to give rise to *infinitely branching*: the transition system allows each term (containing at least one receptionist) to activate an infinite number of transition, at least one for each possible message that can be sent to one of the receptionists. If, for example, a receptionist will be no more able to receive a message (e.g., it is executing an infinite computation) or external actors never send messages to a receptionist, the transition system allows (infinite) useless transitions. One of the most important advantages of the rule *IN* is that it allows the definition of observational semantics (*e.g.*, bisimulation)

that capture interesting aspects of asynchronous communication. Instead we follow the approach of [5], where it is shown that the same oservational semantics can be obtained by eliminating the problem of infinitely branching by slighty modifying the usual (synchronous) observational semantics.

4   Hurried Philosophers

In this section we present a specification of the Hurried Philosophers case study, which aims to establish the expressive power of object-oriented formalisms, testing features such as dynamic object creation and termination; dynamic binding of communication channels (*i.e.*, the partner of a communication is defined when the communication occurs); inheritance (a class of objects may inherit some features from other classes of objects).

To improve the presentation, we introduce some further notations which will be useful in the specification, giving a direct representation of the standard object-oriented concepts of methods, inheritance and attributes. Then, we provide a specification of the traditional (static) version of dining philosophers problem [17]. Finally, we show how dynamic aspects concerning the creation of new philosophers and the deletion of them, can be added to the specification exploiting inheritance.

4.1   Methods, Inheritance and Attributes

To provide a more modular specification of the problem, we add some further syntactic notations and we show how the usual object-oriented concepts of methods, inheritance and attributes can be represented in the algebra of actors.

First of all we define a new boolean expression *otherwise* which can be used as possible guard in a choice:

$$e_1{:}P_1 + e_2{:}P_2 + \ldots + e_n{:}P_n + otherwise{:}P_{n+1}$$

Such a new program is the same as:

$$e_1{:}P_1 + e_2{:}P_2 + \ldots + e_n{:}P_n + \neg(e_1 \vee e_2 \vee \ldots \vee e_n){:}P_{n+1}$$

The second extension we consider helps in writing programs incrementally, exploiting a form of inheritance. Given a behaviour of an actor defined as follows:

$$Beh \stackrel{def}{=} e_1{:}P_1 + e_2{:}P_2 + \ldots + e_n{:}P_n$$

we can consider programs $P_1, P_2, \ldots, P_n$ as methods, where their activation depends on the conditions $e_1, e_2, \ldots, e_n$ on the incoming message. In an object oriented scenario it could be useful to add new features to it, *i.e.*, more branches to the choice. This feature is obtained introducing a new operator for defining behaviours:

$$Beh \stackrel{ext}{=} e_{n+1}{:}P_{n+1} + e_{n+2}{:}P_{n+2} + \ldots + e_{n+m}{:}P_{n+m}$$

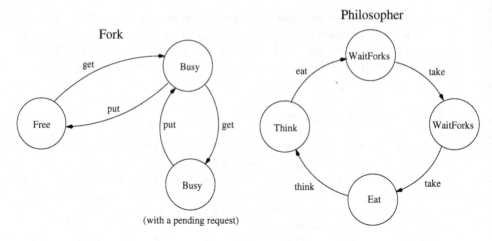

Fig. 1. Behaviours of forks and philosophers.

which redefines the initial behaviour in the following way:

$$Beh \stackrel{def}{=} e_1{:}P_1 + e_2{:}P_2 + \ldots + e_{n+m}{:}P_{n+m}$$

When a behaviour with an *otherwise* guard is extended as above, we suppose no new *otherwise* guards can be added, and the meaning of the *otherwise* is updated taking into account also the new conditions.

Since objects are usually modelled as entities having several attributes, we introduce a notation to represent records as possible structured values in actor expressions. A record r is denoted with $(f_1{:}e_1, \ldots, f_n{:}e_n)$ and the selection of the value in the field f_i is written as $r.f_i$. An operator $\uplus$ which allows to add fields to records is assumed; it is defined as follows:

$$(f_1{:}e_1, \ldots, f_n{:}e_n) \uplus (f_{n+1}{:}e_{n+1}) \stackrel{def}{=} (f_1{:}e_1, \ldots, f_{n+1}{:}e_{n+1})$$

4.2   Static Dining Philosophers

In this section we present the specification of a static version of dining philosophers in which no new philosophers can be added to the table, neither philosophers can leave or change their place on the table. In order to avoid both deadlock and starvation, we use the right-left algorithm presented in [17]. It consists of having two categories of philosophers: *right* and *left* ones, asking for the forks in a different order. A right philosopher asks for the fork on his right first, he waits until the fork is available, and then he asks for the fork on his left. A left philosopher makes the same operations but inverts the order of the requests (*i.e.,* he asks for the left fork first). The philosophers are hurried: everytime a philosopher asks for a fork, if the fork is not free, his request is stored in order to give him the fork as soon as possible.

Table 2. Static dining philosophers.

$Free \stackrel{def}{=}$	
$message.con = get:$	$become(Busy).$
	$send(message.act, (act{:}self, con{:}take))$
$Busy \stackrel{def}{=}$	
$message.con = get:$	$become(Busy, message.act)+$
$message.con = put \wedge state = null:$	$become(Free)+$
$message.con = put \wedge state \neq null:$	$become(Busy, null).$
	$send(state, (act{:}self, con{:}take))$
$Think \stackrel{def}{=}$	
$message.con = eat:$	$become(WaitForks).$
	$send(state.fork1, (act{:}self, con{:}get))$
$WaitForks \stackrel{def}{=}$	
$message.con = take \wedge$	
$message.act = state.fork1:$	$become(WaitForks).$
	$send(state.fork2, (act{:}self, con{:}get))+$
$message.con = take \wedge$	
$message.act = state.fork2:$	$become(Eat)$
$Eat \stackrel{def}{=}$	
$message.con = think:$	$become(Think).$
	$send(state.fork2, (act{:}self, con{:}put)).$
	$send(state.fork1, (act{:}self, con{:}put))$

This algorithm ensures that neither deadlock nor starvation can occur, if at least one right and one left philosopher are present at the table. In fact, under this condition, no circular waiting chains can be reached during the computation (see [17] for more details on the proof of the correctness of the algorithm).

The way we implement the right-left algorithm consists in representing each philosopher and each fork as an actor. We assume a fixed structure for messages which are represented using records containing two fields: *act* indicating the name of the sender of the message, and *con* representing the contents of the message.

An actor representing a fork can be free or busy. The state of the fork f_i can be *null* or p_j indicating that the philosopher p_j is waiting for that fork (hence when the fork f_i will be freed it will be given to p_j) while *null* says that no philosophers is waiting for the fork f_i. When a fork receives a *get* request and it is free it becomes busy, otherwise if it is busy it stores such a request. This is needed to implement a fair policy. See Figure 1 for a graphical representation of the possible behaviours of fork actors.

Initially all the philosophers are thinking. When a philosopher receives a message having contents *eat*, he starts asking for the forks he needs to eat, and, eventually, he starts eating. A philosopher eats until he receives a message having contents *think*, in this case he frees the forks and starts thinking. Note that, we do not consider the source of the messages *eat* and *think*, we only

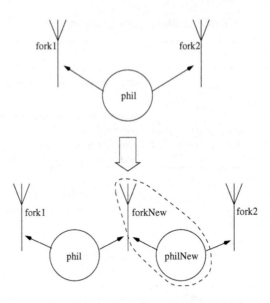

Fig. 2. Insertion of a new philosopher.

define how the philosophers react to them. The state of an actor representing a philosopher is a record having the form: $(fork1{:}f_i, fork2{:}f_j)$ where f_i is the name of the first fork that the philosopher asks for, and f_j the name of the second one. In our solution actors do not know the names of their neighboring philosophers: a philosopher only knows his first and his second fork. See Figure 1 for a graphical representation of the possible behaviours of philosopher actors.

The actor term which specifies the static dining philosophers is the following one:

$$DinPhil \stackrel{def}{=} {}^{f_0}Free_{null} \mid \ldots \mid {}^{f_{n-1}}Free_{null} \mid$$
$${}^{p_0}Think_{(fork1{:}f_0,fork2{:}f_1)} \mid \ldots \mid {}^{p_{n-1}}Think_{(fork1{:}f_0,fork2{:}f_{n-1})}$$

where the actors ${}^{f_i}Free_{null}$ and ${}^{p_i}Think_{(fork1{:}f_k,fork2{:}f_l)}$ represent the i-th fork and the i-th philosopher, respectively.[1] We explicitly require that the philosophers p_0 and p_{n-1} share the first fork, in this way the condition of having at least one right and one left philosopher is satisfied. The behaviours *Free* and *Think* are defined in Table 2 where in every behaviour is omitted the choice:

$$otherwise{:}send(self, message)$$

which resends to the same actor a message which can not be managed from the actual active message handlers.

---

[1] The state *null* of each fork indicate that no philosopher is waiting for it, while the fields *fork1* and *fork2* of the state of each philosopher contain the name of the first and of the second fork that the philosopher needs to eat.

Table 3. Adding new philosophers.

$Heap \overset{def}{=}$	
$message.con = newFork \wedge$	
$state \rangle 0:$	$become(Heap, state - 1).$
	$create(d, Free, null).$
	$send(message.act, (act:d, con:takeNew))+$
$message.con = newFork \wedge$	
$state = 0:$	$become(Heap).$
	$send(message.act, (act:self, con:noNew))$
$Think \overset{ext}{=}$	
$message.con = create:$	$become(WaitNewFork).$
	$send(heap, (act:self, con:newFork))$
$WaitNewFork \overset{def}{=}$	
$message.con = takeNew:$	$become(Think, (fork1:message.act,$
	$fork2:state.fork2)).$
	$create(d, Think, (fork1:message.act,$
	$fork2:state.fork1))+$
$message.con = noNew:$	$become(Think)$

4.3 Adding New Philosophers

In Table 3 we show how it is possible to extend the specification of static dinining philosophers of Table 2 introducing a protocol for creating new philosophers. First of all, we introduce a new heap actor:

$$^h Heap_m$$

which has the responsability of creating a new fork, whenever a new philosopher is added to the table. The maximum number of new forks that can be created is m (the initial state of the heap actor). Thus, if n is the number of starting philosophers, no more than $n + m$ philosophers can be present around the table at the same time. In fact, when the heap creates a new fork, the state of the heap is decremented, and when it reach 0 no further fork can be added.

The process of creation of a new philosopher starts when a philosopher who is thinking receives a message containing the *create* operation. A philosopher can manage this kind of messages only when he does not own any fork.

A philosopher, before creating his new neighbor, asks to the heap for a new fork. If the operation succeeds, the heap creates the new fork, and sends its name to the creating philosopher. Otherwise, the creation fails and the protocol terminates. When a philosopher receives from the heap the name of the new fork, he creates his new neighbor philosopher who is located in between himself and his first fork (see Figure 2). It is interesting to note that every philosopher is able to introduce new neighbors without knowing the name of the actual neighbors.

Table 4. Eliminating philosophers.

$Think \overset{ext}{=}$	
$message.con = kill \wedge$	
$state.phil1 \neq state.phil2$:	$become(Leave).$
	$send(state.phil1, (act{:}self, con{:}leave))+$
$message.con = kill \wedge$	
$state.phil1 = state.phil2$:	$become(Think)+$
$message.con = leave$:	$become(Halt).$
	$send(message.act, (act{:}self, con{:}state))$
$Leave \overset{def}{=}$	
$message.act = state.phil1$:	$become(Leave, state \uplus (stPhil1{:}message.con)).$
	$send(state.phil2, (act{:}self, con{:}leave))+$
$message.act = state.phil2$:	$send(state.fork1, kill).$
	$send(state.phil1, (act{:}self,$
	$\quad con{:}compState(1, self, state$
	$\quad\quad \uplus(stPhil2{:}message.con)))).$
	$send(state.phil2, (act{:}self,$
	$\quad con{:}compState(2, self, state$
	$\quad\quad \uplus(stPhil2{:}message.con))))+$
$message.con = leave$:	$become(Think)$
$Halt \overset{def}{=}$	
$message.act = state.phil1 \vee$	
$message.act = state.phil2$:	$become(Think, message.con)$
$Free \overset{ext}{=}$	
$message.con = kill$:	$send(heap, (act{:}self, con{:}forkDies))$
$Heap \overset{ext}{=}$	
$message.con = forkDies$:	$become(Heap, state + 1)$

4.4 Eliminating Philosophers

In Table 4 we show how it is possible to extend the specification of dining philosophers by introducing also the possibility of eliminating philosophers actually around the table. Also the process of elimination can start only when a philosopher is thinking. When a thinking philosopher receives a message containing the *kill* operation, he starts the protocol for leaving. To support this protocol a philosopher needs to know the names of his neighbors. Hence, we extend the state of each actor with two new fields *phil1* and *phil2* containing the name of the philosophers who share the first and the second fork, respectively.

When a philosopher starts the leaving protocol, he continues only if his neighbors are different, *i.e.*, the philosophers at the table are more than two. If the protocol continues, then the philosopher that is leaving the table asks the neighbors for the name of the forks they use in order to be able to define the new links. When he receives these informations, he eliminates his first fork and he communicates to the neighbors that they have to share his second fork (see Figure 3).

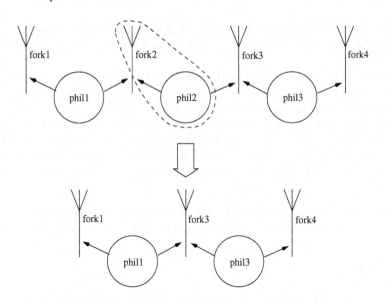

Fig. 3. Elimination of a philosopher.

Moreover, in order to be sure that at least one right and one left philosopher stay at the table, the neighbors will share the same fork as first fork. The function $compState(i, self, state)$ computes the new state of the neighbor i of the leaving philosopher:

$compState(i, self, state) \stackrel{def}{=}$
$rec: = (fork1{:}state.fork2, phil1{:}state.phil_{\bar{i}})$
if $(state.stPhil_i.phil1 = self)$
then $rec: = rec \uplus (fork2{:}state.stPhil_i.fork2, phil2{:}state.stPhil_i.phil2)$
else $rec: = rec \uplus (fork2{:}state.stPhil_i.fork1, phil2{:}state.stPhil_i.phil1)$
return rec

where $\bar{i}$ is used to indicate 2 if $i = 1$, or 1 otherwise.

If a philosopher executing this protocol receives a message having contents *leave*, which means that also a neighbor philosopher is leaving, he aborts the protocol. This protocol guarantees that the definition of the new links between the remaining philosopher and forks is correct, *i.e.,* fairness is still preserved.

5   Conclusion

We have defined an algebra of actors that provides basic object-oriented features. We have presented the specification of the Hurried Philosopher case study which shows the expressive power of the algebra.

As we have already stated, the process algebraic definition that we have given to the actors, allows the reuse of standard techniques used in the field of asynchronous calculy. For instance we have investigated the actor model by using the asynchronous bisimulation of [5] and we found interesting results presented in [12].

Currently the applications of the algebra of actors concern the study of a communication language for multi-agent systems [11] and the specification of the new asynchronous messaging service of CORBA [13].

A number of additional research items still need to be carried out. Among them: a study of how typing and inheritance issues, such as in [1], can be addressed in our algebra; and the definition of a framework for formal reasoning about programs, *e.g.,* following the style of the Hennessy and Milner logic [18].

Acknowledgements: The authors would like to acknowledge discussions with Gul Agha, and to thank the anonymous referees for helpful suggestions.

References

1. M. Abadi and L. Cardelli. An Imperative Object Calculus. *Theory and Practice of Object Systems*, 1(3):151–166, 1995.
2. G. Agha. *Actors: A Model of Concurrent Computation in Distributed Systems.* MIT Press, Cambridge, MA, 1986.
3. G. Agha, I. Mason, S. Smith, and C. Talcott. Towards a Theory of Actor Computation. In *Proc. of CONCUR'92*, volume 630 of *Lecture Notes in Computer Science*, pages 564–579. Springer Verlag, 1992.
4. G. Agha, I. Mason, S. F. Smith, and C. Talcott. A Foundation for Actor Computation. *Journal of Functional Programming*, 7(1):1–69, January 1997.
5. R. Amadio, I. Castellani, and D. Sangiorgi. On Bisimulations for the Asynchronous π-Calculus. *Theoretical Computer Science*, 195(2):291–324, 1998.
6. J. Andreoli and R. Pareschi. Linear Objects: Logical Processes with Built-in Inheritance. *New Generation Computing*, 9(3-4):445–473, 1991.
7. M. Campione and K. Walrath. *The Java Tutorial: Object-Oriented Programming for the Internet.* Addison Wesley, 1996.
8. L. Cardelli. A Language with Distributed Scope. *Computing Systems*, 8(1):27–59, January 1995.
9. P. Ciancarini, R. Gorrieri, and G. Zavattaro. Towards a Calculus for Generative Communication. In *Proc. FMOODS'96*, pages 283–297. Chapmann & Hall, 1996.
10. F. deBoer, J. Klop, and C. Palamidessi. Asynchronous Communication in Process Algebra. In *Proc. LICS'92*, pages 137–159. IEEE Computer Society Press, 1992.
11. M. Gaspari. Concurrency and knowledge-level communication in agent languages. *Artificial Intelligence*, 105(1-2):1–45, 1998.
12. M. Gaspari and G. Zavattaro. An algebra of actors. In *Proc. 3nd IFIP Conf. on Formal Methods for Open Object-Based Distributed Systems (FMOODS)*, pages 3–18. Kluwer Academic Publishers, Feb 1999.
13. M. Gaspari and G. Zavattaro. A process algebraic specification of the new asynchronous corba messaging service. In *Proc. European Conf. on Object Oriented Programming (ECOOP)*, Lecture Notes in Computer Science. Springer-Verlag, Berlin, 1999.
14. C. Hewitt. Viewing control structures as patterns of passing messages. *Artificial Intelligence*, 8(3):323–364, 1977.

15. K. Honda and M. Tokoro. An Object Calculus for Asynchronous Communication. In *The Fifth European Conference on Object-Oriented Programming*, volume 512 of *Lecture Notes in Computer Science*, pages 141–162. Springer-Verlag, Berlin, 1991.

16. S. Khoshafian and G. Copeland. Object Identity. In *Proc. OOPSLA '86*, pages 406–416, September 1986.

17. N. Lynch. *Distributed Algorithms*. Morgan Kaufmann Publishers, San Francisco, CA, USA, 1996.

18. R. Milner. *Communication and Concurrency*. Prentice Hall, 1989.

19. R. Milner. Functions as processes. *Mathematical Structures in Computer Science*, 2(2):119–141, 1992.

20. R. Milner. Elements of interaction. *Communications of the ACM*, 36(1):79–89, January 1993.

21. G. Plotkin. A structural approach to operational semantics. Technical Report DAIMI FN-19, Department of Computer Science, Aarhus University, Denmark, 1981.

22. C. Talcott. Interaction Semantics for Components of Distributed Systems. In *Proc. FMOODS'96*, pages 154–169. Chapmann & Hall, 1996.

23. V. Vasconcelos. Typed Concurrent Objects. In *8th European Conference on Object Oriented Programming*, Lecture Notes in Computer Science. Springer-Verlag, Berlin, 1994.

24. S. Vinoski. CORBA: Integrating Diverse Applications Within Distributed Heterogeneous Environments. *IEEE Communications Magazines*, 14(2), February 1997.

25. D. Walker. Objects in the π–calculus. *Information and Computation*, 116(2):253–271, 1995.

Formal Reasoning about Actor Programs
Using Temporal Logic

Susanne Schacht*

isys software gmbh, Engelbergerstr. 21,
79106 Freiburg, Germany

Abstract. We here present an approach to reasoning about actor programs on the basis of temporal logic. Temporal logic is particularly appropriate for the specification of concurrent programs, but most known temporal logic proof systems for concurrent computations rely on imperative language constructs, ignoring, e.g., the creation of processes and the dynamic configuration of communication channels, which are crucial for actor based programming. We will demonstrate our approach by applying it to a detection algorithm for termination of parts of a computation.

1 Temporal Logic for Actor Computations

Temporal logic is particularly appropriate for reasoning about distributed programs. Whereas operational or denotational semantics are very fine-grained, temporal logic statements can be sufficiently abstract to make useful assertions about the overall behavior of a complex system. Using a temporal logic framework, safety properties (invariants like partial correctness, absence of deadlock, mutual exclusion etc.) and liveness properties (termination, total correctness, cycles) of concurrent programs can be shown without taking implementation details into account [22,5]. One advantage of the use of temporal logic specifications is their incrementality. New features can simply be added, thus decreasing the set of possible models, i.e. implementations. However, most temporal logic proof systems for concurrent computations rely on imperative language constructs. They only take a simple action language as a basis to build the states of a program and ignore the creation of processes and the dynamic configuration of communication channels, issues that are crucial for object-oriented programming, in particular, the actor model of computation [14,1].

We here present an approach to reasoning about actor programs on the basis of temporal logic, called Temporal Actor Logic – TAL. In general, a proof system for a specific program consists of three parts: a basic temporal logic system, a

---

* This work has been done as a part of my dissertation at the Computational Linguistics Lab, Freiburg University, Germany. Thanks to all of the colleagues and especially to Prof. Hahn!

G. Agha et al. (Eds.): Concurrent OOP and PN, LNCS 2001, pp. 445–460, 2001.
© Springer-Verlag Berlin Heidelberg 2001

programming language specific part (the *domain* part) and a program specific part. In Section 2, we introduce the basic temporal logic system. In Section 3, we briefly sketch the actor model and introduce a simple syntax for actor programs. Next, we adjust the general proof system to the description of actor computations in Section 4, thus introducing the domain part of our proof system. We then demonstrate the application of the logical framework – the building of the program part – to a definition of receipt handlers which manage the termination recognition of subcomputations.

2   A Basic Proof System

In this section, we give a basic proof system for discrete and linear[1] temporal logic, following [16]. We will first introduce a set of temporal-logic *future* operators:

"next", $\bigcirc A$,   "A will hold at the next state"
"eventually", $\diamondsuit A$,   "A holds now or at least at one of the following states"
"always", $\square A$,   "A holds now and at all following states"
"until", $A \, \mathcal{U} \, B$, "A will hold up to the first state at which B holds"
"atnext", $A \, \mathcal{N} \, B$, "A will hold at the first state at which B holds"

The semantics of a linear first-order temporal logic language is given by a (Kripke-) structure $\mathbf{K}$ which consists of a structure for the first-order kernel (all formulae containing no temporal operators) and a sequence of variable valuations. $\mathbf{K}_i(F)$ is the valuation of a formula F (in state i) as in FOL and, additionally, for the temporal operators:

$\mathbf{K}_i(\bigcirc A) = \mathbf{t}$ iff $\mathbf{K}_{i+1}(A) = \mathbf{t}$
$\mathbf{K}_i(\diamondsuit A) = \mathbf{t}$ iff $\mathbf{K}_j(A) = \mathbf{t}$ for some $j \geq i$
$\mathbf{K}_i(\square A) = \mathbf{t}$ iff $\mathbf{K}_j(A) = \mathbf{t}$ for every $j \geq i$
$\mathbf{K}_i(A \, \mathcal{U} \, B) = \mathbf{t}$ iff $\mathbf{K}_j(B) = \mathbf{t}$ for some $j \geq i$ and $\mathbf{K}_k(A) = \mathbf{t}$ for every $k, i \leq k < j$
$\mathbf{K}_i(A \, \mathcal{N} \, B) = \mathbf{t}$ iff $\mathbf{K}_j(B) = \mathbf{f}$ for every $j > i$
    or $\mathbf{K}_k(A) = \mathbf{t}$ for the smallest $k > i$ with $\mathbf{K}_k(B) = \mathbf{t}$

A is *valid in* or *satisfied by* $\mathbf{K}$, $\models_{\mathbf{K}} A$, if $\mathbf{K}_i(A) = \mathbf{t}$ for every $i \geq 0$. A is *valid* if $\models_{\mathbf{K}} A$ for every $\mathbf{K}$, i.e. that a valid formula must be true in *every* possible state. B *follows from* A, i.e., $A \models B$ if $\models_{\mathbf{K}} B$ for every $\mathbf{K}$ with $\models_{\mathbf{K}} A$. In contradistinction to the correspondence between the semantic satisfiability $\models$ and the syntactic implication $\rightarrow$, i.e., $A \models B$ iff $\models A \rightarrow B$ in classical logic, in temporal logic implication applies only locally to states, and $A \models B$ is equivalent to $\models \square(A \rightarrow B)$; this can also be written as "$\Rightarrow$" (*entailment*): $\square(A \rightarrow B)$ iff

---

[1]   Branching-time logics (cf., e.g., [10]) are much more complex than linear ones, as their models are graphs. This extra amount of complexity is not needed in our application. Hence, we concentrate on sets of sequences that can be adequately described by linear temporal logic.

$A \Rightarrow B$. Accordingly, $A \models B$ and $B \models A$ is captured by *congruence*: $\Box(A \leftrightarrow B)$ iff $A \Leftrightarrow B$.

The axioms and inference rules below together with all FOL tautologies, build our basic proof system. All As and Bs are formula schemata; axioms can be instantiated with any *wff*, while for the rules' premises instantiations occur with already derived formulae. The soundness of the deduction system, stating that any derivable formula is valid, can be proven by induction over the derivation of A.

$$\neg \bigcirc A \Leftrightarrow \bigcirc \neg A$$
$$\bigcirc (A \to B) \Rightarrow (\bigcirc A \to \bigcirc B)$$
$$\Box A \Rightarrow (A \land \bigcirc \Box A)$$
$$\bigcirc \Box \neg B \Rightarrow A \mathcal{N} B$$
$$A \mathcal{N} B \Leftrightarrow \bigcirc (B \to A) \land \bigcirc (\neg B \to A \mathcal{N} B)$$
$$\forall x \bigcirc A \to \bigcirc \forall x A$$
$$A \to \bigcirc A$$
(if A contains no flexible variable)

modus ponens:
$$\frac{A, A \to B}{B}$$

next:
$$\frac{A}{\bigcirc A}$$

induction:
$$\frac{A \to B, A \to \bigcirc A}{A \to \Box B}$$

3   The Actor Model

An actor system consists of several concurrently processing objects, called actors, which can send messages to each other asynchronously; messages are assumed to be buffered in their receivers' mail queue. Message processing is required to be *fair*, i.e. all sent messages are eventually received and consumed. An actor *program* contains a set of actor definitions as well as some initially created actors and messages sent to them. Actor definitions declare internal variables for actors instantiated from them (the actors' *acquaintances*), and define their instances' reaction to incoming messages (their *behavior*) by way of method definitions. Each instantiated actor has a unique mail address (its *identity*). Upon receiving a message an actor performs a composite action consisting of sending further messages to other actors it knows about (**send**), creating new actors from actor definitions, thus supplying their initial acquaintances (**create**), or specifying a new behavior and/or new acquaintances for itself (**become**). The replacement behavior defined by a **become** action (of which only one per method can be performed) is effective for the next message the actor accepts. Additionally, we allow for local variables, using a **let** expression for binding actors to identifiers.

Several kinds of formal semantics have been proposed for the actor model, usually based on the notion of events (processing of a single message at an actor) and tasks (pending events). Agha's operational semantics [2] assigns a tree of *configurations* related by transitions to each program as its meaning. A configuration consists of a *local states function* and a set of unprocessed *tasks*, where the local states function maps actor addresses to behaviors. Tasks are functions

```
program       ::= actorDef* action
actorDef      ::= 'defActor' actorType (variable*)
                     methodDef*
                  'endActor'
methodDef     ::= 'defMeth' messageKey (variable*)
                     action
                  'endMeth'
action        ::= action; action
                | 'send' actor messageKey (actor*)
                | 'if' boolean 'then' (action)
                  ['else' (action)]
                | 'let' variable 'be' value 'in' (action)
                | 'become' actorType ({variable: value}+)
actor         ::= 'self'| variable |('create' actorType(value*))
actorType     ::= identifier
messageKey    ::= identifier
variable      ::= identifier
value         ::= basicValue | actor
basicValue    ::= 'nil' | integer | boolean | char | real |...
```

Fig. 1. Syntax of a basic actor language

as well, mapping a unique tag (an identifier created in order to distinguish between similar tasks) to pairs consisting of a target actor address and a message. Transitions between configurations are determined by choosing one of several unprocessed tasks for execution: the effects of processing of this task's message at its target are evaluated, giving a set of new tasks, a set of newly created actors and the replacement behavior for the target actor. This semantics depends on the existence of one (of several possible) global view(s) of the system: the configurations describe all actors involved. This kind of semantics as well as the denotational one of Clinger [8] does not allow abstract propositions about *parts* of the system's computations that could be combined. Mechanisms for abstraction over the total ordering of events as well as abstraction over specific details of a configuration are needed to make comprehensive statements about the overall system behavior or that of selected parts.

A more recent approach to actor semantics is conducted in [4] where the λ-calculus-based semantics of functional languages in terms of transitions modeled as reduction steps is extended by integrating actor primitives like send, become, newAddress, and initBehavior, the latter building create. To specify *open* systems, a configuration states *internal* recipient actors which are able to receive messages from outside and *external* actors which do not belong to the actors in the configuration but can be addressed by them. Although partial descriptions of an actor system become possible now, this semantics, on the other hand, looses an abstraction level as message processing is not seen as atomic, but as an application of a function (i.e. the receiver's behavior) to a value (i.e. the message), and is merged with internal transitions. In order to compare our temporal logic to this approach, the reminder of this chapter describes it in depth.

The behavior of an actor $b \in \mathcal{B}$ is the fixed point of a function defined by a λ-abstraction. The processing of a message $m \in \mathcal{M}$ is modelled as the application of a behavior to this message, $\mathrm{app}(b, m)$. The actor syntax can be translated into the λ-based syntax as follows:

```
let x be v in (e)            ::= app(λx.e, v)
let a be (create b) in(e)    ::= app(λa.(seq(initBehave(a, b),newAddress)))
```

Some further definitions: Usual λ expressions e are taken from the set $\mathcal{E}$, actor addresses a from $\mathcal{AI}$. Local states s are elements of $\mathcal{AS} := \{(?_a)|a \in \mathcal{AI}\} \cup \{(b)|b \in \mathcal{B}\} \cup \{[e]|e \in \mathcal{E}\}$. A local states function s_a is defined as an element of $\alpha : \mathcal{AI} \cup \{\_\} \to \mathcal{AS}$. Messages m are in $\mathcal{M}$, behaviors b in $\mathcal{B}$, and tasks (a, m) in $\mathcal{T} := \mathcal{AI} \times \mathcal{M}$.

Let α be a partial function from addresses $(\mathcal{AI} \cup \{\_\}^2)$ to local states $(\mathcal{AS})$, let $\theta \subseteq \mathcal{T}$ be a set of tasks, $\rho \subset \mathrm{Dom}(\alpha)$ a set of recipient actors, and $\eta \subset \mathcal{AI}$ ($\eta \cap \mathrm{Dom}(\alpha) = \emptyset$) a set of external actors. This way, a *configuration* is a quadrupel c of $(\alpha, theta, \rho, \eta)$ written as $\langle \alpha \mid \theta \rangle_\eta^\rho$.

The transition rules for configurations are shown below (we omit the rules for $\langle \mathbf{in} \rangle$ and $\langle \mathbf{out} \rangle$, the rules that concern external actors). The $\langle \mathbf{fun} \rangle$ rule is for internal computation steps that can be done in λ-calculus. The other rules use reduction contexts that incorporate the usual left-first - call-by-value reduction strategy [23]. These are expressions with a "hole" that can be filled with a reducible expression (e.g., e in $R[\![e]\!]$) [11]. $\langle \mathbf{newadd} \rangle$ reduces a reduction context with $\mathbf{newAddress}$ to a new address with an undefined state and to the reduction context that contains this new address. $\langle \mathbf{init} \rangle$ assigns a new behavior to an actor with undefined state. $\langle \mathbf{become} \rangle$ reduces the actor to a new behavior and leaves an anonymous address in the context. $\langle \mathbf{send} \rangle$ adds a new task to θ, and $\langle \mathbf{rcv} \rangle$ reduces an idle actor and one of its tasks to the application of the actor's behavior to the task.

$\langle \mathbf{fun} \rangle \qquad e \mapsto_\lambda e' \Rightarrow \langle \alpha \cup \{[e]_a\} \mid \theta \rangle_\eta^\rho \mapsto \langle \alpha \cup \{[e']_a\} \mid \theta \rangle_\eta^\rho$

$\langle \mathbf{newadd} \rangle \qquad \langle \alpha \cup \{[R[\![\mathbf{newAddress}()]\!]]_a\} \mid \theta \rangle_\eta^\rho \mapsto \langle \alpha \cup \{[R[\![a']\!]]_a, (?_a)_{a'}\} \mid \theta \rangle_\eta^\rho,$
$\qquad\qquad\qquad\qquad\qquad\qquad\qquad\qquad\qquad\qquad\qquad\qquad a' \notin \mathrm{Dom}(\alpha) \cup \{a\} \cup \eta$

$\langle \mathbf{init} \rangle \quad \alpha \cup \{[R[\![\mathbf{initBehave}(a', b)]\!]]_a, (?_a)_{a'}\} \mid \theta \rangle_\eta^\rho \mapsto \langle \alpha \cup \{[R[\![\mathbf{nil}]\!]]_a, (b)_{a'}\} \mid \theta \rangle_\eta^\rho$

$\langle \mathbf{become} \rangle \qquad \langle \alpha \cup \{[R[\![\mathbf{become}(b)]\!]]_a\} \mid \theta \rangle_\eta^\rho \mapsto \langle \alpha \cup \{R[\![\mathbf{nil}]\!]]_\_, (b)_a\} \mid \theta \rangle_\eta^\rho$

$\langle \mathbf{send} \rangle \qquad \langle \alpha \cup \{[R[\![\mathbf{send}(a', m)]\!]]_a\} \mid \theta \rangle_\eta^\rho \mapsto \langle \alpha \cup \{[R[\![\mathbf{nil}]\!]]_a\} \mid \theta \cup \{(a', m)\} \rangle_\eta^\rho$

$\langle \mathbf{rcv} \rangle \qquad \langle \alpha \cup \{(b)_a\} \mid \{(a, m)\} \cup \theta \rangle_\eta^\rho \mapsto \langle \alpha \cup \{[\mathbf{app}(b, m)]_a\} \mid \theta \rangle_\eta^\rho$

---

[2] $\_$ is like an anonymous variable in PROLOG: only its existence is of interest.

4   Temporal Actor Logic

In order to adapt the system from Section 2 to the actor model and to construct the domain part of the proof system, the actor language must provide the elements for building local, non-temporal propositions about pending tasks and the actors involved. These will serve as basic state descriptions. Sequences of states will then be defined by determining the effects of one of the pending tasks. We assume a standard FOL plus identity in order to express propositions about the bindings between identifiers and values (i.e., actor addresses and basic values) and between actors and their definitions. Accordingly, we use unary predicates to indicate whether a term denotes an instance of a certain actor definition and binary predicates named after acquaintance or variable names to indicate an actor's acquaintances. A name can be used as a function symbol, taking the actor as a parameter and denoting the acquaintance's or variable's value, thus forming a term.

For tasks, we use the notation (a, m), with a denoting an actor address and m denoting a unique message. The predicate Task takes three places, two for the task itself and a third one to actually indicate whether the task is an element of the set of unprocessed tasks. This third parameter is a Boolean variable whose value flips from *"false"* to *"true"* (thus denoting an unprocessed task), is kept unchanged until the task is processed, and after being processed flips back from *"true"* to *"false"* (this value will not be changed anymore). Such a behavior guarantees the interpretation of the predicate Task not to change over time in our semantics. This feature is axiomized as follows:

$$\forall (a, m, f) :$$
$$\mathsf{Task}(a, m, \mathit{true}) \wedge \neg\mathsf{Task}(a, m, \mathit{false})$$
$$\wedge (\,\square\, \neg\mathsf{Task}(a, m, f) \vee (\neg\mathsf{Task}(a, m, f)\,\mathcal{U}\,(\mathsf{Task}(a, m, f)\,\mathcal{U}\,\square\, \neg\mathsf{Task}(a, m, f)))$$

The consideration of the third parameter would massively complicate subsequent derivations, and does not at all clarify the more interesting parts of our logic. Therefore, we will use the predicate Task as a two-place predicate sometimes, but keep in mind that the easier notation does not fit to the easier semantics, but can be transformed into a more elaborate, formally fully adequate one.

A state proposition in the underlying logic consists of SP, a set of propositions (i.e., a conjunction of basic propositions) about bindings of identifiers to values (addresses or basic values) and actors to definitions, and TP, which conjoins Task(a, m, *true*) for every (a, m) that is a task. The effects of processing a task turn up as the creation of new tasks and changes affecting the binding properties. A formal definition for the state transition will be given below. Meanwhile, we express the effects of processing a task (a, m) on SP and TP as *effects*$(a, m)(SP, TP)$, or even shorter as *effects*(a, m). *effects*$(a, m)(SP)$ denotes the local state proposition SP after (a, m) has been processed, *effects*$(a, m)(TP)$ denotes the new proposition about tasks to be executed.

The following axiom *(single transition)* states that as long as there are un-processed tasks, exactly one of them, (a, m), is processed at each step and no changes on SP and TP are made other than those caused by (a, m):

$$SP \wedge TP \wedge \exists\, a, m, f : \mathsf{Task}(a, m, f) \leftrightarrow$$
$$\forall\, b, n, g : \mathsf{Task}(b, n, g) \leftrightarrow$$
$$\bigcirc\, (\neg\mathsf{Task}(a, m, f) \wedge (\mathsf{Task}(b, n, g) \vee (a, m) = (b, n)) \wedge \mathit{effects}(a, m))$$

Fairness must also be included as an axiom of the proof system — any pending task will stay pending until eventually it is chosen for execution. Then, in the next state, it will be removed from *tasks* and its effects can be observed. Since no pending task can be cancelled, this axiom *(fairness)* is sufficient to guarantee fair computations.

$$\forall\, (a, m, f) : \mathsf{Task}(a, m, f) \leftrightarrow$$
$$\mathsf{Task}(a, m, f)\, \mathcal{U}\, (\mathsf{Task}(a, m, f) \wedge \bigcirc\, (\mathit{effects}(a, m) \wedge \square\, \neg\mathsf{Task}(a, m, f)))$$

From this, we can conclude:

$$\forall\, (a, m, f) : \mathsf{Task}(a, m, f) \rightarrow \lozenge\, (\mathsf{Task}(a, m, f) \wedge \bigcirc\, \mathit{effects}(a, m)) \qquad (1)$$

since:

$$\mathbf{K}_i(A\,\mathcal{U}\, (A \wedge \bigcirc\, (B \wedge \neg A))) = \mathbf{t}$$
$$\Rightarrow \mathbf{K}_j(A \wedge \bigcirc\, (B \wedge \neg A))) = \mathbf{t} \text{ for } j > i$$
$$\Leftrightarrow \mathbf{K}_i(\lozenge\, (A \wedge \bigcirc\, (B \wedge \neg A))) = \mathbf{t}$$
$$\Rightarrow \mathbf{K}_i(\lozenge\, (A \wedge \bigcirc\, B)) = \mathbf{t}\,\blacksquare$$

To reason about a particular program, existing actors and sent messages as well as the program's actor definitions must be accounted for. The local state propositions about actors are determined by their binding to actor definitions and the values of their acquaintances, both forming the state propositions SP. For tasks (a, m), from a's method definition for m's message key, we take the combined action to determine m's effects. The bindings of m's parameters can be found in SP, as they must have been either acquaintances or variables of m's sender or newly created actors known from the previous steps of computation. If no identifier of an actor is available (e.g., if it is created anonymously), we use a variable which can be considered existentially quantified. With these conventions, $\mathit{effects}(a, m)(SP,\ TP)$ can be defined in a straightforward way. The effects of performing (a, m) on the set of pending tasks, *tasks*, are determined by identifying all **send** actions in the method's definition that are actually performed (depending on a's state). The receivers and messages are combined to tasks and then added to *tasks*. Below, $act(a, m)$ denotes the action that is the body of m's definition at a ("\" denotes the set complement operator).

$$\mathit{effects}(a, m)(\mathit{tasks}) := \mathit{tasks} \setminus \{(a, m)\}\ \cup\ \mathit{newTasks}(act(a, m))$$

with: $newTasks(act(a, m)) :=$

$$
\begin{cases}
\{(b, n)\} & \text{if } act(a, m) = \texttt{send b n} \\
newTasks(\text{act}) & \text{if } act(a, m) = \texttt{let <var> be <actor> in act} \\
newTasks(\text{act}_1) \cup newTasks(\text{act}_2) & \text{if } act(a, m) = \texttt{act}_1 \texttt{ ; act}_2 \\
newTasks(\text{act}_1) & \text{if } act(a, m) = \texttt{if cond then act}_1 \texttt{ else act}_2 \\
& \quad \text{and } \texttt{cond} \text{ is true} \\
newTasks(\text{act}_2) & \text{if } act(a, m) = \texttt{if cond then act}_1 \texttt{ else act}_2 \\
& \quad \text{and } \texttt{cond} \text{ is false} \\
\emptyset & \text{else}
\end{cases}
$$

The proposition TP is defined via $tasks$: $TP := \bigwedge_{(b,n)\in tasks} \mathsf{Task}(b, n)$ and therefore:

$$
effects(a, m)(TP) := \bigwedge_{(b,n)\in effects(a,m)(tasks)} \mathsf{Task}(b, n).
$$

The effects on the local state propositions, SP, concern newly created actors via **create** actions and changes on a via **become**. For any actor b created from a definition D with initial acquaintances aqc_i bound to a_i, $D(b)$ and $Aqc_i(b, a_i)$ are added to SP. Also, we refer to a_i as $acq(b)$, and extend the function as well as the predicate notation to local variables where necessary. All elements of SP concerning a itself are changed according to the occurrence of a **become** action.

$$
effects(a, m)(SP) := SP \cup created(act(a, m)) \cup new(act(a, m)) \setminus old(act(a, m))
$$

with the following definitions of $created$, new, and old, respectively:

$$
created(act(a, m)) := \begin{cases}
\{\mathsf{D}(\mathsf{var}(a)), \mathsf{Aqc}_1(\mathsf{var}(a), a_1), \ldots, \mathsf{Aqc}_n(\mathsf{var}(a), a_n)\} \\
\quad \text{if } act(a, m) = \texttt{let var be create D(a}_1\texttt{..a}_n\texttt{)in..} \\
\quad \text{and analogously for other occurrences of } \texttt{create}
\end{cases}
$$

$$
new(act(a, m)) := \begin{cases}
\{\mathsf{D}(a), \mathsf{Aqc}_1(a, a_1), .., \mathsf{Aqc}_n(a, a_n)\} \\
\quad \text{if } act(a, m) = \texttt{become create D(a}_1\texttt{..a}_n\texttt{)}
\end{cases}
$$

$$
old(act(a, m)) := \begin{cases}
\{\mathsf{D}_{prev}(a), \mathsf{Aqc}_1(a, b_1), .., \mathsf{Aqc}_k(a, b_k)\} \\
\quad \text{if } act(a, m) = \texttt{become create D(a}_1\texttt{..a}_n\texttt{)} \text{ and} \\
\quad \mathsf{D}_{prev}(b_1..b_k) \text{ the previous definition of } a
\end{cases}
$$

For the initial configuration, $tasks$ and SP are determined from the initially created actors and sent messages in the program.

To summarize, a TAL proof system Θ for a program P as in Section 3 is a quadruple $\langle \Sigma^\Theta, Op, R_b, R_p \rangle$, consisting of a signature Σ^Θ, the set Op of the usual first-order and temporal logic operators, and two sets of rules and axioms. The set R_b contains the axioms and rules of the basic proof system (cf. Section 2). The set R_p consists of the rules taken from P by instantiating the axioms for fairness and single transition. This is achieved by evaluating the $effects$ function on the methods (cf. Section 5 for an example). The signature Σ^Θ is defined as (V, K, O, M, F, P) with V the set of variables, K and O constants and operators for arithmetics and set theory, M a set of function symbols of various arities, F a set of unary function symbols, and P a set of unary and binary predicate symbols as mentioned above. M contains the `messageKeys` in P, F contains symbols that appear in P as `variable`, `acquaintance` or `parameter`; P consists of the symbol Task and the symbols that appear as `actorType` and as `acquaintance`. By convention, we write predicate symbols starting with a capital letter and functions with a lower-case letter.

Terms and formulae are inductively defined; terms are: any element of V and K, any element of F and O applied to terms. Any element of M applied to terms is a message term. Formulae are: $\mathsf{Task}(s, m)$, $\mathrm{P}(s)$, $\forall v : G$, $\exists v : G$, $(G \wedge H)$, $(G \vee H)$, $\neg G$, $\Box\, G$, $\bigcirc G$, $G\,\mathcal{U}\,H$ for a term s, a message term m, an element P of P, a variable v and formulae G and H.

These are the inference rules for TAL:

General inference rules:

$$\text{modus ponens: } \frac{A, A \to B}{B} \qquad \text{next: } \frac{A}{\bigcirc A} \qquad \text{induction: } \frac{A \to B, A \to \bigcirc A}{A \to \Box B}$$

Axioms for the Task *switch:*

$$\frac{}{\mathsf{Task}(a, m, \textit{true})} \qquad \frac{}{\neg\mathsf{Task}(a, m, \textit{false})}$$

$$\frac{}{(\Box\,\neg\mathsf{Task}(a, m, f) \vee (\neg\mathsf{Task}(a, m, f)\,\mathcal{U}\,(\mathsf{Task}(a, m, f)\,\mathcal{U}\,\Box\,\neg\mathsf{Task}(a, m, f))))}$$

Single transition:

$$\frac{SP \wedge TP \wedge \exists\, a, m, f : \mathsf{Task}(a, m, f)}{\begin{array}{l}\forall\, b, n, g : \mathsf{Task}(b, n, g) \leftrightarrow \\ \quad \bigcirc\,(\neg\mathsf{Task}(a, m, f) \wedge (\mathsf{Task}(b, n, g) \vee (a, m) = (b, n)) \wedge \textit{effects}(a, m))\end{array}}$$

Fairness:

$$\frac{\mathsf{Task}(a, m, f)}{\mathsf{Task}(a, m, f)\,\mathcal{U}\,(\mathsf{Task}(a, m, f) \wedge \bigcirc\,(\textit{effects}(a, m) \wedge \Box\,\neg\mathsf{Task}(a, m, f)))}$$

$$\text{or: } \frac{\mathsf{Task}(a, m, f)}{\Diamond\,(\mathsf{Task}(a, m, f) \wedge \bigcirc\,(\textit{effects}(a, m) \wedge \Box\,\neg\mathsf{Task}(a, m, f)))}$$

Rules for method definitions:

$$\frac{\mathsf{Task}(a, m, f) \wedge \bigcirc\,\textit{effects}(a, m)}{\begin{array}{l}\bigcirc\,(\neg\mathsf{Task}(a, m, f) \wedge \forall(a_i, m_i) \in \textit{newTasks}(\texttt{action}(a, m))\exists f_i \mathsf{Task}(a_i, m_i, f_i) \\ \quad \wedge\,\textit{created}(\texttt{action}(a, m)) \wedge \neg\textit{old}(\texttt{action}(a, m)) \wedge \textit{new}(\texttt{action}(a, m)))\end{array}}$$

A model for a TAL formula is a modified Kripke structure $\mathbf{K} = (\mathcal{U}, \mathcal{I}, \mathcal{V})$, i.e. a universe, an interpretation of predicate and function symbols and of constants and a sequence of variable assignments $\mathcal{V}_i$, $i = 1..n$, $\mathcal{V}_i(x) \in \mathcal{U}$ for all $x \in V$. We divide the universe $\mathcal{U}$ into actors (identified by their address), messages, and simple values: $\mathcal{U} = \mathcal{AI} \cup \mathcal{M} \cup Bas$, $\mathcal{AI} \cap \mathcal{M} = \mathcal{M} \cap Bas = \mathcal{AI} \cap Bas = \emptyset$. The simple values Bas can be differentiated by types, e.g. Boolean values or integers.

Terms:

$$\forall x \in V: \quad \mathbf{K}_i(x) \quad = \mathcal{V}_i(x)$$
$$\forall k \in K: \quad \mathbf{K}_i(k) \quad = \mathcal{I}(k)$$
$$\forall f \in F: \quad \mathbf{K}_i(f(t)) = \mathcal{I}(f)(\mathbf{K}_i(t))$$
$$\forall o \in O: \quad \mathbf{K}_i(o(\bar{t})) = \mathcal{I}(o)(\overline{\mathbf{K}_i(t)})$$
$$\forall m \in M: \mathbf{K}_i(m(\bar{t})) = \mathcal{I}(m)(\overline{\mathbf{K}_i(t)})$$

Atomic formulae:

$$\forall r \in V_{\text{actor}}, m \in M, f \in V_{\text{bool}}: \mathbf{K}_i(\mathsf{Task}(r, m, f)) = \mathbf{t} \text{ iff } \mathbf{K}_i(f) = \mathbf{t}.$$
$$\forall t \in V_{\text{actor}}, \mathrm{P} \in M: \qquad \mathbf{K}_i(\mathrm{P}(t)) \qquad = \mathbf{t} \text{ iff } \mathbf{K}_i(t) \in \mathcal{I}(\mathrm{P})$$

For all formulae F and G:

$$\forall v \in V: \mathbf{K}_i(\forall v: G) = \mathbf{t} \text{ iff } \mathbf{K}_i(G[v/t]) = \mathbf{t} \text{ for all } t.$$
$$\forall v \in V: \mathbf{K}_i(\exists v: G) = \mathbf{t} \text{ iff } \mathbf{K}_i(G[v/t]) = \mathbf{t} \text{ for a } t.$$
$$\mathbf{K}_i(\neg F) \quad = \mathbf{t} \text{ iff } \mathbf{K}_i(F) = \mathbf{f}.$$
$$\mathbf{K}_i(F \wedge G) = \mathbf{t} \text{ iff } \mathbf{K}_i(F) = \mathbf{t} \text{ and } \mathbf{K}_i(G) = \mathbf{t}.$$
$$\mathbf{K}_i(F \vee G) = \mathbf{t} \text{ iff } \mathbf{K}_i(F) = \mathbf{t} \text{ or } \mathbf{K}_i(G) = \mathbf{t}.$$
$$\mathbf{K}_i(\Box F) \quad = \mathbf{t} \text{ iff } \mathbf{K}_j(F) = \mathbf{t} \text{ for all } j \geq i.$$
$$\mathbf{K}_i(\Diamond F) \quad = \mathbf{t} \text{ iff } \mathbf{K}_j(F) = \mathbf{t} \text{ for a } j \geq i.$$
$$\mathbf{K}_i(\bigcirc F) \quad = \mathbf{t} \text{ iff } \mathbf{K}_{i+1}(F) = \mathbf{t}.$$
$$\mathbf{K}_i(F \,\mathcal{U}\, G) = \mathbf{t} \text{ iff } \mathbf{K}_j(G) = \mathbf{t} \text{ for a } j \geq i$$
$$\text{and } \mathbf{K}_k(F) = \mathbf{t} \text{ for all } k, i \leq k < j.$$

The main application of TAL is to study given actor programs and to show properties about them that are more general than those that can be directly derived from the program code or its operational semantics. One possibility to show that a given actor program P is a model of a TAL formula F is to construct F from P. For this purpose, one has to build the formulae for every actor definition in P and for the initial actors and messages as shown above and combine them by $\wedge$. This is demonstrated next.

5   Example: Partial Termination

Due to its fine granularity, the asynchronous actor mode of message passing does not easily fit complex real-world applications. Especially, synchronization constructs are needed (for suggestions, cf. [12], and [3]). As an example for applying the temporal logic system, we will discuss a programming scheme for 'partial termination'. Instead of detecting the termination of arbitrary processes, we focus on a characteristic subset of computations which are obliged to signal their termination to their initiators without considering other activities of involved actors. We use a set of *acknowledged* messages that form the sub-computation under consideration. Any involved message sends a receipt that may carry unique descriptions called *tags* of further receipts that have to be expected. Special actors, `ReceiptHandlers`, accept the receipts and manage the tags already received and those still to be expected in two separate sets, `tagsRec` and `tagsExp`, respectively, and, finally, send a `terminated` message to their creator `replyDest`. Assuming `ReceiptHandlers` with initially empty sets, a computation process is considered terminated if both sets are empty and if at least one receipt message has arrived (since no spontaneous computation can occur in an actor system). As a safety property, we expect no `terminated` message to be sent (i.e., being an element of *tasks* in the next state) if both sets are not empty. This can be expressed as:

$$\Box(\, \bigcirc (\mathsf{Task}(\mathrm{replyDest(x)}, \mathrm{terminated}) \to (\exp(x) = \emptyset = \mathrm{rec}(x)))) \tag{2}$$

`ReceiptHandlers` can be defined as follows:[3]

```
defActor ReceiptHandler (tagsExp tagsRec tag replyDest)
  meth receipt (rtag furtherTags)
    let exp be ((tagsExp ∪ (furtherTags \ tagsRec)) \ {rtag})
      in let rec be (tagsRec ∪ {rtag}) \ (furtherTags ∪ tagsExp)
        in if ((exp = ∅) and (rec = ∅))
          then (send replyDest terminated; become Inactive)
          else (become ReceiptHandler (exp rec tag replyDest))
  endMeth
  meth generateTag (returnAddress)
    send returnAddress return (tag+1);
    become ReceiptHandler (tag: tag+1)
  endMeth
endActor
```

From the definition, we can derive the following formulae:

$$(\mathrm{ReceiptHandler}(x) \wedge \mathsf{Task}(x, \mathrm{receipt}(t\ set))) \to$$
$$\Diamond\, (\mathrm{ReceiptHandler}(x) \wedge \mathrm{TagsExp}(x, set_1) \wedge \mathrm{TagsRec}(x, set_2) \wedge \mathrm{Tag}(x, n)$$
$$\wedge \mathrm{ReplyDest}(x, y) \wedge \mathsf{Task}(x, \mathrm{receipt}(t\ set))$$
$$\wedge((\exp(x) = (set_1 \cup (set \setminus set_2)) \setminus \{t\} = \emptyset$$
$$\wedge \mathrm{rec}(x) = (set_2 \cup \{t\}) \setminus (set \cup set_1) = \emptyset$$
$$\wedge \bigcirc (\mathsf{Task}(y, \mathrm{terminated}) \wedge \mathrm{Inactive}(x)))$$
$$\vee(\exp = (set_1 \cup (set \setminus set_2)) \setminus \{t\} \neq \mathrm{rec} = (set_2 \cup \{t\}) \setminus (set \cup set_1)$$
$$\wedge \bigcirc (\mathrm{ReceiptHandler}(x) \wedge \mathrm{TagsExp}(x, \exp(x))$$
$$\wedge \mathrm{TagsRec}(x, \mathrm{rec}(x)) \wedge \mathrm{Tag}(x, n) \wedge \mathrm{ReplyDest}(x, y)))))$$

---

[3] We use set-algebraic operators instead of low-level message passing expressions.

$(\text{ReceiptHandler}(x) \land \text{Task}(x, \text{generateTag}(z))) \rightarrow$
$\Diamond (\text{ReceiptHandler}(x) \land \text{TagsExp}(x, set_1) \land \text{TagsRec}(x, set_2) \land \text{Tag}(x, n)$
$\quad \land \text{ReplyDest}(x, y) \land \text{Task}(x, \text{generateTag}(z))$
$\quad \land \bigcirc (\text{Task}(z, \text{return}(n + 1)) \land \text{ReceiptHandler}(x) \land \text{TagsExp}(x, set_1)$
$\quad \land \text{TagsRec}(x, set_2) \land \text{Tag}(x, n + 1) \land \text{ReplyDest}(x, y)))$

After having received a `receipt` message, both sets `exp` and `rec` will be calculated. If and only if for both sets the result equals the empty set, the `terminated` message will be issued. Therefore, the invariant (2) holds. From the definition of `ReceiptHandler` itself no (useful) liveness properties can be derived, since they depend on the respective application of `ReceiptHandlers`.

We will now demonstrate such an application of `ReceiptHandlers` on a simple recursive computation, the printing of the structure and the contents of a tree in a depth-first manner. Fig. 2 shows a graphical representation of the

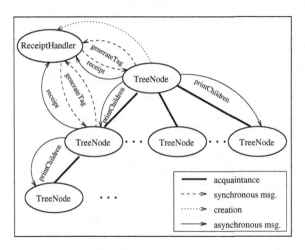

Fig. 2. Printing trees

objects and messages involved. The program code is shown below[4]. All of the messages, except for the generation of tags, operate asynchronously. After a `ReceiptHandler` has been created, each non-leaf tree node's daughters are asynchronously accessed and printed. At each node, for each daughter node a receipt tag is computed and sent to the daughter as a parameter of the `printChildren` message (`rtTag`). With the `receipt` message sent to the receipt handler, the set of all these tags is returned as those which are to be expected (`furtherTags`) and the node's own receipt tag as that tag which is hereby received (`rtag`).

---

[4] Two further syntactical enhancements: **ask** expressions denote rpc-like communications, **for** `<var>` **in** `<collection>` **do** abbreviates an enumeration of the elements of `<collection>`. Both can be expanded to patterns of asynchronously communicating actors.

```
defActor TreeNode(parent children contents)
  meth printAll (printer)
    let tag be 1 in
      let rhActor be
          (create ReceiptHandler(∅ ∅ tag self) in
          (send self printChildren(printer rhActor tag);
          send rhActor receipt(nil {tag})))
  endMeth
  meth printChildren (printer returnDest rtTag)
    send printer print (contents);
    let tagSet be (create Set) in
      (if (children ≠ ∅)
        then (let tag be (ask returnDest generateTag) in
          (for child in children do
            (ask tagSet add(tag);
            send child printChildren(printer returnDest tag))));
        send returnDest receipt(rtTag tagSet))
  endMeth
```

In order to prove the termination we have to show that (a) if the tags transmitted as 'received' are exactly those transmitted as 'expected' the sets will eventually be empty, and (b) given (a), the **terminated** message will be sent, and (c) that (a) and (b) will be guaranteed to happen.

For the first part, we have to prove that the bookkeeping of tags is correct. Assuming a collection of **receipt** messages $r_i(t_i\ S_i)$ with $1 \leq i \leq k$, t_i pairwise distinct, S_i pairwise distinct or empty, and $\{t_1 \ldots t_k\} = \bigcup_{i=1}^{k} S_i$ sent to a ReceiptHandler with initially empty sets $\mathbf{tagsExp}_1$ and $\mathbf{tagsRec}_1$. At each arrival of a message r_i, $\mathbf{tagsExp}_{i+1}$ and $\mathbf{tagsRec}_{i+1}$ are calculated as follows:

$$\mathbf{tagsExp}_{i+1} := (\mathbf{tagsExp}_i \cup (S_i \setminus \mathbf{tagsRec}_i)) \setminus \{t_i\}$$
$$\mathbf{tagsRec}_{i+1} := (\mathbf{tagsRec}_i \cup \{t_i\}) \setminus (\mathbf{tagsExp}_i \cup S_i)$$

$\mathbf{tagsExp}_{k+1}$ will amount to the empty set, since all t_i that became an element of it via S_j, $j < i$, will be eventually excluded, and $\mathbf{tagsRec}_{k+1}$ will be empty, too, since all elements t_i that might have been added have been an element of $\mathbf{tagsExp}_l$, $l < i$, or of S_j, $j \geq i$, and therefore have been removed.

The last two steps are to show the liveness property, i.e. that the sets will eventually be empty and a **terminated** message is really sent to the creator of the ReceiptHandler, which can be described as

$$\Diamond(\text{Exp}(x, \emptyset) \wedge \text{Rec}(x, \emptyset) \wedge \bigcirc (\text{Task}(\text{replyDest}(x), \text{terminated}))) \qquad (3)$$

The methods defined for TreeNodes are formalized as:

$$(\text{TreeNode}(u) \wedge \text{Task}(u, \text{printAll}(pr))) \rightarrow$$
$$\Diamond(\text{Task}(u, \text{printAll}(pr))$$
$$\wedge \bigcirc (\text{ReceiptHandler}(rhActor) \wedge \text{Task}(u, \text{printChildren}(pr\ rhActor\ 1))$$
$$\wedge \text{Task}(rhActor, \text{receipt}(nil\ \{1\})))))$$

(TreeNode(u) $\wedge$ Task(u, printChildren(pr rhActor tag))) $\rightarrow$
$\quad\Diamond$((Task(u, printChildren(pr rhActor tag)) $\wedge$ Children(u, $\emptyset$)
$\qquad\wedge \bigcirc$ (Task(rhActor, receipt(tag $\emptyset$))))
$\quad\vee$(Task(u, printChildren(pr rhActor tag)) $\wedge \neg$Children(u, $\emptyset$)
$\qquad\wedge \bigcirc$ ($\forall v : v \in$ children(u) $\leftrightarrow$
$\qquad\qquad$($\exists$vt : vt $= i \wedge$ vt $\in$ tagSet $\wedge$ Task(v, printChildren(pr rhActor vt)))
$\qquad\qquad\wedge$Task(rhActor, receipt(tag tagSet)))))

The newly generated tags i are assumed to be pairwise distinct, which can easily be ensured by an appropriate name server. Given the above formalizations of `ReceiptHandler` and `TreeNode`, an initial set $tasks = \{(\text{tree}, \text{printAll(prt)})\}$, and TreeNode (tree), we can derive (3) by the following deduction. The single element of $tasks$ is immediately chosen for execution and its effects occur in the next step:

$$
\begin{aligned}
&\text{TreeNode(tree)} \wedge \text{Task(tree, printAll(prt))} \rightarrow \\
&\quad \bigcirc \text{(ReceiptHandler(rh)} \wedge \exp = \emptyset \wedge \text{rec} = \emptyset \\
&\quad \wedge \text{Task(tree, printChildren(prt rh 1))} \wedge \text{Task(rh, receipt}(\textit{nil}\ \{1\}))
\end{aligned} \tag{4}
$$

Now, we can derive the effects of the two new elements of $tasks$:

$$
\begin{aligned}
\ldots \rightarrow& \\
\Diamond(&(\text{ReceiptHandler(rh)} \wedge 1 \notin \text{tagsRec(rh)} \wedge \exp = \text{tagsExp(rh)} \cup \{1\} \neq \emptyset) \\
&\wedge \bigcirc (\text{ReceiptHandler(rh)} \wedge 1 \in \text{tagsExp(rh)}) \\
\vee(&\text{ReceiptHandler(rh)} \wedge 1 \in \text{tagsRec(rh)} \wedge \text{rec} = \text{tagsRec(rh)} \setminus \{1\}) \\
&\wedge \bigcirc (\text{ReceiptHandler(rh)} \wedge 1 \notin \text{tagsRec(rh)} \wedge 1 \notin \text{tagsExp(rh)})) \\
\wedge \Diamond((&\text{Children(tree, } \emptyset) \wedge \bigcirc (\text{rh, receipt(1 } \emptyset)) \in tasks) \\
\vee(&\neg\text{Children(tree, } \emptyset) \wedge \bigcirc (\exists T : A \wedge \text{Task(rh, receipt(1 T)))))
\end{aligned}
$$
$$\tag{5}$$

with $A \equiv \forall c : c \in$ children(tree) $\leftrightarrow$ ($\exists$cTag : cTag $\in$ T$\wedge$ Task(c, printChildren(prt rh cTag))) Transitively from (4) and the first disjunction of (5), we can conclude that whether the $receipt(\textit{nil}\ \{1\})$ or the $receipt(1\ X)$ message arrives first at rh, 1 will be excluded from both sets, $tagsExp$ and $tagsRec$. This holds analogously for all elements of T in the second disjunction of (5) until both sets are empty again ∎

6   Related Work

Our work clearly benefits from previous work that has been done in the field of applying temporal logics to concurrent program specification. Among the temporal approaches, the work of Emerson et al. [7,10] and of Manna & Pnueli [18] is fundamental; TLA [17] and UNITY [6] both are elaborated specification languages based on temporal logics. They all have in common that they concentrate on the temporal part of the specification and do not provide a deep treatment of single states. This strangely neglected topic is, however, crucial for actor programming, since an actor system, as a whole, cannot be described simply by sequences of atomic states or events. Adequate descriptions of actor systems must account for the complexity of objects and connections between objects, instead. The only other approach to actor specifications by temporal logics does not deal with the actor primitive *become* but uses local synchronization techniques to model change of state [9].

Process calculi like CSP [15] or CCS [19] do not allow for changes of the underlying topologies of communication channels as they occur in object-oriented systems as an effect of object creation. The π-calculus [20] and other extensions of process calculi, e.g., the Object Calculus by Nierstrasz [21], overcome these limitations. They could be used for the specification of actor systems, though they are not designed for it, but since they are very fine-grained, this can only be achieved in a rather complicated way.

Initial work on formal actor semantics has been conducted by Greif [13] and Clinger [8]; for a description of the newest operational semantics by Agha et al., cf. Section 3. In contrast to these styles of semantics our work is concentrated on the next higher level of abstraction from computational steps: while our description of state transitions is quite simple – it could easily be exchanged by a more sophisticated one – we focus on the overall development of computations in actor systems.

7 Conclusion

We have shown how a general proof system for discrete linear temporal logic can be tailored to describe the behavior of actor systems. We express the configurations of an actor system at a time step as sets of formulas serving as (temporally spoken) local states for the temporal logic TAL. This specialized proof system has been applied to an extension of the basic actor communication mode that allows for detecting the termination of partial computations. Temporal logics turns out to be a suitable formal framework for describing actor-style distributed computations.

References

1. AGHA, G. *Actors: a Model of Concurrent Computation in Distributed Systems.* MIT Press, Cambridge, MA, 1986.
2. AGHA, G. The structure and semantics of actor languages. In *Foundations of Object-Oriented Languages*, J. de Bakker, W.-P. de Roever, and G. Rozenberg, Eds., no. 489 in LNCS. Springer, Berlin etc., 1990, pp. 1–59.
3. AGHA, G., FRØLUND, S., KIM, W. Y., PANWAR, R., PATTERSON, A., AND STURMAN, D. Abstraction and modularity mechanisms for concurrent computing. In *Research Directions in Concurrent Object-Oriented Programming*, G. Agha, P. Wegner, and A. Yonezawa, Eds. MIT Press, Cambridge, MA, 1993, pp. 3–21.

4. AGHA, G., MASON, I. A., SMITH, S. F., AND TALCOTT, C. L. A foundation for actor computation. *Journal of Functional Programming 7* (1997), 1–72.

5. BARRINGER, H. The use of temporal logic in the compositional specification of concurrent systems. In *Temporal Logics and their Applications*, A. Galton, Ed. Academic Press, London etc., 1987, pp. 53–90.

6. CHANDY, K. M., AND MISRA, J. *Parallel Program Design*. Addison-Wesley, 1989.

7. CLARKE, E. M., EMERSON, E. A., AND SISTLA, A. P. Automatic verification of finite state concurrent systems using temporal logic specifications: A practical approach. *ACM Transactions on Programming Languages and Systems 8*, 2 (April 1986), 244–263.

8. CLINGER, W. D. *Foundations of Actor Semantics*. PhD thesis, Cambridge, MA: MIT, Dept. of Mathematics, 1981.

9. DUARTE, C. H. C. A proof-theoretic approach to the design of object-based mobility. In *Proc. 2nd IFIP Conf. on Formal Methods for Open Object-Based Distributed Systems* (London, 1997), H. Bowman and J. Derrick, Eds., Chapman and Hall, pp. 37–56.

10. EMERSON, E., AND HALPERN, J. 'sometimes' and 'not never' revisited: On branching time versus linear time temporal logic. *Journal of the ACM 33*, 1 (1986).

11. FELLEISEN, M., AND FRIEDMAN, D. P. Control operators, the SECD-machine, and the λ-calculus. In *Formal Descriptions of Programming Concepts III*, M. Wirsing, Ed. Elsevier, Amsterdam, NL, 1986, pp. 193–217.

12. FRØLUND, S. Inheritance of synchronization constraints in concurrent object-oriented programming languages. In *ECOOP '92 – European Conference on Object-Oriented Programming* (1992), O. L. Madsen, Ed., no. 615 in LNCS, Springer, pp. 185 – 196.

13. GREIF, I. G. *Semantics of Communicating Parallel Processes*. PhD thesis, MIT, Dept. of Electrical Engineering and Computer Science, 1975.

14. HEWITT, C., AND BAKER, H. Actors and continuous functionals. In *Proceedings of the IFIP Working Conference on Formal Description of Programming Concepts* (1978), E. Neuhold, Ed., Amsterdam etc.: North-Holland, pp. 367–390.

15. HOARE, C. *Communicating Sequential Processes*. Prentice-Hall, 1985.

16. KRÖGER, F. *Temporal Logic of Programs*. Springer, 1987.

17. LAMPORT, L. The Temporal Logic of Actions. *Transactions on Programming Languages and Systems 16*, 3 (May 1994), 872–923.

18. MANNA, Z., AND PNUELI, A. *The Temporal Logic of Reactive and Concurrent Systems*. Springer, Berlin etc., 1992.

19. MILNER, R. *Communication and Concurrency*. Prentice-Hall, Englewood Cliffs, NJ, 1989.

20. MILNER, R., PARROW, J., AND WALKER, D. A calculus of mobile processes, I/II. *Information and Computation 100*, 1 (1992), 1–77.

21. NIERSTRASZ, O. Towards an object calculus. In *ECOOP'91 – Proceedings of the European Workshop on Object-Based Concurrent Computing* (Berlin etc., Geneva, Switzerland, July 15–16 1992), M. Tokoro, O. Nierstrasz, P. Wegner, and A. Yonezawa, Eds., Springer.

22. OWICKI, S., AND LAMPORT, L. Proving liveness properties of concurrent programs. *ACM Transactions on Programming Languages and Systems 4*, 3 (1982), 455–495.

23. PLOTKIN, G. Call-by-name, call-by-value and the lambda calculus. *Theoretical Computer Science 1* (1975), 125–159.

Flexible Types for a Concurrent Model

Franz Puntigam

Technische Universität Wien, Institut für Computersprachen
Argentinierstr. 8, 1040 Vienna, Austria. `franz@complang.tuwien.ac.at`

Abstract. Subtyping is undoubtedly useful for the support of incremental refinement and reuse of software components, a crucial feature of object-oriented languages. Types and subtyping for concurrent languages are not yet explored sufficiently and tend to be less flexible than desired. We propose a flexible type model for concurrent languages. This model ensures statically that "message-not-understood-errors" do not occur at run-time even if object behavior is changed dynamically.

1 Introduction

The basic concept of object-oriented programming languages is that of objects communicating with other objects [4,22]. An object is a self-contained entity characterized by its identity, state and behavior. Objects are classified according to their behavior into a system of types. We also regard subtyping [5,11] and the related sort of polymorphism as necessary features of object-oriented languages.

In concurrent object-oriented languages it is advantageous to use *active objects* [2,15], i.e. to regard objects as *processes*. The computation models for such languages can be based on, for example, Hoare's CSP model [9], the object-based actor model [1] and process calculi [3] like Milner's π-calculus [13].

In this chapter we explore a combination of the actor model with a process calculus, where active objects communicate by asynchronous message passing. The major contribution is a type model based on two principles: The sender updates his view of the receiver's type after sending a message (1); and each object accepts messages corresponding to all clients' type views in arbitrary interleaving (2). These principles support very flexible and expressive types.

In Sect. 2 we describe how types can help in the development of concurrent software and give a motivation for the type model. Our basic model of computation (without types) is introduced in Sect. 3. Type annotations are added in Sect. 4. We give conditions on which programs are well-typed.

2 Types

Most type-theoretic foundations for object-oriented languages are based on a typed λ-calculus extended by records [5,8]. Subtyping is defined by Wegner and Zdonik's principle of substitutability [22]: "An instance of a subtype can always be used in any context in which an instance of a supertype was expected."

G. Agha et al. (Eds.): Concurrent OOP and PN, LNCS 2001, pp. 461–472, 2001.
© Springer-Verlag Berlin Heidelberg 2001

```
task type Buffer is              task body Buffer is
  entry put(e: in Data);           x: Data;
  entry get(e: out Data);        begin
end Buffer;                        loop
                                     accept put(e: in Data) do x := e; end;
                                     accept get(e: out Data) do e := x; end;
                                   end loop;
                                 end Buffer;
```

Fig. 1. Task Type of a Buffer in Ada

Some approaches consider object states: A type specifies a set of available operations as well as preconditions, postconditions and invariants for the execution of these operations. A server handles requests from clients by executing operations according to (sequential) procedure call semantics. The server's type represents a contract with the clients [12]: If a client requests a service and provides appropriate input, the server returns the promised output. A subtype is a more detailed contract than a supertype.

According to Liskov and Wing, types are partial specifications of object behavior that express how instances can be used safely [11]. A subtype is a more complete specification than a supertype. Complete specifications are, in general, not considered to be types because they include details probably irrelevant to all imaginable uses. On the other side, most-general types inferred from programs are often inappropriate because local modifications easily have global effects, hindering the incremental refinement and reuse.

Type models for active objects are not yet established. We demonstrate what we may expect from a type system by an example using tasks in Ada. Tasks are regarded as active objects that accept the messages specified by entry declarations in task type declarations.

Fig. 1 shows a buffer task as example. Using the information provided by the task type, "b.put(e_1); b.put(e_2); b.get(e_3);" would be a type-conforming piece of code, where b is an instance of "Buffer", and e_1, e_2, e_3 are variables of type "Data". Provided that no other task sends messages to b, the execution results in a deadlock: b expects to receive "put" and "get" in alternation, beginning with "put". The repeated behavior changes of b are not reflected in the type, although clients must know in which ordering messages are accepted.

Types shall specify message orderings in order to prevent such deadlocks[1] or other unexpected reactions of a server. A compiler shall be able to ensure that each client obeys these ordering restrictions and each server can handle messages in all allowed orderings. According to the principal of substitutability, a subtype shall specify messages in all orderings specified by supertypes. We show that such a type system can be developed.

---

[1] It is not our goal to prevent deadlocks in general. But the execution of an active object shall be blocked only if the queue of received messages is empty.

$$(p_1; p_2); p_3 \equiv p_1; (p_2; p_3) \qquad p; \varepsilon \equiv p \qquad \varepsilon; p \equiv p$$
$$(p_1 \parallel p_2) \parallel p_3 \equiv p_1 \parallel (p_2 \parallel p_3) \qquad p_1 \parallel p_2 \equiv p_2 \parallel p_1 \qquad p \parallel \varepsilon \equiv p$$
$$(p_1 + p_2) + p_3 \equiv p_1 + (p_2 + p_3) \qquad p_1 + p_2 \equiv p_2 + p_1 \qquad p + \varepsilon \equiv p$$
$$(p_1 + p_2); p_3 \equiv p_1; p_3 + p_2; p_3 \qquad p + p \equiv p$$
$$x(v_1, \ldots, v_n) \equiv p[v'_1, \ldots, v'_n / v_1, \ldots, v_n] \quad \text{if} \quad x(v'_1, \ldots, v'_n) := p$$

Fig. 2. Equivalence of Process Expressions

3   Model of Computation

An actor is an active object with a unique mail address, a behavior and a mail queue holding received messages. The messages in the mail queue are dealt with in the same sequence as they were received: Depending on its current behavior and the message dealt with, the actor sends messages to acquainted mail addresses, creates new actors and replaces its own behavior. Furthermore, in our model we assume that messages are received in the same (logical) order they have been sent.

In our model we specify the behavior of actors by process expressions. The actions deal with accepting (receiving) and sending messages and creating actors. We assume that a set of message names M, a set of process names X, an infinite set of variables and mail addresses V are given; elements of M, X and V are denoted by m, x and v (sometimes quoted or indexed), respectively. The syntax of process expressions p and process definitions q is given by:

p ::=	ε	empty process (skip)
	$v\$p$	create actor with mail address v and behavior p
	$v.m(v_1, \ldots, v_n)$	send m with concrete parameters $v_1, \ldots, v_n$ to v
	$m(v_1, \ldots, v_n)$	get m from mail queue; formal parameters $v_1, \ldots, v_n$
	$x(v_1, \ldots, v_n)$	apply process x with concrete parameters $v_1, \ldots, v_n$
	$p_1; p_2$	sequential composition; first p_1, then p_2
	$p_1 + p_2$	alternative composition; either p_1 or p_2
	$p_1 \parallel p_2$	parallel composition (merge); p_1 and p_2 interleaved
q ::=	$x(v_1, \ldots, v_n) := p$	define process x; formal parameters $v_1, \ldots, v_n$

A process expression is the empty process, an action (create, send, receive), a process application, or a sequential, alternative or parallel composition. Empty parentheses can be omitted. Among composition operators, ";" has highest and "+" lowest priority. A process definition associates a process expression with a name and parameters. For each $x \in X$ there is exactly one process definition; the set of these process definitions is a program. The reflexive, transitive closure of the rules in Fig. 2 defines an equivalence relation on process expressions: Composition operators are associative and have ε as neutral element; "$\parallel$" and "+" are commutative; "+" is idempotent; ";" is right-distributive over "+". Process applications are equivalent to the applied process expression, where concrete parameters are substituted for formal parameters. ($p[v'_1, \ldots, v'_n / v_1, \ldots, v_n]$ denotes the simultaneous substitution of $v'_1, \ldots, v'_n$ for all free occurrences of $v_1, \ldots, v_n$ in p.) Usually, we do not distinguish between equivalent process expressions.

We require that all alternative compositions are equivalent to process expressions of the form $m_1(v_{1,1}, \ldots, v_{1,n_1}); p_1 + \cdots + m_k(v_{k,1}, \ldots, v_{k,n_k}); p_k$. (Each operand of "+" is either another alternative composition or a process expression headed by a receive-action.)

The example in Fig. 1 is expressed as a program in our formalism by:

$$\text{Buffer} := \text{put}(v_1); \text{get}(v_2); v_2.\text{back}(v_1); \text{Buffer}$$

"Buffer" is a process name, "put", "get" and "back" are message names, and v_1 and v_2 variables. The argument of "get" is supposed to be the mail address of an actor to which the buffer content v_1 is sent on receipt of "get". The cycle of putting an element into the buffer and getting it from there can be repeated as often as needed by applying the last rule in Fig. 2; the above definition of "Buffer" is (by substituting the process expression for the process name) equivalent to this one:

$$\text{Buffer} := \text{put}(v_1); \text{get}(v_2); v_2.\text{back}(v_1); \text{put}(v_1); \text{get}(v_2); v_2.\text{back}(v_1); \text{Buffer}$$

An actor behaving as "Buffer" always contains only a single element. One behaving as "IBuffer" can hold an unbounded number of elements simultaneously:

$$\text{IBuffer} := \text{put}(v_1); \text{get}(v_2); v_2.\text{back}(v_1) \parallel \text{IBuffer}$$

Because of parallel composition, an arbitrary number of "put"-messages is handled. But the number of "get"-messages must not exceed that of "put"-messages.

The next example shows a bounded buffer that returns error messages when boundaries are crossed. The second parameter of "put" denotes the actor to whom error messages and acknowledgments shall be sent.

$$\text{EBuffer} := \text{put}(v_1, v_2); v_2.\text{ok}; \text{FBuffer}(v_1) + \text{get}(v_3); v_3.\text{empty}; \text{EBuffer}$$
$$\text{FBuffer}(v_1) := \text{get}(v_2); v_2.\text{back}(v_1); \text{EBuffer} + \text{put}(v_3, v_4); v_4.\text{full}; \text{FBuffer}(v_1)$$

An unbounded buffer can deal with "put" and "get" in arbitrary ordering. When the buffer is empty, the execution of "get" is delayed as in the Linda coordination model [7]:

$$\text{UBuffer} := ((\text{put}(v_1) \parallel \text{get}(v_2)); v_2.\text{back}(v_1)) \parallel \text{UBuffer}$$

4   Typed Processes

Under some conditions, an actor cannot proceed with its execution: There may be no appropriate alternative handling the next message in the mail queue; the numbers of concrete and formal parameters in process applications do not match; or the addressee of a message does not exist. A strong, static type system has to ensure at compile-time that such errors cannot occur.

The first step in adding a type concept is to annotate all variables with type expressions. We assume that a set Z of type names z is given. The syntax of type expressions t and type definitions d is given by:

$$(t_1;t_2);t_3 \equiv t_1;(t_2;t_3) \qquad\qquad t;\varepsilon \equiv t \qquad\qquad\qquad \varepsilon;t \equiv t$$
$$(t_1 \parallel t_2) \parallel t_3 \equiv t_1 \parallel (t_2 \parallel t_3) \qquad\qquad t_1 \parallel t_2 \equiv t_2 \parallel t_1 \qquad\qquad t \parallel \varepsilon \equiv t$$
$$(t_1 + t_2) + t_3 \equiv t_1 + (t_2 + t_3) \qquad\qquad t_1 + t_2 \equiv t_2 + t_1 \qquad\qquad t + \varepsilon \equiv t$$
$$(t_1 + t_2);t_3 \equiv t_1;t_3 + t_2;t_3 \qquad\qquad t_1;(t_2 + t_3) \equiv t_1;t_2 + t_1;t_3 \qquad t + t \equiv t$$
$$(t_1 + t_2) \parallel t_3 \equiv t_1 \parallel t_3 + t_2 \parallel t_3 \qquad\qquad z \equiv t \quad \text{if} \quad z(t_1,\ldots,t_n) := t$$
$$r_1;t_1 \parallel r_2;t_2 \equiv r_1;(t_1 \parallel r_2;t_2) + r_2;(r_1;t_1 \parallel t_2)$$

Fig. 3. Equivalence of Type Expressions

$$
\begin{aligned}
t ::=\ \ &\varepsilon && \text{most general (empty) type} \\
\mid\ &m(t_1,\ldots,t_n) && \text{type of a receive-action; parameter types } t_1,\ldots,t_n \\
\mid\ &z && \text{application of type } z \text{ (parameter types are implicit)} \\
\mid\ &t_1;t_2 && \text{sequential composition} \\
\mid\ &t_1 + t_2 && \text{alternative composition} \\
\mid\ &t_1 \parallel t_2 && \text{parallel composition} \\[4pt]
d ::=\ \ &z(t_1,\ldots,t_n) := t && \text{define type } z; \text{ parameter types } t_1,\ldots,t_n
\end{aligned}
$$

Type expressions specify which messages are expected to be received by corresponding processes. Sequential, alternative and parallel type compositions relate the expected message receptions to one another. Type expressions can be associated with type names. As a program associates all process names with process expressions, a type program associates each type name with a type expression. The reflexive, transitive closure of the rules in Fig. 3 defines an equivalence relation on type expressions. These rules are essentially the same as those in Fig. 2 except that ";" also is left-distributive over "+" and right-distributive over "$\parallel$", and each parallel composition is equivalent to a specific combination of sequential and alternative composition (r_1 and r_2 denote types of receive-actions).[2]

From now on, p and q denote typed process expressions and typed process definitions. Typed create-actions, send-actions, receive-actions and applications of processes are of the forms $v{:}t\$p$, $v{:}t.m(v_1{:}t_1,\ldots,v_n{:}t_n)$, $m(v_1{:}t_1,\ldots,v_n{:}t_n)$ and $x(v_1{:}t_1,\ldots,v_n{:}t_n)$, respectively. Typed process definitions are of the form $x(v_1{:}t_1,\ldots,v_n{:}t_n) := p$. Except of the type annotations, the model of computation remains unchanged.

It is easy to see how type expressions are related to process expressions: Given a process expression p, we substitute the most general type ε for each send-action and create-action, $m(t_1,\ldots,t_n)$ for each receive-action $m(v_1{:}t_1,\ldots,v_n{:}t_n)$, and z for each process application $x(v_1{:}t_1,\ldots,v_n{:}t_n)$, where $z(t'_1,\ldots,t'_n) := t$ and $x(v_1{:}t''_1,\ldots,v_n{:}t''_n) := p'$ and $t'_i \equiv t''_i$ ($1 \le i \le n$) and t is equivalent to the type expression constructed in the same way from p'. The resulting type expression (as well as each type expression equivalent to it) is a most specific type of p.

The notion of the "most specific type" suggests that a process expression has also other types than this one. These other types are the supertypes of the most specific type. The subtype relationship $\le$ on type expressions (and

---

[2] Reasons for including the additional rules can be found in [18]. In process expressions, similar rules would have undesired consequences.

$$\frac{t_1 \equiv t_2}{t_1 \leq t_2} \qquad\qquad t \leq \varepsilon$$

$$\frac{t'_1 \leq t_1 \quad \ldots \quad t'_n \leq t_n}{m(t_1,\ldots,t_n) \leq m(t'_1,\ldots,t'_n)} \qquad \frac{z(t_1,\ldots,t_n) := t \leq z'(t'_1,\ldots,t'_n) := t'}{z \leq z'}$$

$$\frac{t_1 \equiv t_3 \quad t_2 \leq t_4}{t_1;t_2 \ \leq \ t_3;t_4} \qquad\qquad \frac{t_1 \leq t_3 \quad t_2 \leq t_4}{t_1 \parallel t_2 \ \leq \ t_3 \parallel t_4}$$

$$\frac{t_1 \leq t_3 \quad t_2 \leq t_4}{t_1 + t_2 \ \leq \ t_3 + t_4} \qquad\qquad \frac{t_1 \leq t_3 \quad t_2 \leq t_4}{t_1 \parallel t_2 \ \leq \ t_3;t_4}$$

$$\frac{t_1 \leq t_3 \quad t_2 \leq t_4}{t_1 \parallel t_2 \ \leq \ t_3 + t_4} \qquad \frac{t'_1 \leq t_1 \quad \ldots \quad t'_n \leq t_n \quad t \leq t'}{z(t_1,\ldots,t_n) := t \ \leq \ z'(t'_1,\ldots,t'_n) := t'}$$

Fig. 4. Subtyping

type definitions) is the reflexive, transitive closure of the rules in Fig. 4. A type expression t is a subtype of t' (and t' is a supertype of t) if $t \leq t'$. Trivially, each type is a subtype of a type expression equivalent to itself and of the most general type ε. The type of a receive-action is a subtype of another receive-action only if message name and parameter number are equal and the types of the parameters in the subtype are supertypes of those in the supertype. These contravariant parameter types ensure that each instance of a subtype can deal with all parameters satisfying the constraints expressed in a supertype, according to the principle of substitutability [5,11]. For the same reason, parameter types in type definitions are contravariant.

The left-hand sides of sequential compositions must be equivalent if they are related by subtyping.[3] An example shows the reason: Assume that $t_1;t_2 \leq t_3;t_4$ if $t_1 \leq t_3$ and $t_2 \leq t_4$. Then, it is possible to infer $m_1;m_2 \leq \varepsilon;m_2$ and (equivalently) $m_1;m_2 \leq m_2$ (where m_1 and m_2 are types of receive-actions without parameters). But according to the principle of substitutability, $m_1;m_2$ cannot be a subtype of m_2 because instances of the first type expect to receive m_1 as the first message, while instances of the second type expect m_2 first; a message m_2 sent according to a supertype cannot be accepted by an actor that behaves according to $m_1;m_2$. Therefore, Fig. 4 does not contain this rule. On the other side, $m_1;m_2 \leq m_1;\varepsilon$ and (equivalently) $m_1;m_2 \leq m_1$ does not cause such problems; a message m_1 sent according to the supertype can be accepted by an actor that behaves according to $m_1;m_2$. It is always possible to extend a supertype by adding type expressions at the end, using sequential composition.

In subtypes of alternative and parallel compositions, both sides can be subtypes of the respective parts of the supertypes. Supertypes can be extended by adding alternatives or type expressions specifying additional messages acceptable in parallel. These rules are understandable through the principle of substitutability: All messages sent according to an alternative (or parallel part) of a supertype can be handled by an actor of a subtype with additional alternatives (parallel

---

[3] It is possible to relax this strict condition by using a more complex set of subtyping rules. The smaller rule set was selected in order to keep subtyping comprehensive.

parts). Using these argument, it is easy to see that a parallel composition can be a subtype of a corresponding sequential or alternative composition.

We have to address the question of "what typed processes are well-typed so that 'message-not-understood-errors' cannot occur": Of course, all parameters must be type-conforming, i.e., concrete parameter types must be subtypes of corresponding formal parameter types. Furthermore, each sender of a message must be sure that the receiver can deal with the message. For example, if a variable is of a type $m_1; m_2 + m_3; m_4$, the first message sent to the corresponding mail address must be either m_1 or m_3; the second message must be m_2 if m_1 was sent as first message, and m_4 if m_3 was the first message.

In concurrent systems it is usually not known which sending actor is the first, second, etc. to send a message to a receiving actor. Moreover, different senders may know different types of the receiver. We have to provide precaution against type conflicts caused by the fact that each actor has only partial information about a receiver's behavior. As a solution of this problem we could require that each message accepted as the first one is also accepted in all further execution states. In a rigid system it might be appropriate to simply restrict changes of object behavior so that the same set of messages is accepted in each state. But we can be much less restrictive: If an actor A behaves according to a parallel composition $t_1 \| \cdots \| t_n$ so that each actor A_i $(1 \leq i \leq n)$ that may send a message to A knows only the partial type t_i of A, the actors $A_1, \ldots, A_n$ can independently send messages to A according to their knowledge of A's type; all possible interleavings of received messages can be handled by A.

We demonstrate this use of parallel type composition by a typed version of a previous example:

$$\text{IBuffer} := \text{put}(v_1{:}\text{tElem}); \text{get}(v_2{:}\text{back}(\text{tElem}));$$
$$v_2{:}\text{back}(\text{tElem}).\text{back}(v_1{:}\text{tElem}) \| \text{IBuffer}$$

$$\text{tIBuffer} := \text{put}(\text{tElem}); \text{get}(\text{back}(\text{tElem})) \| \text{tIBuffer}$$

"tIBuffer" is a most specific type of the typed version of "IBuffer". Because "tIBuffer" is a parallel composition of an infinite number of type expressions "put(tElem); get(back(tElem))", each instance of "tIBuffer" can deal with an arbitrary number of clients, each sending first "put" and then "get". An instance returns on receipt of "get" any of the elements in the buffer, not necessarily the one put into the buffer by the actor sending "get". "tIBuffer" can also be regarded as a parallel composition of an infinite number of type expressions "put(tElem); get(back(tElem)) \| tIBuffer" so that each client of an instance of "tIBuffer" can know the most specific type and send messages according to it.

Typed versions of "Buffer" and "Ubuffer" and their most specific types are:

$$\text{Buffer} := \text{put}(v_1{:}\text{tElem}); \text{get}(v_2{:}\text{back}(\text{tElem}));$$
$$v_2{:}\text{back}(\text{tElem}).\text{back}(v_1{:}\text{tElem})); \text{Buffer}$$

$$\text{tBuffer} := \text{put}(\text{tElem}); \text{get}(\text{back}(\text{tElem})); \text{tBuffer}$$

$$\text{UBuffer} := ((\text{put}(v_1\text{:tElem}) \parallel \text{get}(v_2\text{:back}(\text{tElem})));$$
$$v_2\text{:back}(\text{tElem}).\text{back}(v_1\text{:tElem})) \parallel \text{UBuffer}$$

$$\text{tUBuffer} := \text{put}(\text{tElem}) \parallel \text{get}(\text{back}(\text{tElem})) \parallel \text{tUBuffer}$$

It is easy to see that "tUBuffer" and "tIBuffer" are subtypes of "tBuffer", and "tUBuffer" is a subtype of "tIBuffer". Furthermore, "tUBuffer" has the supertypes

$$\text{tPutUBuffer} := \text{put}(\text{tElem}) \parallel \text{tPutUBuffer}$$
$$\text{tGetUBuffer} := \text{get}(\text{back}(\text{tElem})) \parallel \text{tGetUBuffer}$$

Clients who know that a server's type is "tPutUBuffer" can send only "put"-messages, and others knowing that this server is of type "tGetUBuffer" can send only "get"-messages.

So far we used the principle of type updating only implicitly:

On sending a message, the type of the variable representing the receiver is replaced with the receiver's expected type after processing this message.

In the above examples, the variable v_2 is initialized with the mail address of an actor behaving according to "back(tElem)". After sending "back" to v_2, the type of v_2 (implicitly) becomes ε; no further message can be sent to v_2.

Fig. 5 shows a solution for a version of the dining philosophers problem. The types "tF" and "tH" represent forks on the table and in a philosophers hand, respectively. A new philosopher arriving at the table puts a new fork onto the table. When a philosopher leaves, he takes a fork on the table with him. Philosophers send their forks (being on the table) to their left neighbors and receive forks from their right neighbors. Hungry philosophers keep the forks and begin to eat when they have two forks. In order to avoid deadlocks, a single token is passed around the table. When receiving a token, philosophers must send all forks they have to their left neighbors. Also, a thinking philosopher having the token can create a new philosopher as his left neighbor. After receiving initial messages from their creators, new philosophers behave like all other philosophers at the table. Eating philosophers with the token can ask their left neighbors to leave the table ("go"). The neighbors send "done" around the table, indicating that they are ready to go. (The argument of this message tells the asking philosopher the leaving philosopher's left neighbor.) The forks received by leaving philosophers are immediately sent to the left neighbors; leaving philosophers don't deal with other messages. Each philosopher is either not yet initialized (Pn), thinking (Pth), hungry (Ph), eating (Pe), waiting for a "done"-message (Pw) or leaving (Pl). Not yet initialized philosophers are of the type "tPn", leaving philosophers of "tPl", and all other philosophers of "tP".

The example shows the principle of type updating in many places: After sending "up" to a fork of type "tF" the type of this fork becomes "tH"; after sending "init" to a philosopher of type "tPn" the type becomes "tP"; after sending "go" to a philosopher of type "tP" the type becomes "tPl", and so on.

Fortunately, the principle of type updating is statically enforceable. It would not be necessary to associate each use of a variable with a type annotation as we

$$tF := up; tH + take$$
$$tH := down; tF$$

$$tPn := init(tP); tP$$
$$tP := fork(tF); tP + token; tP + go; tPl + done(tP); tP$$
$$tPl := fork(tF); tPl$$

$$Pn := init(x{:}tP); f{:}tF\$F; x{:}tP.fork(f{:}tF); Pth(x{:}tP)$$

$$
\begin{aligned}
Pth(x{:}tP) := \ & fork(f{:}tF); x{:}tP.fork(f{:}tF); Pth(x{:}tP) + \\
& fork(f{:}tF); Ph(x{:}tP, f{:}tF) + \\
& token; x{:}tP.token; Pth(x{:}tP) + \\
& token; x{:}tP.token; y{:}tPn\$Pn; init(x{:}tP); Pth(y{:}tP) + \\
& go(f{:}tF); f{:}tF.take; x{:}tP.done(x{:}tP); Pl(x{:}tPl) + \\
& done(y{:}tP); x{:}tP.done(y{:}tP); Pth(x{:}tP)
\end{aligned}
$$

$$
\begin{aligned}
Ph(x{:}tP, f{:}tF) := \ & \\
& fork(g{:}tF); f{:}tF.up; g{:}tF.up; Pe(x{:}tP, f{:}tH, g{:}tH) + \\
& token; x{:}tP.fork(f{:}tF); x{:}tP.token; Pth(x{:}tP) + \\
& go(g{:}tF); g{:}tF.take; x{:}tP.fork(f{:}tF); x{:}tP.done(x{:}tP); Pl(x{:}tPl) + \\
& done(y{:}tP); x{:}tP.done(y{:}tP); Ph(x{:}tP, f{:}tF)
\end{aligned}
$$

$$
\begin{aligned}
Pe(x{:}tP, f{:}tH, g{:}tH) := \ & \\
& fork(h{:}tF); x{:}tP.fork(h{:}tF); Pe(x{:}tP, f{:}tH, g{:}tH) + \\
& token; f{:}tH.down; g{:}tH.down; x{:}tP.fork(f{:}tF); \\
& \quad x{:}tP.fork(g{:}tF); x{:}tP.token; Pth(x{:}tP) + \\
& token; f{:}tH.down; g{:}tH.down; x{:}tP.fork(f{:}tF); \\
& \quad x{:}tP.go(g{:}tF); Pw(x{:}tPl) + \\
& go(h{:}tF); h{:}tF.take; f{:}tH.down; g{:}tH.down; x{:}tP.fork(f{:}tF); \\
& \quad x{:}tP.fork(g{:}tF); x{:}tP.done(x{:}tP); Pl(x{:}tPl) + \\
& done(y{:}tP); x{:}tP.done(y{:}tP); Pe(x{:}tP, f{:}tH, g{:}tH)
\end{aligned}
$$

$$
\begin{aligned}
Pw(x{:}tPl) := \ & fork(f{:}tF); x{:}tPl.fork(f{:}tF); Pw(x{:}tPl) + \\
& token; Pw(x{:}tPl) \text{ /* should not occur */ } + \\
& go(f{:}tF); Pw(x{:}tPl) \text{ /* should not occur */ } + \\
& done(y{:}tP); y{:}tP.token; Pth(y{:}tP)
\end{aligned}
$$
$$Pl(x{:}tPl) := fork(f{:}tF); x{:}tPl.fork(f{:}tF); Pl(x{:}tPl)$$

$$F := up; down; F + take$$

Fig. 5. Dining Philosophers

do in this chapter for didactic reasons; the compiler can compute this information from type annotations associated with parameters (variable initializations) in received messages and process name specifications.

Whenever a single object shall be used in several places simultaneously, the principle of type updating is not sufficient: A type update of one variable referencing the object is not visible at other variables referencing the object. Therefore, we additionally need the *principle of type splitting* which is applied when a new variable referencing the object (an alias) is introduced:

The type associated with a variable is split along a parallel composition into two parts whenever the variable is used as an argument; one part is associated with the argument, the other with further uses of the variable. This type splitting is repeated for each use of the variable as an argument.

When this principle is obeyed, all clients can independently send messages according to the supertypes of the servers known by the clients. Parallel composition ensures that servers can deal with all messages received from their clients in arbitrary interleaving.

Types like "tIBuffer" and "tUBuffer" can easily be split into any number of types. In these cases, the principle of type splitting does not restrict the usability of objects in any way. But many types cannot be split arbitrarily. Often, a type t can be split only into t and ε. For example, the type of forks "tF" does not support type splitting because each messages accepted by a fork depends on each previous message. In Fig. 5 each fork is always referenced only once. Philosopher types are split when handling "go"-messages: The variable x of type "tP" is sent to the left neighbor and also occurs in the following call of "Pl" with type "tPl". Therefore, "tP" has to be equivalent to "tP ∥ tPl". The proof of this equivalence using the rules in Fig. 3 is rather difficult because of the recursive definition. But it is easy to see that "tP" supports any number of "fork"-messages in arbitrary interleaving with the other messages specified by the type.

All type expressions have to conform to a further rule:

Each type expression $t \neq \varepsilon$ is equivalent to:

$$t \equiv m_1(t'_{1,1}, \ldots, t'_{1,n_1}); t_1 + \cdots + m_k(t'_{k,1}, \ldots, t'_{k,n_k}); t_k$$

where $m_i = m_j$ implies $n_i \neq n_j$ ($1 \leq i < j \leq n$). And for each type expression $t \parallel t'$: If t contains a subexpression $m(t_1, \ldots, t_n); t''$ and t' contains a subexpression $m(t'_1, \ldots, t'_n); t'''$, then $t'' \equiv t'''$.

This rule prevents an undesired kind of (internal) nondeterminism: Suppose that an actor expects to receive two messages of the same name and number of arguments alternatively or simultaneously (because of alternative or parallel composition). When an appropriate message is received, one of the appropriate actions is selected nondeterministically for execution. However, the sender of the message cannot know which one is selected. The sender may assume that a different action is selected and may continue to send messages based on the wrong assumption. If the remaining types after executing the nondeterminate receive-actions are all the same as required by our rule, the assumption that a specific action is selected does not matter.

The type model is quite expressive. In the dining philosophers example, a compiler can ensure that forks receive "take" as the last message when being on the table, and "up" and "down" in alternation. The messages received by philosophers are also restricted as specified by the types. But, for example, it is not possible to express in types (i.e. enforce statically) that at most one token is passed around the table. No type system can prevent such run-time errors.

5   Related Work

Nierstrasz argues that it is essential for a type model to regard an object as a process in a process calculus [15]. He proposes "regular types" and "request substitutability" as foundations of subtyping. However, his very general results are not concrete enough to be useful in practice. Many proposals are more practical, but do not consider behavior changes [6,16,21,10].

The proposal of Nielson and Nielson [14] can deal with behavior changes. Types in their proposal are based on a process algebra, and a type checker updates type information while walking through a process expression. Their type model cannot ensure that all sent messages are understood. But subtyping is supported so that instances of subtypes preserve the properties expressed in supertypes: If a program corresponding to a supertype sends only understood messages, also a program corresponding to a subtype does so.

The process type model presented in this chapter seems to be the first model considering behavior changes, supporting subtyping and ensuring statically that all messages are understood by the receivers [17,18]. An improved type representation [19,20] eliminates some difficulties of the process type model and makes the proposal feasible as extensions of practical programming languages. Work on improving the process type model will be continued.

Combinations of the actor model with process calculi comprise a large class of computation models. These models differ in the way how objects communicate. For example, if the linear sequence of received messages is replaced by a mail-box not introducing an ordering on received messages, the model provides an additional degree of freedom. Consequences of changed computation model semantics on the type model are not yet explored sufficiently. This is important future work.

6   Conclusions

We have shown that type expressions provide a feasible basis for statically typed, concurrent, object-oriented languages. We proposed two principles—the principles of type updating and type splitting. They support very flexible and expressive type systems: The behavior of active objects can change; subtyping is supported; and type-consistency can be checked at compile-time, ensuring that each object understands all received messages.

References

1. Gul Agha. *Actors: A Model of Concurrent Computation in Distributed Systems.* MIT Press, 1986.
2. Gul Agha, Peter Wegner, and Akinori Yonezawa, editors. *Research Directions in Concurrent Object-Oriented Programming.* MIT Press, 1993.
3. J. C. M. Baeten and W. P. Weijland. *Process Algebra*, volume 18 of *Cambridge Tracts in Theoretical Computer Science.* Cambridge University Press, 1990.

4. Grady Booch. *Object-Oriented Analysis and Design with Applications*. Benjamin-Cummings, Redwood City, California, second edition, 1994.

5. Luca Cardelli and Peter Wegner. On understanding types, data abstraction, and polymorphism. *ACM Computing Surveys*, 17(4):471–522, 1985.

6. Simon J. Gay. A sort inference algorithm for the polyadic π-calculus. In *Conference Record of the 20th Symposium on Principles of Programming Languages*, January 1993.

7. David Gelernter and Nicholas Carriero. Coordination languages and their significance. *Communications of the ACM*, 35(2):96–107, February 1992.

8. Carl A. Gunter and John C. Mitchell, editors. *Theoretical Aspects of Object-Oriented Programming; Types, Semantics, and Language Design*. MIT Press, 1994.

9. C. A. R. Hoare. Communicating sequential processes. *Communications of the ACM*, 21(8):666–677, August 1978.

10. Naoki Kobayashi and Akinori Yonezawa. Type-theoretic foundations for concurrent object-oriented programming. *ACM SIGPLAN Notices*, 29(10):31–45, October 1994. Proceedings OOPSLA'94.

11. Barbara H. Liskov and Jeannette M. Wing. A behavioral notion of subtyping. *ACM Transactions on Programming Languages and Systems*, 16(6):1811–1841, November 1994.

12. Bertrand Meyer. Systematic concurrent object-oriented programming. *Communications of the ACM*, 36(9):56–80, September 1993.

13. R. Milner, J. Parrow, and D. Walker. A calculus of mobile processes (parts I and II). *Information and Computation*, 100:1–77, 1992.

14. Flemming Nielson and Hanne Riis Nielson. From CML to process algebras. In *Proceedings CONCUR'93*, number 715 in Lecture Notes in Computer Science, pages 493–508. Springer-Verlag, 1993.

15. Oscar Nierstrasz. Regular types for active objects. *ACM SIGPLAN Notices*, 28(10):1–15, October 1993. Proceedings OOPSLA'93.

16. Benjamin Pierce and Davide Sangiorgi. Typing and subtyping for mobile processes. In *Proceedings LICS'93*, 1993.

17. Franz Puntigam. Type specifications with processes. In *Proceedings FORTE'95*, Montreal, Canada, October 1995. IFIP WG 6.1, Chapman & Hall.

18. Franz Puntigam. Types for active objects based on trace semantics. In Elie Najm et al., editor, *Proceedings FMOODS '96*, Paris, France, March 1996. IFIP WG 6.1, Chapman & Hall.

19. Franz Puntigam. Coordination requirements expressed in types for active objects. In Mehmet Aksit and Satoshi Matsuoka, editors, *Proceedings ECOOP '97*, number 1241 in Lecture Notes in Computer Science, Jyväskylä, Finland, June 1997. Springer-Verlag.

20. Franz Puntigam. Types that reflect changes of object usability. In *Proceedings of the Joint Modular Languages Conference*, number 1204 in Lecture Notes in Computer Science, Linz, Austria, March 1997. Springer-Verlag.

21. Vasco T. Vasconcelos. Typed concurrent objects. In *Proceedings ECOOP'94*, number 821 in Lecture Notes in Computer Science. Springer-Verlag, 1994.

22. Peter Wegner and Stanley B. Zdonik. Inheritance as an incremental modification mechanism or what like is and isn't like. In S. Gjessing and K. Nygaard, editors, *Proceedings ECOOP'88*, number 322 in Lecture Notes in Computer Science, pages 55–77. Springer-Verlag, 1988.

High Level Transition Systems for Communicating Agents

François Vernadat and Pierre Azéma

LAAS-CNRS
Laboratoire d'Analyse et d'Architecture des Systèmes
7 avenue du Colonel Roche F-31077 Toulouse cedex
{vernadat,azema}@laas.fr

Abstract. This paper presents an agent-oriented formalism based on Logic Programming and Predicate/Transition Nets. The problem of moving philosophers is used as application example. A distributed system is considered as a result of composition of agents, which are instances of predefined classes. A main objective is to describe dynamic systems of communicating agents.

1 Introduction

An agent based formalism is introduced by means of a specific example. A main objective is to describe dynamic systems of communicating agents. These agents are active elements: they may communicate with partners, disappear or create new agents, hence dynamic. The agent behaviors are described by Predicate/Transition Nets [7], equipped with extensions to take into account communication and dynamic changes.

An agent is an instance of a class. The class behavior is depicted by a set of transitions. During its evolution, an agent may create new agents. An agent knows its own identifier and class, i.e. self referencing is available.

The communications are synchronous, i.e. sending and receiving are simultaneous: rendez-vous is the basic synchronization mechanism. Communications directly occur between agents according to their acquaintances, in the same way as in Actor language [1]. Communications are synchronous, contrary to Actor language. Multiple synchronization is allowed : Inside a transition an agent may specify a profile of synchronization, that is a set of (one to one) rendez-vous with different partners. Moreover, the communication profile may be computed during the transition evaluation, resulting in a great expressive power.

A case-study is presented in section II. The main purpose is to take into account the distributed constraints, *i.e.* local agents do not know the current global system state. Sections III and IV present respectively architectural concepts and the behavior of the identified components. An overview of the formalism is given in sections V and VI: the synchronization mechanism allowing for dynamic transition merging is detailed. Section VII presents the associated software environment and deals with verification aspects. It is worth noticing

G. Agha et al. (Eds.): Concurrent OOP and PN, LNCS 2001, pp. 473–492, 2001.
© Springer-Verlag Berlin Heidelberg 2001

that the system dynamism, that is the possibility of agent creation and deletion, does not prevent a space exploration, or even an exhaustive search. Section VIII gives some concluding remarks.

2   The Hurried Philosopher: A Distributed Point of View

Consider philosophers sitting around a table. A fork is shared by two philosophers. A philosopher can eat only when he is holding two forks, the left and the right ones. A philosopher may introduce a new philosopher. The newcomer arrives with a fork. Conversely a philosopher possessing a fork may decide to leave the table. This case study has been proposed by Sibertin-Blanc [14].

A main significance of this example comes from the treatment of the dynamism. The model focuses on distributed aspects of the problem, that is each agent has only a local knowledge of the system (specifically a philosopher only knows its neighbours on the table). Information exchange between partners needs explicit communication.

Moreover, the control of the system is distributed among the different partners. Each agent possesses its own local state. There is no specific centralized component in charge of the management of the system nor specific data structure recording global informations shared by the different partners.

2.1   Distributed Management

A major difficulty concerns the on-line arrival and departure of philosophers. The following assumptions are considered:

- The maximal number of philosophers is invariant. More precisely the set of philosopher names is fixed at the initialisation of the system.
- Initially, the system consists of a ring involving all the philosophers. Each philosopher knows the name of the next philosopher on the ring.

In the sequel, a local predicate next(_nextref) will be associated with each site to record the name of the next site. A philosopher has not to know the name of the previous philosopher on the ring.

Some philosophers may not be present on the ring, and their names have to be locally managed, because there is no centralized control.

As a consequence, each philosopher has to know some non present philosophers, in order to introduce them; the names of these philosophers are supplied by local predicate name(_list_of_names). These predicates define a partition over the set of all non present philosophers. A present philosopher may introduce a new philosopher if and only if predicate name, that is the local name subset, is not empty. Complementary, a philosopher, who leaves the table, transmits the list of non-present philosophers he knows to one of its neighbor.

With respect to the topology structure of the table, a philosopher leaving the table has to transmit the identity of its next neighbor on the ring to its predecessor, so that this predecessor updates predicates next, name, that is its local knowledge about the ring.

Figure 1 demonstrates the modification of local data associated with the departure of a philosopher, in the case of a ring composed of several nodes {A,B,C, ... }.

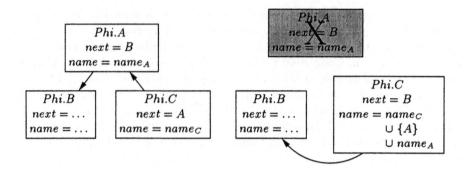

Fig. 1. Departure of a philosopher

When Philosopher A leaves the table, a communication by rendez-vous occurs between philosophers A and C whose purpose is the transmission by philosopher A to C, of the local A values next(B) and name(name_A). Philosopher C consequently updates its local data. Note that Philosopher B is not affected by this transformation of the ring topology.

Figure 2 indicates the modification of local data associated with the entry of a non-present philosopher.

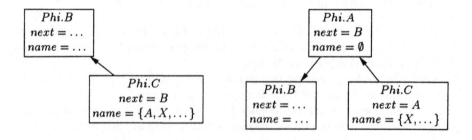

Fig. 2. Introduction of a new philosopher

In this case, external philosopher A is introduced by C. This transformation does not require an explicit communication. Philosopher C, knowing an unused philosopher name (A), creates an A instance of philosopher class. A is created

with an empty set of free references (**name** = ∅) and predicate **next** with value
B (the former successor of philosopher C). Philosopher C changes the successor
value to A and removes value A on its set of unused names.

2.2   Global Architecture

A philosopher corresponds to a single site, and each philosopher is associated
with three agents. The architecture consists of three layers, and there is a specific
agent per layer, which is in charge of a specific aspect of the problem (cf. Fig. 3).

The synchronization between two adjacent agents is carried out by rendez-
vous: this is represented by a black ellipse on Figure.

Agent *phi*, at the application layer, represents the
user behavior,
agent *man*, at the protocol layer, manages forks,
avoids deadlock, etc
agent *net*, at the topology layer, is in charge of
partner communications and topology modifications: predicates **next**, **name** are managed at this
level.

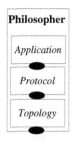

Fig. 3. Architecture of a Philosopher

These three agents represent a philosopher on a single site. In the case of a
ring configuration of philosophers, each agent *net* only knows the identity of one
neighbor.

3   Description of the System Architecture

The former system is now described by means of a PROLOG environment, so-
called VAL. Two types of components are considered: agents and structures.
This section introduces the structures and the agents of the case study.

- **Agents** are active system elements: they have attributes, methods and
behaviors. They send and receive messages, they may disappear and/or create
new agents. In our case, following the architecture developed in the previous
section, three classes of agents are considered: **phi**, **net** and **man**. The behavior
of agents is first presented in the context of our case study in Section 4 and is
formally described in Sections 5 and 6 .

- **Structures** are passive elements, that is without behavior.

- The configuration specifies an instance of the general architecture system
described by the root structure of the hierarchy.

nb: A self-referencing mechanism is available. A component (agent or structure) knows its own identity: attributes @*Class* and @*Ref* allow a component
to reference its own class and reference.

A structure may be viewed as a generic "template" allowing to define the architecture of the system: the list of components (or sub-structures), their hierarchy and the connections among the different components. In this case, three kinds of structure are used:
- structure site defines the three layer architecture of a philosopher
- structure ring defines a ring architecture
- structure hurried defines the whole system.

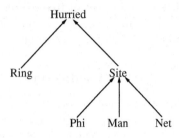

Fig. 4. Description of the Hurried Philosophers System

3.1   Structures Description

Structure Site declares the philosopher architecture. Sub-components are defined by means of clause *structures*. The creation procedure (clause *create*) describes the effect of the creation of instance @ref of structure *site*. This creation is parameterized by the identity of the next site on the ring (predicate *next(_other)*). In this case, the creation of site.@ref consists of the creation of agents phi.@ref, man.@ref and net.@ref. The knowledge of the next site on the ring is only transmitted to agent *net*.

```
structures([phi,man,net]).
-- declaration of sub-components: phi, man and net

private([]).                      -- no local routines

create([next(_other)],_create) :-  -- definition of term _create
     _create_phi = (phi,@ref,[]),
     _create_man = (man,@ref,[]),
     _create_net = (net,@ref,[next(_other)]),
     _create = [ _create_man, _create_phi , _create_net ] .
```

Structure Ring is in charge of the creation of a ring. This structure may be viewed as a generic template in the sense that the class of the component to be created is a sample parameter (_class). The purpose of this template is to create the components of the ring and to initialize each of them with the identity of the next site on the ring. *Ring* does not declare the component _class, the associated declaration is performed in structure Hurried. The creation is parameterized by the set of references (_group) and the class of the component to be created (_class).

```
structures([]).           --  no sub-component
private([create_site]).   --  a local routine declaration

create([class(_class),group(_group)],_create) :-
     create_site(ring,_class,_group,_group,[],_create).
.../...
```

create and create_site are directly written in a prolog syntax. The input parameters, _group and _class, respectively represents the list of references and the class of the elements to be created. Clause create_site produces as output term _create specifying the set of associated creations.

For instance, when _group = {1, 2, 3} and _class = site then produced term _create = {(site, 1, [left(2)]), (site, 2, [left(3)]), (site, 3, [left(1)])}. Each element of this set will be interpreted according to the creation procedure of structure site as viewed in the previous paragraph. For people interested by PROLOG source, here is the code:

```
-- predicate create_site computes the create clauses

create_site(ring,_class,_group,[],_create,_create):-!.
create_site(ring,_class,_group,[_head|_tail],_pc,_create) :-
     others(ring,_entity,_group,[_left,_right]),
     _npc = [(_class,_head,[next(_left)])|_pc],
     create_site(ring,_class,_group,_tail,_npc,_create).
```

nb: *others* is a routine provided by the kernel and computes the identities of adjacent sites according to a specific topology. Several topologies are available: ring, complete graph, n-trees, bus. In the present case, the left neighbor identity is enough to define a ring.

Structure Hurried is the highest structure in the hierarchy. The two sub-structures ring and site are declared. The creation is parameterized by the set of references (_group). The creation consists in the ring creation of site components referenced by _group.

```
structures([ring,site]).   -- component declaration
private([]).   -- local routine declaration : empty in this case

create([group(_group)],[(ring,@ref,[group(_group),class(site)])]).
```

3.2   Agent

An agent declaration (cf Table 1) is similar to a structure definition.

An agent has no sub-component and corresponds to a leaf of the architecture tree. The agent behavior is introduced by clause *body*.

Clause *init* defines the initial state of the created instance. This state may be derived from a parameter list: *init(_list_of_parameters, _initial_marking)).*

The agent behavior is described in Section 4 and 6.

In the case of agent *Phi*, the parameter list is empty, and the initial state is *thinking*. Agent *man* initially is idle and possesses a fork. For agent *net*, the initial marking consists of two predicates. Predicate *next(_other)* supplies a ground value _other, i.e. the identity of the next site on the ring, and predicate *name*([]) defines an (initially empty) list of free references.

Table 1. Agent declaration

AGENT PHI	AGENT MAN	AGENT NET
body(phi).	body(man).	body(net).
private([]).	private([]).	private([]).
init([],	init([],	init([next(_other)],
[thinking]).	[idle,fork]).	[next(_other)],name([])]).

3.3 Configuration

Finally, the configuration file defines a specific instance of the whole system described by the hierarchy of structures. The following configuration describes a system constituted of 3 "philosophers" referenced by $\{1, 2, 3\}$.

```
create([[(hurried,example,[group([1,2,3])])]]) -- Configuration file
```

This leads to the following ring creation:
`(ring,example,[group([1,2,3]),class(site)])`
according to the `ring` creation procedure, the creation of 3 *site* structures is performed: `(site,1,[left(2)])` `(site,2,[left(3)])` and `(site,3,[left(1)])`
Finally, according the `site` creation procedure, the whole system consists of three agents per site.

Table 2. Initial state of the system

phi_1([thinking])	phi_2([thinking]),	phi_3([thinking]),
man_1([fork,idle]),	man_2([fork,idle]),	man_3([fork,idle]),
net_1([next(2),	net_2([next(3),	net_3([next(1),
name([])]),	name([])]),	name([])])

4 Description of Layers

The behavior of each agent class is defined by a specific net. In a standard way, a net may be viewed as a set of transitions. A transition occurrence requires enabling conditions, or preconditions. The transition occurrence establishes consequences, or postconditions. The pre and postcondition variables are constrained by application dependant selectors. A transition depicts either an internal event, without communication, or an external event with (possibly multiple) communication. The agent environment is modified by a transition firing when new agents are created, or when the current agent disappears.

A transition definition consists of several fields, introduced by keywords. *Trans* introduces the transition name.
From (resp. *To*) introduces the precondition (resp. postcondition) set.

Rdv, (rendez-vous) introduces a 4-tuple: {agent class, agent identifier, interaction point identifier, message}. A communication partner is identified by class and reference. The message is exchanged at the interaction point.

Create is a 3-tuple:{agent class, agent identifier, creation parameter list}.

Exit is a boolean. When the value is true, the current agent disappears.

An overview of the formalism is given in Section 6.

4.1   Application Layer

At the upper level, the behavior of agent *phi* is described by the following transitions:

```
Trans req
From  thinking
Rdv   (man,@ref,ip,ask)        Trans leave
To    waiting                  From  thinking
                               Rdv   (man,@ref,ip,stop)
Trans entry                    Exit  true
From  waiting
Rdv   (man,@ref,ip,entry)      Trans new
To    eating                   From  thinking
                               Rdv   (man,@ref,ip,new)
Trans release                  To    thinking
From  eating
Rdv   (man,@ref,ip,stop)
To    thinking
```

Transitions on the left describe the standard problem.

req: agent *phi* issues a request for eating, the rendez-vous is used to keep agent *man* informed of the request. A design choice has been to give the same identifier ($@ref$) to the agents on the same site, and of distinct classes. The *phi* state goes from *thinking* to *waiting*,

entry: when agent *man* agrees, a rendez-vous may occur: agent *phi* starts to eat.

release: agent *phi* stops to eat, and informs manager *man*.

Any transition involves a synchronization with manager agent *man*. A single interaction point (ip) is defined. The message meanings are self explanatory. Transitions on the right deal with dynamic aspects.

leave: agent *phi* leaves the table, this agent disappears (cf **Exit true**), the rendez-vous with agent *man* propagates this decision,

new: agent *phi* introduces a new philosopher, the creation modalities are processed at a lower level.

4.2   Management Layer

The following behavior is introduced: each philosopher possesses a fork and asks the left neighbor for a second fork, via the topology layer, that is by means of a rendez-vous with agent *net*.

```
Trans req                          Trans lending
From  idle                         From  fork
Rdv   (phi,@ref,ip,ask)            Rdv   (net,@ref,previous,give)
To    waitnext                     To    nofork

                                   Trans recovering
Trans ack                          From  nofork
From  waitnext                     Rdv   (net,@ref,previous,release)
Rdv   (phi,@ref,ip,entry),         To    fork
      (net,@ref,next,give)
To    work                         Trans dynamic(_mes)
                                   From  fork
                                   Rdv   (phi,@ref,ip,_mes),
Trans release                            (net,@ref,ip,_mes)
From  work                         Cond  If _mes = stop
Rdv   (phi,@ref,ip,exit),                then _exit = true
      (net,@ref,next,release)            else _exit = fail
To    fork,idle                    TO    fork
                                   Exit  _exit
```

req: agent *man* evolves to state **waitnext**, in synchronization with agent *phi*.

ack: two rendez-vous simultaneously occur: the one with agent *phi*, the other with agent *net*. In this case, the left neighbor agrees to give his fork.

release: agent *phi* stops to eat, this information is propagated in a single step to agent *net*.

In a standard way, agent *man* gives (recovers) a fork to (from) the right neighbor by transition **lending** (**recovering**). It is worth to emphasize that, while a philosopher is performing an *ack*, the partner is performing a *lending*, and vice-versa transition *release* is associated with transition *recovering*. The partner reference is supplied by agent *net*.

Transition **dynamic(_mes)** represents agent *man* counterpart of transitions **stop** and **new** of agent *phi*. Note that a new constraint is added, as a consequence of precondition *fork*, a neighborhood modification is allowed only when the fork is available on the site. The introduced keyword *Cond* declares a (PROLOG) assertion, which has to be fulfilled for the transition to occur. In this example, according to the message value, either *stop* or not, the current agent will either disappear or not.

4.3   Topology Layer

Agent *net* of the topology layer is in charge of communications between adjacent philosophers and of configuration evolutions. Three points are successively considered: site removal, site insertion and message transmission.

Site Removal. Predicate **name** deals with references available for the creation of a new philosopher. This data management is locally performed.

Transitions **leave** and **change_next** are complementary and simultaneously occur between two adjacent philosophers. A philosopher disappearance is decided at the upper (application) layer, a synchronization occurs with the predecessor; two kinds of information are exchanged: the next philosopher identity and the available reference set. On the other side **change_next(_Nnext)**, the partner updates its successor value and the available reference set.

```
Trans leave
From  name(_names),
      next(_next)
Rdv   (man,@ref,ip,exit),
      (@class,_x,ip,
          exit(_next,_names))
Exit  true
```

```
Trans change_next(_Nnext)
From  name(_names),next(_next)
Rdv   (@class,_next,ip,
          exit(_Nnext,_nextnames))
Cond  ground(_Nnext),
          union([_next|_names],
          _nextnames,_Nnames)
To    name(_Nnames),next(_Nnext)
```

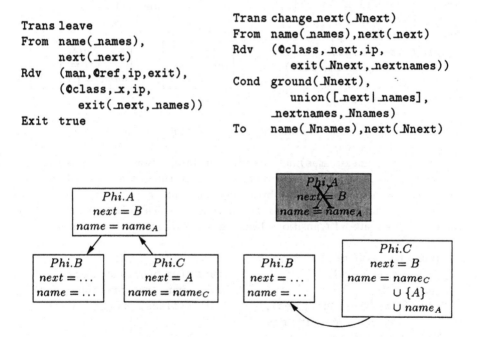

Fig. 5. Management of acquaintances at the topology layer

In transition **change_next**, condition **ground(_Nnext)** (i.e., variable _Nnext has to be a ground term), prevents the dynamic merging of two identical transitions, either **leave** or **change_next**.

Site insertion

The decision to introduce a new partner is local to a site. The site needs an available reference in order to name the new site. The created site takes as "next" value, the old one of the creating site, conversely, the creating site updates the value of predicate **next** by the identity of the created site. The figure on the right indicates the main creation steps.

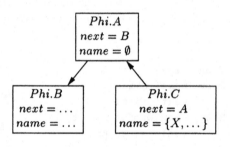

```
Trans   new(_first)
From    name([_first|_tail]),
        next(_next)
Rdv     (net,@ref,ip,new)
Create  (site,_first,
              [next(_next)])
To      name(_tail),
        next(_first)
```

Transmission. Agent *phi* knows only the left neighbor identity: predicate **next** records this knowledge.

Transition **to_next(_mes)** describes the synchronization of two adjacent sites. The condition _next $\neq$ @ref prevents the exit of the last element of the ring (in the case of a ring constituted by a single element, _next = @ref).

Transition **from_previous(_mes)** represents a synchronization initiated by the predecessor site (cf transitions **lending** or **recovering** of agent **man**).

```
Trans to_next(_mes)
From  next(_next)              Trans from_previous(_mes)
Cond  _next ≠ @ref             Rdv   (man,@ref,previous,_mes),
Rdv   (man,@ref,next,_mes),          (@class,_prev,ip,_mes)
      (@class,_next,ip,_mes)
To    next(_next)
```

5   Overview of the Formalism

5.1   Basic Ingredients

Notations: The following elements are considered: C a set of constants, V a set of variables, Π a set of predicate symbols. *Terms* denotes the set of terms associated with $\triangleleft C, V, \Pi \triangleright$. The notion of substitution is defined as usual. A term i is an instance of the term t, iff there exists a substitution σ such that $(t : \sigma) = i$. A ground term is a term without any variable. In the sequel, $G(Terms)$ denotes the subset of the ground terms of $Terms$. The set of identifiers is denoted $Identifier$.

Communications: For communication-structuring purpose, a **topic** is associated with any message. This concept is similar to the concept of interaction point or gate introduced in formal description techniques.

An **elementary synchronization** consists of a quadruple
$<Class, Ref, topic, message> \in Terms^4$ where $Class, Ref$ represents the partner in the synchronization, $topic$ the topic devoted to the synchronization and $message$ the message content.

The formalism allows multiple synchronisations, a **synchronization** is a (finite) set of elementary synchronisations. $Synchronisations$ is the set of all synchronisations, $Synchronisations =_{Def} \mathcal{P}_f(Terms^4)$.

Dynamism: The creation constraint is the name conflict resolution: two distinct instances of the same class must have distinct names. The "initial state" of an instance of agent is specified at its creation.

An **elementary creation** is represented by a triple
$<Class, Ref, Marking> \in Terms^2 \times \mathcal{P}_f(G(Terms))$ where $Class$ represents the class of agent to be created, Ref its reference, and $Marking$ its initial "state" The formalism allows multiple creations, a **creation** is a set of elementary creations. We note $Creations$, the set of all creations.
$Creations =_{Def} \mathcal{P}_f(Terms^2 \times \mathcal{P}_f(Terms))$

5.2   Class of Agent

Transition: Table 3 gives the **transition definition** and its representation.

Table 3. Transition definition

Trans	$name$	$(\in Terms)$
From	$Pre - Conditions$	$(\in \mathcal{P}_f(Terms \setminus V))$
Rdv	$Synchronization$	$(\in V \cup \mathcal{P}_f(Terms^4))$
To	$Post - Conditions$	$(\in \mathcal{P}_f(Terms \setminus V))$
Create	$create$	$(\in V \cup \mathcal{P}_f(Terms^2 \times \mathcal{P}_f(Terms)))$
Exit	$exit$	$(\in V \cup \{true, false\})$
Cond	$condition$	$(\in Prolog\_Expression)$
	$trans$	$(\in Identifier$ identifier of the transition$)$

We note $Transitions$, the set of all the transitions. We assume the existence of an injective map from $Transitions$ to $Identifier$ furnishing an unique identifier for each transition.

A **behavior**, is a (finite) subset of the set of transitions. We note $Behaviour$, the set of all behaviors. $Behaviour =_{Def} \mathcal{P}_f(Transitions)$

Interpretation: To facilitate data manipulation, logic programming (Prolog) is used as a declarative and prototyping language. The interpretation is defined as a set of Horn Clauses (implication with at the most one conclusion). The selector of a condition will be expressed as a Prolog Expression. In the sequel, we note $(cond : \sigma) \models_I true$ to indicate that the instance of a prolog expression $cond$ determined by σ holds according to interpretation I. In the sequel, we assume the existence of a general interpretation allowing for arithmetic operations shared by all the agents.

A **Class of Agent** is composed of a behavior and an interpretation. We note $Agents$, the set of all agents. $Agents =_{Def} Behaviour \times Interpretations$

A **marked instance of agent** is a triple $\langle class, ref, marking \rangle$
$$\in Agents \times Identifiers \times \mathcal{P}_f(G(terms))$$

$\mathcal{C}$, a set of marked instances of agent makes a **Configuration of marked instances of agent** if this set does not contain homonymous agent (instances of the same class sharing the same reference), that is:

$\forall \langle class_1, ref_1, marking_1 \rangle \in \mathcal{C}, \forall \langle class_2, ref_2, marking_2 \rangle \in \mathcal{C} :$
$$\langle class_1, ref_1, marking_1 \rangle \neq \langle class_2, ref_2, marking_2 \rangle$$
$$\Rightarrow \langle class_1, ref_1 \rangle \neq \langle class_2, ref_2 \rangle$$

We note $Configurations$ the set of all configurations. $Configurations$ is a subset of $Creations(\mathcal{P}_f(Terms^2 \times \mathcal{P}_f(Terms)))$

5.3 Systems, Marked System

A **system** is defined as a (finite) set of agent classes. We note $Systems$, the set of all the systems. $Systems =_{Def} \mathcal{P}_f(Agents)$

$\mathcal{C}$, a Configuration of marked instances of agent is **valid** for a system S iff
$$\langle class, ref, marking \rangle \in \mathcal{C} \Rightarrow class \in S$$

We note $Valid\_Configurations(S)$, the set of the valid configurations for system S. Table 2 gives an instance of such configuration.

A **marked system** is finally defined as a pair $(S, \mathcal{C})$ where S is the system and $\mathcal{C}$ a valid configuration for S.

With respect to standard Petri Nets, the following analogies occur: a net corresponds to a system (a set of agent classes) and a marking corresponds to a valid configuration (a set of marked agent instances).

6 Semantics

Notations: In the sequel of the paper, $(S, \mathcal{C})$ denotes a marked system and $< class, ref, marking >$ a marked agent instance of the configuration $\mathcal{C}$. For a transition $trans =< name, \ldots, cond >$ and a substitution σ, we note $(trans : \sigma)$ for $< (name : \sigma), \ldots (cond : \sigma) >$

- Let t, a transition belonging to the behavior of agent class $class$ and $trans$ its identifier, σ a substitution such that $(trans : \sigma)$ is a ground transition instantiation.

$class(ref).trans$ denotes transition identified by $trans$ of the instance ref of the agent class $class$ and $class(ref).trans : \sigma$ the instance of the above transition defined by substitution σ.

$class(ref).trans.field$: σ a specific instantiation of a particular field of the above transition

$$\mathcal{C} \vdash S \xrightarrow{\quad class(ref).trans:\sigma \quad} \equiv \text{for system } S, \text{ configuration } \mathcal{C} \text{ enables}$$
transition $class(ref).trans : \sigma$.

$$\mathcal{C} \vdash S \xrightarrow{\quad class(ref).trans:\sigma \quad} \mathcal{C}' \equiv \text{the firing leads to configuration } \mathcal{C}'$$

Transitions including dynamic aspects are first considered while section 3 deals with general transitions (including communication and dynamism).

6.1   Transition with Dynamism

Enabling Rule:
Let σ such that $class(ref).trans : \sigma$ is a ground transition,

$$\mathcal{C} \vdash S \xrightarrow{\quad class(ref).trans:\sigma \quad} \text{ iff}$$

(1) $< class, ref, marking >\in \mathcal{C}$
(2) $[trans.from : \sigma] \subset marking$
(3) $([trans.to : \sigma] \setminus ([trans.from : \sigma] \cap [trans.to : \sigma])) \cap marking = \emptyset$
(4) $[trans.cond : \sigma] \models_I true$
(5) $[trans.exit : \sigma] \in \{false, true\}$
(6) $([trans.create : \sigma] \cup \mathcal{C}) \in Configurations$
(7) $[trans.create : \sigma] \in Valid\_Configurations(S)$

Corresponding agent instance exists in the current configuration (1), Conditions (2) to (4) are standard [7]: Preconditions are fulfilled (2), Postconditions which are not preconditions are not fulfilled (3), Condition is satisfied in the considered interpretation (4). The three last conditions are relative to the dynamism. Condition (5) only requires that the value of the Exit field is correctly instantiated. Condition (6) ensures that the creations associated with the firing leads to a valid configuration (cf def 5.2) and condition 7) the class of the created agents must belong to system S

Firing Rule If $\mathcal{C} \vdash S \xrightarrow{\quad class(ref).trans:\sigma \quad}$ Then $\mathcal{C} \vdash S \xrightarrow{\quad class(ref).trans:\sigma \quad} \mathcal{C}'$ where
$\mathcal{C}' = (\mathcal{C} \setminus \{<class, ref, marking>\}) \cup New\_marking\_class(ref) \cup [trans.create : \sigma]$

$$New\_marking\_class(ref) =_{Def} \begin{cases} \emptyset & \text{if } [trans.exit : \sigma] = true \\ \{<class, ref, marking'>\} & \text{otherwise} \end{cases}$$

where $marking' = (marking \setminus [trans.from : \sigma]) \cup [trans.to : \sigma]$

The instances of agents specified in the create field of the transition appear in the resulting configuration. If boolean `exit` holds, then the current agent instance disappears, else its marking ($marking'$) is updated in a standard way [7].

Example: Consider transition new(_first) of agent *net* (cf section 4.3) and the state system described in table 4.

Table 4. State of the System

phi_1([thinking])   phi_3([thinking]),
man_1([fork,idle]),   man_3([fork,idle]),
net_1([next(3),name([])]),   net_3([next(1),name([2])])

Tables 5 gives an instantiation of transition new(_first) for agent *man*.3 by the following substitution (@ref/3,_first/2,_tail/[] _next/1).

Table 5. Instance of firable transition

Trans	new(_first)		Trans	new(2)
From	name([_first | _tail]),		From	name([2 | []]),
	next(_next)			next(1)
Rdv	(net,@ref,dyn,new)		Rdv	(net,3,dyn,new)
Create	(site,_first,[next(_next)])		Create	(site,2,[next(1)])
To	name(_tail),		To	name([]),
	next(_first)			next(2)

The firing of the resulting instance of transition leads to the state system[1] described in table 6.

Table 6. State of the system after the firing of transition man(3).new(2)

phi_1([thinking])	phi_3([thinking]),	phi_2([thinking]),
man_1([fork,idle]),	man_3([fork,idle]),	man_2([fork,idle]),
net_1([next(3),	net_3([next(2),	net_2([next(1),
name([])]),	name([])])	name([])]),

6.2 Synchronous Transition

Communication by rendez-vous is now considered. Rendez-vous is interpreted as a transition resulting from the merging of synchronous transitions. A synchronization transition is a set of linked synchronous transitions. In the proposed approach, this merge is dynamic.

The elementary synchronisations of a transition are constraints to be solved. A set of synchronous transitions makes a synchronization transition iff each synchronization constraint is satisfied within the set, all the synchronisations are required because each transition within the set is linked (directly or not) with each other, and finally an agent is involved at most once.

---

[1] The creation of structure site is described in section 3.1

In the sequel, we consider a set E of n distinct (ground) instances of transitions: $Cl_1(ref_1).trans_1 : \sigma_1, Cl_2(ref_2).trans_2 : \sigma_2, ...Cl_n(ref_n).trans_n : \sigma_n$.

Synchronization Transition . • With each synchronous transition is associated a **synchronization set** (the set of its elementary rendez-vous)
$$synchro(Cl(ref).trans : \sigma) =_{Def} \bigcup_{(x,y,z,t)\in[trans.rdv:\sigma]} \{(Cl, ref, x, y, z, t)\}$$
i.e, an elementary rendez-vous is a sextuple $(Cl_1, ref_1, Cl_2, ref_2, suj, mes)$ specifying identity (class and reference) of the two partners, the topic and the message exchanged.

Notion of synchronization set is extended to the set of synchronous transitions. For a set E of synchronous transitions, we associate:
$$SYNCHRO(E) =_{Def} \bigcup_{e\in E} synchro(e)$$
• Two elementary rendez-vous are **complementary** (denoted $rdv_1 \ C \ rdv_2$) if they are of the following form:
$$rdv_1=(Cl_1, ref_1, Cl_2, ref_2, topic, mes) \ \& \ rdv_2=(Cl_2, ref_2, Cl_1, ref_1, topic, mes)$$

Example: Transition `req` of agent *phi* involves the following elementary synchronization: `(man,@ref,ip,ask)`, while transition `req` of agent *phi* involves `(phi,@ref,ip,ask)` leading to 2 complementary rendez-vous:

   `(phi,@ref,man,@ref,ip,ask)`, `(man,@ref,phi,@ref,ip,ask)`.

Definition: A set E of **independent** instances of synchronous transitions makes a **Synchronization transition** iff
(1) $\bigoplus (SYNCHRO(E)) = \emptyset$
   All the constraints specified by E are resolved within E
(2) $F \neq \emptyset$ and $F \subset E \Rightarrow \bigoplus (SYNCHRO(F)) \neq \emptyset$
   E is minimal ensuring that E does not contains any superfluous constraint.
Where the sum ($\bigoplus$) of a Synchronization set is defined as follows:

$$\bigoplus (E) =_{Def} \begin{cases} \emptyset & \text{if } E = \emptyset \\ \bigoplus (F) & \text{if } E = \{rdv_1, rdv_2\} \cup F \text{ and } rdv_1 \ C \ rdv_2 \end{cases}$$

Enabling and Firing Rule. Let $E = \{Cl_1(ref_1).t\_1 : \sigma_1, Cl_2(ref_2).t\_2 : \sigma_2, ...Cl_n(ref_n).t\_n : \sigma_n\}$ a synchronization transition, $\mathcal{C} \models S\overset{E}{\Longrightarrow}$ Iff the two following hold.

1) $\forall i \in [1,n] : \mathcal{C} \models S\xrightarrow{Cl_i(ref_i).t\_i:\sigma_i}$
   each transition of the set is enabled (in the sense of definition 6.1)
2) $Create(E) \in Valid\_Configurations$
where $Create(E) =_{Def} \bigcup_{Cl_i(ref_i).t\_i:\sigma_i \in E}[t\_i.create : \sigma_i]$
   the set of the elementary creations involved in the Synchronization transition makes a valid Configuration.

Firing Rule: For $E = \{e_1, ... e_n\}$, If $\mathcal{C} \models S\overset{E}{\Longrightarrow}$ Then $\mathcal{C} \models S\overset{E}{\Longrightarrow} \mathcal{C}'$
   where $\mathcal{C} \models S\xrightarrow{e_1} \mathcal{C}_1, ... \ \mathcal{C}_{n-1} \models S\xrightarrow{e_n} \mathcal{C}_n = \mathcal{C}'$

As E is composed of independent transitions, the order of transition firing is irrelevant.

Example: Consider C, the state of the system described by table 6.

Let $E = \{e1, e2, e3, e4\}$, the set constituted of the following instances of enabled transitions:

e1 = phi(1).leave:(@ref/1),

e2 = man(1).dynamic:(@ref/1,_mes/stop,_exit/true)

e3 = net(1).leave(@class/net,@ref/1,_refs/[],_next/3,_free/2)

e4 = net(2).change_next:(@class/net,@ref/2,_refs/[],_next/1,_Nrefs/[1],_Nnext/3)

then $C \vdash S \xrightarrow{\quad ei \quad} \forall i \in [1,4]$

SYNCHRO(E) =   { (phi,1,man,1,ip,stop),(man,1,phi,1,ip,stop),
                  (man,1,net,1,ip,stop),(net,1,man,1,ip,stop),
                  (net,1,net,2,ip,exit(3,[])),(net,2,net,1,ip,exit(3,[]))
                  }

and

$\bigoplus_{E} (SYNCHRO(E)) = \emptyset$

Then $C \models S \Longrightarrow C'$ where

$C' = \begin{array}{ll} \{\text{phi\_2([thinking])} & \text{phi\_3([thinking])}, \\ \text{man\_2([fork,idle])}, & \text{man\_3([fork,idle])}, \\ \text{net\_2([next(3),name([1])])}, & \text{net\_3([next(2),name([])])}\} \end{array}$

7   VAL Prototyping Environment

Software environment VAL is based on the described formalism. The ability to describe dynamic systems, i.e. agent creation and suppression, and mobile processes, i.e. addressing by name and logic unification, offers a great flexibility to the designer.

For debugging purpose, a specification may be interpreted step by step. The simulator and the associated graphical user interface provide standard facilities : display of the global state and of the communication queues, list of the enable transitions.

Verification. In order to perform formal verification, the reachable state space may be computed (even in the case of dynamic description). This space may then be analyzed either by algebraic approaches, i.e. observational equivalence [10], or by temporal-logic model checking [4].

For the present example, a verification using behavioral equivalences has been carried out. The state space enumeration is first performed. In the case of three sites, a (deadlock free) state graph has been derived. It consists of 94 states and 258 edges.

To check the mutual exclusion property, the observed events are, for each philosopher, the critical section entry and exit. Weak bissimulation is considered, the resulting quotient graph is represented in Fig. 6.

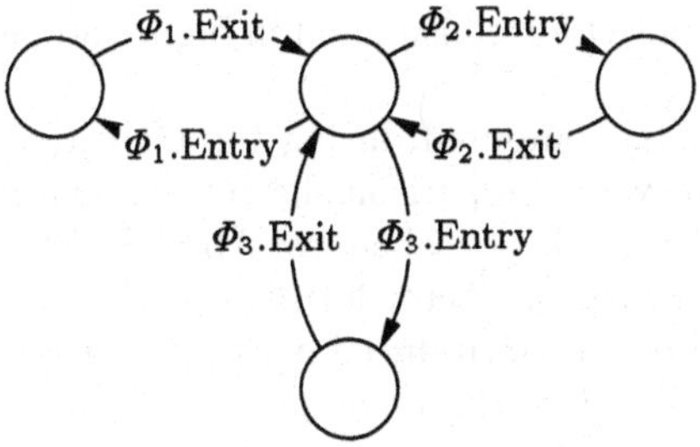

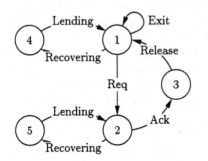

The mutual exclusion property holds because between two consecutive entries in the critical section, the exit of the last philosopher must necessarily occurr.

Fig. 6. Behavioral Verification of the Mutual Exclusion Property

The second verification concerns the local service associated with a single philosopher. More precisely, observed events are those occurring at the management layer. The minimal LTS obtained in the case of safety equivalence [6] is depicted below. Roughly speaking, class 1 corresponds to system states where the philosopher is thinking, class 2 to states where the philosopher is waiting and state 3 where the philosopher is eating. Since this quotient LTS is deadlock free, we can deduce that a philosopher is itself deadlock free.

Another interesting properties may be deduced from a simple LTS analysis:
The departure of a philosopher is only possible in the case where this philosopher is thinking (cf loop exit in class 1)
Starvation of a philosopher is possible: the loop Lending, Recovering, situated in class 2, shows that a waiting philosopher may wait an infinite time before eating.

Fig. 7. Observation of the local service

8   Conclusion

A formalism based on Logic Programming and Predicate/Transition Nets has been introduced for describing dynamic systems by means of communicating agents.

Multiple synchronisations are allowed. Within a transition, an agent may specify a synchronization profile, that is a set of one to one, possibly bidirectional rendez-vous. In Lotos [9], the synchronization operates through gates and does not require any knowledge on the process environment. A process may specify a rendez-vous with others processes on a specific gate without knowing the number of these processes. From this point of view, Lotos multi-rendez-vous is more expressive than the proposed one. However Lotos does not allow to exchange gates as messages; this is allowed in the proposed formalism, and in higher order formalisms such as Π-Calculus [11].

Finally, the communication profile may be computed during the transition evaluation, resulting in greater expressive power. Comparisons have to be made with higher order formalisms such as the Π-Calculus or Chemical Abstract Machine [3]. A first comparison has been made with the Actor formalism [16].

The proposed formalism meets the requirements introduced in the case-study proposal, by C. Sibertin-Blanc [14,13], for the 2nd Workshop on Object-Oriented Programming and Models of Concurrency [19,20]

Local Control: The formalism is mainly devoted to the description of distributed systems, that is agents without common memory or global control, each agent is thus an autonomous entity.

Dynamism is an important feature of this proposal. Agents may join or leave the system at run time. Moreover, dynamism is supported during all the design steps, even the formal verification.

Dynamic Binding holds because the communication partners are defined online, when the communication happens. Moreover, another feature (not demonstrated here) is the dynamic specification of communication profiles, such as a generic pattern.

Inheritance In the current state, the proposal does not allow for inheritance.

Verification The proposed formalism is supported by a transitional semantics; there is no need to translate the description into an other kind of net. The formal verification is conducted directly on the initial description.

A software environment VAL based on this formalism is available. The possibility to describe dynamic systems, through agent creation and suppression, and mobile processes, through addressing by name and logic unification, offer a great flexibility to the designer. The associated execution engine includes all the features of the presented model. The specification may be simulated for debugging purpose or exhaustively analyzed for verification purpose, even in the case of dynamic description. The former environment is presently used in the domain of communications protocols and Computer Supported Cooperative Work [15,18].

References

1. Agha, G.: *Actors: a model of concurrency computation in distributed systems.* MIT Press, Cambridge (Mass.), 1986
2. Banâtre, J.P., Le Métayer, D.: *The Gamma model and its discipline of programming.* Science Computer Programming, Vol. 15, 1990
3. Boudol, G.: *Some Chemical Abstract Machines.* In: J.W de Bakker and All (Eds.), "A Decade of Concurrency: Reflections and Perspectives." LNCS, Vol. 803, Springer Verlag, 1994

4. Emerson, E.A., Srinivasan, J.: *Branching Time Temporal Logic.* LNCS Vol. 354., Springer Verlag, 1989
5. Diaz, M. and all (Eds.) *The formal description technique Estelle.* North-Holland, 1989
6. Fernandez, J.C, Garavel, H., Mounier, L., Rasse, A., Rodriguez, C., Sifakis, J.: *A Tool Box for the verification of Lotos programs.* In: 14th Conf. on Software Engineering. Melbourne, Australia, 1992
7. Genrich, H.J.: *Predicate/Transition Nets.* In: Jensen, K., Rozenberg, G. (Eds.), "High-Level Petri Nets: Theory and Application." Springer Verlag, 1991, pp 3-44
8. Lloret, J.C., Azéma, P., Vernadat, F.: *Compositional design and verification with labelled predicate transition nets.* In: Clarke, E.M., Kurshan, R.P. (Eds.), "Computer Aided Verification" DIMACS, Vol. 3, 1990, pp 519-533
9. van Eijk, P.H.J. and all (Eds.) *The formal description technique Lotos.* North-Holland, 1989
10. Milner, R.: *Communication and Concurrency.* Prentice Hall, 1989
11. Milner, R., Parrow, J., Walker, D.: *A Calculus of mobile processes.* Information and Computation, Vol. 100, pp 1-77
12. Reisig, W.: *Petri Nets: an Introduction.* Monographs on Theoretical Computer Science, Springer Verlag, 1985
13. Sibertin-Blanc, C.: *Cooperative Nets.* Application and Theory of Petri Nets. LNCS Vol. 815, Springer Verlag, 1994
14. Sibertin-Blanc, C.: *The hurried Philosophers.* Case study proposal for the 2nd Workshop on Object-Oriented Programming and Models of Concurrency. 1996
15. Villemur, T., Diaz, M., Vernadat, F., Azéma, P.: *Verification of Services and Protocols for Dynamic Membership to Cooperative Groups.* Workshop on Computer Supported Cooperative Work, Petri Nets and related formalisms, 1994
16. Vernadat, F., Azéma, P., Lanusse, A.: *Actor Validation by means of Petri Nets.* In: " Object-Oriented Programming and Models of Concurrency.", Italy, 1995
17. Vernadat, F., Azéma, P.: *Validation of Communicating Agent System.* In: "Special Issues in Object-Oriented Programming." Max Mühlhäuser (Ed.) dpunkt.verlag, Heidelberg Germany ISBN 3-9200993-67-5, 1996
18. Azéma, P., Vernadat, F., Gradit, P.: *A workflow specification environment.* In: "Workflow Managament: Net-Based Concepts, Models, Techniques, and Tools." A workshop within the XIX ICATPN, Lisbon, Portugal, 1998
19. Agha, G., de Cindio, F. (Eds.): *First Workshop on Object-Oriented Programming and Models of Concurrency.* A workshop within the XVI ICATPN, Turin, Italy, 1995
20. Agha, G., de Cindio, F. (Eds.): *Second Workshop on Object-Oriented Programming and Models of Concurrency.* A workshop within the XVII ICATPN, Osaka, Japan, 1996

Schedulability Analysis of Real Time Actor Systems Using Coloured Petri Nets[1]

Libero Nigro and Francesco Pupo

Laboratorio di Ingegneria del Software
Dipartimento di Elettronica Informatica e Sistemistica
Università della Calabria, I-87036 Rende (CS) – Italy

Voice: +39-0984-494748 Fax: +39-0984-494713
{l.nigro, f.pupo}@unical.it

Abstract. This paper proposes a modular and real-time actor language which addresses timing predictability through a holistic integration of an application with its operating software, i.e., scheduling algorithm. The actor language fosters a development life cycle where the same concepts and entities migrate unchanged from specification down to design and object-oriented implementation. The paper concentrates on the specification and analysis of distributed real-time systems using Coloured Petri Nets. Both functional and temporal properties can be validated by simulation and occurrence graphs in the context of the Design/CPN tools.

Keywords: Actors, Modularity, Real Time, Design/CPN, Temporal Analysis, Occurrence Graphs.

1 Introduction

Many distributed software systems can be represented, abstracting away from time aspects, as a set of asynchronous, autonomous components which interact one to another in order to co-ordinate local activity. The Actor model [1] is a well established computational framework suitable for building open and re-configurable general distributed applications.

In real-time systems time management is fundamental: application correctness depends not only on the functional results produced by computations, but also on the time at which such results are generated. Timing correctness is related to guaranteeing that a given logical behaviour is provided at the right time, not before nor after a due time.

In the last years, Agha et al. [27, 28, 30] have proposed extensions to the actor model in order for it to be applicable to real-time systems. The extensions rely on capturing message *interaction patterns* among actors through a *RTsynchronizer*

---

[1] A preliminary version of this paper was presented at the *First Workshop on The Practical Use of Coloured Petri Nets and Design/CPN*, University of Aarhus, K. Jensen (Ed.), DAIMI PB-532, pp. 271-285, May 1998.

G. Agha et al. (Eds.): Concurrent OOP and PN, LNCS 2001, pp. 493–513, 2001.

construct [27]. RTsynchronizers are declarative in character. They refer to groups of actors and specify timing constraints on the execution of relevant messages. Their concrete application depends on the possibility of ensuring a global time notion in a distributed system. Moreover a suitable scheduling strategy is required in order to guarantee that the timing constraints of messages are ultimately met. A major benefit of the use of RTsynchronizers is modularity. Actors are firstly defined according to functional issues only. Then timing aspects are separately specified through RTsynchronizers which affect scheduling.

This work is concerned with a development of distributed real-time systems using the concepts of actor and RTsynchronizer. However, a variant of the Actor model is actually proposed with the goal of ensuring *time predictability* [33] through a holistic integration of an application with its operating software, i.e., scheduling algorithm. Concurrency among the actors of a same community on a physical processor depends on message processing interleaving and not on over-killing, OS hidden mechanisms. The approach delivers a seamless development life cycle where all the development phases share the same vocabulary of concepts (actors, messages, timing constraints, scheduling control, ...). As a consequence, analysis can start as early as a specification model of a system with its timing information has been derived. As the project is moved to its target architecture and the timing attributes become more precise, the same analysis model can be reused to check the system remains schedulable. The concept is that of an iterative development life cycle where analysis results can directly be used at the design and implementation levels [36].

This paper focuses on the specification, visualisation and analysis of actor systems with timing constraints using Coloured Petri Nets [15, 16, 17].

The possibility of modelling actor systems through high-level Petri nets has been previously investigated. In [31] an equivalence between actors and the formalism of CPN [15] is provided with the goal of formalising actor semantics. The work of Agha et al. described in [3] is concerned with the use of Predicate Transition nets [10] in order to support the visualisation of actor programs. Predicate Transition nets (PrT-nets) and CPN are formally equivalent and can be considered as two slightly different dialects of the same language. But none of these approaches has analysis aims. Moreover, only functional aspects are covered.

In [24] the temporal analysis of object-based real-time systems through TER nets [12] was explored. However, the lack of mechanisms for modularity and compositionality, which are essential in the modelling of complex systems, was among the motivations leading to the choice of CPN and the use of the Design/CPN tools [18] in particular, in the work described in this paper. Besides the high-level character of CPN, Design/CPN supports the hierarchical and modular construction of nets and provides a built-in time notion on the basis of which temporal properties can be verified by simulation and the generation of occurrence graphs.

The CPN-based specification and analysis mechanisms described in this paper can be related to the Timing Constraint Petri Nets (TCPN) [35] and the Communicating Shared Resources (CSR) [11] modelling and verification tools. From TCPN this work inherits the basic goal of using timed Petri Nets for a *quantitative* evaluation of the timing properties and for the *schedulability* problem of a set of real-time objects, i.e., guaranteeing that the system behaviour satisfies its timing requirements. However, whereas TCPNs represent a particular extension of Petri nets which enables the analysis and the adjustment of timing constraints, this work relies on a widely known Petri nets language, CPN, which is semantically extended in order to support the specification and analysis of real-time actor systems.

The CSR is a language for specifying modular and verifiable real-time systems. Both the functional and the temporal aspects can be specified in the CSR Application Language. Timing parameters, e.g., a deadline on the execution of a command, and more general scheduling attributes (priorities) are part of the Configuration Language which specifies also the CSP-like channel connections among the processes and the mapping of processes to resources (e.g., processors) in order to assemble a system "in-the-large". Parameters and attributes of the Configuration Language are critical to ensure that a system is schedulable, i.e., the hard-real-time operations meet their time obligations. For the verification task, the CSR specification and configuration descriptions of a system are first automatically translated into the terms of the Calculus of CSR (CCSR) which is based on a process algebra and provides a Reachability Analyser to validate and possibly to correct iteratively the system timing constraints.

Besides any differences in the language notations, a key distinction between the real-time actor model proposed in this paper and TCPN and CSR is that the specification of an actor system is kept very close to the its design/implementation. The CPN specification is used as a *prototype* which can be executed and analysed for visualising the functional and temporal properties of the system. On the other hand, both TCPN and CSR separate specification from design and implementation. Verification activities are independent from the scheduler algorithm which ultimately will orchestrate the system evolution. This work argues that the analysis and the fulfilment of timing constraints require prototyping the specification of a system *together* with its scheduling strategy.

The structure of the paper is the following. Section 2 describes the proposed variant of the Actor model [19, 25] where reflective actors are used to capture RTsynchronizers. Section 3 exemplifies the application of the actor concepts and timing constraints to a crane example. Section 4 shows a translation of the actor model into Design/CPN. Section 5 discusses the analysis of an actor-based CPN model with a special focus on the timing verification. The schedulability of the crane example is showed. Finally, the conclusions are presented together with directions of further work.

2 An Actor Language for Time Dependent Systems

A modified version of the Actor model [1] was designed [19, 25] to support the construction of real-time applications. It centres on actors directly modelled as finite state machines. As in the Actor model three basic operations are available:

- *new*, for the creation of a new actor as an instance of a class which directly or indirectly derives from a basic Actor class. The data component of an actor includes a set of *acquaintances*, i.e., the known actors to which messages can be sent

- *send*, for transmitting an *asynchronous message* to a destination actor. The message can carry data values. The sender continues immediately after the send operation
- *become*, for changing the current state of the actor, thus modifying the way the actor will respond to expected messages. The processing of an unexpected message can be postponed by storing it into states or data.

Some differences from the Actor model were introduced for achieving timing predictability. Each object no longer has an internal thread. Rather, concurrency is provided by a *control machine* (see later in this paper) which transparently buffers all the exchanged messages among the actors residing on a same processor, and delivers them according to a control strategy. In other terms, concurrency relies on a light-weight mechanism: *message processing interleaving*, which costs a method invocation in an object-oriented language. Only one message processing can be in progress within an actor at any instant in time. Message execution can't be suspended nor interrupted. All of this contributes to a deterministic computation of message response times.

To ensure the real-timeness of an actor system, message execution times should be bounded and short. Therefore, some constructs should be avoided in actors (e.g., recursion, loops without bounded iterations, dynamic data structures, etc.). Coping with heterogeneous message durations requires careful partitioning and modularising a system in subsystems with different time requirements, allocated to distinct processors [19]. A hard-real-time (HRT) subsystem hosts a community of actors with short and homogeneous message response times. A non HRT subsystem can have actors with longer message execution times.

The external environment is supposed to be interfaced by *terminators*, e.g., periodic actors which sense specific external conditions and raise internal messages corresponding to environment stimuli. A terminator synchronously operates the associated I/O device (e.g., a sensor).

Instead of using a single mail queue to every actor, a few message queues (e.g., one) are handled by the control machine. A customisable scheduler hosting timing constraints avoids control machine dependencies from the policies and mechanisms of an operating system.

As in the Shlaer-Mellor method [32], actors can admit *accessor methods* which can be invoked synchronously instead of by asynchronous message passing. Accessor methods return data values of the current actor state. It is the responsibility of the developer to guarantee the consistency of accessor semantics.

2.1 Basic Components

At the system level an application consists of a collection of subsystems/processors, linked one to another by a (possibly) deterministic interconnection network (e.g., CAN bus [14]). A subsystem hosts a group of actors which are orchestrated by a control machine. The basic components are the control machine and the application actors. The control machine is articulated into the following sub-components (see also Fig. 1):

- a *local clock*, which contains a "real" time reference clock for all the actors in the subsystem
- a *plan*, which arranges message invocations along a timeline
- a *scheduler*, which filters messages, applies to them a set of timing control clauses and schedules them on the plan. The scheduler is actually split into two actors: the *input filter* (iFilter) and the *output filter* (oFilter). iFilter is responsible of the scheduling actions of *just sent messages*; oFilter can be specialised to verifying timing violations at the dispatch time of a message
- a *controller*, which provides the basic control engine which repeatedly selects (*selector* block) from the plan the next message to be dispatched, and delivers it (*dispatcher* block) to its receiver actor.

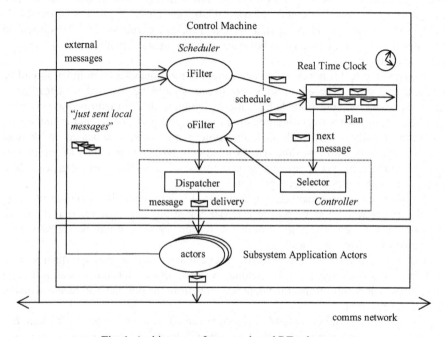

Fig. 1. Architecture of an actor-based RT subsystem.

A control machine is naturally event-driven. Asynchronous messages are received either from the external environment or from within a subsystem, and are processed by the relevant actors. A message processing can generate further messages and so on. From this point of view, the message plan of the control machine can reduce to a FIFO queue.

The time-driven paradigm [26], on the other hand, adds the facility of scheduling messages to occur at certain due times. From this viewpoint the control machine can be customised through programming so as to be application-dependent and sensitive to specific timing models. Different control machines have been implemented, ranging from discrete-event simulation (the time notion is *virtual time*, which gets incremented by the controller of a discrete amount at each message selection and dispatching) to soft/hard real-time. In a soft real-time context, messages can be time-stamped by their *dispatch time*. A scheduled message cannot be delivered *before* its dispatch time. Message selection is regulated by minimum dispatch time. An HRT control machine is depicted in Fig. 1. It will be assumed in the rest of this paper along with a discrete time model. The organisation of the message plan consists of a set of partial ordered *time validity intervals* of message invocations. A *message invocation* includes a message specification (identifier and data contents), its destination actor and two pieces of timing information: *tmin* and *tmax*, *tmin≤tmax*. It is intended that *msg* cannot be dispatched before *tmin* and should be delivered before *tmax* to avoid a timing violation. The selector component chooses the next dispatch message on the basis of an *Earliest Deadline First* (EDF) strategy.

In the architecture of Fig. 1, application actors are not aware of timing nor of *when* they will be activated by a message. Actors are developed according to functional aspects only. Timing constraints are the responsibility of the scheduler component which hosts a set of RTsynchronizers. The approach favours modularity and reusability. A given actor can be re-used within different applications with different timing requirements.

The control machines of a whole system interact one to another in order to fulfil system-level requirements. To this purpose a suitable *interaction pattern* (protocol) [2] is introduced which is responsible of inter-subsystem synchronisation (e.g., real-time barrier synchronisation [30]). For example, in [4] a specific protocol was implemented based on the Time Warp algorithm [9] for synchronising the control machines of a distributed simulation system. An actor system can require both local and global co-ordination and control. A selected subsystem can host a *master* control machine, where the operator, through a friendly and asynchronous graphical user interface, can issue configuration or monitoring commands.

2.2 Timing Constraints

A timing constraint specifies a time interval between the occurrences of a group of causally connected messages. A violation of a timing constraint can require a recovery action to be carried out. The output Filter component in Fig. 1 can purposely be specialised to detect and handle a timing violation by scheduling a message with a suitable time validity interval for an exception handler actor. The following operations are provided for the expression of timing constraints and scheduler programming:

- msg.*cause()*, which returns the identity of the (immediate causal) message msg' whose processing caused the generation of the message msg
- msg.*iTime()*, which returns the time at which the message msg was dispatched
- msg.*deadline()*, which returns the upper limit of msg's time window
- *schedule*(msg, t*min*, t*max*), which schedules on the control machine plan the message msg with the associated time window [t*min*..t*max*], t*min* ≤ t*max*
- *now()*, which returns the current value of local clock.

It should be noted that for the *iTime()* function to be applicable to an incoming network message, a notion of a global time has to be provided [28].

A input/output filter is provided of the timing attributes of a subsystem (actor periods, deadlines, etc.) at the initialisation time. A timing constraint is identified by a given *pattern*, i.e., a boolean expression involving message parameters and local data of the filter. If the pattern is satisfied the corresponding timing constraint is applied, possibly executing a scheduling action.

The following are some common examples of timing constraints. For simplicity, they are expressed in Java syntax. The concept of a "thread of control" [32] is used to model the chained execution of messages triggered by an environment stimulus sensed by a periodic terminator. A thread of control defines the system response to the stimulus. Its processing can be required to terminate within a deadline.

Periodic message
The time clause is associated with an actor (e.g., a terminator) which has a repetitive message (e.g., ReadCO for reading the environment CO level). After being processed, the message is sent again by the actor to itself. Message periodicity is ensured externally to the actor in the iFilter:

```
if( m instanceof ReadCO && m.cause() == m )
    schedule( m, m.cause().iTime()+P, m.cause().iTime()+P );
```

where *P* is the message period, which is data local to the iFilter.

Deadline on a message
A message can be required to be handled within a given deadline *D* measured since the invocation time of its cause:

```
if( m instanceof DeadlineMessage )
    schedule( m, now(), m.cause().iTime()+D );
```

Deadline on a thread of control
A message can be scheduled to occur within the deadline of its cause by the following time clause:

```
schedule( m, now(), m.cause().deadline() );
```

This construction allows to propagate the deadline of a thread among its constituent messages.

Urgent time clause
A message can be scheduled to occur immediately, e.g., for safety conditions:

```
schedule( m, now(), now() );
```

Weak time clause
A weak time clause requiring dispatching a message "as soon as possible" can be achieved by:

```
schedule( m, now(), infinite );
```

Discarding a message
A message can be logically discarded by the following time clause

schedule(m, infinite, infinite).

Finally, it is worth noting that scheduler programming rests on the basic semantics of "*safe progress, unsafe block*" principle [28]. In particular, a scheduler (i.e., iFilter) actor can buffer a message into local data if releasing it (i.e., queuing the message on the Message Plan for subsequent dispatching) can cause violations of constraints in the future. As an example, consider a particular scheduler which is in charge of synchronising the audio and video streams of a multimedia session for the purpose of ensuring a suitable QoS (e.g., lip-synch) [20, 29, 8]. The arriving of an audio message without the corresponding video message implies that the scheduler must store temporarily the arrived message waiting for the associated video message. If the missing message is not received in the admitted *End-to-End* delay, the arrived message can be discarded. On the other hand, if the missing message arrives in the due time, both messages can be released so as to be "simultaneously" played.

3   An Example

The following considers a simplified model of a software controlled crane [5] that can lift and carry containers from arriving trucks to a buffer area, from where they are taken for further handling. To carry the containers, the crane uses a magnetic latching mechanism. Actors can naturally model the crane control system (Crane), the buffer area (BoundedBuffer) and the operator (Console). Terminator actors control the engine, the magnet, and the periodic item detector and left and right position sensor physical devices. The Crane actor understands the following messages:

- *On, Off* (turn on and off the magnet)
- *Loaded, UnLoaded* (raised by the *ItemDetector*)
- *Forward, Backward* (start a forward/backward crane movement)
- *RightStop, LeftStop* (raised respectively by *RightSensor* and *LeftSensor*)
- *Emergency* (stops crane and turns off the magnet in emergency situations).

In addition, Crane exports the accessor functions: *isOn, isOff, Moving, Stationary, isLoaded, isUnLoaded*, which return information about crane current state. The Engine and the Magnet actor classes understand messages for turning their state on and off. The Engine has also a message to setup the direction movement (forward/backward). The ItemDetector senses when a container is loaded into the latching mechanism, and sends a Loaded or Unloaded message to the crane. The Position sensors capture crane limit positions where the engine should be turned off. They transmit a RightStop or LeftStop message respectively.

Starting from a loaded container, first a Loaded message is received by crane. Loaded causes a Forward message to be sent by crane to itself. Processing Forward in turn generates messages for setting up the engine rotation movement and turning it on. Then the crane enters a *MOVING* state where it waits for a RightStop message. After that, the engine is turned off and a Put message is sent to the buffer area. After receiving a reply for put, the crane releases the container by turning off the magnet. It then sends and a Backward message to itself for the backward movement.

The crane system can also be controlled by the operator at the console under the restriction that On or Off requests are rejected while crane is moving. However, an Emergency message has to be processed within an assigned deadline d. Fig. 2 shows an iFilter which schedules the crane system. It is initialised with the relevant timing attributes: period of sensors and emergency deadline. The scheduler behaviour is captured in the *handler* method [25]. To each "just sent message" m the handler applies the relevant time clause.

```
public class iFilter extends Actor{
    long p /*sensors period*/, d /*emergency deadline*/;
    iFilter( long p, long d ){ this.p=p; this.d=d; }
    protected void handler( Message m ){
        if( m instanceof ReadItemDetector || m instanceof ReadLeftSensor ||
            m instanceof ReadRightSensor ) {
            //periodic message
            if( m.cause() == m ) //schedule m according to period p
                ControlMachine.schedule( m, m.cause().iTime()+p, m.cause().iTime()+p );
            else //initialisation
                ControlMachine.schedule( m, now(), now() );
        }
        else if( m instanceof Emergency )
            ControlMachine.schedule( m, now(), now()+d );
        else if( m.sender instanceof Operator || m.cause() instanceof ReadItemDetector ||
                m.cause() instanceof ReadLeftSensor || m.cause() instanceof ReadRightSensor )
            //weak time constraint
            ControlMachine.schedule( m, now(), ControlMachine.infinite )
        else //thread deadline constraint
            ControlMachine.schedule( m, now(), m.cause().deadline() )
    }//handler
}//iFilter
```

Fig. 2. An iFilter for scheduling the crane system.

Self-driving periodic actors send to themselves the first periodic message during initialisation (i.e., within the constructor). Exchanged messages during normal, automated behaviour, are handled by a weak timing constraint with t_{max} being infinite. However, an Emergency thread of control is managed according to its deadline.

4   Modelling Actor Components by CP-Nets

In this section a mapping from actor concepts to CP-nets is described. CPN combine the strength of Petri nets [23], i.e., the possibility of modelling synchronisation and concurrency problems, with the facilities of high-level programming languages which allow the definition of powerful data types and permit the manipulation of their values. Places of a CP-net are typed and tokens carry data values (colours). Transition enabling depends *also* on the token values which can be modified by transition firing. A *binding-element* represents a pair <transition, enabling token values>. CPN ML, an extension of the functional programming language Standard ML [22], is used for colour declarations and net inscriptions. Binding-elements are constrained by input arc inscriptions and transition guards. Output arc inscriptions affect the construction of generated tokens at a transition firing.

Besides the high-level character, CP-nets are provided of abstraction mechanisms such as modularity and compositionality which are essential in the modelling and analysis of complex systems. More in particular, the hierarchical extension of CP-nets [13] allow to relate a transition (and its surrounding arcs and places) to a separate sub-net (*page*) providing a detailed description of the activity associated to the transition. It is possible to construct a large model from a number of pages which are interconnected to one another according to a well-defined semantics which preserve the same formal analysis of the corresponding non hierarchical net model.

CP-nets have also been extended with a time concept [16]. A global clock is introduced that represents the model simulation time. In addition to its value, a token is assigned a time stamp expressing the earliest time at which the token can be removed by a binding-element. To be enabled, a binding-element must be also *ready*, i.e., the time stamps of the enabling tokens must be less than or equal to the current system time. Each generated token gets as the time stamp the occurrence time of the fired transition, that is the current global time. When there is no ready binding-element, the global clock is advanced to the minimum time stamp in the model so enabling at least one transition. A transition can be associated with a delay δ which is the increment to the time stamp of the generated tokens with respect to the transition occurrence time. This means that the produced tokens are unavailable for binding-elements for δ time units.

All these features are integrated into Design/CPN, a graphical environment for the editing, simulation and analysis of CP-nets.

Basic components of an actor system (actors, control machine, scheduler, network protocol, ...) can be 1-to-1 mapped on to CPN pages with well-established interconnection pins. At a first level, a simple description of the system can be given by ignoring component internal details. At a subsequent level a more detailed description of the internal behaviour is specified.

At the highest abstraction level, an actor subsystem is represented by a description page containing the (substitution) transitions *Actors* and *ControlMachine* that respectively model the group of the application actors and the orchestrating control machine. Each transition is refined by substituting it with a page giving the details of the actor group or the internal components of the control machine. Actually, the Actors page is organised as a collection of sub-pages, one per actor. Similarly, the ControlMachine page admits sub-pages associated to its internal components (see section 2.1). The process of substituting a transition with an inner page can be continued until concrete sub-nets are used as substitutions. The refinement process can purposely reuse some pages. For instance, the control machine sub-pages are obvious candidates to be reused into different projects. Moreover, some pages can be shared between different components, e.g., the control machine pages among the various subsystems of a distributed system. Page reuse implies that different page instances are actually used in the model topology. The overall page organisation of a model is summarised in a special Design/CPN page which is the system *hierarchy page*.

Fig. 3 shows the hierarchy page of the CPN model developed for the crane control system. It has a node for each page in the model. An arc between two nodes indicates that the source node contains a transition that is substituted by the page of the destination node (sub-page). Each node is inscribed by the name and number of the corresponding page, while each arc is inscribed with the name of the corresponding substitution transition.

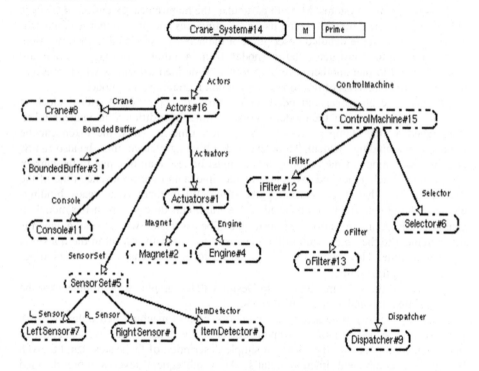

Fig. 3. Hierarchy page of the crane system.

The hierarchy page shows the net structure of the model. It is composed of three parts:

- the System Description part, which consists of the page *Crane_System*
- the Actors part, which consists of eleven pages (*Actors, Crane, BoundedBuffer, Console, SensorSet, LeftSensor, RightSensor, ItemDetector, Actuators, Magnet, Engine*)
- the Control Machine part, which consists of five pages (*ControlMachine, iFilter, Selector, oFilter, Dispatcher*).

The Crane_System is the topmost page in the hierarchy and it is shown in Fig. 4. *Actors* and *ControlMachine* transitions communicate through the interface places CMPIn and CMPOut. Each of these transitions is marked with the HS tag indicating that it is a substitution transition and contains interface information, i.e., the correspondence between super-page *sockets* and sub-page *ports*. For simplicity, the interface information is omitted in Fig. 4.

The page Actors#16 contains the actor sub-pages. Some sub-pages directly model corresponding actors (e.g., the *BoundedBuffer* actor page portrayed in Fig. 5), others abstract a group of logically related actors (e.g. SensorSet and Actuators) that are refined at a further sub-level.

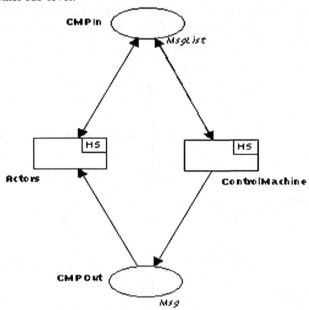

Fig. 4. The Crane_System page.

The BoundedBuffer actor can receive a *Get* or *Put* message and can find itself into one of three states: *Empty*, *Partial* and *Full*. An incoming unexpected message, e.g., a *Put* in the *Full* status, is deferred by storing it in a local queue (token in the DEQ place). Deferred messages are re-scheduled to be received again as soon as the actor changes its current state.

The BoundedBuffer sub-net receives an incoming message in its input place BBP2 and generates new messages in the output place BBP1. These places coincide respectively with CMPOut and CMPIn places in Fig. 4.

Owing to the high-level nature of CP-nets, the actor sub-net is reduced to a single transition which is annotated by the actor worst case execution time. The actor behaviour is captured by CPN ML arc expressions and some local state places. On the other hand, it could equal be possible to derive a more articulated actor sub-net [31, 3] which is capable of displaying more directly states and state transitions of the actor life cycle.

The adopted modelling technique for actors has two advantages. The first one is related to a reduction of the net complexity and therefore to a minimisation of the state explosion problem of the occurrence graph method (see later in this paper). Reductions are among the classical approaches [23] when analysing large systems.

The second one is concerned with a reduction of the "distortion risks" during the translation of the verified model to the target application. This is obtained by expressing actor state transitions and actions in a form very close to the final implementation language. This in turn facilitates an automatic translation of specification to design and implementation [21].

Synchronous accessor methods among the actors of a same subsystem are implemented by shared places in the subsystem page. Each such a place gets a copy of the status of a supplier actor and can directly be read by requestor actors.

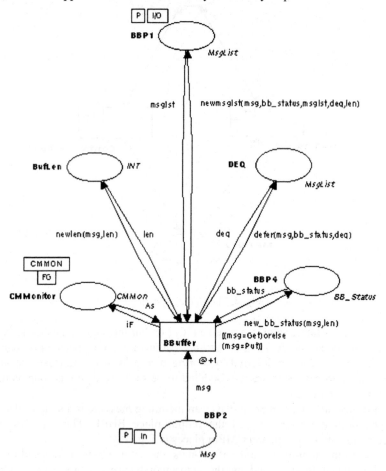

Fig. 5. The Bounded Buffer page.

The Control Machine is modelled by the ControlMachine#15 page shown in Fig. 6.

It contains four substitution transitions (*iFilter, Selector, oFilter* and *Dispatcher*) corresponding to the basic sub-components, and five places two of which (CMPIn and CMPOut) represent the interface with the system actors. The places Plan, CMP1 and CMP2 allow the control machine components to communicate.

The iFilter component is modelled by the page iFilter#12 depicted in Fig. 7.

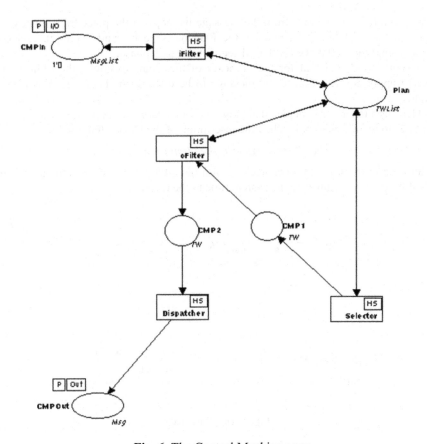

Fig. 6. The Control Machine page.

It contains a transition (iFilter) and four places. At each occurrence of the iFilter transition, the ifilt(...) function is invoked which takes the plan message list from the iFP2 place and the just sent message list from the iFP1 place, applies the time clauses of Fig. 2 and generates a new plan message list. Place LDM holds a copy of the last dispatched message which is used as a repository for inquiring about the causal message and its attributes of a message under scheduling.

The CMMonitor place is a member of a Fusion Global set and contains a token of the CMMon colour set which ensures that scheduling, selecting and dispatching activities in the control machine are strictly sequenced.

The iFP2 place corresponds to the socket place Plan in the page ControlMachine#15. This socket node is assigned also to the Plan port node of the page *Selector#6* which is represented in Fig. 8.

This page models the selection of the next message to be dispatched from the Plan. In this case an EDF strategy is used. First the Plan message list is duplicated into the SP4 and SP5 places. The t_marg() function appends to each message item its *temporal margin*, i.e., the difference between the *tmax* value of the message time window and the current time. Transition ST2 edf() function is responsible of determining in the SP6 place the message with the minimum temporal margin which is removed (ST3 transition) from the message list copy in the place SP4. Finally, the firing of the ST4 transition updates the Plan message list and copies the selected message into the SP9 place for the subsequent dispatch phase.

It should be noted that for the purposes of the Crane example, the oFilter page doesn't apply any function to the message to be dispatched. Therefore, this page is omitted.

The Dispatcher page depicted in Fig. 9 is in charge of delivering the selected message to its relevant actor. The time inscription of the Dispatch transition:

$$@+if\ (tmin>time()) \ then\ (tmin-time()) \ else\ 0$$

allows to timestamp the message token in the DP2 place with the amount of time possibly required to advance the system time to the message *tmin*.

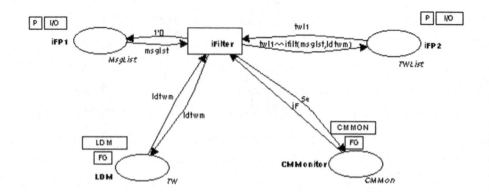

Fig. 7. The iFilter page.

The destination actor receives the dispatched message, processes it by firing its action transition and generates a list of new messages. All these messages get time stamped with the occurrence time of the actor transition augmented by the action duration. After that the control machine resumes its activity into the iFilter component thus starting a new cycle of the control loop.

5   Analysis of an Actor-Based CPN Model

The analysis process of an actor-based CPN model aims at verifying and possibly correcting the specification of timing constraints and action worst-case-execution-times in order to ensure *schedulability* [11, 35]. In particular, the analysis has to assert that every thread is completed within its deadline.

Both informal and formal analysis methods can be applied. The informal analysis can be conducted by simulating the model (*specifications testing* [12]), i.e., by providing an initial marking and then by tracing one or more possible resulting behaviours. By observing these behaviours the analyst can realise whether or not the specified system meets functional or timing requirements. Temporal behaviour can be verified by analysing the multiple concurrent threads especially under peak-loads.

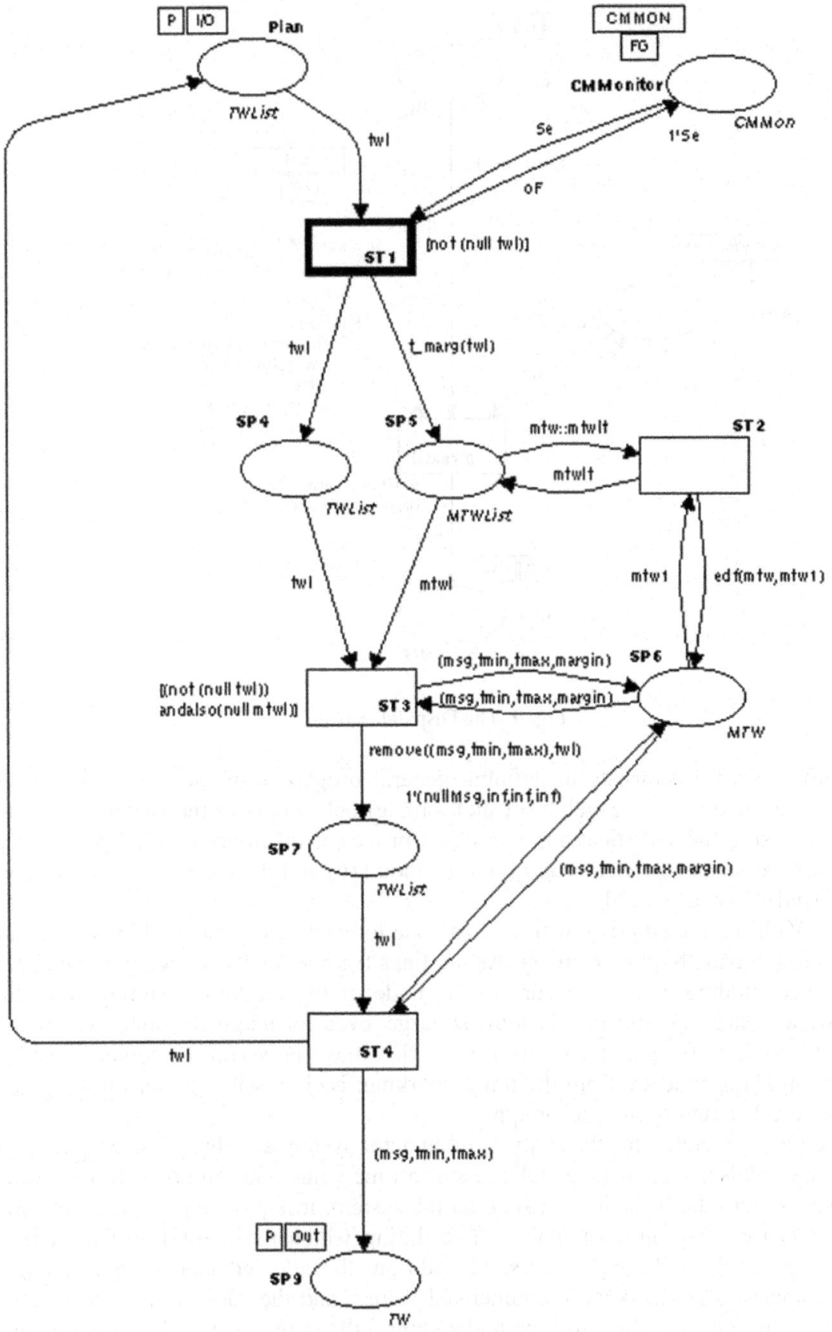

Fig. 8. The Selector page.

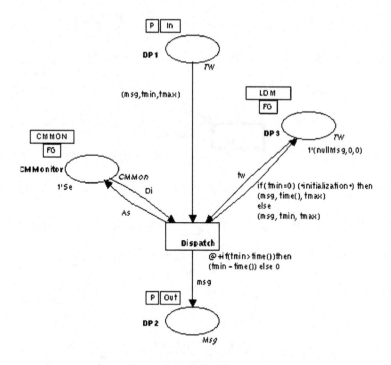

Fig. 9. The Dispatcher page.

Formal analysis consists in defining general properties of the net model which reflect special types of desirable (or undesirable) behaviours of the specified system, and then using the specification to formally prove (or disprove) such properties. For this purpose the *occurrence graph* (OG) method [16] and the Occurrence Graph Tool (Occ Tool) [7] can be used.

An OG is a directed graph which has a node for each reachable marking and an arc for each occurring binding element. An arc links the node of the marking in which the associated binding element occurs to the node of the marking resulting from the occurrence. Such a graph may become so large, even for relatively small nets, that it cannot possibly be generated even with the most powerful computer. Another limitation is dependency from the initial marking: each possible initial marking may originate a different occurrence graph.

The OGs generated for the analysis of an actor system are always timed, i.e., each node represents a system state and contains a time value and a timed marking. Since the typical periodic behaviour of such an RT system, it is possible to reduce the size of an OG by choosing a simulation time during which meaningful system activity (e.g., peak-load conditions) occurs. In addition, the adopted modelling techniques (i.e., colour sets as intervals or enumerated values, and the adopted net structure for actors) purposely contribute to keep under control the state space explosion problem.

All standard dynamic properties for a CP-net can be derived from its OG, e.g., boundedness, home, liveness, and fairness properties. They all can be investigated by available functions in the OG Tool. Moreover ML functions can be written for issuing non-standard inquiries to OG.

For instance, the maximum number of simultaneously enabled transition instances can be found by the MaxTransEnabled function in Fig. 10.

```
fun MaxTransEnabled () : int
    = SearchNodes ( EntireGraph,
                    fn _ => true,
                    NoLimit,
                    fn n => length( OutArcs n ),
                    0,
                    max);
```

Fig. 10. Predictable system behaviour check function.

When invoked on any generated OG for the Crane CPN model, the function returns 1:

MaxTransEnabled ();
1 : int;

thus confirming that the modelled system always evolves in a predictable manner. Indeed, when a system consists of a single subsystem there is always only one transition enabled (belonging either to an actor or to an internal component of the control machine).

Similarly, one such a function can capture the verification of a temporal property in positive or negative form, provided an OG is generated for a suitable simulation period.

To exemplify, it is possible to carry out the verification of properties like the following: "is it always true that for each instance of a given transition firing (representing, e.g., the beginning of a thread) there always (as the OG state space allows) exists an instance of the corresponding transition firing (modelling, e.g., the end of the thread) such that the time distance between them is less than a fixed time interval (e.g., deadline of thread execution) ?"

The negative form of a property can be more immediate. In this case the existence of a single occurrence of a searched event that contradicts the property is sufficient to assert that the property doesn't hold.

In the crane example we have verified (see Fig. 11) that for each occurrence of the *Emergency* event it always (in the generated OG) follows a transition instance corresponding to turning off the magnet (*Magnet* action transition) such that the temporal distance between the two events is less than a deadline d.

When invoked, the EmerDeadln() function always returns an empty list, proving that the list built with Magnet transition instances which are causally connected to an Emergency event and temporally internal to the deadline interval, is never empty:

EmerDeadln();
val it = [] : Node list

The CP-net modelling of an actor system supports an *incremental development* through a modified spiral lifecycle model [36]. Functional and temporal analysis can start as soon as a CPN specification has been produced and be based on the required obligations (periods, thread deadlines, ...) and a preliminary estimation of actor message processing times. As the temporal information about actors get more accurate, i.e., the project is tuned to a physical target architecture, the CPN model can be applied again to check that the system remains schedulable. The key point is that the CPN model closely mirrors the design/implementation models of a system. Therefore, conclusions drawn from CPN analysis can directly be interpreted at the lower levels of development.

```
fun EmerDeadln(): Node list =
  PredAllNodes(fn n =>
     let
        val Cr_Em = StripTime(Mark.Crane_System'CMPOut 1 n);
     in
        if Cr_Em == 1`Cr_Emergency then
              null(PredArcs(EntireGraph,
                       fn a =>
                          if ArcToTI(a) = TI.Magnet'Magnet 1 then
                          ((CreationTime(DestNode(a))-CreationTime(n))<d)
                          andalso(DestNode(a)>n)
                          else false,
                       1))
        else false
     end)
```

Fig. 11. Temporal property check function.

The experimental analysis of the Crane system was performed on a Sun Sparc 5 with 24 MB of physical RAM.

6 Conclusions

This paper describes an actor-based framework suited to the development of real-time distributed systems and shows a possible formalisation in terms of CP-nets. The benefits of the formalisation are the possibility of achieving visualisation and verification support in the context of the Design/CPN tool.

Like in [36, 24] both temporal and functional aspects can be considered since the early stages of a project by using a modified spiral-model design life cycle where specifications can gradually be refined as the development becomes more and more specific. In particular, as the message execution times become more precise, schedulability and timing behaviour can be checked that remain fulfilled. Temporal properties can be analysed by generating and exploring occurrence graphs.

The proposed approach simplifies the transformation of a specified and verified system to its final implementation. Actors and their controlling algorithm can directly be programmed in a object-oriented language like Java [25].

Directions of further work include

- experimenting with the proposed CPN modelling to real-life time-dependent systems
- improving the support for distributed systems by a better exploitation of page reuse and instance management mechanisms, in the presence of deterministic

networks like CAN which assigns unique identifiers (priorities) to network messages to control their transmission delay [34]

- improving property analysis by using timed CTL [6] for expressing queries on the state space of a generated occurrence graph.

The actor-based approach described in this paper is also in current use in the realisation of distributed simulation of cellular networks [4] and multimedia applications [8].

Acknowledgments

Work carried out under MURST project "MOSAICO". The authors are grateful to the anonymous referee whose comments contributed to improving the paper presentation. Thanks are also expressed to CPN group at the University of Aarhus for their support during the experimental work with the Design/CPN tools.

References

1. Agha, G.: Actors: A model for concurrent computation in distributed systems. MIT Press, 1986.
2. Agha, G.: Abstracting interaction patterns: a programming paradigm for open distributed systems. Formal Methods for Open Object-based Distributed Systems, Vol. 1, Najm E. and Stefani J. B. (eds), Chapman & Hall, 1996.
3. Agha, G., Miriyala, S., Sami, Y.: Visualizing Actor Programs using Predicate Transition Nets. *Journal of Visual Languages and Computation*, 3(2), June 1992, pp. 195-220.
4. Beraldi, R., Nigro, L.: Performance of a Time Warp based simulator of large scale PCS networks. *Simulation Practice and Theory*, 6(2), February 1998, pp. 149-163.
5. Bergmans, L., Aksit, M.: Composing synchronisation and real-time constraints, *J. of Parallel and Distributed Computing*, September 1996.
6. Cheng, A., Christensen, S., Mortensen, K.H.: Model checking Coloured Petri Nets: Exploiting strongly connected components. WoDES'96, August 20, 1996. http://www.daimi.aau.dk/designCPN/libs/askctl.
7. Christensen, S., Jensen, K., Kristensen, L.: The Design/CPN Occurrence Graph Tool. User's manual version 3.0. Computer Science Department, University of Aarhus, 1996. http://www.daimi.aau.dk/designCPN/.
8. Fortino, G., Nigro, L.: QoS centred Java and actor based framework for real/virtual teleconferences. *Proc. of SCS EuroMedia98*, Leicester (UK), Jan. 5-6, 1998, pp. 129-133.
9. Fujimoto, R.M.: Parallel discrete event simulation. *Communications of the ACM*, 33(10), 1990, pp. 30-53.
10. Genrich, H. J.: Predicate/transition nets. In *Advances in Petri Nets*, W. Brauer, W. Reisig and G. Rozenberg (eds.), New York, Springer Verlag, 1987.
11. Gerber, R., Lee, I.: A layered approach to automating the verification of real-time systems. *IEEE Trans. on Software Engineering*, 18(9), September 1992, pp. 768-784.
12. Ghezzi, C., Mandrioli, D., Morasca, S., Pezzè, M.: A unified high-level Petri net formalism for time-critical systems. *IEEE Trans. on Software Engineering*, 17(2), February 1991, pp. 160-172.
13. Huber, P., Jensen, K., Shapiro, R.M.: Hierarchies in Coloured Petri Nets. In *Advances in Petri Nets*, Springer Verlag, LNCS 483, 1991, pp. 313-341.

14. ISO/DIS 11898: Road Vehicles, Interchange of digital information, Controller Area Network (CAN) for high speed communications, 1992.
15. Jensen, K.: *Coloured Petri Nets - Basic concepts, analysis methods and practical use.* Vol. 1: Basic concepts. EATCS Monographs on Theoretical Computer Science. Springer, 1992.
16. Jensen, K.: *Coloured Petri Nets - Basic concepts, analysis methods and practical use.* Vol. 2: Analysis methods. EATCS Monographs on Theoretical Computer Science. Springer, 1994.
17. Jensen, K.: *Coloured Petri Nets - Basic concepts, analysis methods and practical use.* Vol. 3: Practical use. EATCS Monographs on Theoretical Computer Science. Springer, 1997.
18. Jensen, K., Christensen, S., Huber, P., Holla, M.: Design/CPN. A reference manual. Computer Science Department, University of Aaurus, 1996. http://www.daimi.aau.dk /designCPN/.
19. Kirk, B., Nigro, L., Pupo, F.: Using real time constraints for modularisation. Springer-Verlag , LNCS 1204, 1997, pp. 236-251.
20. Kouvelas, I.., Hardman, V., Watson, A.: Lip synchronization for use over the Internet: analisys and implementation. In *Proc. of IEEE Globecom'96,* London UK, 1996.
21. Kummer, O., Moldt, D., Wienberg, F.: A framework for interacting Design/CPN and Java processes. In *First Workshop On the Practical Use of CPN and Design/CPN*, University of Aarhus, K. Jensen (Ed.), DAIMI PB-532, 1998, pp. 131-150, http://www.daimi.aau.dk /CPnets/
22. Milner, R., Harper, R., Tofte, H.: *The definition of Standard ML*. MIT Press, 1990.
23. Murata, T.: Petri nets: properties, analysis and applications. *Proceedings of the IEEE,* **77**(4), 1989, pp.541-580.
24. Nigro, L., Pupo, F.: Modelling and analysing DART systems through high-level Petri nets, LNCS 1091, Springer-Verlag, 1996, pp. 420-439.
25. Nigro, L., Pupo, F.: A modular approach to real-time programming using actors and Java. *Control Engineering Practice*, **6**(12), December 1998, pp. 1485-1491,.
26. Nigro, L., Tisato, F.: Timing as a programming in-the-large issue. *Microsystems and Microprocessors*, **20**, June 1996, pp. 211-223.
27. Ren, S., Agha, G.: RTsynchronizer: language support for real-time specification in distributed systems. *ACM SIGPLAN Notices*, **30**, 1995, pp. 50-59.
28. Ren, S., Agha, G., Saito, M.: A modular approach for programming distributed real-time systems. *J. of Parallel and Distributed Computing*, Special issue on Object-Oriented Real-Time Systems, 1996.
29. Ren, S., Venkatasubramanian, N., Agha, G.: Formalizing multimedia QoS constraints using actors. *Formal Methods for Open Object-based Distributed Systems (FMOODS'97)*, Vol. **2**, H. Bowman and J. Derrick (Eds.), Chapman & Hall, 1997, pp. 139-153.
30. Saito, M., Agha, G.: A modular approach to real-time synchronisation. In *Object-Oriented Real-Time Systems Workshop*, 13-22, OOPS Messenger, ACM SIGPLAN, 1995.
31. Sami, Y., Vidal-Naquet, G.: Formalization of the behaviour of actors by coloured Petri nets and some applications. *PARLE '91*, 1991.
32. Shlaer, S., Mellor S.J.: *Object Lifecycles-Modeling the world in states*. Yourdon Press Computing Series, 1992.
33. Stankovic, J.A.: Misconceptions about real-time computing, *IEEE Computer*, **21**(10), 1988, pp.19-19.
34. Tindel, K., Burns, A., Wellings, A.J.: Analysis of hard real time communications. *Real Time Systems*, 9, 1995, pp. 147-171.
35. Tsai, J.J.P., Yang, S.J. Chang, Y.-H.: Timing Constraints Petri Nets and their application to schedulability analysis of real-time system specification. *IEEE Trans. on Software Engineering*, **21**(1), January 1995, pp. 32-49.
36. Verber, D., Colnaric, M., Frigeri, A.H., Halang, W.A.: Object orientation in the real-time system lifecycle. *Proc. of 22nd IFAC/IFIP Workshop on Real-Time Programming*, Lyon, 15-17 September 1997, pp. 77-82.

Control Properties
in Object-Oriented Specifications

A. Diagne

Université Pierre & Marie Curie
Laboratoire d'Informatique de Paris 6 (LIP6)
Thème Systèmes Répartis et Coopératifs (SRC)
4 place Jussieu, F-75252 Paris Cedex 05 France
Phone: (+33)(0)1 44 27 73 65, Fax: (+33)(0)1 44 27 62 86
Alioune.Diagne@lip6.fr

Abstract. Verification and validation are becoming the most important activities in software engineering because of the stronger need to
ensure and enhance quality. The system designers need to ensure the
conformance of models with the functionalities of a target system and
the correctness, safeness and reliability of its operating cycles. Many formal methods have been promoted to support such tasks and Petri nets
formalism seems to be one of the most relevant one to evaluate control
properties. But it lacks of structuration facilities to allow to handle large
scale systems. The aim of this paper is to show that, for some guidelines,
Petri nets can be integrated in usual object-oriented software methodologies in order to evaluate such properties. Hence, one can profit from
contributions of both object and nets to handle quality in systems in a
more satisfactory way.

Keywords: Object Methodology, Petri Nets, Control Properties, Verification & Validation.

1 Introduction

Verification and Validation (V&V) are becoming the most important activities
in software engineering because of the stronger need to ensure and also enhance
quality. The quality assurance had been shifted to the earliest phases of the
software life cycle, thus disabling systematic test along the implementation as
an evaluation procedure. The engineers concerned with system design must be
aware of the specific aspects of a given system like its architecture and expected
properties. These specificities must be represented in a model of the system
and be evaluated before any realization. Evaluation of a model is twofold. A
system designer must be able to answer the couple of questions Boehm asks in
[Boehm 81]:

1. *"Are you building the right product?"*
2. *"Are you building the product right?"*

G. Agha et al. (Eds.): Concurrent OOP and PN, LNCS 2001, pp. 514–533, 2001.
© Springer-Verlag Berlin Heidelberg 2001

The answer to the first question is stated through the activities which ensure that the system or its model is traceable back to customer requirements, it is validation (see [Pressman 92] and [Rushby 93]). The answer to the second question is stated through the activities which ensure that the system or its model correctly realizes specific functionalities, it is verification. V&V encompass so many kinds of properties that the use of one single method is definitely disabled. Among these aspects we stress on control properties and advocate the use on Petri nets for their evaluation. There are some kinds of control properties which can be considered as basic ones on system models because they mean that some *bad control happenings* like deadlocks or starvation do not occur. Some other kinds of properties are specific to a given application domain because they represent what can be considered as *good control happenings* for that domain.

Petri nets is a very relevant formalism to evaluate both basic and specific control properties like liveness, fairness, home state or terminal state, etc. Their main drawback is the lack of structuring facilities that allow one to handle large scale systems. Many extensions of nets aiming to overcome this drawback have been proposed in the literature (see [Buchs 91], [Sibertin-Blanc 94] and [Lakos 95a] as a non-exhaustive list). Almost all of them are based on integration of object-oriented contributions. [1]But in general authors do not worry too much about the cost of integration of such extensions in existing methodologies, a sine qua none condition to the success of Petri nets as industrial-strength formal method. Object paradigm has gained legitimacy to cover life-cycle of systems from requirements analysis to exhibition of a model in a language-independent way and further to its implementation using an object-oriented language. Its suitability comes mainly from the tight correspondence between its construction units (object and class) and the entities of the real world. The seamless use along the life-cycle is an appeal for engineers who only need to win up and manage a restricted set of concepts. In the object-oriented specification area, there are many methodologies, each one organizes and uses the concepts according to a given set of steps and instructions towards production of software. The steps and instructions characterize the methodology. The specificities of a given application domain and the skillfulness of its experts can be caught into reusable design patterns to structure the result of a methodology (see [Gamma 95]). Granularity and complexity of objects can be tuned to model architecture and control aspects at a level that can be evaluated. But in general, there is no built-in facilities in object methodologies that allow one to undertake such evaluation.

Modular and incremental constructions in Petri nets have also been investigated for verification purposes (see [Baumgarten 88], [Vogler 92], [Murata 94], [Valmari 94] and [Schreiber 95] as a non-exhaustive list). In [Baumgarten 88], distributed systems are structured into building blocks on which one has two kind of functional equivalences. Internal and external equivalences and their relationships allow to state about interchanging building blocks and its induced effect on the neighborhood. An extension of external equivalence to failure semantics is achieved in [Vogler 92] where a complete range of materials on equivalence and

---

[1]   The main trends in integrating objects with nets can be found in [Bastide 95].

composition of modules is presented. In [Murata 94], a notion of equivalence is presented which allows the authors to build reachability graph in a hierarchical way and to extend liveness results to the resulting compact representation of the graph. This equivalence is based on the notion of I-O similarity which very closed to the previous external equivalence. In [Valmari 94], the author proposes an equivalence based on failures and divergence semantics extended with a notion of "chaos". This equivalence allows to condense the environment of a component while verifying it. The condensation keeps the information about the failures and divergences of the environment. In [Schreiber 95]), a notion of observation which implies external equivalence is used. Building blocks have observable places containing the requests they accept and results they compute consequently. The equivalence is based on the capability to compute the same "functions" i.e. the same transformations from observable markings.

We choose to adapt the notion of functional equivalence presented in the previous works to a model more dedicated to object-orientation. We focus on the definition of a component model and afterward, we specialize functional equivalence for our purpose. Another aim of this paper is to show that, for some guidelines, we can integrate contributions of Petri nets in usual object-oriented methodologies in order to undertake the evaluation of software models. Petri nets allow a system designer to verify and validate models. State-exploration techniques allow to perform verification while simulation allows to animate the model and henceforth involve end-users and system-owners in its validation. This paper is organized as follows: the next section describes how a given design methodology can be constrained in order to support introduction of nets for evaluation. The section 3 shows the way to extract from the result of the design methodology control aspects and how to translate them into nets. In section 4, we propose a method to undertake evaluation of control properties for a whole system or parts of it.

2   Design Methodology

Engineers often use CASE[2] environments to model systems according to procedures enforced by their application domain. A CASE environment is an integration of a set of tools that implement a methodology. Each tool is dedicated to support one or many steps of the design methodology. A very important class of tools is the one supporting the conceptual modeling activities. These tools are used by engineers to catch the relevant information in the application domain into computer-based models. We assume in this paper that conceptual modeling tools are object-oriented ones.

2.1   Object-Based Design Concepts

Object-oriented conceptual models are based on two sets of concepts:

---

[2] Computer Aided Software Engineering.

1. basic concepts like *modularity, encapsulation* and *factorization* of similar objects into *classes*,
2. elaborated concepts like *inheritance* between classes and *advanced composition mechanisms* to build (sub-)systems from objects.

Basic concepts enhance the decomposition of a system or its model into interacting components. Each component encapsulates data and bears operations (also called services or methods) an is called an object. Data is represented as attributes of object (instance variables) or on classes (class variables). The data determines the state of the object. An operation is a set of access/modification instructions on the encapsulated data that has semantics meaning for the object and its environment. Operation invocation may have side effects (persistent or not) on the state of the object.

The elaborated concepts give facilities to enhance reuse and to formalize interactions between objects. Composition mechanisms allow the system designer to put control on inter-objects communication. The basic composition mechanism is the functional one. With functional composition, an object is viewed a generic parameterized entity that can be activated by receiving arguments which are bound to parameters. In order to undertake formal activities, we must emphasize the notion of object compatibility. Each object is characterized by the sequence of messages it accepts and the other objects are expected to respect such sequences as contracts. For instance, an object modeling a file server will accept first an open operation on an existing file, then one can repeat the read or write operation and finishes by close. Such composition mechanism allows a system designer to avoid propagation of faulty events. It is used in [Nierstrasz 93], [Diagne 96a] and [Sa 96] and is a valuable way to formalize the interface of objects that was usually regarded as a record of functions. An object can have many accepted sequences, each of them then determines a given class of clients.

2.2   Control and Architecture in Object-Based Specification

The use of the design concepts is determined by the specification methodology. A specification is description of *what* is desired from the system to model and sometimes *how* it must be realized. In classical CASE environments where formal methods are not involved, the accent is put on readability because specification stands first of all for repository and transmission of information on a model between concerned engineers. The description must be unambiguous to support this couple of needs. In presence of formal methods, the specification must obey some other constraints. The information must be organized and represented in a way that it can be processed by the formal methods.

In the purpose of control properties evaluation, three aspects need to be enhanced:

1. The *"how to realize"* i.e. one must put in the model the information on the operating cycles of the target system and its components. To fulfill this need, we need more information about the dynamic of the objects than the axiomatic

pre and post-conditions for the methods. The limitations of pre and post conditions is that they are not history sensitive (see [Matsuoka 93]). With pre and post conditions we can not ensure that an object will not receive a message it can not *yet* or *no longer* handle (see [Puntigam 97]). Objects are mainly active entities and their possible behaviors must be known by their environment which must stick on that as contracts. Behaviors are implicit in the dynamic models and need to be made explicit (see [Rumbaugh 91] and [Rumbaugh 97]).

2. The explicit expression of expected control properties on the model. Expression of properties is often missing in classical object-oriented specification. However, it is a rather good way to enforce engineers *"to think formal"*. There is no universal good or bad properties and the best way to build a specification with such properties is first to determine which they are. For instance a system with a terminal state can be considered as system with one admissible deadlock (the terminal state) and therefore reaching this state is quite a *good property*. It must hence be clearly stated in the specification and verified afterwards.

3. The explicit expression architecture of the system. This last need is harder to fulfill. The architecture of a system is made of static and dynamic links between its components. The static links are often not directly related to the control. They are often integrity constraints expressed through the class diagrams. The dynamic links are made of roles assigned to components and the rules governing the interactions. The architecture of a system, at less in its dynamic part, helps to understand, formalize and evaluate the synchronization problems which are often tough to tackle. In a method like OMT, it can be deduced from the dynamic model. This model hides representation of the composition mechanism that can be highlighted by restriction to couples of interacting objects.

2.3   Outcome of an Object-Based Specification

The specification procedure must be oriented to represent the information necessary to run evaluation of control properties. The object-oriented methodology is not so much burdened by this fact; it can be achieved by design-patterns (see [Gamma 95]). The information we need can be represented even like overloading annotation on the model. This information can be extracted when the model reaches a satisfactory level of maturity as another model dedicated for evaluation (see figure 1). The information is extracted from the different diagrams (eventually overloaded) used in analysis and design and organized to be relevant for the activities we are concerned with. For verification and validation of control properties, one must enhance the expression of behaviors and interactions as shown in figure 1. The verification and validation environment gives results that are traced back to the analysis and design environment to tune the specifications.

The level of maturity is measured according to the evaluation one wants to undertake by asking two questions:

1. *Is the information relevant to my evaluation sufficiently supplied and correctly organized?*
2. *Is there expressed any property to evaluate on the model?*

The first question is answered by statements about the completeness and the coherence of the models. Such statements can be issued through simulation of the models if they are executable or by a prototyping approach (see [Diagne 96b]). Our Petri net model can also help for that purpose (see 3.2 below). The second question has at least an implicit answer. There are always some properties like deadlock freeness, liveness, etc. that are worth to check on a specification.

The evaluation model is a flat one without any elaborated object concept like hierarchy or inheritance. In our concern, it only expresses the control in the system as well as properties we might evaluate on it. The sequences of messages accepted by an object determine its interaction pattern against which all requests must be checked (see figure 1). The messages that are correct towards this interaction pattern execute the part of the behavior corresponding to the right method. The role of an object acting as a client facing a server is determined by the special sequence of authorized messages it uses. For instance clients of a file server can act as readers or writers. The extraction of the evaluation model from the OO model can be automatic for a given methodology or performed as hand-work by design engineers. Anyway, it must be made as much formal as possible to avoid drift between the two models that can lead to mismatch in results interpretation. The evaluation model is tailored to contain only what

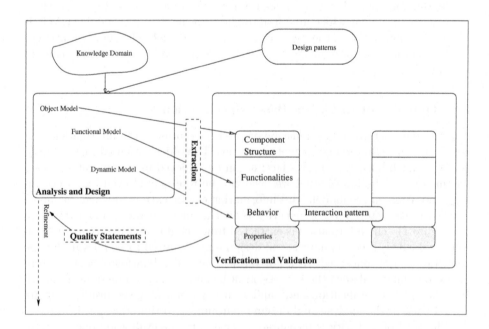

Fig. 1. Outcome of an OO Specification.

is relevant to be considered for control modeling. Hierarchy, inheritance and polymorphism are up to now not worth to be represented. They are not the most important aspects that can influence the control properties. They are more relevant as reuse mechanisms in the object-oriented methodology.

2.4 From Objects to Nets

The evaluation model in figure 1 is the entry point of the Petri nets evaluation environment. Petri nets are equivalent to high level programming languages and all control structures can be represented (see [Jensen 90] and [Heiner 92]). So it is rather easy to transform each object of an evaluation model into modular colored Petri nets.

2.5 Control Expression in Petri Nets

We present hereafter how the basic control structures can be modeled in Petri nets. This is to give hints and directions to someone who wants to translate software statements into nets. The main structures used to express control are:

1. *modification* of the value of a variable or a parameter of a method or an attribute of an object,
2. *conditionals* which can be *if-then-else* or *case statement*,
3. *loops* which can be *for* or *while* statement,
4. *sequence* of previous actions to build methods.

A general transformation principle is the following "*Attributes, variables parameters are mapped to places and actions are translated to transitions. The colors domains of the places are determined by the type of the entities they model. Valuated arcs are added to model consumption/production of values by actions from/into the right places*". This kind of transformation can be based on the method developed in [Heiner 92].

This method is based on the fact that Petri has the expressiveness of a programming language. We must express the control in objects in a procedural way instead of an axiomatic way with pre and post-conditions or a descriptive one in natural language. So the transformation to nets can be achieved with a good level of precision. Once again, it is a way to make engineers construct their application in order to apply formal validation and verification. They must therefore prepare their models to support such activities.

2.6 An Object-Based Colored Petri Net Model

Integration of object paradigm contributions with Petri nets in an active research topic. Authors have proposed Petri nets extensions to integrate modularity and encapsulation, refinement in design phase, architectural aspects like client-server in interactions and elaborated concepts like inheritance, polymorphism and hierarchy. Some authors propose new object-based net models without worrying about support or integration with existing methodologies. Up to now, these models are not successful for industrial use because of the adaptation needed for engineers. Some other try to support a design methodology. A general observation about the trials to integrate nets with object methodologies is the trade-off that exists between supporting object contributions and proving power of resulting nets. Authors generally make the choice to implement all aspects of an object methodology on top on nets (see [Bruno 95] and [Lakos 95b]). Net models must therefore be made complex to support the many steps and the organization of object-oriented concepts of a given methodology. Our choice is to integrate nets in a CASE environment implementing a methodology as a tool dedicated to evaluate control properties. So we are not loaded down neither with the methodological aspects nor with every thing not relevant to express and evaluate control. We now define the net model which is also presented in [Diagne 96a], and used within a multi-formalisms approach in see [Diagne 96b].

Now we present the *OF-CPN (Object Formalism Colored Petri Net)* model. In the remainder of the section Γ is a set of elementary color sets. An elementary color set is a finite set of elements called colors. A color domain can be an elementary color set or a cartesian product of countably many such elementary color sets. Let us give some preliminary definitions.

DEFINITION 1 (PRELIMINARIES)

1. $\Gamma^n = \underbrace{\Gamma \times \ldots \times \Gamma}_{n\ times}$ and $\Gamma^\star = \bigcup_{n \in I\!N} \Gamma^n$,

2. C_γ is the set of all the constants of the elementary color set γ, V_Γ the set of variables over the elementary color set γ and $Symb_\gamma = C_\gamma \cup V_\gamma$. For a color domain $\gamma_1 \times \ldots \times \gamma_n \in \Gamma$, $C_{\gamma_1 \times \ldots \times \gamma_n} = C_{\gamma_1} \times \ldots \times C_{\gamma_n}$, $V_{\gamma_1 \times \ldots \times \gamma_n} = V_{\gamma_1} \times \ldots \times V_{\gamma_n}$ and $Symb_{\gamma_1 \times \ldots \times \gamma_n} = Symb_{\gamma_1} \times \ldots \times Symb_{\gamma_n}$.

3. For a given set S, Bag(S) is the set of multi-sets over S. Roughly speaking, a multi-set is a set where elements may occur several times. A multi-set over a set S is formally denoted $\sum_{s \in S} x(s).s$ where $x(s) \in I\!N$. Multi-sets can be equipped with addition, subtraction and multiplication by an integer. The empty multi-set is denoted $\langle \rangle$.

4. As usually, $^\bullet x$ and $x^\bullet$ are the pre and post sets of a place or a transition in a Petri net. If S is a set, $^\bullet S$ and $S^\bullet$ are the union of pre and post sets of elements of S.[3]

Definition and characteristics of an OF-CPN. Now we give the definition and characterization of our modular Petri net model.

DEFINITION 2 (OF-CPN)
An OF-CPN is a 7-tuple (Net, P_{acc}, P_{res}, P_{snd}, P_{get}, $\Im_{acc-res}$, $\Im_{snd-get}$) where:

1. Net is a colored Petri net (P, T, Dom, Pre, Post, Guard, M_0) with:

---

[3] Presentation of Petri nets can be found in [Murata 89].

(a) P is the set of places and T the set of transitions and $P \cap T = \emptyset$,

(b) Dom: $P \cup T \longrightarrow \Gamma^*$ defines the color domains for places and transitions,

(c) Pre and Post define respectively the backward and forward incidence color functions: Pre, Post: $P \times T \longrightarrow Bag(Symb_{Dom(P)})$,

(d) Guard defines the guards on transitions:
$$\forall\, t \in T,\ Guard(t)\colon Bag(Sym_{Dom(t)}) \longrightarrow \mathcal{B} = \{True, False\},$$

(e) M_0 is a marking for Net i.e. $\forall\, p \in P,\ M_0(p) \in Bag(C_{Dom(p)})$,

2. $P_{acc} \subset P$ such that $\forall\, p_{acc} \in P_{acc}$, ${}^\bullet p_{acc} = \emptyset$ and $M_0(p_{acc}) = \langle\rangle$,

3. $P_{res} \subset P$ such that $\forall\, p_{res} \in P_{res}$, $p_{res}{}^\bullet = \emptyset$ and $M_0(p_{res}) = \langle\rangle$,

4. $P_{snd} \subset P$ such that $\forall\, p_{snd} \in P_{snd}$, $p_{snd}{}^\bullet = \emptyset$ and $M_0(p_{snd}) = \langle\rangle$,

5. $P_{get} \subset P$ such that $\forall\, p_{get} \in P_{get}$, ${}^\bullet p_{get} = \emptyset$ and $M_0(p_{get}) = \langle\rangle$,

6. the sets P_{acc}, P_{res}, P_{snd} and P_{get} are pairwise disjoint,

7. $\Im_{acc-res}\colon P_{acc} \longrightarrow P_{res}$ is a bijection such that:

   (a) $\forall\, (p_{acc}, \Im_{acc-res}(p_{acc})) \in P_{acc} \times P_{res}$, $\forall\, t_n \in {}^\bullet(\Im_{acc-res}(p_{acc}))$,
   $\exists\, t_1 \ldots t_{n-1} \in T\colon t_1 \in p_{acc}{}^\bullet$ and $t_i{}^\bullet \cap {}^\bullet t_{i+1} \neq \emptyset$ for $1 \leq i \leq n-1$,

   (b) $\forall\, (p_{acc}, \Im_{acc-res}(p_{acc})) \in P_{acc} \times P_{res}$, $\forall\, t_1 \in p_{acc}{}^\bullet$, $\exists t_2 \ldots t_n \in T : t_n \in (\Im_{acc-res}(p_{acc}))^\bullet$ and $t_i{}^\bullet \cap {}^\bullet t_{i+1} \neq \emptyset$ for $1 \leq i \leq n-1$,

   (c) $\forall\, (p_{acc}, \Im_{acc-res}(p_{acc})) \in P_{acc} \times P_{res}$, $p_{acc}{}^\bullet \cap {}^\bullet(\Im_{acc-res}(p_{acc})) = \emptyset$,

8. $\Im_{snd-get}\colon P_{snd} \longrightarrow P_{get}$ is a bijection such that:

   (a) $\forall\, (p_{snd}, \Im_{snd-get}(p_{snd})) \in P_{snd} \times P_{get}$, $\forall\, t_n \in {}^\bullet(\Im_{snd-get}(p_{snd}))$,
   $\exists t_1 \ldots t_{n-1} \in T\colon t_1 \in p_{snd}{}^\bullet$ and $t_i{}^\bullet \cap {}^\bullet t_{i+1} \neq \emptyset$ for $1 \leq i \leq n-1$,

   (b) $\forall\, (p_{snd}, \Im_{snd-get}(p_{snd})) \in P_{snd} \times P_{get}$, $\forall\, t_1 \in p_{snd}{}^\bullet$, $\exists t_2 \ldots t_n \in T : t_n \in (\Im_{snd-get}(p_{snd}))^\bullet$ and $t_i{}^\bullet \cap {}^\bullet t_{i+1} \neq \emptyset$ for $1 \leq i \leq n-1$,

   (c) $\forall\, (p_{snd}, \Im_{snd-get}(p_{snd})) \in P_{snd} \times P_{get}$, $p_{snd}{}^\bullet \cap {}^\bullet(\Im_{snd-get}(p_{snd})) = \emptyset$.

An *OF-CPN* is a Petri net with some special subsets of places (P_{acc}, P_{res}, P_{snd} and P_{get}) called the *interface places*. P_{acc} is the the set of *accept places* holding the tokens modeling requests accepted from the environment. P_{res} is the the set of *result places* holding the tokens modeling results issued for requests accepted from the environment. The bijection $\Im_{acc-res}$ ensures the correspondence between incoming requests and outgoing results.

P_{snd} is the the set of *send places* holding the tokens modeling requests sent to the environment. P_{get} is the the set of *result places* holding the tokens modeling results for requests sent to the environment. The bijection $\Im_{snd-get}$ ensures the correspondence between outgoing requests and incoming results.

The *interface transitions* are those in $Int_{obs} = (P_{acc}^\bullet \cup {}^\bullet P_{res} \cup {}^\bullet P_{snd} \cup P_{get}^\bullet)$. Firing these transitions consumes or produces tokens in the interface places. We assume that there is a *naming facility* mapping the interface places to different names. The interface transitions are named according to the names of places they are connected with.

The points (7) and (8) of the definition give an operational semantics to *OF-CPN*. They allow to model an operation as a sequence of elementary actions. Point (7a) ensures that every transition producing an outgoing result belongs to a possible sequence containing a transition that consumes an incoming request. Point (7b) states the symmetrical assertion. Point (8a) is similar to point (7a) for outgoing requests and incoming results. Point (8b) is the symmetrical of point (8a). Point (8c) ensures that the computation of request is not immediate i.e. one can not send a request and expect the result by the same transition. This point has equivalent (point 7c) for incoming requests. This operational semantics is a structural one. Behavioral operational semantics is also ensured by verification (see 3.1).

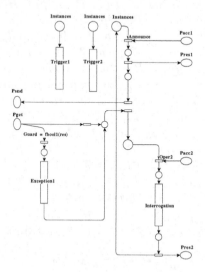

Fig. 2. Intuitive Example of an OF-CPN.

The *OF-CPN* in figure 2 models a component which has two operations: an *Interrogation* and an *Announce*. The *Interrogation* accepts a request in the place *Pacc2* and computes a results sent back in *Pres2*. The *Announce* accepts a request and issue immediately after an acknowledge. The request is processed after and causes another one to be sent to a remote component in the place *Psnd*. The result of this last request is got from the place *Pget* and it may trigger an *Exception*. The interrogation is available once the announce has been called. Beside this two operations, the component has two triggers which can be executed in parallel with them. They have a precondition governing their execution.

DEFINITION 3 (COMPOSITION OF OF-CPNs)
Two OF-CPNs O_1 and O_2 can be combined if there is a mapping
$\zeta: P_{snd}(O_1) \cup P_{get}(O_1) \longrightarrow P_{acc}(O_2) \cup P_{get}(O_2)$ verifying:

1. $\zeta(P_{snd}(O_1)) \subset P_{acc}(O_2)$ and $\zeta(P_{get}(O_1)) \subset P_{res}(O_2)$,
2. if $\zeta(p)$ is defined then $\zeta(\Im^{O_1}_{snd-get}(p))$ is also defined,
3. $\forall p \in P_{snd}(O_1) \cup P_{get}(O_1)$, $\zeta(\Im^{O_1}_{snd-get}(p)) = \Im^{O_2}_{acc-res}(\zeta(p))$,
4. $\forall p \in P_{snd}(O_1) \cup P_{get}(O_1)$, $Dom(\zeta(p)) = Dom(p)$.

In the previous configuration, O_1 is the *client* and O_2 is the *server*. We can build a composite of the two *OF-CPNs* and merge each place p with $\zeta(p)$ if the mapping is defined. The resulting *OF-CPN* is denoted $O_1 \oplus O_2$. These merged places are dropped out from the interface places of the composite *OF-CPN*. This operation can be performed for one server (resp. one client) and many of its clients (resp. many of its servers). Such constructions allow to build *ad-hoc* composite components and sub-systems. Its this way we can validate a given scenario involving many objects.

3   Validation and Verification

The V&V is run in a Petri net environment on a set of OF-CPNs (see figure 1). The aim of this work is to take profit from the natural modularity in object orientation and avoid to build a global net from OF-CPNs by some composition mechanism (places or transitions fusion or arcs addition). We promote a method in which V&V focus on a given object and consider an abstraction of its environment (see figure 3). The abstraction of an object is built according to the encapsulation equivalence presented in section 4. The abstraction of the environment is a set of black boxes able at their interfaces to accept requests and issue results without modeling into full details their internal control flow.

3.1   Encapsulation Equivalence

Encapsulation is one of the main concepts promoted by object paradigm. It states that an object is characterized by *what* it realizes rather than *how* it operates to realize it. Specialized to our modular nets, this statement means that an OF-CPN is characterized by *what happens in its interface places*. From the environment, one can hide all actions that do not modify the interface places. We define an equivalence based on that principle to build for a given component a reduced abstraction that is used by its environment. The next section is dedicated to the equivalence relation.

Occurrence Graphs

DEFINITION 4 (OCCURRENCE GRAPH)
An occurrence graph is defined by a 5-tuple $OG = (V, E, T, L, v_0)$ where $v_0 \in V$ is called the initial vertex and (V, E) is a finite labeled directed graph in which each vertex in V is reachable with a path from v_0.
$L : E \longrightarrow T$ is a mapping from the set E of edges to the set T of labels.

The definition of an occurrence graph is similar to the one of an R-graph given in [Murata 94]. E is a subset of $V \times T \times V$ and L is then the projection on T. It is obvious that the reachability graph of a Petri net is an occurrence graph where the set of labels is the set of occurrences of transitions of the net.

We construct the occurrence graph of an OF-CPN by overloading the places of the sets P_{acc} and P_{res} and then building its reachability graph with that extended initial marking. Tokens in the interface can be tuned to model the environment of the component and the hypotheses it assumes on it. Here we just overload these places by all their possible markings. The goal is to have a graph in which one can see how the object can react to the interactions that can be sent by its environment. This reachability graph may be very complex because of the interleaving of requests. But we can have some strategies to reduce it. For instance, we can consider that objects have a parallelism level that determines the number of requests they can operate at a time. Therefore, we consider from the whole reachability graph only states verifying:

1. $\sum_{p \in P_{acc}} (| M(p) | + | M(\Im_{acc-res}(p)) |) \geq \sum_{p \in P_{acc}} | M_0(p) | -K,$
2. $\sum_{p \in P_{snd}} (| M(p) | + | M(\Im_{snd-get}(p)) |) \geq \sum_{p \in P_{snd}} | M_0(p) | -K,$

where M(p) is the number of tokens in place p at the considered state, $\Im_{acc-res}$ and $\Im_{snd-get}$ are the bijections given in the definition of an OF-CPN and K is a constant that defines the level of parallelism. This statement means that at most K requests are being operated each time. If K is 1, we have sequential objects. If $K \geq 1$, the equations (1) and (2) can be refined by:

1. $\forall p \in P_{acc}, | M(p) | + | M(\Im_{acc-res}(p)) | \geq | M_0(p) | -K_m,$
2. $\forall p \in P_{snd}, | M(p) | + | M(\Im_{snd-get}(p)) | \geq | M_0(p) | -K_m,$

where K_m is the parallelism level for each method and $K_m \leq K$. In the remainder of the paper, we assume that K_m is 1 for sake of simplicity. All the results can be obtained without this restriction. We consider that objects are entities which operate concurrent activities but each operation is sequential (i.e. one request being processed for each operation at a given time). The restriction of the reachability graph with respect to statement (1 & 2) is another reachability graph in which some interleavings are discarded because they are not worth to be considered. Such interleavings correspond to cases of parallelism that do never occur in an implementation of the system.

Encapsulation Equivalence for Occurrence Graphs. The *E-equivalence* in occurrence graphs is based on a kind of *I-O similarity* defined by Murata and Notomi (see [Murata 94]). *Two vertices are E-equivalent for the environment if their past and future produce the same changes on interface places, which changes must respect the request-reply correspondence.* For the purpose of a formal definition of *E-equivalence*, we define two sub-sets of characterized vertices of an occurrence graph:

1. $V_{IN} = \{w \in V$ such that there is an edge terminating at w and labeled by an element of $T_{interface}\}$,
2. $V_{OUT} = \{u \in V$ such that there is an edge outgoing from u and labeled by an element of $T_{interface}\}$.

V_{IN} (resp. V_{OUT}) is the set of vertices reachable by consuming tokens from (resp. producing tokens in) the interface places. V_{IN} is called the post-synchronization set of the graph and V_{OUT} its pre-synchronization set.

For a given vertex v, we define its post-synchronization set and pre-synchronization set like following:

1. the restriction of the post-synchronization for a vertex v is called its *out-set* denoted $v_{IN} = \{w \in V_{IN}$ such that $\exists\, \sigma \in (T \backslash T_{interface})^\star$ verifying $w[\sigma \rightsquigarrow v\}$[4] if this set is not empty and $\{v\}$ otherwise,
2. the restriction of the pre-synchronization for a vertex v is called its *in-set* denoted $v_{OUT} = \{u \in V_{OUT}\ \exists\, \sigma \in (T \backslash T_{interface})^\star$ verifying $v[\sigma \rightsquigarrow u\}$ if this set is not empty and $\{v\}$ otherwise.

DEFINITION 5 (THE ENCAPSULATION EQUIVALENCE)
Two vertices v^1 and v^2 of the occurrence graph of an OF-CPN are E-equivalent if and only if:

1. *they have the same in-set ($v^1_{IN} = v^2_{IN}$) and out-set ($v^1_{OUT} = v^2_{OUT}$),[5]*
2. *if an edge ending at a state in v_{IN} is labeled by a synchronization transition $t_1 \in p_{acc}^\bullet$, $p_{acc} \in P_{acc}$, then each path from each state in v_{OUT} contains an occurrence of $t_2 \in {}^\bullet(\Im_{acc-res}(p_{acc}))$,*
3. *if an edge ending at a state in v_{IN} is labeled by a synchronization transition $t_1 \in p_{snd}^\bullet$, $p_{snd} \in P_{get}$, then each path from each state in v_{OUT} contains an occurrence of $t_2 \in {}^\bullet(\Im_{snd-get}(p_{snd}))$.*

Two equivalent vertices are linked by paths, each of which contains at least two synchronizations with the environment, one modeling a request or result received and the other modeling either the result delivery of the previous request or another request issue. In some cases, we are in presence of nested incoming and outgoing requests, e.g. an object that needs to send a request in order to achieve another one. In the remainder of the paper, the equivalence class of a vertex v is noted $v^\sim$.

E-equivalence is stronger then *I-O similarity* presented in [Murata 94]. *I-O similarity* corresponds to the condition (1) given in the definition 5 which means that if $v_{IN} \cap V_{IN} \neq \emptyset$, $\forall v_{in} \in vIN, \exists v_{out} \in v_{OUT}, \sigma_{internal} \in (T \backslash T_{interface})^\star$ such that $v_{in}\,[\sigma_{internal} \rightsquigarrow v_{out}$. *E-equivalence* takes into account the nested requests and allows one to look for preservation of the request-reply correspondence. This strong condition (2 & 3) of definition 5 is enforced to check interaction faults (see 3.1). If it is not true, then the request-reply correspondence is broken.

Some equivalence classes must be distinguished from the others. They are the one which contains (post or) pre-synchronizations i.e. $\{v^\sim$ such that $v^\sim \cap V_{IN} \neq \emptyset$ or $v^\sim \cap V_{OUT} \neq \emptyset\}$. They model events occurring at the interface of the objects.

---

[4] σ leads from w to v.
[5] The in-set and the out-set are noted v_{IN} and v_{OUT} in the remainder of the definition.

We call them synchronizing classes. The others which are in the set $\{v^\sim$ such that $v^\sim \cap V_{IN} = \emptyset$ or $v^\sim \cap V_{OUT} = \emptyset\}$ model internal events that are related by the precedence relation of the flow control. We call them internal classes. For all classes, we are interested in what we call entry-state (resp. exit-state). *An entry-state of a class is a state which is not reachable from any other state of the class. An exit-state of a class is a state such that there is no other state in the class reachable from it.* For synchronization classes, the entry and exit-states are unique. For a given class, each couple of entry and exit state is linked by a class encapsulated path, i.e. a path which only visits states of the class. Such paths do not involve the environment and can be hidden except for classes which verify $v^\sim \cap V_{IN} \neq \emptyset$ and $v^\sim \cap V_{OUT} \neq \emptyset$. For such classes, the encapsulated paths are relevant for the environment.

DEFINITION AND ALGORITHM 6 (THE E-EQUIVALENCE QUOTIENT)
If OG is an occurrence graph, its quotient by the E-equivalence relation $OG^\sim = (V^\sim, E^\sim, T^\sim, L^\sim, v_0^\sim)$. $OG^\sim$ is built as follows:

1. $V^\sim = V_{/E-equivalence}$, $E^\sim = \emptyset$ and $T^\sim = \emptyset$,

   (a) $\forall\, v^\sim \in V^\sim$, if $v^\sim \cap V_{IN} \neq \emptyset$ and $v^\sim \cap V_{OUT} \neq \emptyset$, then
   $E^\sim = E^\sim \cup \{(v^\sim, t_{IO}(v^\sim), v^\sim)\}$; $T^\sim = T^\sim \cup \{t_{IO}(v^\sim)\}$,

   (b) $\forall\, (v^\sim, w^\sim) \in V^\sim \times V^\sim$ such that $v^\sim \cap V_{IN} \neq \emptyset$; and $w^\sim$ is an internal class, so that $v^\sim [t \rightsquigarrow w^\sim$ then $E^\sim = E^\sim \cup \{(v^\sim, t_{IN}(v^\sim, w^\sim), w^\sim)\}$; $T^\sim = T^\sim \cup \{t_{IN}(v^\sim, w^\sim)\}$,

   (c) $\forall (v^\sim, w^\sim) \in V^\sim \times V^\sim$ such that $v^\sim$ is an internal class; $w^\sim \cap V_{OUT} \neq \emptyset$, and $\exists t \in (T \setminus T_{interface})$ so that $v^\sim [t \rightsquigarrow w^\sim$ then
   $E^\sim = E^\sim \cup \{(v^\sim, t_{OUT}(v^\sim, w^\sim), w^\sim)\}$; $T^\sim = T^\sim \cup \{t_{OUT}(v^\sim, w^\sim)\}$;

   (d) $\forall (v^\sim, w^\sim) \in V^\sim \times V^\sim$ such that $v^\sim$ and $v^\sim$ are internal classes, and $\exists t \in (T \setminus T_{interface})$ so that $v^\sim [t \rightsquigarrow w^\sim$ then $E^\sim = E^\sim \cup \{(v^\sim, \tau, w^\sim)\}$; $T^\sim = T^\sim \cup \{\tau\}^6$,

2. $v_0^\sim$ is the equivalence class of v_0.

   This reduced occurrence graph is an exhaustive model of the observable behavior of objects. Furthermore, it models the internal transitions that cannot be hidden by *E-equivalence*. They must be validated as autonomous actions or interaction failures for the object. One drawback of this reduction method is that it hides divergent traces (see [Valmari 94]). A divergence occurs in an element $v^\sim \in V^\sim$ such that there is $v_1, v_2 \in v^\sim$ with $(v_1, v_2) \in E$ and $(v_2, v_1) \in E$. A divergence trace is a trace in which the object can loop forever, executing internal actions. To avoid skipping this problem, we propose a version of the method which shows full internal details of a given object (see section 3.2 below). Nevertheless, the existence of divergence traces can be checked before computing the reduced occurrence graph to select objects to consider in full details.

---

[6] τ denotes internal actions as in process algebras

Interaction Failures. An interaction failure is a path of the occurrence graph that corresponds to one of the two following cases:

1. a path σ has an edge $t \in p_{acc}^{\bullet}$ where $p_{acc} \in P_{acc}$ and σ has no edge in $^{\bullet}(\Im_{acc-res}(p_{acc}))$,
2. a path σ has an edge $t \in {}^{\bullet}p_{res}$ where $p_{res} \in P_{res}$ and σ has no edge in $(\Im_{acc-res}^{-1}(p_{res}))^{\bullet}$,
3. a path σ has an edge $t \in {}^{\bullet}p_{snd}$ where $p_{snd} \in P_{snd}$ and σ has no edge in $(\Im_{snd-get}(p_{snd}))^{\bullet}$,
4. a path σ has an edge $t \in p_{get}^{\bullet}$ where $p_{get} \in P_{get}$ and σ has no edge in $(\Im_{snd-get}^{-1}(p_{get}))^{\bullet}$,

An interaction failure is either *"a path in which an object does not send (resp. receive) a result for an accepted request (resp. for a sent request)"* or *"a path in which an object receives (or sends) an unexpected result without any previous request sent (or received)"*. Interaction failures model starvation when waiting a result or what is called honesty and discretion of objects in [Sibertin-Blanc 93]. They correspond to pathes that can not be reduced by *E-equivalence*. They must be evaluated on the initial occurrence graph. Once interaction failures evaluated, one can focus on reduced occurrence graph to evaluate a whole system of objects.

The Minimal Representation of an Occurrence Graph. The advantage of occurrence graph is the reduction that can be performed on them using the *E-equivalence*. Once this reduction is done, it is valuable to consider the reduced model instead of the initial one. We propose a way to have a minimal Petri net representation of an occurrence graph in order to support symbolic simulation and prototyping. The minimal representation is a net which only model events on interface places.

DEFINITION AND ALGORITHM 7 (THE MINIMAL REPRESENTATION)
Consider $OG^{\sim} = (V^{\sim}, E^{\sim}, T^{\sim}, L^{\sim}, v_0^{\sim})$ the E-equivalence quotient of an occurrence graph $OG = (V, E, T, L, v_0)$ built from the OF-CPN $O_{net} = (Net, P_{acc}, P_{res}, P_{snd}, P_{get}, M_0)$. The minimal net representation of $OG^{\sim}$ is the net $O_{net}^{\sim}$ built as follows:

1. *the set of places of $P(O_{net}^{\sim})$ is initialized to the interface places of O_{net} and the set of transitions of $T(O_{net}^{\sim})$ is initialized to $\emptyset$,*
2. $\forall v^{\sim} \in V^{\sim}, P(O_{net}^{\sim}) = P(O_{net}^{\sim}) \cup \{p_{v^{\sim}}\}$,
3. $\forall v^{\sim}, w^{\sim} \in E^{\sim}, T(O_{net}^{\sim}) = T(O_{net}^{\sim}) \cup \{t_{(v^{\sim},w^{\sim})}\}$,

   (a) $\forall p \in P_{interface}(O_{net}), Pre(p, t_{(v^{\sim},w^{\sim})}) = M_{v^{\sim}}(p)$ *where $M_{v^{\sim}}(p)$ is the marking of p in the entry-state of $v^{\sim}$,*
   (b) $\forall p \in P_{interface}(O_{net}), Post(p, t_{(v^{\sim},w^{\sim})}) = M_{w^{\sim}}(p)$ *where $M_{w^{\sim}}(p)$ is the marking of p in the exit-state of $w^{\sim}$,*
   (c) $Pre(p_{v^{\sim}}, t_{(v^{\sim},w^{\sim})}) = 1$ *and* $Post(t_{(v^{\sim},w^{\sim})}, p_{v^{\sim}}) = 1$.

The abstraction of a component is another one just accepting requests and giving results without modeling the internal processing necessary for that purpose. For the environment, a component and its abstraction provide the same behavior. Abstractions are interesting in order to validate a set of objects while focusing on a given one among them (see figure 3). One just considers the whole net for the interesting object and abstractions for the others. This approach is very valuable when using Petri nets to validate and verify a large number of interacting objects. It then allows one to avoid managing the growth of the size of nets with the number and complexity of objects. The interaction failures of a component can be check on its reduced occurrence graph. Once validated, we can use the minimal representation of a component to validate and verify systems where that component is involved.

3.2   Object-Based Verification and Validation

Object-based V&V is promoted to take profit from the natural structuration and modularity of objects. It is also a way to enhance reuse of valied objects. The principle of our method is the one shown in (see figure 3). When validating and verifying an object, a system designer needs minimal information about its environment. The current object relies on what its environment supplies in order to guarantee its own functionalities (see [Colette 93]). Therefore, one does not need to consider full representation of the environment, only an abstraction suffices. To compute such an abstraction for a given object of the environment, we overload the places of P_{acc} and P_{res} by expected incoming requests and results. We then compute the occurrence graph of the component which can be reduced as shown in section 3.1. It makes sense to consider first the net modeling the abstractions of all objects in a system instead of the set of all nets modeling all objects. This net gives high level information on the system modeling both control and data flows. In order to go into details on some aspects of that information, we will need some kind of viewpoint from an object. From an object, we therefore consider a viewpoint on the system which is constructed by interfacing the net modeling that object with minimal representations of the other objects. This viewpoint can support deeper evaluation of the concerned object.

System Analysis. The main idea here is to prove that results which would be obtained from the global net built from all the objects can be deduced from the one built from their minimal representation. We consider in the remainder of the paper a system of n ($n \geq 2$) objects $(O_i)_{i \in [1,..,n]}$ and $(O_i^{\sim})_{i \in [1,..,n]}$ their minimal representations. Each of the objects has an initial marking $M_0(O_i)$. We denote $\bigoplus_{i \in [1,..,n]}(O_i)$ the net built from the objects by place fusion and conserving the initial marking of objects. $\bigoplus_{i \in [1,..,n]}(O_i^{\sim})$ is the net built from the minimal representations with the initial markings of objects. $OG(\bigoplus_{i \in [1,..,n]}(O_i)$ and $OG(\bigoplus_{i \in [1,..,n]}(O_i^{\sim}))$ design the respective occurrence graphs of $\bigoplus_{i \in [1,..,n]}(O_i)$ an $\bigoplus_{i \in [1,..,n]}(O_i^{\sim})$.

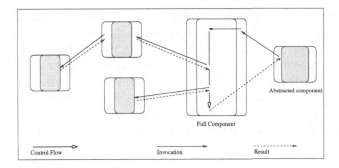

Fig. 3. Object-based V&V.

As *E-equivalence* is stronger than *I-O similarity*, we can use the abstraction $OG(\bigoplus_{i\in[1,..,n]}(O_i^\sim))$ to validate the reachability of states of the system according to the results demonstrated by Murata and Notomi in [Murata 94]. Interaction failures also can be validated on $OG(\bigoplus_{i\in[1,..,n]}(O_i^\sim))$ which costs less to compute than $OG(\bigoplus_{i\in[1,..,n]}(O_i))$. If the reduced occurrence graph has no interaction failure, then the global one and henceforth the system do not have interaction failure. The terminal vertices in the occurrence graph $OG(\bigoplus_{i\in[1,..,n]}(O_i^\sim))$ must be analyzed. Each of such vertices has restrictions on the occurrence graph of the object which are terminal ones. So the deadlocks can be checked into details by analyzing the faulty objects. For this purpose, we also avoid to build the global occurrence graph $OG(\bigoplus_{i\in[1,..,n]}(O_i))$. We build the occurrence graph for the faulty object interfaced with the minimal representations of the others.

Object Analysis. The analysis at object level allows one to go into details of a given object when it appears faulty in the analysis of interactions or terminal vertices. We consider therefore the object and an abstraction of its environment. The viewpoint from O_1 is denoted $O_1^+ = O_1 \oplus (\bigoplus_{i\in[2,..,n]}(O_i^\sim))$ with the initial marking of O_1. It is built by place fusion between the whole net modeling O_1 and the minimal representations of the other objects. The corresponding occurrence graph is noted $OG(O_1^+)$. We consider a weak encapsulation equivalence noted *WE-equivalence* on the set $(O_i)_{i\in[1,..,n]}$ which is the *E-equivalence* on $(O_i)_{i\in[2,..,n]}$ In other words, the equivalence is not considered on object O_1 because we want to keep it entirely. With $OG(O_1 \oplus (\bigoplus_{i\in[2,..,n]}(O_i^\sim)))$, we can go into details of one (or even many) object model(s). If the object O_1 presents an interaction failure, we can examine in detail the sequence(s) leading to that interaction failure. It is a kind of local debugging on the high symbolical level of specifications. This per-object V&V is a way to enhance reuse.

Validation and Model Execution. Specification simulation is a key activity for validation (see [Mellor 94]). It allows the experts (the ones of formal methods and the ones of application domains) to involve end-users and system owners into validation. Simulation also can be used to execute specification models in a symbolic way. The interest of building minimal representations here is to run simulation without going into details. Tracing a model of a system back to requirements can be made at a higher level of abstraction. The net $\bigoplus_{i\in[1,..,n]}(O_i^{\sim})$ can support the simulation of the whole system. The net $(O_1 \oplus (\bigoplus_{i\in[2,..,n]}(O_i^{\sim})))$ can support the simulation from an object viewpoint. It shows into details the actions happening in the object O_1 with an abstraction of its environment. This kind of simulation can be used to run symbolic debugging on specifications.

4   Conclusions

This paper presents a method whose aim is to integrate Petri nets in object-oriented methodologies in order to validate and verify control properties. The method is based on analysis of occurrence graph with a reduction technique. It focuses on validation and verification of interaction between objects. The reduction technique allows one to build smaller nets to handle in order to manage large libraries of objects. The method aims to integrate Petri nets in an object-oriented methodology by means of design patterns. It allows one to manage a library of objects and to build (sub-)systems in a bottom-up approach with many abstraction levels. It is a way to enhance reuse of valid objects. The design patterns must be tailored according to the application domain and we can not be prescriptive on them as the paper does not consider any domain in particular. The reduction method allows to enhance reuse and management of objects libraries. Once an object is validated and verified, we can henceforth use its reduction in systems where it is reused. The reduced occurrence graph is a valuable basis to evaluate temporal logic formula on the observable behavior of a system or its components.

Acknowledgements

All the proofreadings done on the first versions of this paper are acknowledged here. Anonymous reviewers have also helped by their comments and suggestions.

References

Bastide 95.   Bastide R. , *Approaches in Unifying Petri Nets and the Object-oriented Approach*, In Proceedings of the First Workshop on Object-oriented Programming and Models of Concurrency, Torino, June 1995.

Baumgarten 88.   Baumgarten B., *"On Internal and External Characterizations of PT-net building block Behaviour"*, In Advances in Petri Nets, 1988, Rozenberg G. Ed., Springer Verlag, 1990, LNCS vol. 340, pages 44-61.

Boehm 81.   Boehm B., *"Software Engineering Economics"*, Prentice Hall, 1981.

532   A. Diagne

Bruno 95.   Bruno G., Castella A., Agarwal R. & Pescarmona M.-P., *"CAB :
                 An Environment for Developing Concurrent Applications"*, In
                 Proceedings of ICATPN'95, Torino, Italy, June 1995, De Miche-
                 lis G. & Diaz M. Eds., Springer Verlag, LNCS vol. 935, pages
                 141-160.
Buchs 91.   Buchs D. & Guelfi N., *A Concurrent Object-oriented Petri Nets
                 Approach for System Specification*, In Proceedings of the Inter-
                 national Conference on Application end Theory of Petri Nets,
                 Aarhus, Denmark, 1991, LNCS vol. 524.
Colette 93.   Colette P., *"Application of the Composition Principle to Unity-
                 Like Specification"*, In Proceedings of TAPSOFT'93, Orsay,
                 France, April 1993, Gaudel M.-C. & Jouannaud J.-P. Eds., LNCS
                 vol. 668, pages 230-242.
Diagne 96a.   Diagne A. & Estraillier P., *"Formal Specification and Design
                 of Distributed Systems"*, In Proceedings of FMOODS'96, Paris,
                 France, March 1996, Najm E. & Stefani J.-B. Eds, Chapman &
                 Hall, pages 325-340.
Diagne 96b.   Diagne A. & Kordon F., *"A Multi-Formalism Prototyping
                 Approach from Conceptual Description to Implementation of
                 Distributed Systems"*, In Proceedings of IEEE International
                 Workshop on Rapid System Prototyping'96, Greece, Porto
                 Caras, Thessaloniki, June 1996, IEEE Comp. Soc. Press, Num.
                 96TB100055, pages 102-107.
Gamma 95.   Gamma E., Helm R., Johnson R. & Vlissides J., *"Design Pat-
                 terns : Elements of Reusable Object-oriented Software"*, Addison
                 Wesley Professional Computing Series, 1995.
Heiner 92.   Heiner M., *"Petri Net Based Software Validation, Prospects and
                 Limitations"*, Technical Report TR92-022, GMD/First at Berlin
                 Technical University, Germany, March 1992.
Jensen 90.   Jensen K., *"Coloured Petri Nets : A High Level Language for
                 System Design and Analysis"*, In Proceedings of the Advances in
                 Petri Nets 1990, Rozenberg G. Ed., Springer Verlag, 1990, LNCS
                 vol. 483, pages 342-416.
Lakos 95a.   Lakos, C. A., *"From Coloured Petri Nets to Object Petri Nets"*, In
                 Proceedings of ICATPN'95, Torino, Italy, June 1995, De Michelis
                 G. & Diaz M. Eds., Springer Verlag, LNCS vol. 935, pages 278-
                 297.
Lakos 95b.   Lakos C. A. & Keen C. D., *"An Open Software Engineering En-
                 vironment Based on Object Petri Nets"*, Technical Report TR95-
                 6, Computer Science Department, University of Tasmania, Aus-
                 tralia.
Matsuoka 93.   Matsuoka S. & Yonezawa A. *Analysis of Inheritance Anomaly
                 in Object-Oriented Concurrent Programming Languages*, In Re-
                 search Directions in Concurrent Object-Oriented Programming,
                 Agha G., Wegner P. & Yonezawa A. Eds., MIT Press, 1993, pages
                 107-150.
Mellor 94.   Mellor S. J. & Shlaer S., *"A Deeper Look at .. Execution and
                 Translation"*, In Journal of Object Oriented Programming, 7(3),
                 1994, pages 24-26.
Murata 89.   Murata T., *"Petri Nets : Properties, Analysis and Applications"*,
                 In the Proceedings of the IEEE, 77(4), April 1989, pages 541-580.

Murata 94.   Murata T. & Notomi M., *"Hierarchical Reachability Graph of Bounded Nets For Concurrent Software Analysis"*, In Transactions IEEE on Software Engineering, 20(5), May 1994, pages 325-336.

Nierstrasz 93.   Nierstrasz O., *"Regular Types for Active Objects"*, In Proceedings of the ACM OOPSLA'93, Washington DC, USA, October 1993, pages 1-15.

Pressman 92.   Pressman R. S., *"Software Enginnering : A Practitionner's Approach"*, Mc Graw Hill 1992.

Puntigam 97.   Puntigam F., *"Coordination Requirements Expressed in Types for Active Objects"*, In Proceedings of ECOOP'97, Jyväskylä, Finland, June 1997, Akşit M. & Matsuoka S. Eds., LNCS vol. 1241, pages 367-388.

Rumbaugh 91.   Rumbaugh J., Blaha M., Premeralani W., Eddy F. & Lorensen W., *"Object-Oriented Modeling and Design"*, Prentice Hall, 1991.

Rumbaugh 97.   Rumbaugh J. & Booch G., *"Unified Method for Object-Oriented Development"*, Version 1.0, Teaching Documentation Set, Rational Software Corporation, Santa Clara, USA, 1997.

Rushby 93.   Rushby J., *"Formal Methods and the Certification of Critical Systems"*, Technical Report CSL-93-7, Computer Science Laboratory, SRI International, Menlo Park CA 94025, USA, December 1993.

Sa 96.   Sa J., Keane J. A., & Warboys B. C., *"Software Process in a Concurrent, Formally-based Framework"*, In Proceedings of the IEEE International Conference on Systems, Man and Cybernetics, Beijing, China, October 1996, pages 1580-1585.

Schreiber 95.   Schreiber G., *''Functional Equivalence of Petri Nets"*, In Proceedings of ICATPN'95, Torino, Italy, June 1995, De Michelis G. & Diaz M. Eds., Springer Verlag, LNCS vol. 935, pages 432-450.

Sibertin-Blanc 93.   Sibertin-Blanc C., *"A Client-Server Protocol for the Composition of Petri Nets"*, In Proceedings of ICATPN'93, Chicago, Illinois, USA, June 1993, Ajmone Marsan M. Ed., LNCS vol. 691, pages 377-396.

Sibertin-Blanc 94.   Sibertin-Blanc C., *"Cooperative Nets"*, In Proceedings of ICATPN'94, Zaragoza, Spain, June 1994, Valette R. Ed., LNCS vol. 815, pages 471-490 .

Valmari 94.   Valmari A., *"Compositional Analysis with Bordered Places Subnets"*, In Proceedings of ICATPN'94, Zaragoza, Spain, June 1994, Valette R. Ed., LNCS vol. 815, pages 531-547.

Vogler 92.   Vogler W., *"Modular Constructions and Partial Order Semantics of Petri Nets"*, Springer Verlag, LNCS vol. 625.

A Cooperative Petri Net Editor

R. Bastide†. C. Lakos‡, and P. Palanque†

†University of Toulouse I, France
‡University of Tasmania, Australia

Introduction

This case study presents a standard example for evaluating formal specification approaches that combine a formal model of concurrency (such as Petri nets) with the object-oriented approach. You are expected to exercise your formalism of choice on this problem in order to demonstrate how well it deals with a system of reasonable size.

We describe how the structure of a problem can be modeled in a formalism and how component reuse can be incorporated.

It is desirable to highlight or emphasize the way that the structure of the problem can be modeled in your formalism, and the way that reuse of components can be incorporated.

Statement of the Problem

The system being studied is a piece of software allowing cooperative editing of hierarchical diagrams. These diagrams may be the work products of some Object-Oriented Design methodology, Hardware Logic designs, Petri Net diagrams, etc. Note that if the diagrams happen to coincide with the formalism being tested, care must be taken to clearly distinguish between the two.

One key aspect of this problem is that the editing software should cater to several users, working at different but connected workstations, and cooperating in constructing the one diagram. In the Computer Supported Cooperative Work (CSCW) vocabulary, such a tool could fall in the class of synchronous groupware (each user is informed in real time of the actions of others) allowing for relaxed WYSIWIS (What You See Is What I See): each user may have their own customized view of the diagram under design, viewing different parts of the drawing or examining it at a different level of detail.

A second key aspect of this problem is that the editor should cater to hierarchical diagrams, i.e. components of the diagram can be exploded to reveal subcomponents.

A simple coordination protocol is proposed to control interactions between various users:

- Users may join or leave the editing session at will, and may join with different levels of editing privileges. For example, a user may join the session merely

G. Agha et al. (Eds.): Concurrent OOP and PN, LNCS 2001, pp. 534–535, 2001.

to view the diagram, or perhaps to edit it as well. (See below.) The current members of an editing session ought to be visible to all, together with their editing privileges.
- Graphical elements may be free or owned by a user. Different levels of ownership should be supported, including ownership for deletion, ownership for encapsulation, ownership for modification, and ownership for inspection. Ownership must be compatible with a user's editing privileges.
- Ownership for deletion requires that no one else has any ownership of the component - not even for inspection.
- Ownership for encapsulation requires that only the owner can view internal details of the component - all other users can only view the top level or interface to the component.
- Ownership for modification allows a user to modify attributes, but not to delete the component.
- Ownership for inspection only allows a user to view the attributes.
- Only ownership for encapsulation can persist between editing sessions. Note that this ownership is tied to a particular user, not a particular workstation; all other ownership must be surrendered between sessions.
- Ownership for inspection is achieved simply by selecting the component.
- Other forms of ownership (and release) are achieved by an appropriate command, having first selected the component.
- The level of ownership of a component is visible to all other users, as is the identity of the owner.
- The creator of an element owns it for deletion until it is explicitly released.

Treatment of the Case Study

We have described above the basic requirements of the example system in an informal, incomplete (and possibly even inconsistent) manner. One of the purposes of the case study is to explore the extent to which a formal specification may further the completeness of these requirements. One may deviate from those requirements if necessary, to highlight some feature of the approach you propose, but in such a case the reasons for this deviation should be precisely stated.

This case study may be undertaken at whichever level is appropriate for the approach (e.g. to deal only with requirements engineering, or with the software architecture), or to focus on only one precise aspect of the problem that seems most relevant (e.g. with the technical details of the necessary network communications, or on the high-level protocol between users). However, an effort should be made to make scope of the treatment precise.

The Hurried Philosophers

C. Sibertin-Blanc

University of Toulouse I/ IRIT, France

Introduction

The purpose of this case study is to test the expressive and analytic power of languages merging Petri nets and concepts of the Object-Oriented approach. Regarding the expressive power, the main tested features are:

- local control: each object plays its own Petri net, without regard to the state of other objects;
- dynamic instantiation: while the system is running, new objects may appear and some may disappear;
- dynamic binding: the partner of a communication is defined when the communication occurs, and not by the model (or at "compile time");
- inheritance and polymorphism: a class may inherit some features from other classes, and a communication happens according to the actual class of the partner.

Statement of the case

This case is a refinement of the well known dining philosophers example. Some philosophers are sitting around a table. Each of the philosophers has his plate, and he shares one fork with his right neighbor and another with his left neighbor.

1. When a philosopher does not have his left fork, he can ask his left neighbor for this fork, and the same holds for his right fork;
2. When a philosopher has his right fork and has received a request for this fork, he must give it to his right neighbor, and the same holds for his left fork;
3. A philosopher can eat only when he has both forks;
4. Some philosophers have the additional capability enabling them to introduce a new guest; when a new philosopher arrives at the table, he has one plate and one fork coming from a heap of forks;
5. Some philosophers (possibly the same ones as described above) have the additional capability to leave the table, or to ask their right (or left) neighbor to leave; when a philosopher leaves the table, he goes with his plate and one fork, and he puts the fork into the fork heap;
6. Each philosopher behaves in such a way that the whole system is fair: philosophers eat in turn;
7. At all times, there is at least two philosophers at the table;

G. Agha et al. (Eds.): Concurrent OOP and PN, LNCS 2001, pp. 536–537, 2001.

8. Additional properties may be considered, such as: a philosopher must eat at least once before he leaves, and so on

Examples of interesting properties to prove would be:

- The system never deadlocks;
- If, at the initial state, the fork heap contains a finite number of forks, the system is bounded;
- At all times, a fork is either in the hand of a philosopher or in the fork heap.

A solution of this case may be found in the following paper: "Cooperative Nets," ATPN'94 (Chicago), LNCS 815.

Author Index

Lecture Notes in Computer Science

For information about Vols. 1–1944
please contact your bookseller or Springer-Verlag